B. S. Sokolov A. B. Iwanowski (Eds.)

The Vendian System

Vol. 1 Paleontology

With 126 Figures

Springer-Verlag
Berlin Heidelberg New York London
Paris Tokyo Hong Kong Barcelona

Prof. Dr. Boris S. Sokolov
Prof. Dr. Andrew B. Iwanowski
Paleontological Institute
Academy of Sciences USSR
Profsoyuznaya 123
117868 Moscow, USSR

Translated from Russian by:

Tatjana I. Vasiljeva
Reveka Y. Sorkina
Raisa V. Fursenko
VSEGEI, All-Union Research
Institut of Geology
Srednij prospekt 74
199026 Leningrad, USSR

Title of the original Russian edition: Vendskaja sistema 1, istoriko-geologičeskoe i paleontologičeskoe obosnovanie Paleontologija

ISBN-13: 978-3-642-73974-3 e-ISBN-13: 978-3-642-73972-9
DOI: 10.1007/978-3-642-73972-9

Library of Congress Cataloging-in-Publication Data. Vendskaia sistema. English. The Vendian system / B. S. Sokolov, A. B. Iwanowski, eds. p. cm. Translation of : Vendskaia sistema. Contents : v. 1. Paleontology. – v. 2. Regional geology. Vol. 2 edited by B. S. Sokolov and M. A. Fedonkin. ISBN-13: 978-3-642-73974-3 1. Geology, Stratigraphic – Precambrian.
I. Sokolov, Boris Sergeevich, 1914– . II. Iwanowski, A. B. (Andrew Borisovich)
III. Fedonkin, M. A. (Mikhail Aleksandrovich) IV. Title. QE653.V4313 1990 551.7'1–dc20
89-21944

Softcover reprint of the hardcover 1st edition 1990

Typesetting: International Typesetters Inc., Makati, Philippines
2132/3145(3011)-543210 – Printed on acid-free paper

Preface

The two-volume monograph "Vendian System: Vol. 1 Paleontology, Vol. 2 Regional Geology" is a synthesis of studies carried out over 30 years and devoted to the crucial problem of stratigraphic geology and the history of organic evolution at the Precambrian-Phanerozoic boundary. The studies showed the necessity in recognizing a new stratigraphic system – the Vendian (for the first time in the last 100 years).

Volume 1 deals with different concepts concerning the nature and importance of the Precambrian-Cambrian period, the history of recognition of the Vendian (first as a complex and then as a system) and correlative units of the Upper Precambrian. A detailed promorphological analysis of the Vendian Metazoa is used as a basis for recognition of some new major taxa of multicellular organisms, and is followed by a comprehensive systematic description of all major groups of organic life in the Vendian which include: unique fauna of multicellular, non-skeletal organisms (from the world's richest localities along the Zimny and Letny Coasts in the White Sea area, the Dniester Basin of Podolia in the Ukraine, and Olenek Uplift, as well as peculiar localities of the Metazoa in the Urals); trace fossils, multicellular megascopic algae, actinomycetes and organisms of uncertain systematic position, various microfossils (acritarch and filamentous forms); calcareous algae from the uppermost Vendian; stromatolites and microphytolites. All the described forms are illustrated in the corresponding plates. References are given in Volumes 1 and 2.

The Editors

Contents

List of Contributors

1. Becker, Yurij Rafailovich VSEGEI, All-Union Research Institute of Geology, Srednij prospekt 74, SU – 199026 Leningrad
2. Volkova, Nina Andreevna Geological Institute of the USSR Academy of Sciences, Pyzhevskij pereulok 7, SU – 109017 Moskow
3. Hermann, Tamara Nikolaevna Institute of Precambrian Geology & Geochronology, USSR Academy of Sciences, Makarova Embankment 2, SU – 199164 Leningrad
4. Gnilovskaya, Marina Borisovna Institute of Precambrian Geology & Geochronology, USSR Academy of Sciences, Makarova Embankment 2, SU – 199164 Leningrad
5. Dolnik, Tatjana Aleksandrovna East Siberian Research Institute for Geology, Geophysics, and Mineral Resources, Dekabrskih sobijtij 29, SU – 664026 Irkutsk.
6. Korolyuk, Irina Nikolaevna Institute of Geology and Mining of Oil Resources, Academy of Sciences, and Ministry of Oil Industry, Fersmana 50, SU – 117312 Moscow
7. Krylov, Jgor Nikolaevich Geological Institute of the USSR Academy of Sciences, Pyzhevskij pereulok 7, SU – 109017 Moskow
8. Luchinina, Veronika Akberovna Institute of Geology and Geophysics, Siberian Branch USSR Academy of Sciences, SU – 630090 Novosibirsk
9. Pyatiletov, Valerii Gennadievich Institute of Geology and Geophysics, Siberian Branch USSR Academy of Sciences, SU – 630090 Novosibirsk
10. Ragozina, Alla Leonidovna Palaeontological Institute of the USSR Academy of Sciences, Profsoyuznaya 123, SU – 117868 Moskow
11. Rudavskaja, Valeria Vladimirovna All-Union Oil Research Geological Prospecting Institute, Liteyny prospect 39, SU – 192104 Leningrad
12. Sivertseva, Irina Aleksandrovna Geologicheskij fakultet (Department of Geology) Leningradskij Universitet, SU – 199164 Leningrad
13. Sidorov, Anatolii Dmitrievich Institute of Geology and Mining of Oil Resources, Academy of Sciences and Ministry of Oil Industry, Fersmana 50, SU – 117312 Moskow
14. Sokolov, Boris Sergeevich Palaeontological Institute of the USSR Academy of Sciences, Profsoyuznaya 123, SU – 117868 Moscow
15. Fedonkin, Michael Aleksandrovich Palaeontological Institute of the USSR Academy of Sciences, Profsoyuznaya 123, SU – 117868 Moscow

16. Shenfil, Viliam Yurievitsch Institute of Geology and Geophysics, Siberian Branch USSR Academy of Sciences, SU – 630090 Novosibirsk
17. Yakshin, Michael Semionovitsch Institute of Geology and Geophysics, Siberian Branch USSR Academy of Sciences, SU – 630090 Novosibirsk
18. Jankauskas, Tadas-Rimas Vladimirovich Lithuanian Geological Research and Survey Institute, Sevcenkos 13, SU – 232600 Vilnius

List of Acronyms

CGM	Central Geological Museum
GIN	Geological Institute of the USSR Academy of Sciences
IGGD	Institute of Geology and Geochronologya of the Precambrian
IGiG	Institute of Geology and Geophysics
IGN	Institute of Geological Sciences of Ukr.SSR Academy of Sciences
IGiRGI	Institute of Geology and Exploration of Combustible Minerals
IGU	Irkutsk Geological Survey
ISC	Interdepartmental Stratigraphic Committee of the USSR
LGU	Leningrad State University
LitNIGRI	Lithuanian Research Geological Prospecting Institute
PIN	Paleontological Institute of the USSR Academy of Sciences
VNIGRI	All-Union Petroleum Research Geological Prospecting Institute (Leningrad)
VNIGNI	All-Union Research Geological Prospecting Petroleum Institute (Moscow)

Introduction

The concept that the base of a normal, sedimentary cover of the Russian (= East European) Platform contains deposits older than Cambrian, in the sense of English geologists, has been controversial for a long time. Primarily, it concerns "barren" beds of the well-known "blue clays" and underlying basal sandstones composing the slopes of the Baltic Shield (Schmidt 1897; Lamansky 1897; Bubnoff 1926; Scupin 1928 and others, see Sokolov 1952). However, only the deep drilling on the Russian Platform, started in the late 1940s, has encountered a section between the oldest exposed Lower Cambrian and crystalline basement and has shown that the cover is underlain not by the Caledonides (see Arkhangelsky 1940), but by an earlier unknown, thick succession of marine terrigenous rocks.

The age and stratigraphy of the succession led to many debates known from both Soviet and Western literature. Some workers (the majority) considered these terrigenous deposits as the older beds of the Lower Cambrian (Bruns 1957 and many others), others treated it as a platform equivalent to the entire Precambrian Riphean group of the Urals (Shatsky 1952a,b), still others recognized a new major stratigraphic unit equal in rank to the Cambrian (Sokolov 1952). Naturally, the complex has been compared to the Sinian or the first Precambrian system of the Paleozoic group in the sense of Grabau (1922) and Mazarovich (1947).

The latter statement was based mainly on a paleogeographic account of new data on the pre-Devonian deposits of the Russian Platform, prepared by the author and Aleksandrova during the period of 1948–1952 (Sokolov and Aleksandrova 1952). The oldest deposits of the sedimentary cover of the platform were subdivided into two major complexes: the upper Baltic Complex (with a characteristic assemblage of Lower Cambrian fauna) and the lower, Vendian Complex (containing, as known at that time, only microfossils, scarse trace fossils and problematica, later named Beltanelloides). The author had no doubts about the Precambrian age of the Vendian Complex (or Series). This opinion was shared by Shatsky. However, we proposed quite a different interpretation for the stratigraphic rank of a unit (system or group = erathem) and its correlation with the entire section of the exposed Riphean of Gornaya (Mountain) Bashkiria. Hence, different nomenclature was used (Sinian system or Riphean group), and a different attitude towards the historical-geological relationship between the Redkino and Valdai Series (Vendian) and Baltic Series (Cambrian) was formed, concerning also the importance of the Precambrian climatic minimum (tillites and cold-water sediments during the first historical phase) and the high level of differentiation of Precambrian life for the Vendian, which was considered as deposits accumulated during a particular system and period.

The history of further developments in studies of the Vendian of the Russian Platform and its equivalents in the Urals, Siberia, in the Arctic regions and in Central Asia is well known. These developments have been widely reported in publications on stratigraphy, micropaleontology (and since the mid-1960s in works on paleontology of the Metazoa and Metaphyta), lithology and facies analysis, paleogeography and tectonic processes and mineral resources, primarily, hydrocarbons and ore deposits. We will dwell upon the most important of them. The new birth of the concept of the Sinian terminated by the Vendian Complex has been widely accepted in the USSR and since the 1950s has been used in state mapping. It seems to be the best reflected in general statements on the geological history of "pre-Sinian platforms" of Eurasia (Sokolov 1958).

However, it soon became evident that the Sinian of China and that of the USSR are in fact very similar and contain stromatolite assemblages of the same type (Vologdin 1962, and others); their stratigraphic extent and age correspond to the entire Upper Proterozoic or Upper Precambrian. Based on earlier and new data, the lower boundary of the Riphean in area type, the South Urals, was placed at 1650 ± 50 m.y. (Keller and Krasnobaev 1983). The base of the Sinian suberathem, as named by Chinese geologists between 1975–1980 (The Sinian System of China, Sinian Suberathem in China 1980) is dated now at 1900 m.y. at the base of the Changcheng system or at 1650 m.y. at the base of the Nankou system, which is assigned to the "paleophyte erathem", an approximate equivalent to the Riphean of the USSR (Zhao 1983). In 1982, the Stratigraphic Committee of the People's Republic of China deleted the Sinian Suberathem and it was replaced by several new systems; the name Sinian system was retained only for the upper system at 800 ± 50 m.y. = 570–600 m.y. Thus, the duplication in usage of the term "Sinian" was abolished by this act. Now isotope ages provide a basis for recognizing the Sinian as a particular era with the Vendian period in its upper part (Harland et al. 1982). The Riphean was ranked as an era as well. Actually, Riphean deposits form a special eonothem, corresponding to an eon, which is important for the entire history of the Earth. A classical Sinian in the Jixian type section (North China, Tianjin, Jixian) may be ranked as a standard of similar importance (Sinian Suberathem.... 1980).

In the USSR, the relationship between the Riphean and Vendian has been interpreted in a different way, however, since the early 1960s they were assigned to the Upper Proterozoic. This was mainly due to a comprehensive study by Keller (Upper Precambrian 1963; Keller 1968). However, some workers still regarded the Vendian as the terminal (fourth) division, differing in extent to the Riphean group and at the same time they retained the name terminal Riphean or Vendomian (Keller 1966a,b, 1973 and others; Chumakov and Semikhatov 1981, and others) and the Vendian s.str. (Yakobson and Krylov 1977 and others), whereas others consistently preferred an independent Riphean and Vendian with quite different extents (Sokolov 1964a, 1972a, 1976a, 1980 and others). The former put forward mainly the similarity between stromatolite and microphytolite assemblages, which is quite natural for these biosedimentary products of prokaryotic life activity; they now form in a similar way and are strongly morphologically and ecologically dependent. They were hypnotized in part by the concept of the Riphean group, including the Asha group of the Urals (it is now a proven paleontological equivalent of an incomplete Vendian section), reflecting a tectonically single formation stage in geosynclinal development (Shatsky 1945).

The Riphean and Vendian, as separate stratigraphic units, were accepted at the first All-Union Meeting on the Upper Precambrian stratigraphy of Siberia (Anon .... 1962) and since then, the Vendian (Yudoma Complex, Yudomian) is a general term clearly differentiated from the Riphean complexes of a transgressive cover composed mainly of carbonate deposits directly underlying the Lower Cambrian of the Siberian Platform (All-Union Symposium .... 1965; Equivalents .... 1975; Khomentovsky 1976).

Special emphasis was put on the Vendian-Cambrian pretrilobite boundary beds. They contain the oldest Early Cambrian fauna of diverse skeletal fossils which provided a basis for recognizing a new Tommotian stage (Rozanov et al. 1966) and furthering discussion on the Precambrian-Cambrian boundary problem (Cowie and Rozanov 1973, 1982; Sokolov 1974c, 1977b, 1980; Rozanov 1976; Rozanov and Sokolov 1980, 1982 and others). In fact, it defined a Vendian (Yudoma Complex) – Tommotian boundary, or to be more exact, the Nemakit-Daldyn Horizon (stage) – Vendian and Sunnagin s.l. Horizon (Lower Cambrian) boundary. The Phanerozoic (biological) approach to typification of system boundaries was successfully applied and became the basic approach in the multiyear activity of the Project 29 Working Group of IGCR for defining the Precambrian-Cambrian boundary.

On the Russian Platform, since the earlier studies carried out by the author, the Vendian-Cambrian boundary was placed at the boundary between the Kotlin Horizon (Vendian) and the Rovno Horizon of the Baltic Series (Lower Cambrian). However, the discovery of an extremely rich sabelliditid association in the Nemakit-Daldyn Horizon and in its equivalents in Siberia (Sokolov 1968), the association typical of the Rovno Horizon (Lower Baltic) in the European part of the USSR showed that they are coeval and that the Vendian-Cambrian boundary was at a different level on the Russian and Siberian Platforms. Subsequent records of similar vendotaenid associations from these horizons have confirmed this conclusion (Luchinina et al. 1978 and others) and have necessitated the allocation of the Rovno Horizon into the Vendian.

As mentioned above, from the very beginning the Vendian has been ranked as a system, although it has been called a series or complex. Menner and Streis (1971), emphasizing the importance of the Vendian geological events, ranked it even higher. The first publication in which the Vendian was recognized as a special system was Sokolov (1964), recently it has been mentioned by Keller (1983b) who discussed the priority of system (period) names, such as Vendian, Sinian – in the new sense of Chinese geologists – as mentioned above (The Sinian System of China 1976) and Ediacarian (Cloud 1968; Cloud and Glaessner 1982). As a very old division of the Cambrian, the latter was reported even earlier (Termier and Termier 1960). However, the author has also widely used the term "protosystem" (Keller 1966a,b), i.e. a system in relation to the Phanerozoic (Sokolov 1972a, 1980). When the Stratigraphic Committee of the USSR discussed major stratigraphic units for the Precambrian of the USSR, it formally retained the term Vendian as a

Precambrian unit (independent of the Riphean) of the standard stratigraphic scale for the Upper Proterozoic (Keller and Shulga 1978; Aksenov et al. 1978; Semikhatov 1979; Keller and Krasnobaev 1983). In this book the Vendian is considered as an independent geological system.

Strictly speaking, units of a system rank have been formally accepted only for the Phanerozoic or Phanerozoic eonothem, and the problem concerning the systematic position for a new system has arisen only recently due to recognition of the Vendian, Sinian s.str. and Ediacarian systems. The latter two, as will be shown later, are not synonymous, differing in extent and geochronological boundaries, viz. 650 ±10 (even 680), 800–850 and 640–670 m.y. respectively for the base of the Vendian (Semikhatov 1979; Keller and Krasnobaev 1983), Sinian s.str. and Ediacarian, on the basis of two dates (Jenkins 1981; Cloud and Glaessner 1982). The upper boundary, however, is to be considered the most important (which is or may be paleontologically substantiated) since it has and may provide paleontological evidence and hence may be correlated due to an adequate biostratigraphic standard chosen for the lower boundary of the Cambrian. The Vendian (based on the same data) is dated at 570 ± 20 m.y. (it commonly includes the Rovno or Nemakit-Daldyn Horizons). The Sinian s.str. and Ediacarian are dated at 560 ± 10 and 540–550 m.y. respectively. The latter quite tentative due to the absence of Lower Cambrian beds in Australia, and stratigraphic equivalents of the Vendian of the USSR have also not been recorded.

However, the duration of the period is equal to or longer than similar major periods of the Phanerozoic eon. However, this alone does not show that the Vendian, Sinian or Ediacarian are lacking Phanerozoic units. Such a conclusion requires the evaluation of the main trend of a general geological process: i.e. whether a certain period starts or terminates a major phase in geological history. However, with regard to the first statement, paleontological characterization of the period, complete biochronological and ecosystematic succession of biotas, diverse fauna and flora, and the level of differentiation and radiation become twice as important. With this in mind, the author (Sokolov 1972a, 1980) has defined the Vendian as a special stage or period (system) in the history of the Earth. As an antithese an attempt was made to relate the Vendian to the Riphean using mainly a continuous sequence of stromatolite buildups associated with the Precambrian carbonate facies.

The Vendian undoubtedly marks a new stage in geological history called the Vendian-Paleozoic or Vendo-Cambrian stage by structural geologists. Now we know that its fossils are very diverse and reflect such a high degree of evolutionary differentiation that the idea of placing the Vendian into the Criptozoic, as proposed by Chadwick in 1930, seems absurd. The earlier concept of the Proterozoic is quite comprehensive and reflects some important ecosystem restructurings, and in this context the Vendian holds a unique position. Hence, its biological and stratigraphic interpretation is difficult. Nominally, it is still the Precambrian, but due to differentiation of life it is already the Phanerozoic, however, its peculiar stages of development and particular groups of organisms allow the recognition of a separate unit and a unique evolutionary stage near the Proterozoic-Paleozoic boundary. In this context it is quite a separate system whose final position in a standard stratigraphic scale must be determined in terms of a general philosophy of geological time, space and evolution of life systems.

Abundant paleontological discoveries were of fundamental importance for the understanding of the Vendian period and its organic life. They were discussed at two All-Union symposia on Precambrian and Early Cambrian paleontology (held in 1965 and 1976), with special emphasis on the largest new occurrences of non-skeletal Metazoa of Ediacarian or, to be more exact, Vendo-Ediacarian type, recorded in Podolia, Ukraine, in the White Sea area, on the Olenek Uplift, and in many other regions of the Russian and Siberian Platforms, and in the Urals (Sokolov 1965a, 1972a, 1975, 1976a; Zaika-Novatsky et al. 1968; Keller et al. 1974; Paliy 1976; Fedonkin 1978, 1980a,b, 1983b and others), as well as megascopic plants of vendotaenid flora (Gnilovskaya 1971, 1976, 1979 and others; Ishchenko 1982). Micropaleontological studies begun by Timofeev (1952, 1959, 1973a and others) and Naumova (1951 and others) have been widely extended and specified (Volkova 1962, 1981; Aseeva 1968–1982; Paškevičiene 1980 and others). Volume I provides the most complete account of all Vendian fossil groups of the USSR, especially of the Russian and Siberian Platforms.

Figure 1 shows the evolutionary pattern of major fossil groups through the Vendian and Early Cambrian. Here, we see peculiar features of the time distribution of the non-skeletal Metazoa, as if terminating in the Redkino age (middle Vendian or late Early Vendian), of trace fossils of benthic organisms, of vendotaenian flora, of abundant microfossils often of uncertain systematic position and of phytolites typical of carbonate facies of the Siberian Platform. In the post-Laplandian epoch, we can recognize three other biostratigraphic units, viz. the Redkino, Kotlin and Rovno Horizons.

Therefore, a four-partite subdivision is generally accepted for the Vendian-type area. In terms of nomenclature they are stratigraphic horizons and regional stages with consistent physical and especially paleontological features. Geographically, a succession of all horizons between the Upper Riphean (Pachelma Horizon and its equivalents) and Lower Cambrian (Tommotian, Lontova Horizon of the Russian Platform) remains persistent over a vast area. The Laplandian or Varangerian Horizon (widely reported as the Volyn Series), containing young tillites of this group, was formed in the course of gradual warming of climate, and the initiation of an extensive post-Riphean transgression, which followed a global glaciation, resulted in an important change of the ecosystem. The Redkino Horizon crowns the Early Vendian and yields an extremely rich biota of the non-skeletal Metazoa of Ediacarian type. The Kotlin and Rovno Horizons form the Upper Vendian, best recorded at present in the USSR and partly in China. A general synthesis of the Vendian is presented in the Conclusion to Volume II (see Chap. "The Vendian period in the history of the Earth", Sokolov 1984). The synthesis was prepared for the 27th Session of the International Geological Congress, held in Moscow in 1984 (Sokolov 1984).

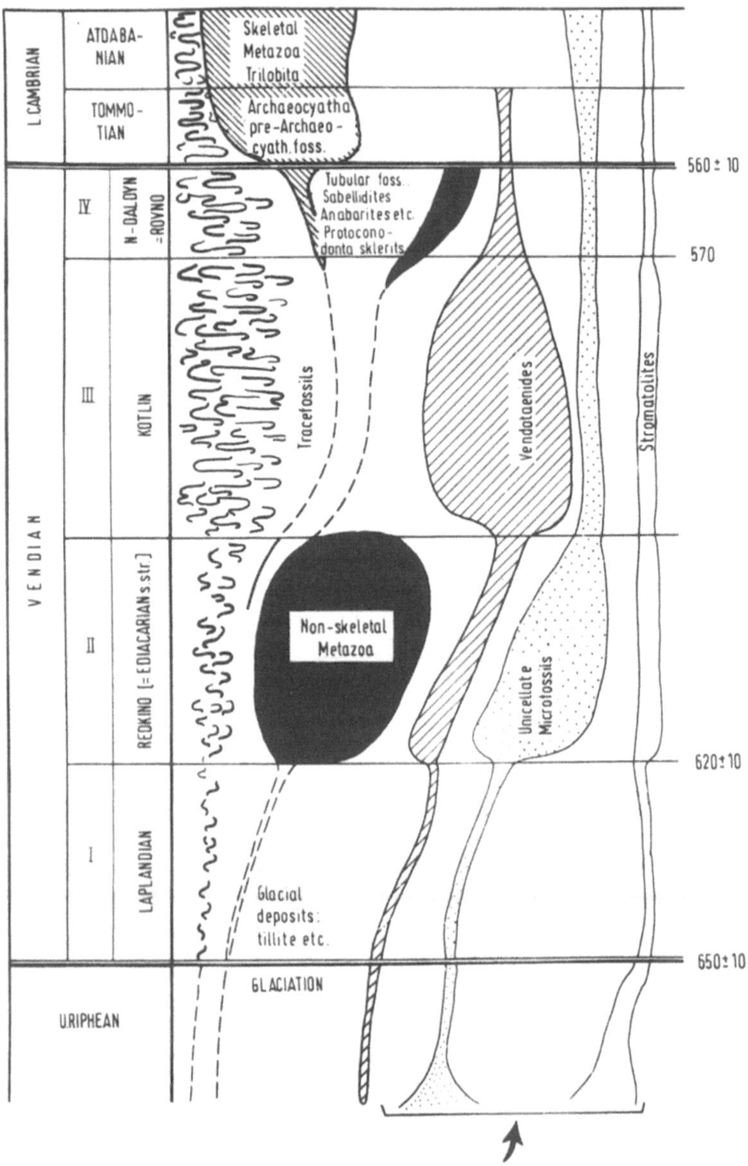

Fig. 1. Development of major groups of organic life in the Vendian and Early Cambrian

Chapter 1

Non-Skeletal Fauna of the Vendian: Promorphological Analysis

1 Introduction

This chapter is devoted to the discussion of major groups of Late Precambrian multicellular fossils, which are the earliest known page of the metazoan fossil record. Below, attempts are made to reveal some of the most important morphological features of the oldest invertebrates, typical only of the Vendian fauna, to better understand the metazoan evolutionary trends.

The present chapter emerged from the study of the rich collections of Precambrian fossils collected by the author from the largest localities in Eurasia, primarily from the south eastern White Sea area, the Dniester Basin of Podolia (Ukraine) and the Olenek Uplift. Material, notable for its excellent preservation and surpassing all the earlier findings in abundance and diversity, helped to widen our knowledge of Vendian fauna, to delineate some new phenomena, typical only of the Vendian, and to appraise their value and scale. On the other hand, study of the organisms, previously known from publications of foreign, mainly Australian, paleontologists, has enabled us to check some long existing concepts, to reveal the weak and strong points of their arguments and to draw other, more substantiated conclusions.

An important feature of Vendian fauna, in contrast to faunas of the subsequent periods of the Phanerozoic, is the absolute dominance of soft-bodied organisms, i.e. animals without any mineralized skeleton. First, tubular forms of mineral skeletons appeared only at the close of the Vendian; stratigraphically, they are recorded at the level of the Nemakit-Daldyn Horizon, Siberia, and probably, a little earlier in the Nama Group, Namibia. But these problematic and non-varied forms are hardly to be considered as typical of the Vendian. It seems more natural to regard them as precursors of the "explosion" of the skeletal Metazoa at the very inception of the Cambrian. The soft-bodied, multicellular animals, known as Ediacaran (or Ediacarian), Vendian-Ediacaran or Ediacara-type fauna, should be undoubtedly regarded as the most dramatic and peculiar feature of the organic world of the Vendian.

The biological uniqueness of the Vendian fauna, apart from other features typical of the organization of its specimens, is represented by the absence of biochemical mechanisms of secretion of mineralized skeletons. This feature, common to the Vendian fauna as a whole, produced some unexpected effects, which are extremely important in terms of methodology. The most important of these can be summarized as follows.

The soft-bodiedness of Vendian metazoans made all megascopic organisms more or less equal with respect to preservation potential. The Vendian can be regarded as the only period in the paleontological record of the metazoans, when all the animals were equal in the face of history. Since the Cambrian, the history of the organic world has been mainly represented by skeletal organisms. This is probably the reason why the paleozoology of the Phanerozoic is basically that of skeletal organisms. The study of Vendian fauna provides us with a unique opportunity to adequately delineate the general panorama of the animal kingdom, although the preservation potential of the soft-bodied animals is determined not only by mechanical properties of the tissue of the organism, but, in great part, by its size, morphology, ecology and taphonomy. Allowances for these features can be helpful in avoiding some misinterpretation.

Another importance effect of soft-bodiedness is the possibility, rare for paleozoologists, to study imprints of their soft tissues and organs. The diversity of preservation degree of non-skeletal organisms demonstrates not only their external forms, but also their internal structure. For instance, it is not unusual to find composite moulds showing the finest canals of the gastrovascular system, gonads or tentacles in the Vendian medusae. Therefore, some new methods of comparative morphology and comparative anatomy can be widely used in the study and interpretation of the fossils.

The excellent preservation of casts and moulds of the Vendian soft-bodied organisms suggests that during the Late Precambrian in many groups of metazoans tissues became denser due to their evolution and differentiation. The mass appearance in the Cambrian of skeletal fauna, especially those groups whose skeletons formed from scleroproteins, may be considered as a direct continuation of the process. Evidence suggests that at that time some Vendian organisms could already secrete fairly dense forms of scleroprotein. The most vivid example is furnished by fine, ridgelike forms of *Redkinia spinosa* Sokolov (Plate 88) which is composed of chitinous matter; the forms were described from the Redkino Horizon, Valdai Superseries of the Russian Platform (Sokolov 1977a). Similar matter was isolated from argillite of the Mogilev-Podolian Series in the Dniester Basin of Podolia by Aseeva (Velikanov et al. 1983). Besides, preparations, made by acritarchologists, commonly contain fragments of opaque, chitinous objects, which are believed to be of animal rather than vegetable origin. All these lines of evidence suggest that the biochemical mechanism of secretion of scleroproteins had already been developed in some groups of Vendian metazoans (Sokolov 1977).

Gigantism is one of the typical features of the Vendian fauna. This phenomenon seems especially striking because the succeeding skeletal fauna of the Early Cambrian metazoans is represented by small forms. Gigantism was primarily characteristic of the most primitive groups of coelenterates, an abundant group of petalonams, flat leaf-shaped worms of the family Dickinsoniidae, and colonial polyps. Some medusoids from the Vendian of the Russian Platform, South Australia and Newfoundland attain 1 m in diameter. Some *Dickinsonia* specimens were also 1 m long; one specimen of *Charniodiscus*, preserved in the Geology and Geophysics Department, University of Adelaide, is 1.2 m high. Taking into account that, along with large forms, minute, 5–10-mm-long organisms had survived, the appearance of megascopic metazoan fossils may be anticipated in the paleontological record immediately after the emergence of multicellular organisms.

It should be noted that gigantism is primarily typical of those groups of metazoans, whose direct ancestors are unknown from the Phanerozoic. Warm and vast Vendian epicontinental seas produced immense masses of such primary food resources as phytoplankton, which formed the basis of the diet for most metazoans, as well as bacteria and primitive organisms which, in turn, served as the basis for many trophic chains. Therefore, the statement that the Vendian was one of the first or even the first epoch of the flourish of metazoans appears to be valid. But why do only some groups exhibit gigantism? One of the assumptions relates this phenomenon to the ecogenetic or phylogenetic termination of the evolution of these groups of metazoans (Sokolov 1977a). In fact, we cannot find direct descendants of these giants.

The Vendian fauna, as a whole, is characterized by high morphological differentiation reflecting taxa of high rank which evolved in the course of the earlier diversification of the main metazoan lineages. At this stage of differentiation, species diversity appears to have been insignificant in the great majority of groups. Only the most primitive groups of invertebrates, particularly gigantic forms, could have a high species diversity.

As yet no reliable data on the diversification of taxa of high rank in the Vendian are available: it seems as if every type and class appeared suddenly. To explain this phenomenon it must be assumed that all the main types of symmetry and architectonics of invertebrates had developed very rapidly during a geologically short time span. At the earlier stages of evolution of metazoans, when their organization may have been relatively low, each new character may have been very important in terms of taxonomy and responsible for any trends of a taxon. However, many of these characters may have not been vital for the function and physiology of the organisms themselves. Therefore, some forms of Vendian metazoans could be virtually more primitive than later invertebrates with which they are compared. Hypothetically, the process of rapid realization of the main versions of invertebrate architectonics seems more probable immediately after, or even in the process of, the development of multicellularity, i.e. during a relatively short time span. This assumption is indirectly supported by the following lines of evidence: about 60% of classes of marine multicellular animals appeared during Vendian-Cambrian time, and only 5% classes of multicellular animals appeared in the time interval from the Permian to the present. Thus, the assumption that metazoans may have had a very long history of evolution in the Precambrian has no solid basis at present. The following morphological analysis of the main groups of Vendian multicellular animals points to an extreme primitivity and archaism pertinent to many groups and rather suggests the late appearance of metazoans.

Despite the excellent preservation of some Vendian metazoans, their classification and systematic interpretation can pose severe problems: thus, the interpretation of some taxa varies over a wide range. In this study the problems are approached by one of the methods of comparative anatomy, namely, the comparative analysis of promorphology or the body-plan analysis of Precambrian non-skeletal metazoans.

Some of the reasons which make us resort to the analysis of promorphology are as follows. First, in lower metazoans symmetrical features vary widely and thus allow us to recognize taxa of low rank. A diversity of symmetries exhibited by specimens of the class Hydrozoa provides an illustrative example. Therefore, the analysis of recent coelenterates is impossible without invoking data on their symmetry. Second, the number of characteristic morphological features that can be used to establish the systematic position of Vendian metazoans is small. For example, in the system of recent coelenterates one must use such features as the presence and structure of cnidoblasts, the position of gonads in ectoderm and endoderm, etc. Unfortunately, fossils do not always exhibit these structures. And the last, but not the least reason to employ the promorphological method is the fact that the existing classificatory system of Metazoa cannot incorporate many Precambrian fossils. Numerous attempts to compare the Vendian soft-bodied organisms with extant invertebrates and even with fossils of later epochs are most often disputable. Hence, it would be more promising to consider the Vendian fauna, not as ancestors of the Phanerozoic fauna, but as descendants of the older Precambrian metazoans, still unknown to us. And, if in this case we consider the Vendian soft-bodied organisms as evolving in Eocambrian time, then we obtain a system not associated with the later metazoans. Therefore, the only objective approach to be used is the promorphological method whose efficiency was brilliantly proved by Beklemishev in his work "Principles of comparative anatomy of invertebrates" (1964).

Below we shall discuss and compare the main body plans of Vendian invertebrates, based on the concept of symmetry as the most precise and developed technique in the study of comparative anatomy. We shall try to reveal principal structural features of the oldest known Metazoa and, as a result, to give a more accurate description of similar and differing characters, which arise from phylogeny.

2 Promorphology of Vendian Radialia

Vendian fauna, in the case of megascopic multicellular animals, was no doubt dominated by coelenterates, such as solitary and colonial polyps and medusae. The Vendian can be characterized as a time of explosive development of coelenterates (Sokolov 1976b). In fact, it was precisely this type of invertebrate that displays the greatest abundance, and species and ecological diversity in the Vendian. Coelenterates constitute up to 67% of 1500 specimens of soft-bodied metazoans, collected in the Flinders Ranges area at Ediacara, South Australia (Glaessner 1972, 1979b). Various coelenterates account for 70% or more of the total number of described species of Vendian biota from the White Sea area (Fedonkin 1980c, 1981a,b). The predominance of coelenterates in the Vendian metazoans can be considered to some extent as an artefact. Vendian coelenterates usually have a greater body size. They contain abundant and various polyps, which deformed the sediment while still in life position, thus producing structures capable of preservation and fossilization. We can exclude the possibility of protracted transport of the dead organisms and can assume a rapid burial process at the habitat without considerable post-mortem alterations. This could be beneficial for the preferential preservation of coelenterates. But not only the large size of individuals and colonies increased the possibility of imprinting coelenterates in the sediment. The capacity of cnidoblasts to preserve their function for some time after the death of an organism may have protected it from carnivores, which could live on other organisms, thus excluding them from the paleontological record at the very beginning; hence, the greater number of coelenterates as compared to other groups of metazoans. On the other hand, a fair amount of small-sized invertebrates, as compared to the dominating coelenterates, suggests that the former may have provided food for the latter. This is quite possible, if we keep in mind that all living coelenterates are predators. Their Precambrian ancestors may well have terminated the trophic chains, not suffering themselves from predators and scavengers. If we accept the widespread statement that the skeletization by many groups of metazoans in the Cambrian could be regarded as an adaptability response of animals to active predation, then the soft-bodiedness of most Vendian multicellular animals implies the lack of active, mobile predators in the Vendian Sea. If active predators did exist in the Vendian, they do not appear to have been macrophages (as well as the vast majority of other Vendian metazoans). Therefore, the necessary and sufficient reaction of possible victims (especially, sedentary forms) may have been an increase in body size. In any case, no impression of a body damaged by predators or scavengers has been found among Vendian fossil metazoans as yet.

The low weight of infauna in benthic communities inhabiting the Vendian Sea was conductive to the intactness of buried organisms and structures formed by their bodies in the sediment (later, imprints, moulds). Taken together, these factors may have contributed to the preferential burial and preservation of coelenterates, and, as a result, we cannot rule out the possibility of some overstatement of their proportion and, hence, species diversity of this phylum as compared to other metazoan groups. However, such an overstatement, if any, hardly changes the general pattern. There are localities which contain exclusively non-skeleton coelenterates; the most interesting of these is the Conception Group, Avalon Peninsula, Newfoundland. Numerous tuffs suggest frequent, catastrophic burials of animals in the shallow-water environment, but, nevertheless, as yet none of the numerous stratigraphic levels has yielded any fossils other than coelenterates. All the paleogeographic and geochronological data suggest that this locality is the oldest (Anderson 1972). This fact, along with the high diversity of Vendian coelenterates, supports the phylogenetic models, which regard coelenterates as the oldest metazoan phylum.

The great evolutionary distance separating Vendian soft-bodied animals from recent Metazoans, as well as large gaps in the fossil record of non-skeletal cnidarians, as a whole, make the elucidation of the systematic position and phylogenetic relations of the Precambrian metazoans rather difficult. Attemps to place the Vendian animals within taxonomic schemes of post-Vendian living taxa are usually disputable, in spite of the fact that the placement of certain groups of Vendian metazoans into classes and orders of the system of living coelenterates was based on the consistency principle, but not on the principle of full correspondence, whose usage is excluded because of the character of paleontological material. Nevertheless, most groups do not fit the present system of coelenterates. Hence, the classification of Vendian material, which will help to understand the relationships between certain groups of coelenterates, primarily for the Vendian, is necessary.

Unlike Enterozoa, grouped into Bilateria, all coelenterates form a particular division of the animal kingdom, Radiadia. Radial symmetry is the most important taxonomic character of coelenterates. Various forms of radial symmetry largely control the extent of systematic taxa of low rank. The diversity of various kinds of radial symmetry in coelenterates may indicate that the evolution of this phylum led to the attainment of various forms of radial symmetry. This is evident even if one has only a slight idea of the highly diverse types of symmetry in the living Cnidaria. For instance, some of the most primitive hydroids contain forms displaying extremely archaic types of symmetry, the variable multiradial. This type of symmetry, with a rotation axis of variable high order, is typical of planula larvae of most hydroids, in Octocorallia of the Anthozoa and a few other groups (Beklemishev 1964). A monaxonic, heteropolar form with a symmetry axis of variable high order is an initial form of symmetry, common to all multicellular organisms. But, the overwhelming majority of hydroids exhibit symmetry of some definite, almost always even order, usually four-fold. Finally, some forms demonstrate incipient bilateral symmetry against the background of radial symmetry. Symmetry of odd order, viz. three-, five-radial and so on, known in hydroids and, in general, in coelenterates, is rather the exception.

The four-fold (or tetramerous radial) symmetry dominates the Scyphozoa, and in this respect higher medusae represent a separate, much more homogeneous class. Only some specialized forms of the class, primarily belonging to the suborder of rhizostoms (Discomedusae Rhizostomeae), show a deviation from the four-fold symmetry or even from the regular radial symmetry. Unlike the Scyphozoa, higher polyps, or Anthozoa, exhibit a less perfect radial symmetry, which is, as a rule, incomplete and does not embrace the entire body. Superficially, all anthozoans are more or less radial, as reflected both in the shape of the body and in the distribution of tentacles, whose number can be divided by 6, 8, 10 or 12. However, their internal structure displays a strong tendency towards bilateral symmetry. The symmetry of some coelenterates can vary widely due to the combination of its different types.

Precambrian Radialia are also highly diverse, though they differ in some symmetry characters from living and fossil animals of this type. Below is a description of the morphology of medusoids and solitary polypoids with particular emphasis on their symmetry. Based on symmetry, Vendian Radialia can be subdivided into three main groups: first, numerous forms with a symmetry axis of infinitely high order; second, forms with a finite, but indefinite number of antimeres or radially arranged elements, whose number increases with growth; and third, forms with a constant order of symmetry.

2.1 Forms with a Symmetry Axis of Infinitely High Order

The most abundant and diverse group of Vendian coelenterates consists of extremely simple forms, lacking any morphologically expressed radial elements, which, if any, are few and unordered. This is a monaxial, heteropolar symmetry of infinitely high order.

2.1.1 Undifferentiated Forms

The structurally simplest organisms of this group are sedentary *Nemiana simplex* Palij and *Beltanelliformis brunsae* Menner. *Nemiana* are usually preserved on the lower surfaces of rock slabs as convex discs with margins bending towards the centre. As a rule, they occur in groups or colonies and completely cover bedding planes for a distance of many metres. Locally, they are so closely spaced that they deform each other and form a kind of polygonal mesh (Paliy 1976). Casts show that the organisms has calyx-like bodies without any appendages (Fig. 2). The surface of their bodies was smooth, without any signs of sculpture. The relief of casts suggests the presence of well-developed mesogloea, which performed supporting functions. The oral side, sometimes preserved on some casts, has a small rounded cavity in the centre, a sort of a mouth. As yet, no specimens demonstrating that *Nemiana* had any radial elements (septa, tentacles, etc.) have been found. The largest specimens of *N. simplex* exhibit the finest concentric furrows on the aboral side of the body. These structures probably suggest the presence of an annular musculature in the basal part of the polyp.

Beltanelliformis brunsae Menner could be regarded as similar to *Nemiana* in some structural elements and in mode of burial. They also occur in abundance as convex discoid casts with a flattened central portion and fine concentric rugae around the periphery. Margins of the largest specimens bend towards the centre of the oral side. This species was described as the aboral side of the bell of the medusoid organism (Keller et al. 1974b). However, the study of the character of preservation, based on mass material, rather suggests that these organisms were sessile, i.e. solitary polyps (Fedonkin 1981a) with a developed annular musculature, particularly around the periphery of the aboral side. No radial elements have been observed and no certainly assignable oral side is known.

Nemiana and *Beltanelliformis* are the most primitive organisms among the known Precambrian metazoans. They show a monaxial heteropolar symmetry with a rotation axis of infinitely high order; this type of symmetry is also typical of many other Vendian Radialia. But unlike the latter, the forms in question exhibit no signs of concentric organization, which, as shown below, is characteristic of Precambrian coelenterates. The symmetry and body plan of *Nemiana* and *Beltanelliformis* are probably similar to the hypothetical gastrulalike ancestor of all multicellular animals, conventionally placed at the base of the phylogenetic tree of Metazoa. This

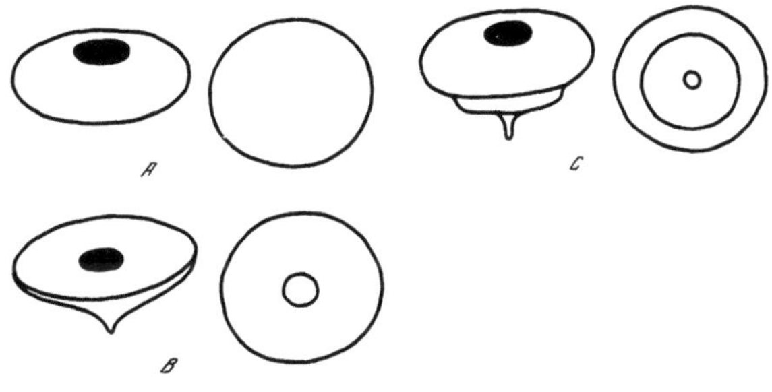

Fig. 2A-C. The most primitive forms of coelenterates of the class Cyclozoa. **A** *Nemiana simplex* Palij; **B** *Tirasiana disciformis* Palij; **C** *Paliella patelliformis* Fedonkin. *Left*, reconstruction; *right*, aboral side

structure is actually the most primitive and, possibly, the earliest. Comparable symmetry is rare in the recent cnidarians and occurs, in particular, in epalpate hydroid Protohydra, Hybrida (Beklemishev 1964, vol. 1, Fig. 15). It is tempting to regard *Nemiana* and *Beltanelliformis* as direct descendants of the earliest multicellular organisms, commencing the evolution of eumetazoans. The two genera are worthy of being classified as a separate taxon of higher rank, since other Vendian Radialia, with a symmetry axis of infinitely high order, are marked by a concentric organization of the discoid body. This could be regarded as a further step in the realization of spatial possibilities of the monaxial, heteropolar symmetry with a rotation axis of infinitely high order.

2.1.2 Forms with a Concentric Differentiation of the Body

The concentric body plan appears to be predominant in Vendian coelenterates, which can be placed into at least three groups. The first group exhibits a relatively regular concentric structure of wide rings on the aboral side of the bell. The second group includes medusoids with a prominent, regular concentric body plan. In both groups the number of concentric zones increases in ontogeny. The third group represents forms with a constant, but small number (two or three) of concentric rings, preserved throughout the lifetime of the organism. Some organisms of the above groups have radial morphological elements as well, but they are extremely unordered and their number varies.

2.1.2.1 Oligocyclic Forms

This group is most vividly exemplified by some species of the genus *Cyclomedusa* Sprigg. Organisms in this genus are medusoid in shape (Fig. 3). Their aboral side has several to many near-concentric rugae which indicate, by their excentricity, a general conical shape for the centre or even most of the body. Many but not all of these have radial structures of varying degrees of coarseness. The conical, rugose, flexible body indicates that it did not swim like a medusa, with the muscles reacting against a mesogloeal "skeleton". On the contrary, the rugae suggest radial contractility, and it seems reasonable as a contractile animal only if attached in life by the apex. This projecting apex is preserved on most of the *Cyclomedusa* species, thus suggesting that it was relatively resistant. Thus, the probability is that, like the *Aurelia* scyphistome and *Stephanoscyphus* (Chapman 1966; Wade 1972a), *Cyclomedusa* developed (or retained in a phyletic sense) chitin at the point of

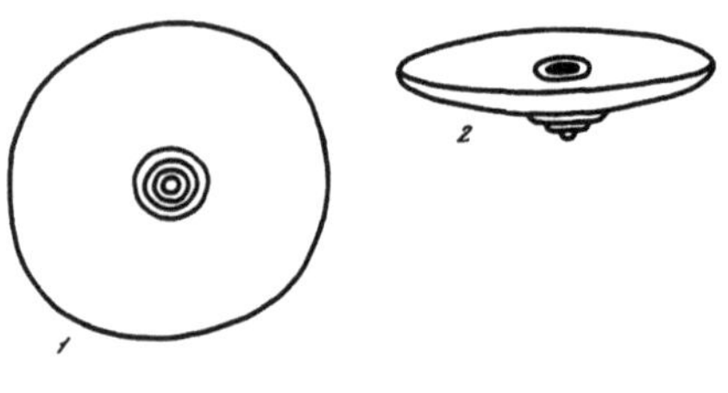

Fig. 3. *Cyclomedusa plana* Glaessner and Wade. **1** Aboral side; **2** reconstruction

attachment. The fixed mode of life of *Cyclomedusa* is also suggested by the state of preservation of fossil material (with a few exceptions, the aboral side is usually well preserved) and by the mode of reproduction of these organisms: at least some species had reproduced asexually by simple longitudinal fission. This mode of reproduction was observed in *Cyclomedusa plana* (Glaessner and Wade 1966) and, probably, *C. davidi* (Wade 1972a). It is of interest that this type of asexual reproduction is a peculiar feature of the class Anthozoa. Most Hydrozoa and Scyphozoa had lost the ability for asexual reproduction; a freely moving mode of life had made high demands on the integration of the organism, hence, the much higher level of organization of Scyphozoa as compared to polyps. Therefore, only rare hydrozoans, for instance, Aequoraeidae, some Eucopidae and Cunina larvae may have reproduced by longitudinal fission (Beklemishev 1964). Some modes of asexual reproduction were also recorded in other groups of Vendian coelenterates with a concentric body plan; hence, it would be more reasonable to regard these forms as polypoids.

Fine, straight furrows interpreted as gastrodermal canals are discernible on the aboral side of some forms, in particular, *C, davidi* Sprigg and *C. radiata* Sprigg. On *C. davidi,* the radial furrows are very numerous and variable. They never extend directly to the central peak (attachment organ), but cross the entire body from the second ring furrow outside the central peak to the margin. No regularity in the addition of furrows has been observed but, if well expressed, they tend to maintain a relatively constant spacing except on the outer ring or rings where new furrows are rarely inserted. These are interpolated between older furrows in forming the radial structure. On other specimens there may be few or no furrows. A similar increase of furrows (Fig. 4) is demonstrated by *C. radiata* as well (Wade 1972a). For example, in *C. plana* Glaessner et Wade which demonstrates some annular zones only in the centre, the large, smooth bell bears radial furrows that could be regarded as fortuitous creases on the mesogloea formed in the process of burial. In general, the question whether the radial furrows are equated to gastrodermal canals can hardly be considered solved if one takes into account the following two factors. First, the aboral side bears concentric rugae and they are basically interpreted as a mesogloeal relief. But in this case, the question may be raised as to why the radial furrows cannot reflect the sculpture of the aboral side of the bell, especially as they are quite straight and unbranching, and new radial furrows, emerging on the outer rings, are interpolated between older furrows, but do not detach from them. In general, these features are not typical of canals of the gastrovascular system. The second factor is associated with the evidence that in medusae (all morphological characters suggest that *Cyclomedusa* is closely related to the Scyphozoa, in spite of the fixed mode of

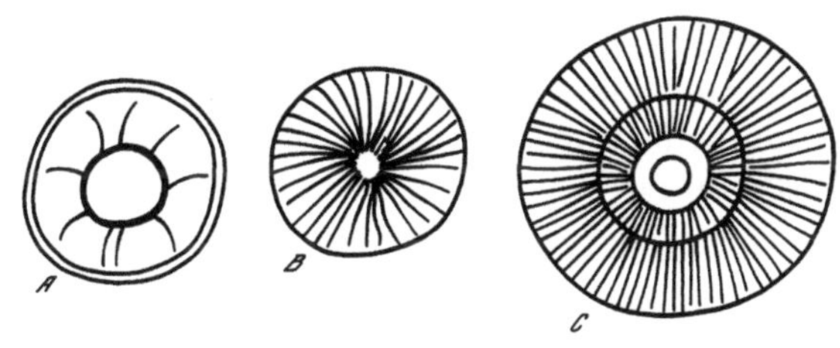

Fig. 4A-C. Body plan of Vendian medusoids with irregular radial symmetry (monoaxial, heteropolar symmetry of variable order). **A** *Medusinites asteroides* Sprigg; **B** *Irridinitus multiradiatus* Fedonkin; **C** *Cyclomedusa radiata* Sprigg

life) canals of the gastrovascular system run nearer to the oral surface; from the top, the canals are covered by the exumbrella mesogloea, which can be very thick on large organisms. Had the *Cyclomedusa* had the canals and had they occupied the same position as that on the living medusae, it can hardly be expected that they would have "imprinted" through a thick layer of the solid mesogloea. Moreover, all the available lines of evidence suggest the presence of the mesogloea on the *Cyclomedusa*. The alternatives are: (1) if the *Cyclomedusa* had had gastrovascular canals, they could be only on the exumbrella, but this appears to be highly improbable; (2) radial furrows of the *Cyclomedusa* correspond to the radially dissected relief of the mesogloea surface from the aboral side. The latter seems more probable. Although this question seems to be rather particular, in this case it defines the level of organization of Cyclomedusae as the most typical representative of Vendian coelenterates, which is quite important. Moreover, in the study of fossil organisms, so remote from us in time, it seems methodically justified to put forward as many alternatives and hypotheses as possible, and to compare the fossil and the modern forms rather than to seek similarities in rare and, at present, uncharacteristic phenomena, since the norm itself can produce a hypnotizing effect on a researcher.

Since the architectonics of *Cyclomedusa* is typical of many groups of Vendian coelenterates, fairly heterogeneous organisms have been placed into the genus. Further investigations will possibly help to better recognize species and to establish new genera of this unwieldy, all-embracing group. The morphological diversity of fossils often shows only the multiformity of preservation of non-skeletal organisms, whereby the negligence of this factor could lead to exaggerated synonymy.

A small specimen of *C. cf. plana* from the Zimny Coast of the White Sea is possibly the oral side, where a smooth subumbrella with a mouth in the form of several cups inserted one into the other is visible in the centre (see Fedonkin 1981a, Plate III, Fig. 1). The absence of concentric rugae or rings on the outer side of the bell indicates that this zone was incapable of radial contractility. This suggests that *C. plana* was not attached in adult life. Presumably, swimming with the aid of a disc edge was a possibility, or its near-flat shape may have enabled it to lie free on the seafloor, as some modern medusae do. These two modes are not mutually exclusive (Wade 1972a).

Wade regards *Cyclomedusa* as a possible ancestral type of chondrophore (Chondrophora, Hydrozoa), a functional intermediate between normal, attached Hydrozoa and the pelagic forms with a float (pneumatophore). *Cyclomedusa* may represent a hydrozoan form sharing common ancestry with the Chondrophora. This assumption is confirmed by the following evidence: (1) its low conical form thins to an outer edge which is an extension of the coenosarcal disc edge rather than a separate outer flange which the medusae possess; (2) its internal radial structure resembles the radial gastrodermal canals of young *Porpita* in shape and position; (3) the probability that it was adherent by the apex suggests the secretion of chitin where the float rediment appears in the development of recent chondrophores (Wade 1972a).

2.1.2.2 Polycyclic Forms

Some Vendian organisms show a fairly regular concentric body plan combined with a bilateral body plan in some cases. Vendian *Eoporpita medusa* Wade exhibit a purely concentric body plan (Wade 1972a; Fedonkin 1981a). These are rather large solitary polypoids, circular or elliptic in shape. The oral side of the relatively flat body carries two series of club-shaped "tentacles" or polypides. They form several outer whorls (dactylozooids) of near-constant length, and several inner whorls (gonozooids) of inwardly reducing side, encircling a single, large central cone (gastrozooid). The aboral surface is smooth except for very fine radial striae with a small, low, central dome. Radial striae may be somewhat related to gastrodermal canals (Wade 1972a). The float, which is rather rarely preserved, is delicate, with numerous narrow, annular chambers enclosing a small, circular, central chamber. The smooth, very finely striate surface is believed to represent the fleshly surface layer on the aboral side of the float.

The structure of the float in *Eoporpita* is strongly annular and bears no close resemblance to that in the Ediacaran bilateral chondrophores *Ovatoscutum concentricum* Glaessner et Wade and *Chondroplon bilobatum* Wade (Wade 1971), which were resistant enough to form external moulds. The chamber sutures are much less sharply defined in most specimens of *Eoporpita* and this tends to substantiate the much softer material of the float suggested by the usual preservation as casts.

The shape of the float is a critical consideration in classifying chondrophores. *Eoporpita* appears to have had narrow circular chambers with a few radial creases and no definite bilateral symmetry. Not only does it differ from the other Ediacaran chondrophores, but also from Paleozoic Porpitidae (Moore and Harrington 1956). Vendian *Eoporpita* is morphologically closer to the Recent *Porpita*, but there are some differences. The tentacles or dactylozooids of *Eoporpita* are invariably simple, club-shaped and similar in size. Polypides of the Recent *Porpita* are variable and morphologically complex, and through the growth of *Porpita* younger whorls are acquired outside the older whorls. The whorls of gonozooids are also supplemented by the outward growth in *Porpita*, whereas the inner whorls of the inner series of tentacles are smaller in *Eoporpita*. The fact that the inner and outer whorls of *Eoporpita* are similar in shape suggests a lack of polymorphism and functional differentiation of the whorls. It is of interest that as yet no polymorphism of polyps has been observed in Vendian organisms, which could be interpreted as colonial coelenterates. Functional differentiation of polyps in coelenterates may have developed much later.

It seems unlikely that the same family of coelenterates could have survived the dramatic Phanerozoic history without suffering serious variations. Nevertheless, Glaessner (1979b and others) believes that this may be attributed to the mode of life of Porpitidae which, along with other chondrophores, now constitute the pleuston of tropical seas. They may have been abundant in Vendian time since they inhabited the ocean surface layer and spread with the current and wind. A quiet habitat medium, rich in food resources, might have contributed greatly to the survival of the family. This is also suggested by the species homogeneity of

Porpitidae. The only factor which might have led to the complete disappearance of the group was predation (Glaessner 1979b and others).

In the author's opinion, there are strong arguments against the assignment of *Eoporpita* to the Chondrophora and to the Hydrozoa in general. First, *Eoporpita* and the Recent *Porpita* differ greatly in the mode of whorl development (see above). Second, *Eoporpita* exhibit a very simple and uniform structure of polypides (or tentacles!), and conversely, the *Porpita* are complex in structure, particularly in the structure of gonozooids, and are marked by morphological and functional differentiation of polypides. Third, the largest of the Recent *Porpita* are an order of magnitude smaller in size than *Eoporpita*.

Fossil chondrophores with tentacles or polypides are usually rare. The only form which was placed into the family Porpitidae is *Paropsonema* Clarke (Silurian-Devonian), but it has distinctly branching tentacles (Ruedemann 1916; Moore and Harrington 1956; Chapman 1966). The Lower Devonian form *Palaeonectris discoides* Rauff of the family Vellelidae has dactylozooids, simple in shape, with blunt ends, i.e. similar to those of *Eoporpita*. One specimen of this species bears a thick ridge across the body, interpreted as a sail.

The family Chondroplidae, known from moulds of the float, holds a unique position among multicellular organisms of the Vendian (Wade 1971). This family had bilaterally symmetrical, flat floats consisting of fine, concentrically curved tubular chambers which narrowed from one end of the axis to the other, leaving a notch with the apex pointed towards the centre of the body and oriented along the axis. In one of the forms the notch is seen in the region of the narrow ends of the tubular chambers, but in another form it is discernible at either end of the axis. The first form, *Ovatoscutum concentricum* Glaessner et Wade, has preserved the signs of concentric structure better than the latter, *Chondroplon bilobatum* Wade (see Fig. 12).

Ovatoscutum represents a shield-shaped or oval disc with a slightly marked axis of bilateral symmetry (Glaessner and Wade 1966; Wade 1971; Fedonkin 1981b). A small oval chamber, aligned along the axis and concentrically embraced by long, narrow chambers, is situated at the centre. Chamber breadth is greatest adjacent to the axis at one end, and at the other end, at some distance from the centre, the chambers are progressively less embracing, leaving a triangular notch in the peripheral part, where the chambers are narrowest. The notch is oriented along the axis with the apex pointed towards the centre, but not reaching it.

Chondroplon is much larger in size (the largest specimen is about 19 cm long and 24 cm wide) and consists of two rounded lobes (Wade 1971). The axis of the body is strongly marked due to a ventral groove and probably a blunt dorsal keel; the initial chamber is relatively large and rather elongate; early chambers embracing the initial one are annular, later chambers are progressively less embracing, thus leaving a notch at one end of the axis (proximally), where the chambers are the broadest. The latest chambers form a notch at the opposite end of the axis (distally), where the chambers are minimum in breadth and length and are arranged transversally. The relief of the mould suggests that the chambers were higher than broad.

According to Glaessner (1979b), the similarity between the *Ovatoscutum* and *Chondroplon* floats, on the one hand, and the pneumatophores of the living

Porpitidae, on the other hand, is so impressive and, more important, specific, that there is no necessity at all to know the morphology of the polyps in order to place these Vendian forms into the class Hydrozoa.

Paleozoic deposits yielded some putative chondrophores. *Archaeonectris benderi*, which may be most reliably assigned to chondrophores, was reported from the Ordovician or Early Silurian beds of Jordan (Huckriede 1967). *Protolyella*, abundant in the Lower Cambrian and previously interpreted as a medusa of uncertain systematics (Moore and Harrington 1956), was later assigned to the chondrophores (Glaessner 1972). It is quite possible that *Velumbrella czarnoskii*, described from the upper Lower Cambrian (or from the lower Middle Cambrian, Bednarczyk 1970) of Poland and interpreted as the earliest trachymedusa (Stasinska 1960; Scrutton 1979), may well belong to the same order as well. This form has rigid, radially divergent ribs and such a regular outline that it is an elastic float rather than a medusa (Glaessner 1979b). Despite the burial of *Velumbrella* in a relatively mobile substrate and its randomly oriented imprints, they show no deformation common to small and thin medusae, but rather resemble relatively solid discs. *Discophyllum* Hall from the Middle Ordovician of North America (Walcott 1898) and *Parapsonema* Clarke from the Upper Devonian (Clarke 1902; Ruedemann 1916) of North America are also placed in the family Porpitidae. *Plectodiscus* Ruedemann from the Devonian of North America (Ruedemann 1916) and the genus *Palaeonectris* Haeckel from the Devonian of West Germany are placed into the family Velellidae. Against the background of such a diversity (with a possibly in complete fossil record) two extant genera of chondrophores, namely, *Porpita* and *Velella,* look like relics of an old and once diverse group of Hydrozoa which, in Glaessner's opinion, more than 600 m.y. ago discovered the ocean surface layer as an ecological niche, rich in food, and adapted to it. The diversity of this group drastically decreased at the close of the Paleozoic, probably due to aggravated competition and predation in the ocean surface layer (Glaessner 1979b). If the above interpretation is correct, then chondrophores could be regarded as the only group of Vendian Metazoa with direct descendants in modern fauna. This seem unlikely, since this is not the case with other groups. Recently, it has seemed rather easy to distinguish the Hydrozoa and Scyphozoa among Precambrian coelenterates and assign most fossils to the existing orders (Fundamentals... 1962). However, the state of preservation of soft-bodied organisms does not always allow the recognition of diagnostic characters, on which the systematics of living coelenterates is based. And, second, Precambrian organisms may have been void of taxonomic characters, which serve as the basis for the establishment of orders of modern fauna. Nevertheless, the study of Precambrian fauna provides extremely valuable information which can contribute greatly to the solution of some basic problems relating to phylogeny.

A morphological entity of Vendian chondrophorelike animals is worthy of consideration. The concentrically regular type float of *Eoporpita* seems more primitive, as compared to others, and can be considered as initial for chambered chondrophores. From this type of architectonics of the float, the type of symmetry exhibited by *Ovatoscutum*, could be developed by means of insignificant transformations, first along the diameter, leading to the formation of bilateral symmetry, and then across. The first stage of transformation may yield the plan of structure of a transitional form whose fossil will probably be found one day. After the appearance of the plane of bilateral symmetry, further transformation (second stage) is associated with the region of thin ends of chambers on either side of the plane, which could result in the appearance of a notch. Thus, we have found a new manner of development of bilateral symmetry and segmentation from radial symmetry. But, unlike those workers who develop Articulata from Radialia with a finite number of radial elements (Beklemishev 1964), we believe in an earlier divergence of these groups: a bilateral articulate animal, possibly, irregularly articulate at the first stage, but still segmented, can be developed directly from a form with a bilateral symmetry of infinitely high order and concentric body plan. We can obtain convincing results if we continue our transformations and fix them at stages in agreement with real organisms.

The float of *Ovatoscutum* can be easily transformed to that of *Chondroplon* by a further increase of bilaterality, and a decrease of concentricity due to an increase and unfolding of the notch (see Fig. 12). There the transformation proceeds in the direction perpendicular to the plane of symmetry, considerably continuing and enhancing the tendencies which had resulted in the emergence of *Ovatoscutum*. Thus, we obtain a normal, bilaterally symmetrical organism. Of particular interest is the inference that a slight, further transformation would produce the symmetry of Dickinsoniidae, i.e. flat, leaf-shaped organisms whose organization has long been a subject of controversy. Nevertheless, these bilateral, segmented forms are not true articulates yet, but only an extremely primitive branch of Metazoa, probably without any descendants in Phanerozoic history.

A simple approach of morphological transformations to study Vendian forms was taken from the brilliant book by D'Arcy Thompson "On growth and form" (1942). It vividly demonstrates the value of the euristic method of transformation, whose history can be followed far back in time. The method is particularly effective in comparing closely related forms, as it provides better knowledge of the trend and character of evolutionary transformations resulting in their divergence. However, the same method applied to the study of fossil organisms of uncertain nature (to say nothing of the degree of affinity) can help to better classify related organisms into groups.

The above transformations of body plan of Vendian organisms do not mean at all that *Eoporpita* is an ancestor of *Ovatoscutum* and, hence, that of *Chondroplon*, but they do demonstrate a possibility of a natural evolutionary transformation of the most primitive, concentric type of symmetry into the bilateral symmetry typical of articulates, showing homonomous metamerism. This way, with landmarks recorded by Vendian non-skeletal Metazoa, appears substantiated and, therefore, probable. As to chondrophores, they all derived from a primitive form with monaxial, heteropolar symmetry, a rotation axis of infinitely high order and concentric organization. With regards to Precambrian organisms, *Eoporpita medusa* Wade is the closest to this hypothetical form, although other groups of Vendian Metazoa contain regular concentric forms, in particular, *Cyclomedusa minuta* Fedonkin, *C. delicata* Fedonkin (Fedonkin 1981a) and *Kullingia concentrica* Glaessner (Foyn and Glaessner 1979), as well as *Kaisalia mensae* Fedonkin. These forms exhibit a regular concentric organization primarily reflecting the character of growth. In other words, concentric forms are stages of growth, and, hence, the change in morphology as the organisms grows, due to addition of new rings. However, there is a matrix which can be assumed even in these polyconcentric forms, but it is most purely expressed in such a characteristic form as *Ediacaria flindersi* Sprigg and some others: i.e. the division of the discoid body into central and peripheral zones. Concentric zones of such polymeric forms as *Kullingia* and *Kaisalia* could be equated to concentric soft septa which add to the surface of the gastric cavity and mechanical strength of the body.

2.1.2.3 Forms with a Constant Number of Concentric Rings

This group exhibits a different body plan, but the same type of monaxial, heteropolar symmetry with a rotation axis of infinitely high order. Like the above group, this one is characterized by a concentric organization, but the number of concentric rings is, as a rule, small and remains constant throughout the lifetime of the individual. This suggests a distinctly differentiated structure of the body and a possible specialization of some parts, i.e. a higher general organization, as compared to polymerous (polyconcentric) forms. *Ediacaria* are usually preserved as low exumbrella casts in the form of a disc embraced by a flat, smooth and broad ring (Fig. 5). The disc is often slightly elevated above the outer ring with a slight peak in the centre. The radius of the central disc is from one-quarter to three-quarters of the total. A deep, rounded groove separating the disc from the annular zone could indicate the edge of the gastric cavity. Outside this groove, fine, radial furrows or striae are sometimes preserved. The subumbrella structure, which is rarely preserved, demonstrates a small central mouth without any appendages, and numerous fine, marginal tentacles.

Ediacaria is the largest Vendian medusoid. The diameter of its bell can exceed 0.5 m. But small forms are also common. This may be explained by the attached or fixed mode of life. Sedentariness of the form is inferred from the character of reproduction. Mass occurrences provide unquestionable evidence for some types of asexual reproduction in *Ediacaria*, such as external budding, longitudinal fission, multiple fission as its modification, and stolonate reproduction. Asexual reproduction of modern coelenterates in much more typical in sessile forms than in freely moving medusoids, regardless of some exceptions. Therefore, the mode of reproduction and the character of preservation of *Ediacaria* are most likely indicative of sedentariness.

Ediacaria appear to have been relatively flat, step-conical forms attached by the apex or merely lying on the seafloor, aboral side downwards. It is of interest that

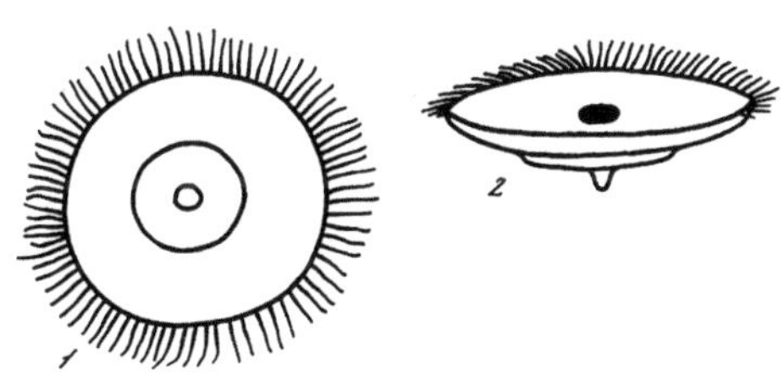

Fig. 5. *Ediacaria flindersi* Sprigg. **1** Aboral side; **2** reconstruction

Tirasiana disciformis Palij, extremely similar to *Ediacaria*, provide evidence for the following phenomenon. In relatively weathered rocks, *Tirasiana* casts are locally ferruginate, i.e. they have a brown film, whereas the slight, central peak is usually marked by fresh truncation, suggesting its projection into underlying rocks. This probably indicates that *Tirasiana* had a small appendage by which they attached to the seafloor.

Binary fission of *Ediacaria* (and *Tirasiana*) is associated with the development of a septum and two attachment apexes on either side. Such a simple, longitudinal fission is likely to start directly with the development of a pair of attachment apexes, i.e. the fission initiates in the oldest part of the body, corresponding to the initial evolutionary stage of the organism. The septum separating the apexes progressively deepens and the apexes diverge. At first, the body becomes elliptical, and then eight-shaped, then the bells separate. In some cases both individuals remain side by side. Multiple fission, whereby the body is divisible into three or four segments at once (Velikanov et al. 1983), is occasionally observed as well.

Longitudinal fission as a mode of asexual reproduction is typical of Vendian coelenterates and is observed, in particular, in *Cyclomedusa plana* Glaessner et Wade, *Paliella patelliformis* Fedonkin, *Nimbia occlusa* Fedonkin, *N. dniesteri* Fedonkin, *Tirasiana disciformis* Palij, *Ediacaria flindersi* Sprigg and others. However, the preservation in one species of some modes of reproduction, irrespective of individual growth, is quite a primitive feature, suggesting that no particular mechanism was responsible for the process. If budding and fission could be regarded as similar processes, then stolonate reproduction may be considered as a somewhat different type. A long, stemlike stolon starts in the centre of the parent individual and terminates with a small daughter individual at its distal end. In general, the stolon is invariable in width from start to end (1–2 to 6–7 mm) and 15 cm and greater long.

The Ediacarian form can be interpreted in the following way: the central disc appears to really correspond to the gastric cavity, whereas the outer flat ring may have served for passive capture of organic detritus suspended in water. This is one reason why *Ediacaria* and similar flat, conical polyps contain the largest forms: passive capture of organic detritus or plankton from water, with minimal energetic expense of movement, must have been beneficial, and the fixed mode of life was optimal in the Vendian Sea, which was rich in plankton, void of active filtrators and predators and without any hard struggle for the substrate. This may be considered as one of the causes responsible for gigantism, primarily typical of sessile forms.

Ediacaria and some other forms with a rigidly fixed number of concentric rings may be marked by fine, radial furrows, mostly confined to a relatively wide zone of the outer ring. Examples are furnished by *Ediacaria flindersi* Sprigg, *Medusinites asteroides* Glaessner et Wade and *Paliella patelliformis* Fedonkin. The radial structures are not present in all specimens, but under favourable conditions they look like abundant, straight, unbranching rugae or ridges. In *Ediacaria*, the number of radial furrows on the wide outer ring increases towards the periphery, not by branching, which is very little, but usually by interpolating new furrows between older ones. This method of adding furrows is known in some *Cyclomedusa* (Wade 1972a). These radial furrows are surface structures rather than gastrovascular canals. However, the state of preservation suggests that these relief forms are associated not with the surface of the body, but rather with that of the mesogloea.

In forms with a fixed number of concentric rings, the central disc was equated to the gastric cavity, and metabolites are likely to have been distributed within the body directly through the mesogloea which performed, apart from supporting and protective functions, a distribution function to a larger degree than in the modern Cnidaria. In this group, the gastric cavity extended laterally and had neither septa nor other large structures, which were able to increase its surface, necessary for fast assimilation. If *Ediacaria* and *Cyclomedusa* used a large bell for passive capture of planktonic "rain" with further transportation of food to the central oral cavity, then only a very high plankton biomass and low density of other consumers of seston in the pelagic region and on the seafloor could provide suitable conditions for these large, flat, conical polyps. Therefore, shelf zones were the most favourable for benthic forms in the Vendian. Such were conditions set up in shallow and warm epicontinental seas as a result of post-glacial transgression.

The main conclusions from our study of Vendian coelenterates with a symmetry axis of infinitely high order can be summarized as follows.

1. Forms of coelenterates with a monaxial, heteropolar symmetry of infinitely high order generally belong to the most primitive, though flourishing group of Vendian coelenterates and metazoans as a whole.
2. This group exhibits such common features as type of symmetry, concentric body plan, the lack of well-developed radial elements of the gastric cavity, the lack of radially arranged, branching gastrovascular canals and the presence of different modes of asexual reproduction, particularly simple, longitudinal fission.
3. All these features suggest a primitive and archaic character of the group, and a monophyletic origin of its representatives. However, homogeneous though it may seem, as early as the Vendian the group had been differentiated enough to be divided into three subgroups on the basis of differing body plan, character of growth, and, possibly, modes of reproduction.
4. The first subgroup consists of organisms with, or even without, irregular concentric structure. The symmetry and the body plan of the latter are similar to the hypothetical gastrulalike ancestor of all multicellular organisms, usually placed at the base of the metazoan phylogenetic tree. This structure is the most primitive among Vendian coelenterates and, possibly, the earliest in the history of the phylum. Comparable symmetry is rare in the modern cnidarians and known only in the larval stage and in adults of apalpate hydroid Protohydra.
5. The second subgroup includes medusoids with a prominent, regular concentric body plan, in which the number of concentric rings increases in ontogeny. This subgroup also contains organisms similar, in the shape of the body and in mode of life, to those of the first subgroup, which are marked by an irregular concentric organization. However, this subgroup includes new forms with initial bilateral symmetry, apart from a regular concentric body plan. These forms allow the derivation of bilaterally segmented animals from

the forms with concentric body plan and a symmetry axis of infinitely high order. Organisms of the second subgroup are extremely close to the living chondrophores, but do not belong to the order Hydrozoa.

6. The third subgroup of coelenterates with a symmetry of infinitely high order is composed of forms with a constant, but small number of concentric rings, which remain invariable throughout their lifetime. This group is characterized by the functional differentiation of different concentric rings, thus suggesting a relatively higher organization of this subgroup, as compared to other taxa of the group.
7. In fact, each of the three subgroups demonstrates a combination of symmetry types and different body plans, in particular, radial symmetry of infinitely high or uncertain order with concentric organization, or bilateral symmetry with concentric body plan.
8. Among organisms with a symmetry of infinitely high order and a concentric organization (concentric body plan) the sedentary forms are the most common; pleustonic and planktonic animals are rare. This predominance of one ecological type in the group of the most primitive Coelenterata may provide evidence for the primary nature of a polypoid stage (Fig. 6). An anisotropic environment on the seafloor and a spectrum of various habitats that was wider than in the pelagic zone could have promoted such a diversity of forms. Similarity in symmetry in sedentary polyps and pleustonic organisms suggests their genetic affinity and may reflect the history of their assimilation on the water surface.
9. Vendian coelenterates with a symmetry of infinitely high order show one of the ways of attaining diversity within this type of symmetry. This results in the concentric zonation of the body. It is followed by the polymerization and ordering of concentric rings, then by oligomerization in combination with radial symmetry of uncertain order and bilateral symmetry.
10. The peculiarity of body plan, growth, reproduction and ecology of the above discussed organisms with a monaxial, heteropolar symmetry of infinitely high order and concentric body plan, along with their probable monophyletic origin, allow us to classify this group as Cyclozoa, a separate class of Coelenterata (Fedonkin 1983a, 1984).

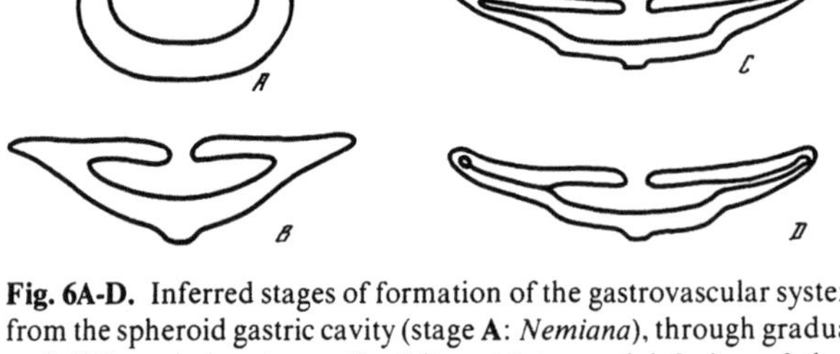

Fig. 6A-D. Inferred stages of formation of the gastrovascular system in the phylogeny of coelenterates from the spheroid gastric cavity (stage **A**: *Nemiana*), through gradual flattening (stage **B**: *Cyclomedusa*) and differentiation (stage **C**: *Ediacaria*) to partial fusion of the lower and upper surfaces of the endoderm, resulting in the formation of a plexus, straight radial or other canals according to the mesogloeal relief (stage **D**)

2.2 Forms with a Symmetry Axis of Uncertain Order

As mentioned above, we have discussed along with coelenterates showing a pure concentric body plan, the forms which combine concentric and radial organizations, mainly represented by tentacles, polypides and mesogloeal relief. These forms are nothing but a combination of two types of body plan, namely, a concentric and radial organization. Moreover, they combine two types of symmetry: a concentric symmetry with a rotation axis of infinitely high order and a radial symmetry with a rotation axis of uncertain order. Unlike the symmetry of infinitely high order, variability in the order of symmetry is determined by a finite, but uncertain number antimeres (tentacles, gonads, etc.).

As a whole, Vendian coelenterates yield many forms showing the concurrent presence of concentric and radial symmetry. This suggests historical links between the concentric and radial body plans and, on the other hand, it also implies a conventional boundary between forms with concentric body plan and animals marked by the combination of a radial and relict concentric body plan. These or those kinds of concentric organizations are known in some modern groups of coelenterates, however, radial organization almost always predominates. Nevertheless, a large proportion of corals and medusae show no signs of concentric body plan and evolve by attaining a radial body plan which, in many respects, proved to be more advanced.

Among Vendian Radialia with a symmetry axis of variable order, *Lorenzinites rarus* Glaessner et Wade and *Hiemalora* (= *Pinegia*) *stellaris* Fedonkin are structurally the simplest. These are relatively small, low polyps with radiate lobes. A single specimen of *Lorenzinites* was reported from the Pound Quartzite of South Australia (Glaessner and Wade 1966); the status of the taxon is still uncertain (Glaessner 1979b). *Hiemalora* (Fig. 7) was found at two richly fossiliferous localities on the Russian Platform (Fedonkin 1980b, 1981a,b) and on the Olenek Uplift.

The state of preservation of *Hiemalora* suggests that the genus had medusoid and polypoid generations. Besides, we cannot rule out the inference that in relatively early ontogeny, *Hiemalora* may have been polypoids and became puberal medusoid in later stages, when they dwelt in the pelagial and performed a habitating function. *Hiemalora* are occasionally preserved as casts showing a projecting disc much lower than the plane bearing tentacle imprints. Tentacles are rather thick, some slightly widening distally. Probably, these are casts of sessile forms. But more common are relatively thin, flat moulds and casts showing larger

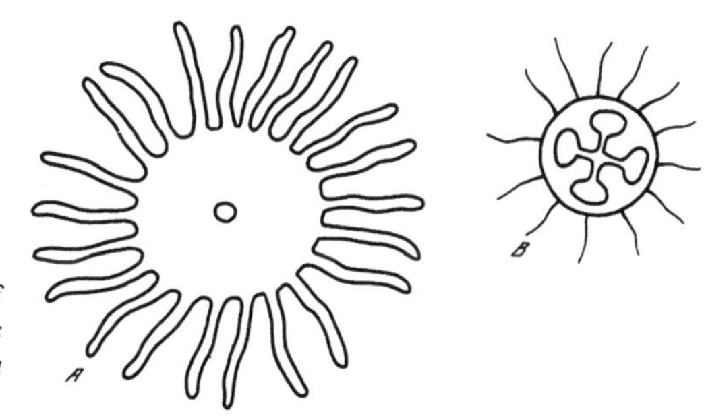

Fig. 7A,B. Some typical specimens of Vendian plankton. **A** *Hiemalora stellaris* Fedonkin; **B** *Stauridinia crucicula* Fedonkin

discs and thinner tentacles; this state of preservation may reflect the medusoid stage. None of a great many well-preserved specimens exhibits any structures of the central part of the disc, except for a small, round mouth in a few specimens. This can be easily explained: the centre of gravity of medusae is, as a rule, shifted into the mesogloeal region, towards the animal pole and, therefore, dead organisms usually swim and fall to the seafloor with their head downwards. Therefore, casts showing the aboral sides of medusae are more common. Polyps are also preserved mainly as casts showing the aboral sides; this is determined by the life position of animals in the sediment.

The absence of any signs of a radial body plan in the polyp body or in *Hiemalora* bell suggests a wide gastric cavity, cyathiform and discoid in shape respectively, in polyp and medusa. The thickness of tentacles suggests the continuation of the gastric cavity into the tentacles. The mode of reproduction of the organisms is unclear, but it could be both asexual and sexual. One of the *Hiemalora* specimens demonstrates a possible longitudinal fission of the bell, i.e. the imprint resembles an ellipse with radiate tentacles, which tapers down at the ends of the short diameter. The specimen was found in the Khatyspyt Formation, Olenek Uplift (Khorbusuonka River). The other specimen from the Vendian deposits, Zimny Coast of the White Sea, is of interest as it bears numerous random bosses in the disc region. These bosses could well be primitive reproductive organs, although the position of the reproductive organs on most living coelenterates is strictly ordered. However, the ordered position of gonads on modern *Cnidaria* is governed by that of the gastric canals. If *Hiemalora* had no canals, which is quite probable, gonads could have a random position. The bosses may be also interpreted as larvae occupying a broad chamber or cavity. A similar phenomenon is observed in *Eleutheria*, the genus of Hydromedusa, which is quite atypical at present (Hyman 1940). This medusa secretes both eggs and sperm in the wall of the epidermal brood sac located above the stomach (Fig. 126A in Hyman 1940, p. 423). *Planula* larvae developed in the same sac. *Hiemalora* might well have a similar sac.

The growth of *Hiemalora*, which can be inferred from series of individuals from small juveniles to largest adults, indicates that marginal tentacles are added while the organism is growing, but their number is not necessarily divisible by any particular value, as in most living cnidarians. However, in living cnidarians, particularly in many *Leptolida* (Athecata, Hydrozoa) the hydranth also does not bear a strictly fixed number of tentacles irregularly scattered over the head. Such a structure is known in Zanclea and some other Corynoidea, Clava, Myriothele, Hydra, and some others (Beklemishev 1964; Ivanov et al. 1981). In living cnidarians, this is the lowest type of symmetry. However, Beklemishev (1964) was right when he wrote that this is not a dissymmetry known at the end of lineages. This is the case when symmetry is not "already absent", but "still absent". This is an uncertain arrangement that precedes a true radial symmetry. In *Hiemalora*, like in the simplest hydroids, tentacles only disorganize the symmetry, but do not determine its order. If we ignore the tentacles, then *Hiemalora* would have a monaxial symmetry with a rotation axis of infinitely high order, as in many Vendian coelenterates and living Protohydra.

Radial symmetry of variable order is also inherent in other forms of coelenterates, even in relatively highly organized forms, for instance, *Hallidaya brueri*

Wade (Wade 1969). These small, discoid medusae were found in the uppermost Precambrian deposits of Central Australia; they reflect a relatively late stage of scyphozoan radiation during the Late Precambrian (Jenkins 1981). The main reason for placing *H. brueri* into the class Scyphozoa is due to a rather complex system of radial gastric canals which diverge from a zigzag furrow around a central gastric cavity and branch dichotomously at least two times towards the margin of the bell. It is not quite clear how the number of radial canals vary as the medusa grows, but as for rounded "nuclei" grouped around the stomach, 3 to 13 are known in different specimens. The nuclei may also be interpreted either as medusoid buds (but, then, why is their later development not known?), or as gonads, perhaps, those in which the embryos are incubated during early development. This is suggested by the "behaviour" of the nuclei in impressions as firm, discrete bodies; in specimens in which they are very numerous, they somewhat overlap. There is also the possibility that a central mouth is represented by the star-shaped slits.

A variable number of reproductive organs are observed in *Elasenia aseevae* Fedonkin (Fig. 8) from the Vendian of Podolia, Ukraine (Velikanov et al. 1983) and, possibly, in *Armillifera parva* Fedonkin from the Vendian of the Zimny Coast, White Sea (Fig. 8), if arcuate structures on the bell of this minute medusa were really equated to gonads (Fedonkin 1980b).

Evmiaksia aksionovi Fedonkin, similar to the above forms in general body plan, has a wide outer ring, but the centre is embraced by ovate structures with radially oriented long axes (Fig. 9). These structures are interpreted as gonads rather regularly spread along the margin of the gastric cavity (Fedonkin 1983a, 1984). We are not sure of the number, but it increases as the organism grows, irrespective of any order of symmetry.

Disorder in radial symmetry is better seen on reproductive organs as the most dense structures on the body of medusae and is better preserved in fossils, especially in small individuals with a thin mesogloea. Disorder is also observed in a system of radial canals, for instance, in *Rugoconites* Glaessner et Wade (Glaessner and Wade

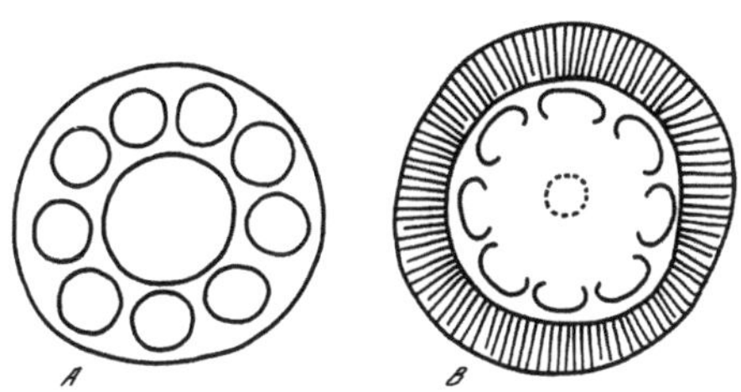

Fig. 8A,B. Body plan of Vendian medusae with the characteristic radial symmetry of variable order. **A** *Elasenia aseevae* Fedonkin; **B** *Armillifera parva* Fedonkin

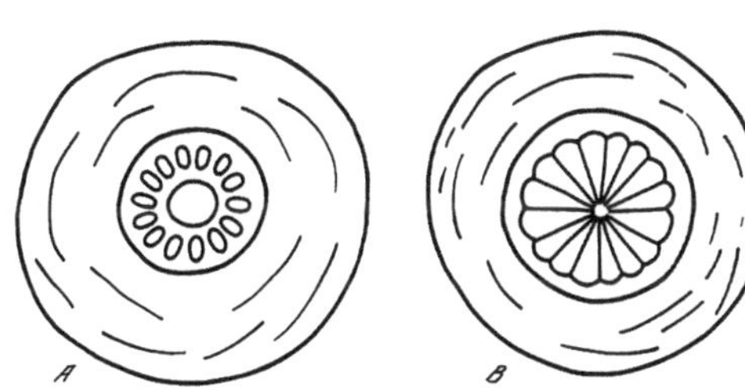

Fig. 9A,B. Body plan of some Vendian medusae with the characteristic symmetry of variable order. **A** *Evmiaksia aksionovi* Fedonkin; **B** *Bonata septata* Fedonkin

1966; Wade 1972a). These low, conical medusae have a few radial canals which diverge from a small polygonal centre in the region of the stomach and branch dichotomously. The margin of the bell, which is usually preserved as a ring, bears multiple, fine tentacles. They do not start from the very edge, but a little closer to the centre, although it is not quite clear on which side of the marginal part of the bell, either on the exumbrella, like some recent Hydrozoa, for example, Trachymedusae, or on the subumbrella, as inferred from the state of preservation.

The number of radial canals diverging from the stomach cannot be accurately counted, as these medusae had the thickest mesogloea in the central part and, therefore, the canals could not produce imprints in the sediment. But examination of the best preserved central parts of some specimens including the stomach region shows some 10 to 20 canals. Specimens with a few initial radial canals have a strongly branching gastrovascular system (canals branch dichotomously up to five times, compared to three times in other specimens). An other feature which could also be regarded as a primitive character is the disordered branching of radial canals and even their coalescence, resulting in the formation of something similar to a network. No annular canal is probably present, while numerous, thin and parallel distal ends of radial canals pierce the marginal part of the bell.

Among the modern medusae, Proboscidactyla (Hydrozoa) seems to bear a strong resemblance to Rugoconites, but the coalescence of canals and the formation of a network represent a rare teratological phenomenon in the living Hydrozoa (Wade 1972a). Not only simple branching canals, but also the absence of the annular canal are common to these forms. An analogy is probably to be sought in rhizostoms. For instance, Cassiopeia (Rhizostomeae) has a rather dense gastrovascular system, which fills the space between straight radial canals, and has no annular canal (Fig. 173A in Hyman 1940, p. 526). However, unlike both Proboscidactyla and Cassiopeia, Vendian forms display a variable number of radial gastrovascular canals even in the area where they leave the stomach.

A number of other forms, for example, *Bonata septata* Fedonkin (Fedonkin 1980b, 1981a), show disordered radial structures or antimeres. These rounded medusoids consist of two concentric rings (see Fig. 9). The outer ring is rather flat, wide, with rare fortuitous concentric wrinkles suggesting that the marginal zone of the bell is thin and fragile. The inner ring has more relief, is fairly convex and consists of radial, wedge-shaped lobes with rounded, distal and acute proximal ends not reaching the centre. The wedge-shaped lobes may be equated to chambers lying between radial septae in the gastric cavity of the organism and thus *B. septata* could be interpreted as a lineage of the Scyphozoa and/or Anthozoa. The lobes could be also interpreted as gonads. In any case, the lobes are added as the organism grows (Vendian deposits of the Zimny Coast, White Sea area, yield specimens displaying 11–19 lobes, but a larger number cannot be ruled out).

Vendian medusoids often demonstrate a combination of radial and concentric body plan. Some of them were described under the section dealing with Radialia. They show a symmetry of infinitely high order, for instance, *Eoporpita* (concentric float and concentric series of radial polypides) or *Ediacaria* whose radial structures are represented by fine marginal tentacles and a fine radial relief of the aboral side of the mesogloea.

One of the forms of Vendian medusoids, namely *Mawsonites spriggi* Glaessner et Wade, also shows the combination of concentric and radial body plan, but the concentric body plan preserved only in the central part of the bell looks like a relict and is forced out by the radial body plan (Glaessner and Wade 1966). In life, this form presumably represented a dome-shaped disc becoming steeply conical near the centre. The central conical part is smooth-walled, truncated and surrounded by a concentric series of large, irregular bosses which increase in size outwardly. The peripheral zone consists of irregular lobes separated by deep furrows, which branch in some specimens. The periphery of the bell is lobate.

Mawsonites combine features of two other Vendian coelenterates. The concentric structure of the central part of *Mawsonites* resembles that of *Cyclomedusa plana* Glaessner et Wade, but the series of bosses resemble series of polypides in *Eoporpita medusa* Wade. The lobate peripheral part of the bell suggests a similarity with some forms of the Scyphozoa, especially with those of Rhizostomeas and Coronata.

The following generalizations have emerged from the study of Vendian Radialia with a symmetry axis of uncertain order:

1. Forms of coelenterates with monaxial, heteropolar symmetry and with an axis of uncertain, but finite order comprise a large and diverse group among Vendian coelenterates. Rotation symmetry of variable order is reflected in the whole organism as well as in different organs (marginal tentacles, canals of the gastrovascular system, gonads, etc.). In some specimens this type of symmetry is reflected in the character of mesogloeal "relief". Forms with this type of symmetry were, in the Vendian, more widespread than now or even in the Paleozoic.
2. The main peculiarity of the representatives of this group of Radialia is an increase in the order of rotation symmetry in ontogeny, that is, new antimeres are added while the organisms are growing, but no order or succession in their formation have been observed. This feature appears to be an extreme primitive character, however, it could be regarded as a necessary stage in the evolution of coelenterates towards the radial symmetry of a stable order, i.e. to those types of symmetry which were mainly typical of Phanerozoic cnidarians.
3. Despite the primitive mode of formation of antimeres, this group shows, on the whole, a much higher organization than Radialia with an axis of symmetry of infinitely high order. The members of the group with variable symmetry usually have a well arranged system of tentacles and a different and complicated system of gastrovascular canals which branch dichotomously several times outwardly and sometimes form a plexuslike network. The system of reproductive organs (gonads) in many forms of this group indicates the prevalence of sexual reproduction; asexual reproduction of this group of Radialia is rare. This seems to indicate that in the level of organization the Vendian coelenterates with symmetry of variable order is much higher than Radialia with symmetry of infinitely high order (class Cyclozoa), although the combination of symmetry types, characteristic of the first and second groups, suggests a phylogenetic relation between them.

4. The peculiarity of this group of Radialia and, in particular, the most important promorphological phenomenon, typical of all its representatives, namely, the radial symmetry of variable order allows us to consider the group as a special class, Inordozoa, in the system of coelenterates.

2.3 Vendian Coelenterates with Radial Symmetry of Stable Order

Coelenterates with a stable order of radial symmetry are not as numerous among the Vendian Radialia as representatives of other groups with a more primitive organization and archaic symmetry. This proportion, obviously opposite to that observed in recent coelenterates, indicates that the ordering of radial symmetry in the form of stabilization of the number of antimeres was a relatively late development, representing a specific and main trend in the evolution of Coelenterata. Actually, in many characteristics the species of Precambrian Radialia with an axis of symmetry of stable order are more similar to living forms than other Vendian forms. Nevertheless, in this group we observe characters which are neither typical of the coelenterates nor of the metazoans in general.

One such peculiarity is a three-fold symmetry. This order of symmetry is extremely rare in recent coelenterates, and in other groups of Metazoa it is always a secondary phenomenon. This symmetry was, however, widespread in the Vendian.

2.3.1 Three-Fold Symmetry

Three-fold symmetry is distinctly demonstrated by *Albumares brunsae* Fedonkin (Keller and Fedonkin 1976; Fedonkin 1981a). These small medusae are trilobate in structure. Delicate gastrovascular canals, three on each lobe, divide dichotomously four times each towards the periphery. There are over 100 short, very thin, marginal tentacles. Three relatively short ridges arranged between the lobes narrow radially. The ridges are probably equated to gonads, since a similar structure is typical of some modern medusae. Elongate gonads, radially oriented, are observed in many hydroids, particularly Phialidium (Leptomedusa) and Homoeonema (Trachymedusa), but on these the gonads are arranged along radial canals. The position of the gonads between the lobes in *Albumares* suggests an analogy with the Scyphozoa which usually furnish examples of rotational displacement of systems of antimeres relative to one another, when the order of symmetry, similar in both, is completely preserved. In so doing, antimeres of one system occupy the position in the interradial zone of another, for instance, the system of oral arms and gonads in *Aurelia* (Fig. 17a, Hyman 1940, p. 523) are displaced with respect to one another by 90°, when both systems show a four-fold symmetry.

The lobate structure of the bell, the complex branching of the gastrovascular canals and the character of gonad arrangement of *Albumares* primarily resemble those of the Scyphozoa, but no three-fold symmetry is known in its normal specimens. However, here taxonomic problems arise, and in this context the question may be raised as to whether the order of radial symmetry is an argument of the same (or greater) taxonomic value as other characters, allowing the placement of *Albumares* in the class Scyphozoa. The problem has some aspects which cannot be avoided by classification. First, the significance of this or that character to the normal, vital activity of the organism. Wide variations in radial symmetry of coelenterates seem to suggest that a radially symmetrical animal is slightly affected by the order of symmetry it has, as a change in the order of symmetry induces no critical changes in the relations between the animal and the horizontally isotropic habitat. The number, size and position of vitals, mainly controlled by the order of symmetry, appear be much more important. Therefore, the survival of the species would be determined not so much by the order of symmetry, as by the structures, one way or another associated with the order of symmetry. Modern coelenterates show a substantial variation in the number of antimeres and in the order of symmetry within the limits of individual variability (Thompson 1942).

The above medusa (*Albumares*) with a three-fold radial symmetry seems to be related to *Anfesta stankovskii* Fedonkin. The discoid body of the latter is divided by three ridges, probably equated to gonads, which were sausage-shaped and diverged radially, but their outer ends did not reach the periphery of the bell. Unlike *Albumares*, the bell of *Anfesta* is discoid, but not trilobate, though it stems from the rotational, three-fold symmetry. The discoid bell of *Anfesta* consists of 18 narrow, major lobes. The distal ends of the lobes are notched into two minor lobes, each of which is dissected in turn by a small, radial furrow into narrow, smoothly sharpened marginal lobes carrying minute tentacles. It is of interest that the proximal ends of the major lobes orient towards the nearest radius coinciding with the axis of gonads. Only the lobes arranged directly above the gonads and along the bisectrix of the interradius orient radially. This suggests that *Anfesta*, like *Albumares*, is based on a trilobate structure, but in the former it is obscured by numerous, narrow lobes and the discoid bell. *Anfesta* have neither an annular canal, nor radial gastrovascular canals, although the lobes might have been hollow and appendages of the cavities could have reached the marginal lobes. This type of distribution system is more primitive than the actual gastrovascular canals. As a whole, the character of arrangement of lobes in *Anfesta* resembles the structure of the gastrovascular system in some Scyphozoa, for instance, *Aurelia*. Resemblance can also be seen in the following features. Major radial lobes of *Anfesta* arranged above gonads and minor lobes branching from either side of the major lobes resemble, in general, perradial branching canals of *Aurelia*, and lobes, arranged along the bisectrix of the interradius in *Anfesta* may be equated to adradial unbranching canals of *Aurelia*. However, such a comparison can only be valid for the organization of the distribution system.

Although *Anfesta* and *Albumares* could be considered as related forms, thin, dichotomously branching canals of *Albumares* substantially differ from the lobes of *Anfesta*. Nevertheless, Vendian medusae have a form whose distribution system is intermediate in structure between primitive lobes and more advanced branching canals.

Skinnera brooksi Wade (Wade 1969), a medusa with a pronounced three-fold symmetry, like *Albumares* and *Anfesta*, but differing in internal structure, was described from the uppermost Precambrian of Central Australia. The central, relatively smooth area may be equated to the stomach, which is surrounded by three, symmetrically arranged, large, semicircular inner depressions, interpreted by Wade as gastric pouches, with a concave curve towards the axial region. A network of paired canals connects the three inner pouches to the centre on one side and to 15 secondary pouches in a ring outside them. The paired canals connecting the secondary pouches to the stomach are well seen in the spaces between the primary (gastric) pouches, but where the paired canals cross the area of primary pouches they are not seen in the imprints and, hence, it is not clear whether the secondary pouches are connected to the stomach through the primary pouches, or whether the paired canals bypass the primary pouches and attach directly to the stomach. Between the depressions corresponding to the discernible secondary pouches are also other linear, radial structures which may be equated to radial canals passing from the stomach directly to the periphery of the bell. No ring canal is observed.

Of 26 composite moulds of the medusa found up to now some specimens show a similar type of symmetry: 3 primary pouches and 15 secondary pouches. However, other specimens have three, but not five, secondary pouches, compared to one primary. This provides evidence for the dominant radial symmetry with a rotation axis of the third order. The combination with a rotation symmetry of higher order, probably divisible by three, is possible due to an increase in the number of secondary pouches.

According to Wade (1969), "the shape of the internal system of spaces and canals in *Skinnera* closely parallels the gastrovascular system of a medusa of the scyphozoan grade of complexity".

Inner primary pouches connected to the regions of the stomach by paired canals are considered as the direct homologue of the scyphozoan gastric pouches, but none of the living medusae of the class Scyphozoa has pouches similar to those of *Skinnera*. The system of secondary pouches attached to the stomach by paired canals, in fact, does not differ from that of the primary (inner) pouches, suggesting that the primary and secondry pouches constitute a single physiological system, namely, the gastrovascular system which, however, has no morphological analogues in the present Coelenterata. The system of secondary pouches with paired canals leading to the stomach and with intermediate unbranching canals could be regarded as a homologous plexus of Semaeostomeae and Rhizostomeae, but with a more uniform plexus development. On the other hand, the system of secondary canals could be regarded as an intermediate stage on the way from the primitive ring sinus, which may have been typical of the simplest organisms, to thin branching gastrovascular canals. The other and, possibly, earlier stage of the formation of the gastrovascular system in the Coelenterate is represented by the lobate structure of *Anfesta*, though these two ways might have been unrelated as successive stages, however, they represent independent ways in the formation of the gastrovascular system. As for the three major depressions in the central part of the bell of *Skinnera* ("inner pouches" according to Wade 1969), however, a different interpretation is possible: they may be regarded as gonads which are, as a rule, the most dense organs in small medusae.

To complete the characterization of the Coelenterata with the unique three-fold symmetry let us dwell on a problematic organism, whose interpretation presents particular difficulty. This is *Tribrachidium heraldicum* Glaessner (Glaessner and Daily 1959; Glaessner and Wade 1966; Fedonkin 1981a,b 1982). The disc-shaped organism is preserved as impressions (negative hyporelief) or as casts (positive epirelief). The former demonstrates a steeply sloping peripheral margin with three raised brachia (arms) radiating at equal angles within (Fig. 10). The arms (according to Glaessner) curve clockwise (in artificial casts representing the bodies as deposited) to join the periphery of the disc where they taper. A small, central Y-shaped groove which, in Glaessner's opinion, may represent the mouth, is rarely seen in the centre between the arms.

Attached to each arm on its convexly curved side, is a small, raised area ("bulla"). The distal seven-tenths of each arm bears short, stout "tentacles" on the outer side and tip. Fine, long, straight or gently curved linear structures (furrows) may extend from the crest and concave side of each arm across the tips of the adjoining arms. With a hint of doubt Glaessner considers this side oral.

Both the Australian material and the collection from the Vendian of the Zimny Coast, White Sea area, contain specimens demonstrating a different type of preservation. These are casts on the overlying silty slabs: they are very flat, disc-shaped and rather rare. The centre has three arcuate grooves with their narrow extremities oriented towards the centre of the disc (see Fig. 10). The outer tips of the arcuate grooves are somewhat wider and deeper due to rounded and broadened terminals. The disc surface bears tiny radial furrows which enter the arcuate grooves. The periphery has no tentacles. The disc edge is smooth (Glaessner and Wade 1966; Fedonkin 1981b).

These features suggest that this is a turned-up oral side. The arcuate grooves with dilated terminals may be equated to the mouth slits, although we cannot rule out that only dilated terminals correspond to cavities which served food intake. In this case, the tiny radial furrows falling into the arcuate grooves may be interpreted as a microrelief promoting transport of food particles from the entire area of the bell to the mouth cavities, possibly with the aid of the ciliated epithelium.

If the side with the three arcuate grooves is an oral one, then concave imprints on the lower surface of siltstone or sandstone slabs, described by Glaessner, should be regarded as casts of the internal mould. This mode of preservation is rather widespread, for example, fine medusae *Albumares*, similar to *Tribrachidium* in size, are preserved in the form of a concave mould on the base of rock slabs, however, this imprint displays readily discernible gonads and branching gastrovascular canals. Therefore, with due regard to the possibility of the impression of some internal structure and the probability that the oral side of *Tribrachidium* is represented by a different mode of preservation, the following assumptions can be

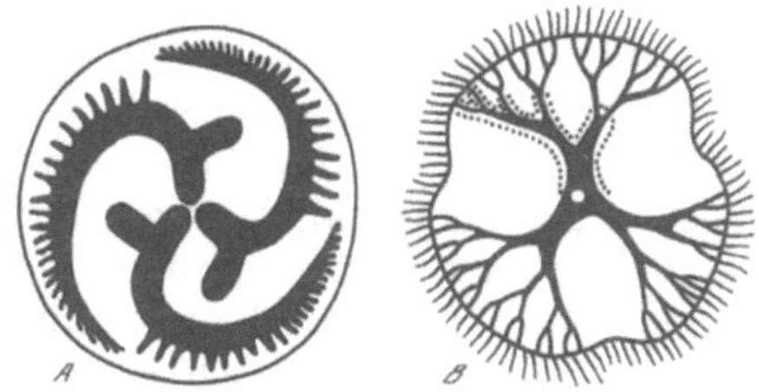

Fig. 10A,B. Comparison of the structure of arms of *Tribrachidium* with the gastrovascular system of a hypothetical medusa (*dots* show the area from where the arm of *Tribrachidium* is produced)

made. The "arms", bearing short "tentacles" at their distal ends and furrows branching into interbrachial spaces near the centre, could be regarded as the distribution system, similar in organization to the gastrovascular canals. The arms are not closed up in the centre, but exhibit rounded, proximal ends which form a Y-shaped groove, the mouth, according to Glaessner.

On the whole, the organism could be conceived as a sessile form which passively caught organic detritus from water. Food particles, caught by the disc, were transported with the aid of the ciliated epithelium through the narrow furrows to the three mouth slits, from where food reached the three gastrovascular canals at once, to their proximal, large diverticula and then to numerous distal branches, radially oriented towards the disc edge; the branches may be regarded as tentacles. Large diverticula ("bullae") could well be interpreted as the gonads.

If such an interpretation proves true, then *Tribrachidium* is to be considered as an organism at the coelenterate level of organization without any affinity among the living Coelenterata either in the common type of symmetry or in the structure of separate organs. However, the Vendian forms are known to contain some species which, though somewhat different, still could constitute, together with *Tribrachidium*, a taxon of high rank in the system Coelenterata: these are radially symmetrical organisms with a three-fold symmetry (*Albumares, Anfesta, Skinnera*). This specific type of symmetry, not typical of other groups of Vendian and living Metazoa, could serve as indirect evidence for the placement of *Tribrachidium* into the Coelenterata.

It should be mentioned that some years ago *Tribrachidium* was still considered as a possible representative of the deuterostomatous lineage of echinoderms. The shape of the "arms" and the tri-slit structure of the "mouth" suggest some resemblance of *Tribrachidium* to certain Edrioasteroidea. The finest radial furrows extending from the crest and the concave curve of each "arm" were regarded as tubular feet or respiratory organs (Glaessner and Wade 1966). The resemblance between *Tribrachidium* and Edrioasteroidea was also strengthened by the occurrence of tri-radial mutants among Echinodermata, for instance, a specimen of the genus *Pentremites* May (Blastoidea) from the Mississipian beds of North America (Macurda 1964). The symmetry of juvenile Edrioasteroidea was adduced as a crucial argument (Bell 1976). Bell described the ontogeny of some representatives of four families from three orders of Edrioasteroidea and noted a strong similarity in their ontogeny in which he recognized seven stages. In their early development, Edrioasteroidea were distinctly tri-radial, however, it is hardly possible to consider this symmetry as a true rotation symmetry of the third order, because in fact it is based on bilateral or biradiate symmetry; the pentaradial symmetry, attained later during ontogeny, formed by bifurcation of the right and left lateral primary axes of growth. Hence, an appeal to tri-radial symmetry of Edrioasteroidea appears to be unsound. Moreover, echinoderms could well have bilateral ancestors since their larvae are bilaterally symmetrical.

In recent papers, Glaessner and some of his followers discounted statements suggesting a relationship between *Tribrachidium* and echinoderms and left the question of the systematic position of the genus open (Glaessner 1971a, 1979b; Jenkins 1981). The curved arms with tentacles strongly resemble lophophores, however, comparison of *Tribrachidium* and Lophophorata is confined to the statement of the resemblance. It should be mentioned that no tri-radial lophophores are known. They are always bilaterally symmetrical and develop from a bilaterally symmetrical horseshoe-shaped rudiment.

The above concept of Tribrachidium as an organism at the coelenterate level of organization appears to be the least ambiguous, although the organism itself remains unique in many respects. Apart from the three-fold symmetry and the unusual character of the gastrovascular system, the presence of three mouths also seems strange. However, the latter is not quite rare among Coelenterata. In adult specimens of some Scyphozoa, in particular, in rhizostoms (for example, *Rhizostoma pulmo*) the mouth tightens and its function is performed by multiple tiny pores in creases of stomatic lappets. This type of oral cavity is typical of forms which feed on fine planktonic organisms sucked, with water, into the gastric cavity. Of interest is the formation of stomatic brachia in a rhizostom, Mastigias, which have a normal, four-lobate shape at early stages of development. As the organism grows, the lobes branch at the ends and their edges start converging. At the final stage the edges of the stomatic lappets converge fully, (a longitudinal groove, also branching twice, remains in place). Mouth cavities, 16 in number, thus remain at the lobe ends (Fig. 173 C-E in Hyman 1940, p. 526).

The assumption that the "arms" of *Tribrachidium* are a homologue of the branching gastrovascular canals of *Albumares* could be supported by a speculative transformation of the curved canals of *Tribrachidium* from the branching canals of *Albumares*, though it is also true that the initial form should possess three, but not nine, initial canals. The transformation of the dichotomously branching canal should proceed by curving in a clockwise direction with a simultaneous reduction of the first (proximal) left-hand offshoot down to a short appendix. All the left offshoots of the first right canal branch should also undergo such a reduction, thus transforming into the "arm" of *Tribrachidium*.

This purely speculative transformation possibly provides evidence for the above homologue and, hence, for the interpretation of *Tribrachidium* as a representative of Coelenterata. However, there is still a very interesting point. It remains difficult to explain how the three-fold symmetry had formed in these organisms and why it had not been attained in cnidarians in spite of its obvious simplicity: i.e. three rays and three radial directions, the least number of radial elements able to adequately react to the isotropic medium in a plane of these radii. In principle, radially arranged organs of these forms are not antimeres in the strictest sense of the word, because it is impossible to draw the plane of symmetry between them. The emplacement of these may have followed no perfect regularity common to all Radialia, when all opposite antimeres emplace simultaneously. In tri-radial medusoids, the identical radial elements may have emplaced, so to say, on an invisible concentric matrix. These concentric rings are represented, for example, by lines connecting points of branching of radial canals of the same order in *Albumares*, i.e. the points of first branching, second branching, etc.

A unique, three-fold radial symmetry makes the above group sufficiently homogeneous in the sense of phylogeny and allows us to consider it as a taxon of relatively high rank. It is possible to recognize it as a coelenterate class. It may also include the oldest skeletal organisms known as the Augustiochreidae which appeared at the close of Vendian and became widespread in the Early Cambrian

(Sysoyev 1965; Valkov 1968, 1982; Valkov and Sysoyev 1969, 1970; Missarzhevsky 1969, 1974; Abaimova 1978).

Twenty species of Augustiochreidea, assigned to 12 genera and 4 families have been described up to now. Valkov, who studied this group in detail, recognized a new subclass, Augustomedusae Valkov, which includes one order, Augustiochreida Valkov et Sysoyev, 1969 (Valkov 1982).

The Augustiochreidae possess tests characterized by internal cavities with a circular, circular-triangular, tri-lobate or star-shaped (with the number of ends divisible by three), hexagonal cross-section. The sculpture of the internal cavity is complicated by wrinkles, furrows, ribs and spines. The test itself is cylindrical or conical; in many Augustiochreidae the surface of the test is ornated with rings, lines of growth, etc.

The major structural features of the Augustiochreidae skeleton are: (1) three-radial symmetry; (2) small size combined with closeness of initial parts; (3) bilamellar walls; (4) the presence of an internal cavity, the presence of triangular stomatic lappets and communicating two-walled septae; (4) chevronlike sculpture; (6) a type of transverse segmentation. They also exhibit the ability for strobilation and transverse, as well as possible longitudinal, fission (Valkov 1982). The above features suggest some relation of the Augustiochreidae to the Conularia discussed below. The placement of the Conularia into the class Scyphozoa is rather widely accepted (Fundamentals of paleontology 1962), however, some workers classify them in a class of their own.

The above medusae with the three-fold symmetry, whose complex structure strongly resembles the organization of the living scyphozoans may be placed into the same subclass as the Augustiochreidae. In this case such a unique type of symmetry as three-fold symmetry makes this relationship rather valid. The affinity of these organisms with the class Scyphozoa remains uncertain, since a three-fold symmetry is not characteristic of this class of cnidarians. A four-fold symmetry dominates scyphozoans. In this respect scyphozoans, or higher medusae, constitute a single group. Some deviation from four-fold symmetry or even from regular radial symmetry is observed only in some specialized forms, mainly, of the rhizostoms suborder (Discomedusae, Rhizostomeae) (Beklemishev 1964). Therefore, comparing the Vendian medusoids and the Augustiochreidae with the class Scyphozoa two possibilities are to be borne in mind: either to widen the diagnosis and the composition of the class to include organisms with the tri-radial symmetry, or to recognize a separate coelenterate class, the Trilobozoa with an organization similar to that of the class Scyphozoa in many respects, but with a different symmetry. With respect to their symmetry this makes it as coherent a group as the scyphozoans. The second possibility appears to be less controversial.

2.3.2 Four-Fold Symmetry

As a whole, the Vendian coelenterates with four-fold symmetry are less common that organisms with three-fold symmetry, although, at least from the Late Paleozoic to Recent, this type of symmetry dominated many groups of coelenterates.

The simplest form among the Vendian coelenterates of this group showing radial symmetry with a rotation axis of the fourth order is *Staurinidia* Fedonkin (Fedonkin 1983a). These are relatively small medusae rarely larger than 10 mm in diameter (see Fig. 7). The state of preservation (negative hyporelief) suggests that the medusae possessed a convex bell, a relatively small stomach, if any, from which four straight, unbranching gastrovascular canals diverged outwardly. Rather large, heart-shaped structures, which may be equated to gonads, are arranged at the ends of canals. Neither on outward continuation of the radial canals, ring canal or velum were observed, probably because of the poor preservation. The bell edge carries fine tentacles.

On the whole, *Staurinidia* strongly resemble the living medusae, *Obelia geniculata* (Leptomedusa) in body plan. The latter have a wide, flat bell, 2–3 mm across, and a rather weakly developed, sometimes even rudimentary, sail. The centre of the subumbrella bears a stomatic pedicle, hanging down and ending with a mouth. Four radial gastrovascular canals start from the small central stomach and enter into the ring canal running along the bell edge. The lower side of the bell bears four gonads arranged along the radial canals. The gonads are rather large and more or less regular, and spherical in shape (Ivanov et al. 1981, p. 221). The relief of the gonads in *Staurinidia*, as in *Obelia*, suggests their arrangement at the subumbrella side. Irrespective of the fact whether *Staurinidia* had a ring canal or not, we could regard this form as in close affinity to the most primitive hydroid medusae. In fact, there is no serious reason against assigning *Staurinidia* to the class Hydrozoa.

Among the Vendian coelenterates with four-fold symmetry *Conomedusites lobatus* Glaessner et Wade holds a unique position (Fig. 11). This genus was recognized on the basis of the study of three small fossils from the Late Precambrian Pound Quartzite, South Australia (Glaessner and Wade 1966). About two dozen specimens, including those from the Vendian beds of the southeastern White Sea area and Podolia, are known at this time (Fedonkin 1981a; Velikanov et al. 1983). Specimens of this monotypical genus are preserved as small convex or flat conical casts composed of four almost equal lobes, delimited by four sharply defined radial grooves that deepened outwardly, separating the adjacent lobes. A vast internal zone, delimited in each of the four sectors by an arcuate groove, has a low conical shape, with a pointed apex: this probably could be equated to a theca. Post-mortem deformations suggest that the theda had been soft and flexible. Beyond the theda there is a marginal flange of the soft body; the marginal flange is rarely preserved, only in very large specimens. Tiny marginal tentacles can be observed in some

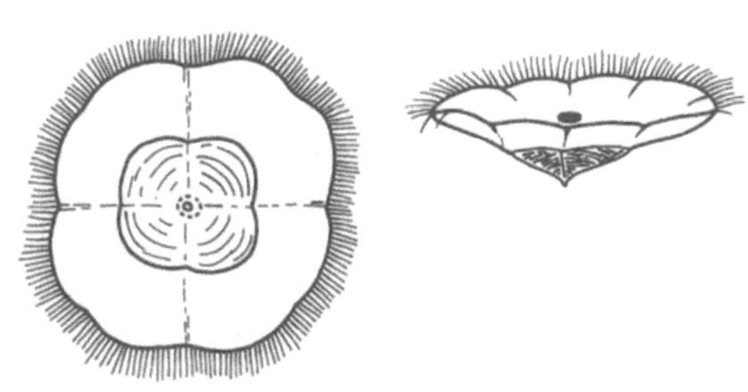

Fig. 11. Aboral side and reconstruction of *Conomedusites lobatus* Glaessner et Wade

specimens. Radial grooves, generally straight and sharply defined, suggest the presence of septal ridges (Glaessner 1971a).

In type of symmetry and low conical shape, *Conomedusites* strongly resemble *Conchopeltis alternata* Walcott from the Middle Ordovician of Minnesota, USA (Fig. 22 in Moore and Harrington 1956; Fundamentals of paleontology, 1962, p. 189). However, *Conomedusites* have neither multiple radial furrows nor prominent concentric wrinkles typical of conchopeltic theca. The structure of the soft body of the latter is still unknown. Four main radial grooves of *Conomedusites* could be interpreted as septal ridges of the theca. The peripheral margin separating the ridges, especially in the largest specimens, may be further subdivided by a shallow indentation resulting in the formation of eight lobes around the body margin. A similar tendency is displayed by the coiled outer edge of *Conchopeltis (Glaessner 1971a). The discovery of Conomedusites* is important because it is the first and so far the only known specimen of Precambrian Coelenterata with an outer theca extremely typical of the living forms, but absent in the great majority of Vendian Coelenterata.

The *Conchopeltis* Walcott, comparable with *Conomedusites*, was placed into the suborder Conchopeltina Moore et Harrington, order Conulariida Miller et Curley, subclass Conulata Moore et Harrington, class Scyphozoa Goette (Moore and Harrington 1956). Such a systematic position of *Conomedusites* is principally based on the notion of Kiderlen (1937) who places the Conularia into the class Scyphozoa. Now it is a widespread belief that conulates are scyphoids. Study of Stephanoscyphus supports this assumption (Werner 1966, 1967a,b). Kozlowski's critical comments stimulated the search for new lines of evidence for a possible affinity between the Conulata and the Scyphozoa (Glaessner 1971a). However, this discussion is beyond the scope of the present study. It should only be noted that the study of the initial growth stages of hydroid polyps shows that the basal parts of the thecae of some species (Fig. 1a in Manton 1941) bear a striking resemblance to the apical parts of Conularia Sowerby (Figs. 10, 19 in Kozlowski 1968).

In any case the Conulata are regarded now under Cnidaria, however, their affinity with scyphozoids could not be so closely inferred elsewhere (Kiderlen 1937; Chapman 1966; Werner 1966, 1967a,b). The essentially tetragonal cross-section of the cone-shaped, elongate pyramidal test is a very important character preserved in the Conulata for almost 350 m.y. of their evolution (Conulariida, s.s., persisted from the Middle Cambrian to the Triassic). It is tempting to regard the septae displayed by many Conulata as the homologue of four or eight teeth in the tube of Stephanoscyphus, however, it can be hardly proved. Complicated stomatic lobes which continue the test sides in the Conulariida are not known in other Cnidaria, but could be functionally attributed to the tetragonal cross-section. As yet, no evidence of strobilation in the Conulata has been obtained, and their reproductive cycle may have profoundly differed from that of the Coronatida (Scyphozoa), comparable to the Conulata. Many authors believe that the Conulata were sedentary at the beginning of the vital cycle, but adults could swim with the oral end down, though they could hardly move in water. More likely is that most of the Conulata were sessile forms which cannot be interpreted as medusoids.

An attempt is made to consider the Conulariida as ancestral to scyphoids in the following way: in one of Conulata lineages the stomatic lobes reduced and the organism thus transformed into a form similar to the scyphistoma, which then attained strobilation ability (Chapman 1966, p. 73). However, paleontological and comparable anatomical data available at present are not sufficiently strong to prove this assumption. In a similar way the Stephanoscyphus could also have evolved from the Conulata (Werner 1966, p. 343), but this assumption is primarily based on the elongate form of the Stephanoscyphus theca. However, the circular cross-section of the latter more closely resembles that of Byronia Matthew (Upper Precambrian of British Columbia, Middle Ordovician of Poland) interpreted as a scyphopolyp (Kozlowski 1968).

Glaessner believes that there is no other alternative than to consider the Conulariida as a peculiar and very conservative lineage of the Cnidaria (Glaessner 1971a, p. 11). This lineage is traceble in time starting from the Late Precambrian and deduced from the group of organisms including *Conomedusites* and Conchopeltis. The evolution of scyphoids through the scyphistoma, as well as through the Stephanoscyphus could also be postulated from this lineage.

The morphological and functional relation of scyphoids precludes from the interpretation of Stauromedusae as being originally polypoids, however, it is rather a modification of a life cycle more common to scyphoids (Chapman 1966; Thiel 1966). Flat cones of Conchopeltina (this group comprises both Precambrian *Conomedusites* and long pyramids of the Conulariida) essentially differ from any polyps of the juvenile and adult stages of the living Scyphozoa, therefore, these groups are not akin to the scyphoids. On the other hand, the structure of Conchopeltina shows that this is a group relating the Scyphozoa to the Hydrozoa. This assumption is based on the comparison of the chitinous theca of Stephanoscyphus, the thecae of Conchopeltis and *Conomedusites*, and the chitinous float of chondrophores. Some authors (Delsman 1923; Mackie 1959) show that the first chitinous chamber of the float in chondrophore larvae is equated to the chitinous theca in actinula of Hydrozoa (Tubularia). Chitinous walls of the float chamber in larvae of *Velella* Lamarck form a flat cone which persists up to the Rataria stage when the float becomes octolobate in shape. The chitinous wall starts closing around the first chamber and the successive chambers, added already concentrically, thus form a flat disc-shaped float. This structure is circular in Porpita Lamarck, elongate-elliptical in *Velella*, and distinctly bilaterally symmetrical in *Ovatoscutum, Chondroplon* and other genera of the fossil chondrophores with elliptical floats. Along with the symmetry of the float, chondrophores demonstrate other differences which can be ascribed to family characters, for instance, the central pore in Porpitidae and the velum in Velellidae.

In terms of the formation of the conical theca in different coelenterates, we can propose a different interpretation for the oldest tubular forms or elongate-conical fossils which appeared at the very close of Vendian and were abundant in the Lower Vendian (Tommotian... 1969). The fossils composed of calcium carbonate or phosphate are generally regarded as worm tubes or molluscan shells, but we cannot rule out the affinity of the oldest elongate-conical fossils with the Coelenterata.

To complete the description of forms displaying four-radial symmetry we must mention two further medusoids from the Precambrian of Iran and France. Some casts of small medusae *Persimedusites chahgazensis* Hehn et Pflug were described from the Late Precambrian Esfordi Formation, Central Iran (Hahn and Pflug

1980). Peculiar features of the species are as follows. The oral region is slightly elevated and occupied by four stomatic tentacles. The subumbrella side is divided roughly into 20 radial lobes separated by deep furrows. The radial lobes start from the oral region and continue to the bell periphery where so-called lobes with tentacles at their outer ends appear between their distal parts. The tentacles have rounded, inflated bases. The above features of *Persimedusites* allow the assignment to the scyphoids, in particular, to the Semaeostomeae.

Ichnusina cocozzi Debrenne et Naud, whose imprints were found in black shales of the essentially terrigenous San Vito Formation in southwestern Sardinia (Debrenne and Naud 1981), also belong to the scyphoids. These Precambrian medusae exhibit a combination of four- and eight-fold radial symmetry. Four elliptical ridges radiating from the centre may be equated to the oral arms and, possibly, gonads. Eight lobes branch dichotomously two times outwardly and form a branching network of relatively wide gastrovascular canals. Very short, paired tentacles fringing the bell, in fact, result from terminal bifurcation of third order lobes.

2.3.3 Six-Fold Symmetry

Only one form of Vendian medusae, *Pomoria corolliformis* Fedonkin displays six-fold symmetry (Fedonkin 1980b). These are small medusae whose subumbrella bears two crowns of closely spaced tentacles. The order of the rotation axis is determined by the symmetry of a small manubrium which is flower-shaped in plan outline: a round mouth and semicircular processes around the outer edge. Tentacles of the stomatic crown are knit in bundles; marginal tentacles are numerous and sharpened. The number of tentacles cannot be counted, as they have not been strictly established.

The following generalizations have emerged from the study of Vendian coelenterates with a stable order of rotation:

1. Inspite of the fact that the forms with a stable order of radial symmetry are relatively rare among the Vendian Radialia, they already show considerable diversity, both in symmetry and in promorphology of the main organ systems.
2. Three-fold forms are the most common and displays a great variation in body plan. Forms with a rotation axis of higher order are less common.
3. Forms with a stable order of rotation display a great diversity of distribution systems represented by single and paired, as well as branching or unbranching radial canals. No ring canal has been found, which may be ascribed to poor preservation. In one form, the bell is divided into narrow branching lobes in which canals are probably present. Such a body plan could be considered as original for forms with thin branching gastrovascular canals.
4. Unlike the forms with a symmetry of infinitely high and variable order, Vendian Radialia with a stable order of symmetry do not show any sign of asexual reproduction. Their reproductive organs (gonads) demonstrate a certain diversity both in shape (circular, ellipsoidal, semicircular, heart-shaped) and in arrangement.

5. Organisms with a stable order of symmetry display the most complex and advanced distribution and reproductive systems. In many respects they are close to the living cnidarians, especially to the scyphoids. It is quite possible that the scyphoid lineage is represented here by the only form with a theca, i.e. *Conomedusites,* which could be regarded as the Precambrian form of the Conulata. The parallel lineage in the Vendian is represented by tri-radial Radialia from which the Augustiochreidae may have derived at the end of Vendian to the onset of Cambrian. The group of tri-radial medusae could be regarded as a separate class, the Trilobozoa.

2.4 Conclusions

The promorphological study of Vendian Radialia opens the possibility of understanding the earliest evolutionary stages of Coelenterata.

The widest distribution in the Vendian of medusoids with a symmetry axis of infinitely high order and with a concentric body plan allows us to consider this group as an independent, major phylogenic branch of early coelenterates. The taxonomic rank of the group seems very high, at least no lower than a class. Therefore, it seems reasonable to place numerous coelenterates with a concentric body plan and symmetry axis of infinitely high order into a separate class, the Cyclozoa, under the Coelenterata. In the earlier evolutionary stages of Radialia, the symmetry typical of the class Cyclozoa is believed to have been dominant. In the Vendian, however, we already see other evolutionary trends in Coelenterata which became dominant in the Phanerozoic. Therefore, the Vendian may be considered to a great extent as the time of the phylogenic closure of the group of coelenterates with a symmetry axis of infinitely high order and concentric body plan. In the benthic and pelagic regions, they were forced out by more highly organized radial cnidarians.

A number of forms demonstrate the concurrent presence of concentric and radial body plans with an obvious prevalence of the latter. This suggests a historical relation between the concentric and radial organizations, and the forms themselves could be regarded as indicative of evolution from a primitive concentric to a more advanced radial body plan. The advancement of radial organization is primarily explained by the attainment of one more degree of freedom in the direction parallel to the axis of symmetry.

Although in the living cnidarians the axis of rotation symmetry of variable order is extremely rare and with good reason is considered as a primitive and archaic character; many groups of Vendian coelenterates were dominated by this type of symmetry. The abundance and diversity of Vendian coelenterates with a radial symmetry of variable order suggest that this type of symmetry represents an important and, probably, necessary stage in the evolution of Radialia.

A search for analogies among the Vendian coelenterates with a symmetry axis of uncertain order and the living forms of this type indicates that many structures in organ systems typical of the modern Cnidaria began to take shape as early as the Precambrian, although judging from their primitive character, the Vendian record of coelenterates represents, in many respects, the early phase in the evolution of many groups of this type.

Organisms exhibiting a radial symmetry of variable order, thus representing the Vendian stage in the evolution of coelenterates, should probably be classified as a separate taxon, although it is uncertain how uniform it would be in terms of phylogeny. The morphological diversity of this group suggests a considerable divergence of the main lineages of Radialia as early as the Vendian, and in this context the symmetry of variable order represents a natural and even necessary stage in the evolution of all the Coelenterata with a radial body plan.

This could eliminate one of the contradictions in the theory proposed by A. Sedgwick (1884) who derived the Articulata from highly specialized Radialia (Anthozoa) with a fixed order of symmetry. The same transformation can be performed from a more primitive form of Radialia, which has an axis of symmetry of variable order. This leads us to the promorphology of those Vendian proarticulates which display unordered segmentation or alternated segments (plane of gliding reflection). This type of symmetry is shown by some Vendian segmented, bilateral organisms and probably reflects one of the modes of emplacement of metamerism in the Articulata.

Analysis of the promorphology of Vendian Radialia also provides another possibility of the emplacement of symmetry in the Articulata which had not been predicted theoretically. This way could be realized by transformation from forms (Fig. 12) with a symmetry of infinitely high order and regular concentric body plan (*Cyclomedusa* or *Kullingia* type) through bilaterally symmetrical forms with relics of concentric organization in the early stages of individual development (*Ovatoscutum* type and then *Chondroplon* type) to bilaterally symmetrical segmented forms (*Dickinsonia* type). This possibility might have been realized in the very early stages of metazoan evolution.

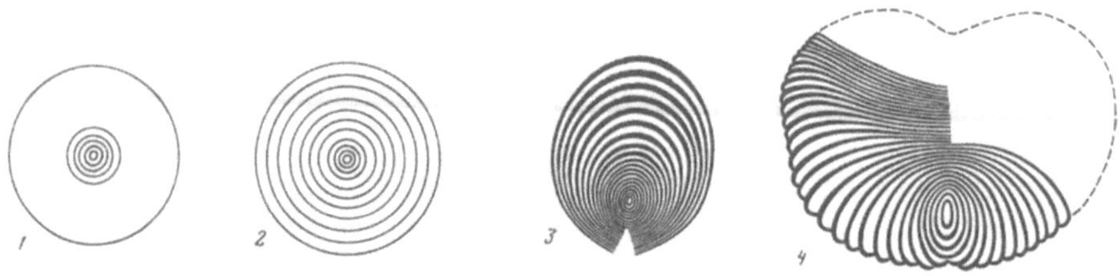

Fig. 12. Forms with concentric body plan. **1** *Cyclomedusa plana* Glaessner et Wade; **2** *Kullingia concentrica* Glaessner and forms exhibiting concentric organization and bilateral symmetry; **3** *Ovatoscutum concentricum* Glaessner et Wade; **4** *Chondroplon bilobatum* Wade

3 Promorphology of Vendian Bilateria

Animals composing the division Bilateria in the system Metazoa demonstrate, as a rule, a well-defined bilateral symmetry. In the modern animal kingdom this group is more abundant and diverse as compared to Radialia. However, as shown above, the Bilateriato Radialia ratio was principally different among Vendian macroscopic, multicellular organisms: paleontological data suggest an overwhelming prevalence of Radialia in the Vendian. This fundamental paleontological observation supports the idea of the affinity of the most primitive groups of Bilateria to Coelenterata (Beklemishev 1964).

The abundance, diversity and large dimensions of Vendian coelenterates suggest that they could be regarded as the oldest type of Eumetazoa. The Vendian life, sternly dominated by Radialia, supports the idea that during the earlier, possibly very short stage of evolution of multicellular animals, radially symmetrical organisms completely dominated (the animal world consisted only of Radialia!). This apparently fantastic supposition is based on retrospective extrapolation, but it is also supported by the Vendian fossil record. One line of such paleontological evidence is the oldest assemblage of Precambrian fossils from the Conception Group, Avalon Peninsula, Newfoundland: this assemblage consists exclusively of solitary and colonial coelenterates, i.e. Radialia. In such cases a possible incompleteness of the fossil record is usually invoked. But it is difficult to conceive that of two groups of soft-bodied metazoans only radially symmetrical animals have been preserved. Along with numerous imprints of coelenterates, simple crawling trails, assigned to possible Bilateria, have been recently reported from the Conception Group (Anderson and Conway Morris 1982). However, in the study of widely known examples of burrowing solitary and colonial polyps extreme care in interpreting the oldest trace fossils must be taken. The fact that the bulk of Vendian Bilateria was represented by very small forms partly explains why their remains are absent or rare in the oldest Vendian strata.

If we assume that the statement of a relatively late appearance of Bilateria is true, then the time of their appearance was not far from that of Radialia. The problem concerning the evolution of bilaterally symmetrical organisms from radially symmetrical forms necessitates, first of all, the promorphological analysis of all known Vendian Bilateria. Such a study and the subsequent elucidation of the group structure as a whole will help to outline the major, earlier unknown lineages, because the Vendian Bilateria are so far the oldest organisms in the fossil record of this metazoan division.

To start the account of Vendian bilaterally symmetrical metazoans it should be recalled that many genera of the group are monotypical and are mainly represented by only a few specimens. Hence, some correction will be introduced into the concepts on the morphology of certain Vendian organisms as new occurrences are found, however, the general body plan, or architectonics, of the animals is usually rather distinct.

Superficially, Vendian Bilateria are dominated by diverse segmented forms, though now not all of these can be classified under Articulata. There are some grounds to recognize a large group close to Articulata and possibly representing an important link on the way to the latter. However, the discussion of segmented forms should be preceded by that of rare Vendian unsegmented organisms.

3.1 Unsegmented Forms

Platypholinia pholiata Fedonkin may well be the simplest form among Vendian Bilateria. They occur as quite flat, leaf-shaped imprints about 70 mm long and 30–35 mm wide. The body tapers more towards one end with the slightly bifid terminal part from which linear structures run inwardly as a poorly visible, narrow median ridge and subparallel furrows are found either side. The structures start from the narrow end, reach roughly the centre of the body or somewhat farther, and die out. The wider end of the leaf-shaped body is smoothly rounded. The outlines are very even, the imprint surface is extremely smooth.

Only three specimens of *Platypholinia* were reported from the upper part of the Zimmy Coast section, White Sea area (Zimnie Gory). They are rather featureless and hence difficult to interpret. The forms could be regarded as fragments of some large, intricately differentiated organism. In this case the narrow end of the body could correspond to the point of attachment. However, the material under study has yielded no attachment organs. Moreover, as yet no organisms whose parts resemble *Platypholinia* have been found in the Vendian. If we interpret this problematic form as a whole organism, then the narrow end could be equated to the anterior or oral end, as the bifid end suggests the presence of a small terminal mouth. Linear structures in the axial part of the body, particularly the narrow, long ridge running from the bifid terminal part could be regarded as a throat, whereas thin, long wrinkles could be equated either to nerve trunks or bands of longitudinal musculature with the aid of which the organism may have been able to draw in the anterior end of the body. The structural features of the digestive system of *Platypholinia* are uncertain. Neither anus nor reproductive organs are discernible.

Additional material will probably help to recognize new structural features. For now, *Platypholinia* seem to be rather primitive organisms and enigmatic in many respects. Data at hand do not allow the placement of *Platypholinia* as taxa of higher organization than the turbellarians. On the other hand, the terminal mouth is not typical of modern Plathelminthes, being more characteristic of nematodes. However, the leaf-shaped form and simple structure of *Platypholinia* place it closer to turbellarians. But the *Platypholinia* probably represents a more primitive organism, as its body plan is easily deduced from the polyp with the most primitive body plan and a symmetry axis of infinitely high order. The fact that *Platypholinia* lacks metamerism is indicative of its initial state. Unlike flat worms, the primary axis, connecting the animal and vegetative ends of the body, may have not changed in *Platypholinia*, therefore, its assignment to the type Plathelminthes is rather tentative.

Another Vendian form, *Vladimissa missarzhevskii* Fedonkin can be more definitely interpreted as a separate organism. Thus far there is only one well-preserved specimen (from Bed 11, Zimnie Gory) demonstrating a leaf-shaped, flat, fairly thick body, sharpened at the anterior and posterior ends. The anterior-posterior orientation of this relatively large organism (about 45 mm long, 32 mm wide) is evidenced by the direction and change of dimensions of the oblique incision of the body edge and by the sharp end of the body. Four pairs of rounded bosses are aligned in two longitudinal rows in the middle part of the body.

Thus, this is an obvious bilaterally symmetrical animal, markedly elongate along the symmetry plane, with two rows of symmetrically (bilaterally) arranged bosses at either side. The bosses and the character of the oblique marginal incision, rather coarse in the anterior part of the body, somewhat coarser in the centre and gradually dying out towards the posterior sharp end, suggest an incipient metamerism in *Vladimissa*.

Comparison with modern organisms suggests some similarity, though superficial, of *Vladimissa* to the turbellarians: the flattened leaf shape of the smooth, undifferentiated body without any appendages supports the similarity. If we develop the interpretation of *Vladimissa* as forms related to the turbellarians, then four pairs of rounded bosses in the middle part of the body may be equated to the reproductive organs, particularly the seminal vesicles, rather large ova or ovular capsules. In this context the closest type of symmetry and incipient metamerism are displayed by modern Planaria Mesostoma ehrenbergii (Fig. 268 in Ivanov et al. 1981, p. 305). These large rectal turbellarians, about 15 mm long, are distinguished by large, reposed ova aligned in two rows at either side of the rectum. The above similarity is obviously insufficient to place *Vladimissa* under rhabdocoelides, nevertheless, placement of this Vendian form into the turbellarians appears to be relatively valid. It is of interest that *Vladimissa* exhibit incipient metamerism of the entire organism: the body exhibits the arrangement of reproductive organs and marginal incision. This type is known as an order metamerism extremely characteristic of the most primitive of the living Bilateria, such as flatworms (Beklemishev 1964).

It should be noted that quite flat, small, smooth, leaf-shaped imprints are abundant in Vendian beds. The fossils lack median structures, but, nevertheless, they display a relatively regular bilateral symmetry, anterior-posterior orientation; these are probably impressions of the oldest flat worms.

The above findings are of interest because no fossil turbellarians have been described as yet. The unique conditions of burial and preservation of Vendian non-skeletal organisms and occasional findings of planarians look promising for filling the gap in the fossil record of the whole phylum (if only due to findings from the Precambrian deposits).

A cast of a wormlike organism, *Protechiurus edmondi,* found in close association with *Preredinium simplex* (Gürich) in the Kuibis Quartzite of the Nama Group, Namibia, was recently described by Glaessner (1979a). The cast (length 74 mm, maximum width 19 mm) shows that the organism had a cigar-shaped body. The spatulate, narrow anterior end is presumed to be the proboscis. A slight constriction separates the proboscis from the trunk which is somewhat widened at posterior end. The greatest width is behind the midpoint of the body. The posterior end is wide and rounded. Prominent longitudinal ridges on the surface are interpreted as representing eight longitudinal muscle bands. *Protechiurus* is considered a representative of the Echiuroidea. The form of the body and the eight longitudinal muscle bands allow a comparison of *Protechiurus* with some forms of the family Echiuridae. If this comparison is confirmed by further studies and new findings, it will be of considerable significance.

The Echiuroidea are related to Annelida, but whether they are ancestral to or derived from annelids by the reduction of segmentation or evolved from common ancestors is not clear (Stephen and Edmonds 1972).

The first fossil representatives of the phylum Echiura were described recently (Jones and Thompson 1977) from the Pennsylvanian of North America (northeastern Illinois) and, therefore, the evolution of the group is uncertain. The Precambrian findings not cast light on the origin and evolution of the Echiura, but they are of importance as an indication of a long-term evolution of the conservative group of marine, benthic, unsegmented, burrowing worms.

3.2 Segmented Forms

The great majority of Vendian Bilateria (in addition to three forms of uncertain systematic position) are articulate or metamerous organisms. This interesting fact deserves further consideration. All the segmented, bilateral Vendian organisms had their growth zone at the posterior end and, hence, could be regarded as representatives of Articulata. However, unlike living and known Phanerozoic articulate animals, many forms of Vendian articulates do not exhibit an opposite, but rather an alternate arrangement of metameres. This appears to be a unique phenomenon, very important for understanding the evolution of articulates as a whole.

The point is that the overwhelming majority of articulates show metamerism due to the simple axis of motion. Movement of the animal along the axis for the length of a metamere (a step of translational symmetry) results in coincidence of the animal with itself. Analogous forms are known from the Vendian as well. Nevertheless, many Precambrian Metazoa display a combination of translational and reflection symmetry due to the plane of glide reflection (Shubnikov 1940, 1951). In this case each subsequent metamere is a mirror reflection of the previous one and thus right and left metameres alternate lengthwise over the whole body of the animal. With this type of metamerism separate metameres per se are dissymmetrical. Of the living organisms glide metamerism is noted in the strobile of the cestodes *Tatria mathevossianae*, Cyclophillidea (Fig. 93 in Beklemishev 1964; Vol. 1, p. 185). This type of metamerism is found in the structure of muscles and in the divergence of spinal nerves of the lancelet. Glide metamerism is a common feature in colonial polyps and plants.

3.2.1 Polymerous Segmented Forms

The genus *Dickinsonia* Sprigg is the most abundant among Vendian segmented Metazoa. Gigantic forms about 1 m in length can be encountered among specimens of the genus. The genus has been studied in detail on mass material from the Pound Quartzite, South Australia (about 400 specimens). Presently, five species of *Dickinsonia* are recognized: *D. costata* Sprigg, *D. elongata* Glaessner et Wade, *D. tenuis* Glaessner et Wade, *D. lissa* Wade and *D. brachina* Wade (Sprigg 1947; Glaessner and Wade 1966; Wade 1972b). Four of the above species, namely, *D. costata, D. tenuis, D. elongata* and *D. lissa* were recorded in the Vendian of the Russian Platform (Keller and Fedonkin 1976; Fedonkin 1978; 1981a, 1983b).

Dickinsonia exhibit a prominent bilateral symmetry of a very thin, flat body, ovate or elongate, and leaf-shaped. The state of preservation suggests that even large specimens did not exceed 3 mm in thickness. Actually, segmentation is clearly seen on the entire body is prominent both on the ventral and dorsal sides. Only in *D. costata* segments become substantially smaller towards the posterior end, while in all the other species, which have shorter and more numerous segments, the length of segments (measured parallel to the axis of the worm) remains more or less constant. The median line, from which the segments start, is usually quite distinct on the dorsal part of the body and hardly seen on the ventral side.

Of the five species of *Dickinsonia, D. costata* are more abundant and, hence, their morphology is better known. Let us now discuss the structure of this form in detail. These worms are flat and elongate-ovate to wider than long. The anterior end is usually convex but rarely indented; the posterior end is almost always notched at the axis. The segmentation is clearly marked on both the dorsal and ventral surfaces, but the former has more relief, since segments here are delimited by rather deep furrows. Segment size diminishes towards the posterior, frequently to several times smaller than the more anterior segments. Throughout, the segments are individually longest at their free edges (tips) and shortest at their centres. The most anterior body segment is fused around and in front of the V- or U-shaped smooth region, interpreted as a prostomium. Other anterior segments curve forwardless progressively (as follows from the centrally-tapering segment shape). More posterior segments slope progressively more sharply in the same direction and decrease substantially in size. The most posterior pair of structures appear to be shorter and broader than the adjacent half-segments, and, in Wade's opinion, may be equated to anal lobes, similar to those of *Spinther arcticus* Sars (Manton 1967; Wade 1972b).

Segment surfaces present two main appearances: (1) they are delimited by furrows continuous from one side of the body to the other (in small specimens these may be indistinct near the centre); this type of surface is interpreted as ventral. (2) They are clearly interrupted (even in small specimens) by an axial structure. This side is considered as dorsal (Wade 1968, p. 262, 263), and the segments themselves are interpreted as "notopodial-elytral" ridges (Wade 1972b, p. 174).

In one of the specimens the small elliptical structure, possibly representing the mouth (or a wrinkle of the everted pharynx), is seen at the rear end of the first pair of segments (Plate 5, Fig. 5 in Wade 1972b). All specimens of the ventral side have been examined for any evidence of neuropodia, but results remain inconclusive. Nevertheless, Wade believes that neuropodia did exist in *D. costata*.

The presence of intestinal caeca, which branch dichotomously around the pharynx, perceived by Glaessner and Wade in one of the partly decayed specimens of *D. costata*, appears to be inconclusive (Plate 101, Fig. 4 in Glaessner and Wade 1966). The other specimen has unbranched ridges in the mid-body region at right angles to the axis. These faint, segmented ridges show no regular arrangement and may be due to the roughness of the microrelief of the sediment, where the animals had been buried, or fortuitous deformation of the thin, gentle body, which is not uncommon.

Some specimens of *D. costata* display a kind of annular structure as if appearing through the segmented surface. Some specimens demonstrate considerable post-mortem contraction of the body. Minor crenations of the segments (similar to secondary segmentation) are seen on some rare individuals. Hence, it is legitimate to assume the presence of basic polychaete musculature of the circular, longitudinal and septal muscles in *Dickinsonia* (Runnegar 1982).

Classification of *Dickinsonia*, like that of other Precambrian multicellular animals, was not unambiguous. At first the monotypical family Dickinsoniidae was placed into the order Dickinsoniidae and the class Dipleurozoa of the Coelenterata (Moore and Harrington 1956; Fundamentals . . . 1962). This was based on a series of misunderstandings starting from Sprigg's (1947) faulty reconstruction of *Dickinsonia* as "symmetrical across both in longitudinal and transverse planes".

But soon Glaessner (1959, 1961, 1962; Glaessner and Wade 1966) established the close relationships between *Dickinsonia* and spongicolous annelids of the genus *Spinther*. Like recent *Spinther, Dickinsonia* are characterized by the anterior body segment being fused around and in front of the prostomium. This resemblance predetermined the logic of further investigations of *Dickinsonia* and led to Wade's (1972b) suggestion to use the order Dickinsoniida for both Dickinsoniidae and Spintheridae. Glaessner (1979b) proposed to accomodate the family Dickinsoniidae in the order Amphinomorpha in Clark's (1969) system of Polychaeta. As a result, some important conclusions on the evolution of annelids were inferred. Thus, Wade indicated that *Dickinsonia* and *Spinther* owe their similarity to a common ancestor. Since *Spinther* are aberrant polychaete, phylogenic relationships suggested by Wade infer the presence of various "cylindrical" annelids before the appearance of, and simultaneously with, *Dickinsonia*, but this suggestion is poorly supported by paleontological evidence. Furthermore, during early ontogeny *Spinther* pass through the stage of a "normal" polychaete and only when they reach 10 mm in length do they become rounded and preserve this form in further development (Manton 1967; Runnegar 1982). Taken together, these factors suggest that the discoid body of *Spinther* and some other traits are adaptive in character and associated with an extoparasitic life habit of the animal.

As for *D. costata*, no stage of a typical polychaeta has been found in their individual development, although fossil material enables the ontogeny of the form to be traced over wide limits. The smallest specimen of *D. costata*, 4 mm in diameter, has a circular, flattened body on which seven segments are discernible (Plate 5, Fig. 8 in Wade 1972b). These and other, primarily anatomical differences indicate that the resemblance mentioned between *Dickinsonia* and *Spinther* is due to convergence, and that the Dickinsoniidae are an independent evolutionary branch of Bilateria.

In recent works some authors stated that *Dickinsonia* had been segmented organisms at the annelid level of organization (Conway Morris 1979; Runnegar 1982). Runnegar believes that *Dickinsonia* might have had a hydrostatic skeleton (true coelome), a metamerously arranged dorsoventral muscles, a longitudinal, transverse and, possibly, diagonal musculature, a relatively resistant collagenic basal membrane, caeca with lateral lobes and a well-developed circulatory system. The presence of the small everting proboscis used to gather detritus was inferred at the ventral side. The statements made by Runnegar are primarily based on the data of Wade (1972b) and on comparative biometric analysis. Runnegar estimated that *Dickinsonia* could live in seawater containing one-ninth of the present content of oxygen dissolved in water. This suggests a low content of free oxygen in the atmosphere for the Late Precambrian.

Recently, the concept of an extremely primitive structure of *Dickinsonia* has become prevailing. Some workers, including the author, permit comparison between *Dickinsonia* and the turbellarians (Termier and Termier 1976; Paliy et al. 1979; Conway Morris 1981; Fedonkin 1981a, 1983a,b). We have certain grounds for such a comparison.

Those who studied *Dickinsonia* disregarded two important factors. First, it was disregarded that at the earliest stages of their development *D. costata* had an ideal, round form of the flat body. Moreover, a relict radial symmetry (often called biradial symmetry) is clearly seen in the arrangement of a few segments. The modern turbellarians can also show radial symmetry, particularly in the structure of the nervous system (Beklemishev 1964; Ivanov and Mamkaev 1973). The other important factor is that metamerism of *Dickinsonia* could not be regarded as strictly ordered, since the left and right metameres are not arranged opposite from each other, but alternately. This is better seen in the region of the largest anterior segments on the dorsal side. Such metamerism is seen not only in *D. costata*, but in *D. elongata* as well. Such an arrangement appears to be an extremely primitive character, quite untypical of articulates.

Comparison of *Dickinsonia* with the turbellarians seems more promising, but some ambiguity still remains. Let us first consider the evidence for the placement of the Dickinsoniidae into Annelida. The main argument is that the segmentation of the body resembles homonomous metamerism of polymerous annelids. However, we cannot unambiguously state that metamerism embraces the entire body and all organ systems. It is possible that only surficial covers were metamerized (Termier and Termier 1976). This could be attributed to the quest for an increase of the surface through which the organisms breathed. The statement of Runnegar that a circulatory system of the coelome type was necessary for normal oxygen supply of such a large body is not the only possible explanation. Living gigantic turbellarians, rare compared to the abundance of small planarians, suggest a relatively wide range of dimensions and body mass for the animals at this level of organization; most species are 0.4 to 5 mm long, but polycladids and terrestrial planarians can attain considerable dimensions. The largest turbellarians, *Bipalium javanum*, is 60 cm long (Ivanov and Mamkaev 1973; Porfirieva 1977).

Dickinsonia differ from Annelida in having imperfect metamerism displayed by an alternate arrangement of metameres and, hence, cannot be accommodated in Annelida. If such metamerism had embraced the entire body of *Dickinsonia*, which is evidenced by an insignificant thickness of the body, then metameres could be interpreted as supporting the intestinal diverticula, which branch from the main intestine both oppositely and alternately, whereby the ends of these branches reach the marginal parts of the body. Obviously, the intestine of such large organisms as *Dickinsonia* performed, in addition to a digestive, also a distributive function.

Beklemishev in his book "Principles of comparative anatomy of invertebrates" (1964, vol 2) under the description of flatworms pointed out that

"When the parenchyma is dense and the circulatory system is absent, the organs and tissues could be supplied, provided that they were not too far from the intestine. In other words, physiologically, the intestine of large Platoda is a decentralized digestive system, similar to the intestine of Coelenterata".

This suggestion is believed to be also true for *Dickinsonia*. Furthermore, some evidence suggests homology between the digestive system of *Dickinsonia* and the gastrovascular system of primitive coelenterates and, particularly, those which display a symmetry axis of infinitely high or variable order. Unexpectedly, this possibility supports the concept of the origin of metamerism in annelids from antimerism ("cyclomerism") of Coelentarata, proposed by Sedgwick (1884), and at the same time, removed the major contradiction, or the paradox, of this concept: an attempt to derive annelids not from primitive, but from highly specialized forms, such as Anthozoa. This problem has been recently discussed elsewhere (Fedonkin 1985). There is the possibility to derive the architectonics of *Dickinsonia* from radially symmetrical animals showing concentric organization, discussed under "Radialia". This example is provided by some Vendian metazoans from *Eoporpita* and *Kullingia* through *Ovatoscutum* to *Chondroplon* and then to *Dickinsonia* (Fedonkin 1983a). The question calls for additional studies, especially in the last link, as the transition from *Chondroplon* to *Dickinsonia* requires the presence of a well-defined median ridge which may be equated to the main intestine, but in *Chondroplon* the left and right metameres meet to form a narrow, median furrow, although Wade (1971) suggests the presence of a "dorsal keel" in *Chondroplon*.

The resemblance between *Dickinsonia* and the turbellarians pertains not only to similar digestive systems, but also to the peculiar form of the body. This similarity is most prominent in those *Dickinsonia* species whose flat body is elongate and leaf-shaped and which consists of a large number of segments, e.g. up to 340 in *D. elongata*. Some of these forms, like Polycladida and Tricladida, exhibit undulation in the thinner, marginal parts of the body. At the same time, we have no grounds to recognize any structures in this entirely segmented body which could be interpreted as the prostomium, peristome or anal lobe. The presence of an anus in *Dickinsonia* is questionable. Therefore, *Dickinsonia* could be classed as homonomously metamerized, undifferentiated animals. The anterior-posterior orientation caused by the presence of the growth zone at the posterior end somewhat resembles the character of growth of some solitary corals, especially Fungia. This similarity could be much more deeply rooted than seems at first glance.

The family Dickinsoniidae accommodates two genera, namely, *Dickinsonia* Sprigg (five species) and the monotypical genus *Palaeoplatoda* Fedonkin. The Dickinsoniidae is probably an independent lineage in metazoan evolution, simultaneously realizing such fundamental promorphological characters as bilateral symmetry and metamerism. Yet we are not in possession of data on the mode of reproduction of the Dickinsoniidae. Fossil material yields neither signs of reproductive organs nor events of asexual reproduction. Above we discussed the characters of the family not permiting placement of the Dickinsoniidae into the Articulata as we see them now. Although some obvious parallels between the turbellarians and the Dickinsoniidae may be drawn, the Dickinsoniidae differ primarily in the display of clearly defined, homonomous metamerism.

The most important general promorphological characters of the group are as follows:

1. Flat and thin, bilaterally symmetrical body.
2. Large number of similar segments or polymerous, homonomous metamerism.
3. Growth zone situated at the posterior end.
4. Alternate arrangement of metameres.

Even these architectonical features would be sufficient to classify the Dickinsoniidae as an independent and specific branch in metazoan evolution on the way to Bilateria and, simultaneously, to Articulata. Further study of this and some other groups of Vendian bilaterally symmetrical forms will probably provide insight into the origin of the two main lineages of multicellular animals or, at least, help to verify some hypotheses of the origin of Articulata.

It is difficult to determine to what extent the lineage of Dickinsoniidae was progressive and advanced: so far, representatives of this group are known only from the Vendian. The systematics and the taxonomic rank of Dickinsoniidae are not quite clear. Following Moore and Harrington (1956), this group could be placed in the class Dipleurozoa, but not Coelenterata. Their promorphology suggests that the Dickinsoniidae are not true Articulata, however, as yet no close analogies have been shown among Plathelminthes either. The author is prone to consider the Dickinsoniidae as a special lateral branch diverging relatively late from the Radialia (at the level of cyclomerous organisms with concentric organization or at the level of animals with radial symmetry of infinitely high order), hence some promorphological relics of radiated multicellular organisms. It is most likely that the Dickinsoniidae represent an independent stock of Bilateria and a taxon of higher rank, equated probably to the phylum Articulata in the system of the animal kingdom proposed by Beklemishev (1964, Vol. 1).

3.2.2 *Sabelliditids*

One of the most typical groups at the Precambrian/Cambrian boundary and, at the same time, one of few groups of animals that survived probably one of the most dramatic events in the history of life evolution on the Earth, are representatives of the order Sabelliditida Sokolov (Sokolov 1965a, 1967). Fossil sabelliditids occur as tubular integuments. These solitary tubes, made of chitinous matter, are usually thin, very long and elastic. The walls of the tubes can be either very thin or thick, often fibrous in structure, with smooth terminal portions or irregularly wrinkled transversely. They exhibit no segmentation of the annelid type or fusellar elements. In one of the groups the tubes consist of funnel-shaped, narrow rings tightly inserted one into the other; the funnels have regular, annular, marginal fringes thus imparting distinct articulation. The tubes vary in colour from lustrous black and dark brown to translucent yellow and nearly colourless.

The order Sabelliditida incorporates two families. The family Sabelliditidae Sokolov, known from the Late Vendian and Early Cambrian, includes several genera: *Sabellidites* Yanichevsky 1926; *Parasabellidites* Sokolov 1967; *Paleolina* Sokolov 1965a. The family Saarinidae Sokolov 1965 also occurring in the Vendian-Cambrian interval includes two genera: *Saarina* Sokolov 1965 and *Caluptrina* Sokolov 1965 (Sokolov 1965a, 1967, 1972; Krandievsky et al. 1968).

For a long time sabelliditid tubes have been compared with those of sedentary annelid worms. However, Sokolov pointed out that sabelliditids differ from sedentary polychaetes in having quite different proportions of tubes; no annelid segmentation; no muff made up of ooze particles around the tube integument; negligible conical tapering of tubes and some other traits. Some resemblance of sabelliditid tubes to those of pterobranchians was also noted, though differences between them are rather substantial; sabelliditids lack any fusellar structure of integuments and features relevant to colonial animals.

In 1965, B.S. Sokolov advanced the hypothesis relating the Sabelliditida to the present Pogonophora. The similarity between these groups is evidenced by the following characters:

1. Organic (chitinous) nature of zooid tubes and similar physical properties.
2. Cylindrical shape of tubes and the absence of articulation and branching typical of annelids.
3. Gradual transition from wrinkled to smooth portions of the tubes.
4. Extremely elongate tubes.
5. "Collars" in some representatives both of Sabelliditida and Pogonophora, which terminate the upper edge of the tube segments ("funnel").

On these grounds Sokolov interprets Sabelliditida as fossil Pogonophora. The fossil record of the Pogonophora is relatively poor. Presumable pogonophore tubes are known from Ordovician erratic boulders in Poland (Kozlowski 1967); Silurian beds of the Taimyr Peninsula (Sokolov 1972a); Lower Pliocene nodules of the Antwerp Basin, Belgium (Heinzilin 1965); and Oligocene beds of Oregon, USA (Adegoke 1967). The Precambrian/Cambrian boundary beds are extremely rich in sabelliditid fossils and all occurrences are related to sediments accumulated in non-agitated shelf environments or epiplatform seas, i.e. in areas of intense colonization and severe ecological competition in the Early Cambrian, at the beginning of the expansion of archaeocyatheans and trilobites. According to Sokolov (1972), sabelliditids were forced to the deeper sea zones, where their slow evolution resulted in such descendants as modern Pogonophora. This interesting hypothesis has been recently verified by a detailed study of the skeleton ultrastructure of sabelliditid and extant pogonophores (Urbanek 1979). Four tubes, 0.5 to 1 mm across, of *Sabellidites* sp. and a tube of modern pogonophore *Zenkevitchiana longissima* Ivanov, were examined. Ultrastructural examination of the sabelliditid tube, carried out for the first time, shows that the wall of the tube consists of two homogeneous layers, outer and inner, and a median laminar part. Specific wrinkles on the surface of the sabelliditid tube are formed exclusively by the outer, almost homogeneous, layer. The walls of the tubes of all studied pogonophores show a fully laminar structure, and wrinkles, visible on their surfaces, are made up of numerous tissue layers. According to Urbanek, the data obtained on the skeleton structure of both groups are inconclusive. Some similarity found in their submicroscopical anatomy provides a very tenuous line of argument phylogenetically connecting Sabelliditida and Pogonophora, but does not fully rule out their possible affinity.

3.2.3 Oligomerous Segmented Forms

Vendian Bilateria, whose symmetry is characterized by a plane of glide reflection, are dominated by forms with a small number of metameres. One of the most peculiar groups is the family Vendomiidae Keller (Keller and Fedonkin 1976), although its volume needs some revision. This group includes five monotypical genera: *Praecambridium* Glaessner et Wade, *Vendia* Keller, *Vendomia* Keller, *Pseudovendia* Boynton et Ford and *Onega* Fedonkin.

The salient characteristics of *Praecambridium* are an oval, disc-shaped body, an axial anterior lobe followed by three to five pairs of small lateral lobes which taper to pointed, outer ends and are directed towards the posterior end. All lobes appear to be confluent in the axial region behind its mid-point. The anterior axial region may have broadened, small, lateral extensions. The axial lobe of the best preserved specimens bears another pair of small lobes laterally. The axial lobe is rounded at the anterior end and tapers slightly backwards.

The interpretation of the form is difficult. The lateral lobes observed on the surface of these minute fossils were tentatively considered as segmental muscle impressions in a thin and probably rather soft shell (Glaessner and Wade 1966). After comparison with the genera *Cambridium* and *Discinella* (or *Mobergella*) the difficult question of the status of *Praecambridium* was still undecided.

Thirteen specimens of *Praecambridium* have been found up to now. The study of fossil material led Glaessner and Wade to the conclusion that the lobes are segmental and that the number of segments increases with growth, from an early stage with three, to later stages with four and five segments. The segments decrease markedly in size towards the posterior margin, where a small triangular terminal area is enclosed laterally by the last well-defined segmental lobes. As stated above, in such cases the grain size of the rock matrix, where the impression was preserved, makes more detailed observations of this feature impossible.

Portions of these minute organisms usually less than 5 mm in length are also peculiar. The anterior half of the body is occupied by the axial lobe, and the posterior half is occupied by lateral lobes. The first segment (terms proposed by Glaessner and Wade) has a peculiar winglike shape as it fits around the tapering end of the anteromedian lobe. The following segments are chevron-shaped. Three-segmented individuals vary in length from 2.8 to about 4.7 mm; the specimens having four or five segments reach 5 mm and more in length.

The new interpretation of *Praecambridium* was based on the assumption that the anterior, horseshoe-shaped region, including the anteromedian lobe, represents the head which is followed by a series of body segments. The apparent addition during growth of new segments at the posterior end is typical of the Articulata. The recognition of the head area is also in agreement with the assignment of *Praecambridium* to the Articulata.

There are at least two reasons against annelid relationships of *Praecambridium*. One is the relative size of the head region; it does not reach this relative size in any known annelid. The other is the absence of any lateral projections from the body which could represent parapodia. Admittedly, some projections may have been situated in *Praecambridium* at the ventral side, but its ventral aspect is unknown.

Possible arthropod relationships of *Praecambridium* are strengthened by a resemblance between it and certain protaspis stages of larval trilobites. However, this resemblance is a general one and sufficient differences exist to make it clear that *Praecambridium* is not a trilobite protaspis. The main difference is the absence of an axis, which is defined by lateral furrows and is divided into segmental rings. This difference appears to be of importance. Similarities are seen in the size, the oval shape, the existence of an axial anterior lobe, which has some characters of a glabella of the head shield of trilobites, and a small number of pleural segments which decrease rapidly in size towards a minute triangular posteromedian area. The distance from the anteromedian edge of the anterior lobe to the anterior margin of the disc equals the length of this lobe. The anterior margin of the disc appears to be crenulated in some specimens. The impression gained from the thorough examination of certain specimens is that of branching ridges extending from the sides of the anterior lobe to the margin where they form crenulation. This peripherally branching, surface sculpture of the head region may be the expression of an underlying digestive (or distribution) system inferred in agnostid trilobites (Öpik 1961). Thus, *Praecambridium* appears to be related to early trilobites or chelicerates or both (Glaessner and Wade 1971). Representatives of these groups are already known from the Lower Cambrian. Their "distant affinity" is contrasted with the "lack of affinity between these classes and the other Arthropoda" (Manton 1969) and hence it is not unlikely that Trilobitomorpha and Chelicerata derived from a common ancestor which had some of the essential morphological characters of *Praecambridium*. In one of his recent works, Glaessner (1979b) suggested that *Praecambridium* resembles primitive Trilobitomorpha or Merostomata. However, the range of comparisons could be much wider.

Birket-Smith (1981a) compared *Praecambridium* to an anaprotaspis of *Elliptocephala* and *Olenus*, a juvenile polychaete *Autolitus fasciatus* and to an embryonic *Limulus mollucanus* (Chelicerata, Merostomata). Comparison of this Vendian organism with larvae appears to be valid. The character of *Praecambridium* growth suggests that the three first segments are initial and that this form is juvenile. Of interest is the comparison of *Praecambridium* with an embryonic *Limulus mollucanus* which develops from a rather large egg (Iwanoff 1933). The limulus stage with four segments (one cephalic and three subsequent) lasts much longer than any of the subsequent stages having a larger number of segments. During this rather lengthy evolutionary stage, the embryonic cells become markedly specialized and the embryo grows. Therefore, three segments next to the cephalic segment could be interpreted as initial. *Praecambridium* seems to exhibit a similar pattern: the stage of three segments next to the head lobe is the initial one, and the growth of the body is maximal in this stage.

A similarity in the structure and evolution of *Praecambridium* and *Limulus* may suggest affinity between the former and Chelicerata, in particular, and Merostomata. Moreover, the phylogenic affinity between Trilobata and Chelicerata, especially Merostomata, has been repeatedly pointed out (Størmer 1942; Bergström 1973, 1979). Hence, *Praecambridium* can probably be an intermediate form between juvenile polychaetes and trilobites; a Vendian juvenile *Spriggina* or some form very closely related to *Spriggina* (Birket-Smith 1981b).

Below we discuss the representatives of the family Springginidae Glaessner 1958 in more detail; these are peculiar Precambrian animals with an unusual combination of characters typical of arthopods and polychaetes. It should be noted that we have to elucidate a very important question concerning *Praecambridium*, namely, whether the arrangement of their left and right segments is opposite, as presumed up to now, or alternate, i.e. staggered. The point is that another Vendian form, *Vendia sokolovi* Keller, which appears to be closely related to *Praecambridium*, possesses conspicuously alternate segments (or a plane of symmetry of glide reflection). *Vendia* was described from the core of the Yarensk borehole (depth 1552 m, Vendian Valdai Superseries) but its systematic position is not clear. Most workers refer *Vendia* to arthopods, particularly to Trilobitomorpha (Sokolov 1965b; Keller 1969; Glaessner and Wade 1971; Glaessner 1979b); but a comparison was made with the pennatulids (Krylov 1972; Termier and Termier 1968); it was also thought to be the lateral aspect of a medusa (Tommotian. . . 1969).

Vendia is preserved as a flat, ovoid impression (the holotype is 14 mm long and up to 8 mm wide). The impression shows a flat, undivided body. The anterior or cephalic area is half-moon or horseshoe-shaped. The median ridge is clearly defined. Five segments or lateral branches, diverging outward and backward, are alternately arranged on either side of the median ridge.

It is clear from a comparison between *Praecambridium* and *Vendia* that they are very closely related and that they differ only at the generic level. *Praecambridium* is much smaller, reaching less than half the length of *Vendia* and slightly more than half its width, with an equal number of segments. The anterior lobe ends at two-sevenths of the length of *Vendia* but extends almost to one-half the length of *Praecambridium*. The apparent staggering of the lateral "outgrowths" from the median ridge of *Vendia* is also found in some specimens of *Praecambridium* (Glaessner and Wade 1971), but the mentioned authors consider this the result of a slight, oblique distortion. In a similar way they explained the symmetry in *Vendia*. It is difficult to agree with this suggestion since the staggering of the lateral segments (lobes, "outgrowths") is known not only in these two forms, but it is also very common in segmented Vendian Bilateria. The existence of a plane of glide reflection in the architectonics of these forms cannot be regarded as a purely Vendian phenomenon, since it is also observed in the structure of some physiological systems of the living Bilateria. However, the apparent staggering of metameres must have been much more common in Vendian bilaterally symmetrical animals and, primarily, this may be ascribed to the process of initial radiation of main stocks of Metazoa, particularly, to the radiation and appearance of the Articulata.

It may be appropriate to recall here that the problem of the origin of the Articulata or that of metamerism of the Annelida remains debatable. Beklemishev (1964) formulated three major concepts, put forward primarily in the last century, which explain the origin of metamerism in annelids through: (1) derivation from antimerism of the Coelenterata; (2) metamerized differentiation and ordering of organs; (3) suppression of transverse fission (strobilar concept). None of these concepts can be accepted now as classical, but each contains a grain of truth. A detailed discussion of these concepts is beyond the scope of this work, but the study of the promorphology of Vendian soft-bodied fauna is believed to lend strong support to this or that concept. First of all, this holds true for the theory of the origin of metamerism in annelids from "cyclomerism" (antimerism) of the Coelenterata. This theory, proposed by Sedgwick in 1884, has since been supported (partly or completely) by many workers. Sedgwick thought that annelids originated from ancestors similar to the Anthozoa. He accepted the development of the mouth and the anus of annelids from the siphonoglyphs of the closed slit mouth, and thus the ventral side of annelids became homologous to the oral side of the Anthozoa body. The body became elongate, the radial chambers of the intestine closed and developed into coelomic pouches. The tentacles of polyps produced extremities. The posterior growth zone in annelids is homologous to the only growth zone of Ceriantharia. In brief, these are fundamental tenets of this simple and elegant theory. According to Beklemishev, this theory can partly be applied to ectodermic metamerism of the larval body, but not to metamerism of the coelom or its derivatives, which is associated with the post-larval body. There is a paradox in this concept, which Beklemishev (1964) attributed to the attempt to derive annelids not from primitive, but from highly specialized forms, such as the Anthozoa.

Analysis of the promorphology of the Vendian Metazoa unexpectedly offers us paleontological arguments supporting Sedgwick's concept and at the same time removing the above contradiction, or paradox. Moreover, the promorphology of some Vendian non-skeletal organisms reveals the most important stages in the development of real metamerism which, although not predicted theoretically, seem necessary. One of them is the stage of unordered, disconformable metamerism characterized by a plane of symmetry of glide reflection. This evolutionary stage of the Articulata could not be lengthy, since the staggering of metameres is imperfect for actively moving animals (to a lesser degree for organisms moving by peristaltic waves running along the ventral side of the body and to a greater degree for organisms with appendages). However, the stage appears to have been a quite natural and even necessary phase, if metamerism is derived not from the Anthozoa, but from less specialized and in all respects more primitive forms of Radialia with an axis of symmetry of infinitely high or variable order. This level of development is known in many groups of Vendian Radialia. The forms having an axis of symmetry of infinitely high or variable order may be assumed to have absolutely dominated Radialia and the Metazoa, in general, at the earliest evolutionary stages of multicellular organisms. Deduction of metamerism of the Articulata from these primitive Radialia must inevitably have led to the form with the unordered arrangement of metameres, similar to *Vendia* (Fedonkin 1985). In this respect it is worthy to cite Beklemishev, who wrote:

"Thus, from the metamerous, disconformable and imperfect arrangement of individual systems of organs which could be assigned to the prototype and possibly to the initial form of annelids and molluscs, two series of forms diverge: in polychaetes – towards enhancement of metamerism, conformity of metamerism of different systems of organs, extending metamerism to new systems of organs, appearance of general metamerism of the body; in molluscs – mainly towards elimination of metamerism due to oligomerization or complete disappearance of metamerously arranged organs. It is quite probable that both groups evolved in the same direction as well. In any case the main tendency of the development of metamerism is believed to have resulted from the mode of movement which had become characteristic of each group" (Beklemishev 1964, Vol. 1).

The above theoretical digression enables us in a different way to consider segmented Vendian organisms, similar to *Vendia*. Obviously, those are not yet the true Articulata in their present sense, but representatives of those stocks which evolved into the true articulate Bilateria. As to the level of development of different systems of organs, these animals were extremely primitive. Thus, *Praecambridium* and *Vendia* segments may well be equated to lateral intestinal diverticula performing a distribution function. The axial anterior lobe in *Praecambridium* may be equated to the pharynx and small intestine (something similar to this is also observed in *Vendia*, and its median ridge may correspond to the caeca). It is possible that just these organisms represent the lineage of a prototype of annelids and molluscs, which Beklemishev wrote about, and hence present natural difficulties in their interpretation (for instance, *Praecambridium* was first compared to molluscs, then to arthropods and polychaetes). In may respects this group does not fit the classification of the living Metazoa, but apparently represents a separate, ancient lineage of the Bilateria, which led to the true Articulata. Organisms of this group may be classified under the Proarticulata, which would reflect the archaic type of organization of these animals (Fedonkin 1983a).

The Valdai Superseries (Letny Coast, White Sea area) yields another species of Proarticulata, namely, *Onega stepanovi* Fedonkin, superficially resembling the larval stages of some trochophoral animals, but, to a greater extent, larvae of arthropods and onychophores. *Onega* has a flat, elliptic body, in the central part of which there is a distinctly elevated, segmented zone represented by a crescent-shaped cephalic area followed by five paired lobes separated by a deep, axial groove. There are five specimens, 4 to 6 mm long. The number of segments increases with growth from three to five respectively, probably reflecting the ability to increase the number of segments as the specimen grows; this ability is extremely typical of annelids, arthropods and some other groups of metamerous animals. So far we cannot interpret *Onega* as a larva, since we were unable to recognize any transitional stages from the known specimens to larger ones, but such a probability cannot be ruled out. We cannot relate *Onega* to any larger metazoan forms known from the Valdai Superseries.

Very close in morphology to the arthropods, and to trilobites in particular, is *Vendomia mennery* Keller (Keller and Fedonkin 1976). Over one-third of the total length of the drop-shaped body of the organism (4 mm) is occupied by the broad, horseshoe-shaped head with two depressions, probably equating to eyes. Six paired segments, decreasing in size posteriorly, arise from the median ridge, resembling the rachis in trilobite. Due to its proportions, particularly the extremely large head, *Vendomia* resembles larval trilobites. In this respect it is appropriate to recall the hypothesis of Termier and Termier (1968, 1976) who proposed the classification under the name nymphoids all the metazoans whose adults do not transcend the stage, otherwise known as the larval stage. One of the characters of larval, soft-bodied animals is the complete or almost complete lack of a mineralized skeleton. According to the above authors, the Vendo-Ediacaran fauna was rich in nymphoids, i.e. adults whose body plan resembled larvae of the later known Metazoa.

Another possible representative of the Precambrian Arthropoda is *Parvancorina minchami* Glaessner (Glaessner 1958). In one of his recent works Glaessner (1979a), based on the examination of 60 specimens, shows the possibility of interpreting these fossils in terms of growth, nutrition and movement of the animal. *Parvancorina* occur as small shieldlike impressions enclosing an anchor-shaped ridge. Curved, relatively large and long structures, interpreted as anterior appendages, diverge obliquely from the anterior lateral arcs (anterolateral ridges) of the anchor-shaped ridge. Numerous filiform structures, regarded as undivided appendages, arise from the median ridge. At present, the shieldlike body is described as a carapace with a faint marginal, raised rim and distinctly elevated, anterolateral and median, smooth ridges. The carapace contains tergites of all the somites of the body, but the carapace itself is unsegmented. The appendages arise from the median zone in regular rows, thus suggesting the metamerous structure of the body under the carapace. Five pairs of relatively stout anterior appendages, which must have been articulated, are followed by up to 20 pairs of posterior, undivided, filiform appendages. Placement of *Parvancorina* into the Arthropoda is based not only on the presence of the peculiar carapace, but on some resemblance with the Paleozoic Trilobitomorpha: *Marella* Walcott, *Mimetaster* Gürich and *Vachonisia* Lehmann. The two latter genera are known from the Lower Devonian, and *Marella* was reported from the Middle Cambrian (Bergström 1979). These genera form the group Marellomorpha (Trilobitomorpha), although some controversies do not allow their classification in one class. Thus, *Marella, Mimetaster* and *Vachonisia* have respectively, only two, three and four pairs of appendages in the head area, although their number in each of the present classes of Arthropoda is more or less constant. However, a widely ranging number of appendages in major arthropod groups may suggest a wider variation in this respect among the earlier arthropods, therefore, a precise number of appendages in the oldest representatives of this group could not be considered as such as important taxonomic character as in the living groups.

Glaessner noted some similarity between the anterolateral ridges in *Parvancorina* and the cephalic (genal) spines in *Marella splendence* Walcott, but the most striking among them are the resemblance of the carapaces of *Parvancorina* and *Vachonisia rogeri* (Lehmann) and a similarity in the differentiation of limbs on the fourth and fifth anterior (oblique) stout appendages followed by numerous (up to 80 in *Vachonisia*), undivided, posterior (transverse) appendages. The systematic position of *Parvancorina* is still uncertain, but available data allow the placement of the genus and the family Parvancoronidae Glaessner close to the point of divergence of the ancestors of Trilobitomorpha in the Crustacea and Chelicerata, but not among direct ancestors of any of these lineages (Glaessner 1979b).

Of great interest is the family Sprigginidae Glaessner. Representatives of this family, such as *Spriggina floundersi* Glaessner, *S. borealia* Fedonkin and *Marywadea ovata* (Glaessner) are fairly abundant in the Pound Quartzite, South Australia (Glaessner 1958, 1976; Glaessner and Wade 1966), but rare in the Vendian of the USSR (Fedonkin 1981a). Sprigginidae show some resemblance to the living pelagic annelids of the family Tomopteridae, but the horseshoe-shaped head more closely parallels some Trilobitomorpha (Glaessner 1958, 1960). Reconstruction of the crustacean ancestors shows a similarity to *Spriggina* (Hessler and Newman 1975). However, it is hardly possible to classify Sprigginidae under the arthropods, because the head of the representatives of the family does not have an appropriate number of segments bearing appendages, furthermore, the segments of the trunk are not clearly articulated and (in *S. floundersi* and *Marywadea*) end with acicular setae. The presence of a simple pharynx in *Spriggina* and a pair of simple teeth in *Marywadea* suggests some relationship to the living polychaete Phyllodocemorpha (Glaessner 1976). In other respects Sprigginidae do not resemble living annelids. A considerable cephalization of these organisms, atypical of polychaetes, and probably an essential sclerotization of epidermic tissues of the head, imply a transitional character of the family Sprigginidae in the evolution from the annelids to the arthropods (Glaessner 1976; Stanley 1976; Termier and Termier 1976).

Birket-Smith (1981b) has rather thoroughly studied *Spriggina* using multiple photographs of latex moulds of fossils (about 30 photographs of each specimen taken in variously arranged lighting with further drawings of details and reconstruction).

The body of *Spriggina* was found to be distinctly segmented, but, unlike the arthropods, segments are not confluent. The dorsal surface is covered by a row of cuticular, caudally overlapping plates. The ventral side of the body bears paired segmental plates. The third and subsequent segments have dorsal and ventral pairs of annelid lateral appendages. The first, strongly modified, horseshoe-shaped segment has paired dorsal depressions, possibly eyes, but no appendages. The ventrally arranged oral cavity occurs under the small prostomium. The second segment is also highly altered, and bears atrophied dorsal, but enlarged ventral appendages which are presumed to be antennae. According to Birket-Smith, *Spriggina* cannot be placed either into the Annelida, or into the Arthropoda in their present sense, despite some similarity to the annelids of the family Hesionidae (double, lateral rows of annelid appendages). Modification of the front segment of *Spriggina* is quite consistent with the concept of the segmentation of the head shield in trilobites (Størmer 1942): the first segment, bearing eyes, and the second segment, bearing antennae, are confluent, while other leg-bearing segments are later additions. Thus, Birket-Smith arrived at the conclusion that trilobites may have derived if not from *Spriggina* proper, then, at least, from one or some very close forms.

The author has recently found two species of unusual organisms of uncertain affinities. One of them, *Bomakellia kelleri* Fedonkin, is a rather large, trilobitelike animal with a half-moon-shaped head region and a spindle-shaped trunk bearing ovoid, flat, lateral appendages. The metamerous structure of the body is emphasized not only by the lateral appendages, but also by two rows of bosses arranged at the dorsal side of the body. The cephalic region strongly resembles in outline the head shield of trilobites, and the axial part of the former can be equated to a large, smooth glabella. Lateral appendages resemble pleura, but it is noteworthy that each appendage has a very regular, two-pennate relief in plan outline. The appendages increase in size from the anterior towards the posterior rounded end, enclosing longitudinal linear structures of uncertain nature. The trunk is not divided into axial rings, as in trilobites. It is clearly seen that many characters of *Bomakellia* do not allow us to propose an affinity either with the Trilobitomorpha, or with the Crustacea, although the lateral appendages with two-pennate relief resemble epipodites and pre-epipodites of some crustaceans. It is appropriate to recall here that some marine polychaetes have pennate or pectinate branchiae. The appendages of *Bomakellia* may have performed, along with the breathing function, a moving function: the animal could actively swim using the appendages. This form might have possessed a thin carapace as suggested by a narrow ridge along the distal ends of the lateral appendages.

The other organism, *Mialsemia semichatovi* Fedonkin, could be regarded as a form related to *Bomakellia*, possibly belonging to the same family, although the forms substantially differ in structure and proportions. *Mialsemia* is ovate and tends to be wider than long. The narrow, smooth, cigar-shaped body, without any signs of segmentation, bears two rows of lateral, opposite appendages. The appendages are long and petaloid, exhibiting a characteristic two-pennate sculpture. The largest appendages occur in the middle part of the body and decrease in size towards both ends of the trunk. The poorly preserved head is half-moon-shaped. A distinct lateral ridge protruding beyond the distal ends of the appendages suggests the existence of a carapace.

The above forms are represented by single specimens and, hence, new finds would possibly help to recognize some new structural details. Now their arthropod (early trilobite or chelicerate) affinity is quite evident. The presence of characters in common with *Spriggina, Bomakellia* and *Mialsemia* allows an assignment to the class Paratrilobita of the Arthropoda.

In summary, the following conclusions can be drawn on the Vendian Bilateria.

1. If we assume that the fossil record of the Vendian Metazoa reflects adequately the proportions of major groups in the Late Precambrian animal world, then during the Vendian the bilaterally symmetrical animals were in the minority, as compared to the radially symmetrical ones. This is partly explained by ecological and taphonomic controls. However, such a disproportion may be chiefly ascribed to the earlier diversification of the Radialia, as compared to the Bilateria.
2. At the same time, the large diversity of body plans, displayed by Vendian Bilateria, reflects a wide rage of types of organization and degrees of advancement of separate lineages of this large group.
3. Bilaterally symmetrical Vendian animals are sharply dominated by flat forms, among them variably segmented organisms, including polymerous and oligomerous forms, showing signs of homonomous and heteronomous metamerism, along with typical unordered segmentation and symmetry of glide reflection.
4. Lack of wormlike animals with a circular or elliptical cross-section implying the development of strong musculature and, possibly, the coelom may suggest a small proportion of coelomates among the Vendian Bilateria.
5. Drastic predominance of segmented forms among the Vendian Bilateria may suggest that in many metazoan lineages the origin of bilateral symmetry and segmentation (metamerism) were interrelated or conjugate processes.
6. Most Vendian Bilateria display features atypical or multicellular animals of later epochs. The most important among these is the plane of symmetry of glide reflection occurring in the extremely primitive Bilateria, especially in polymerous forms exhibiting homonomous metamerism, and in oligomerous Bilateria with conspicuous cephalization. The plane of symmetry of glide reflection could be regarded as a fundamental character reflecting an important stage in the evolution of metamerism in various lineages of the Metazoa. This type of symmetry, atypical of most Phanerozoic multicellular animals, can be considered as paleontological evidence supporting the hypothesis of a possible derivation of metamerism in the Articulata from the radially symmetrical organisms. Here again, we must emphasize that the plane of symmetry of glide reflection in the Vendian Bilateria may indicate the probable extremely early origin of the bilaterally symmetrical segmented forms from the radially symmetrical animals with an axis of rotation of variable order.
7. Thin, leaflike forms of the family Dickinsoniidae show the lowest level of organization among segmented forms. Analysis of symmetry of the Dickinsoniidae and comparison of their body plan with that of the Coelenterata, Turbellaria and Annelida suggest that the Dickinsoniidae are an independent

lineage of the oldest Bilateria. The possibility to derive the symmetry of the Dickinsoniidae from the symmetry of the most archaic Radialia with an axis of rotation of infinitely high order (concentric organization) or with an axis of variable order (radial organization) indicates that this group is very primitive. The Dickinsoniidae is a possible branch of the Bilateria separated from the Radialia earlier than others. It is very difficult to decide whether they have succeeded or not, but judging from the fact, apart from the Dickinsoniidae, that other more highly organized Bilateria existed in the Vendian, the Dickinsoniidae seem to have evolved as a blind alley which possibly became extinct as early as the Vendian: true Dickinsoniida or similar forms are not known either from the Late Vendian or from the Cambrian.

8. Animals with a flat, undivided body, a large cephalic region with an axial lobe followed by segments, belong to an independent and rather specific group of the Vendian Belateria. The segments are opposite or alternate (staggered). This group, represented by the family Vendomiidae, shows a tendency to develop a separate, large, cephalic region, which is a fairly advanced feature. However, the plane of symmetry of glide reflection in the arrangement of segments, and the flat, ovoid, undivided body of these animals may suggest an extremely archaic type of the group, possibly a blind alley, one of Nature's ventures on its way towards the true Articulata.
9. A number of fairly specific groups of Vendian Bilateria is difficult, if possible at all, to place under the existing Metazoa. Representatives of these groups often display a combination of characters, separately relevant phyla drifted apart during later epochs. The horseshoe- or half-moon-shaped head, resembling the head shield of some trilobites, and the body of the annelid type provide a good example in the families Sprigginidae and Bomakellidae. Such forms may be regarded as transitional or ancestral to the groups that become widely separated in the course of subsequent evolution.
10. Among the extremely specific groups of the Vendian Belateria there are lineages that could well continue into the Phanerozoic. This concerns some Arthropoda, rare early Annelida and Sabelliditidae.

4 Colonial Forms and Petalonamae

The Vendian beds contain a considerable number of colonial organisms, which, as a whole, are not as diverse as Phanerozoic colonial forms. The Vendian was dominated by frondlike or pennate organisms (Fig. 13) bearing resemblance, probably only superficial, to modern alcyonaria and pennatularia (Anthozoa). The colonies are not always fully preserved, more common are impressions of their basal parts (attachment disc or the base of the main axial polyp) or structures, similar to Vendian *Pseudorhizostomites howchini* Sprigg, on the base of beds. They are preserved on the lower bedding plane of sandy slabs as furrows which radiate outwardly from a centre of variable shape. As they radiate, they split and become shallow. No outer edge is generally observed. Vertical splitting across the fossils indicates disturbances in bedding above the centre of *Pseudorhizostomites*, or the

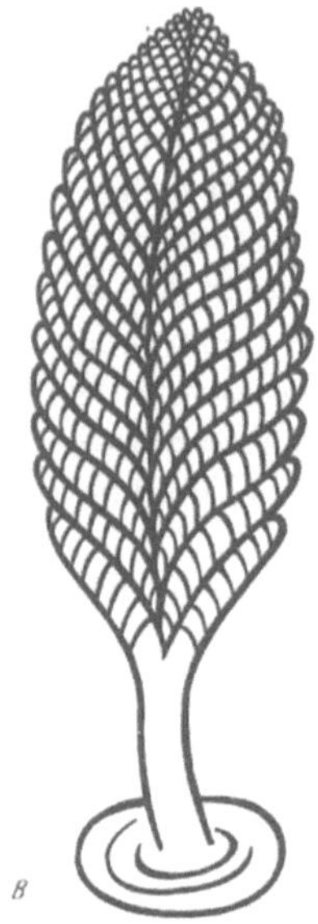

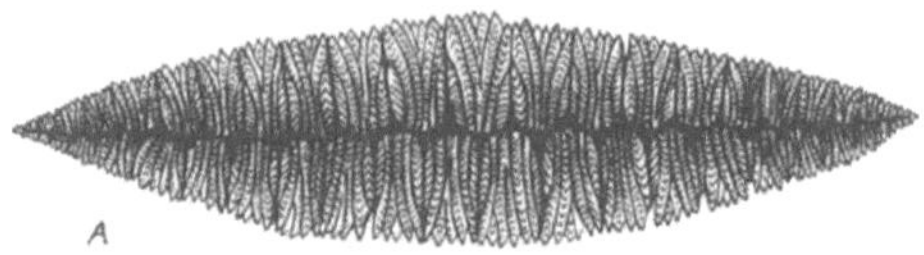

Fig. 13A,B. Some typical colonies of Coelenterata from the Vendian Sea: **A** spindle-shaped colony from the Conception Group, Newfoundland (pelagic form); **B** *Charnia masoni* Ford

presence of structures in the form of vertical flutes or very narrow funnels filled with massive rock. All seem to have resulted from the decay of buried organisms down to consolidation of the sediment with further subsidence of the overlying sediment and flow of the underlying sediment (Wade 1968). However, these fossils can be regarded differently. The slabs on which *Pseudorhizostomites* occur are often rich in different, sometimes very thin impressions and casts of soft-bodied animals, and thus *Pseudorhizostomites* may have been formed by the decay of any of these organisms. Judging from the dimensions of *Pseudorhizostomites*, mostly about 5–7 cm in diameter, the decayed organism must have had appreciable dimensions and volume, however, only one specimen of over 80 fossils, described by Glaessner and Wade (1966), possessed a definite outer edge, a sharp groove preserved around half its circumference. None of dozens of fossils of this species recorded from the Zimny Coast, White Sea area, and from Podolia, Ukraine, possessed outer edges. The above suggests that *Pseudorhizostomites howchini* or, at least, some fossils of this species could be impressions of hydrorhiza, i.e. rhizoid appendages with the aid of which the organisms attached to the substrate (axial adherence). Hydrorhiza are known in living colonial polyps, and *Pseudorhizostomites* closely resemble hydrorhiza of the colonial hydroid polyps, Campanularia (Hyman 1940).

Hydroids could have been much more abundant in the Vendian than it seems now. For example, many Late Precambrian, still unnamed, forms, particularly some colonial forms, from the Conception Group (Avalon Peninsula, southeastern Newfoundland) may well belong in the Hydrozoa.

Frondlike colonial polyps, *Charnia masoni* Ford, were described from the Charnwood Forest, England, and *Charniodiscus concentricus* Ford, diagnosed as a "disclike organism", was described as a separate species, because only in one specimen were "frond and disc apparently associated" (Ford 1958, 1963). Jenkins (Jenkins and Gehling 1978) recently established that in English fossil material not

Charnia, as Ford suggested, but a distinct type of frond is associated with the disclike impression (*Charniodiscus*). He concluded that therefore the entire specimen became the holotype of *Charniodiscus concentricus* Ford and concluded also that this frond showed the characters of *Arborea arborea* Glaessner et Wade, later described from the Pound Quartzite, which consequently became a synonym of *Charniodiscus*. This particular case is cited here to cast light on the taxonomic relationships between the long known and still used names of one and the same Precambrian species.

Charnia and Charniidae Glaessner are generally abundant in Vendian beds of the world (Russian Platform, Siberia, England, Newfoundland). *Charnia* are usually interpreted as colonial polyps resembling, in the form of a colony, living marine pennate organisms, soft corals of the order Pennatularia, subclass Octocorallia (Glaessner 1961; Glaessner and Wade 1966). And, although Ford compared *Charnia* with siphon algae, particularly with Himanthalia and Cymopolia, he did not discount alternative coelenterates affinities of *Charnia* and other rangeids (Ford 1958; Evans et al. 1968). Strictly speaking, too little is known to justify the allocation of *Charnia* not only to Octocorallia, but also to Anthozoa in general. One of the ways to solve the problem of *Charnia* and Rangea affinities would be a detailed study of the structure of separate zooids of these colonies.

Furthermore, two-pennate polyp colonies of uncertain affinities are also known. They contain forms in which the basal part of the primary polyp is elongate and ellipsoid in shape (*Vaizitsinia sophia* Sokolov et Fedonkin), or forms with attachment discs. However, pelagic colonial polyps, for example, elongate, leaflike colonies from the Conception Group, Avalon Peninsula (Newfoundland, Canada), also existed in the Vendian. Tree- or bushlike colonies are rare.

It should be noted that Charniidae differ from living Pennatularia in having a disclike base (Glaessner 1959). The base of the primary polyp in the latter is elongate, pyriform or sausage-shaped, and the disclike basal expansion is known in the representatives of the other order of Anthozoa Octocorallia, in Gorgonaria.

Apart from the attached pennate and less common fanlike colonial organisms, pelagic colonial forms were also abundant in the Vendian. The most interesting are elongate, spindle-shaped colonies, up to 30 cm long, which consist of tightly-spaced lateral branches, with a staggered arrangement and bearing numerous minute polyps, occurring on the branches as grains in the ear. Other colonies resemble a large wormlike body, up to 40 cm long, from which a series of tubular outgrowths detaches towards one side. These forms are long known from the Conception Group, Newfoundland, but have not yet been described.

Analysis of data on Vendian colonial forms allows the following features to be noted: large dimensions of many colonies; insignificant diversity of their morphology; lack of functional differentiation among zooids of the colony. If should be added that we have no reason to assume that the above mentioned colonial forms may be allocated into higher animals than coelenterates. However, it is possible to determine the status of these organisms in the Coelenterata only if the structure of a zooid or the shape and composition of sclerites or spicules are known, provided that some Vendian colonial organisms had had them.

The Vendian fauna incorporates many species whose systematic position remains uncertain, despite the very prominent morphology of these organisms.

Among them, first of all, are forms grouped into the Petalonamae (Pflug 1970a,b, 1972a-c), but it is a fairly heterogeneous group of organisms of uncertain affinities. For instance, representatives of the families Rangeida Glaessner and Charniidae Glaessner bear a close resemblance to pennate colonies of the modern soft corals Pennatulacea, but specimens of the family Pteridinidae Richter (*Pteridium simplex* Richter, *P. nenoxa* Keller) and those of the family Erniettidae Pflug (13 genera and 28 species, partial synonyms) are quite unique, possibly typical of the Vendian exclusively.

Organisms of this group have some features in common, particularly leaflike or bag-shaped bodies and primary segmentation or ribbing, lacking any secondary differentiation of the segments. In addition to Pteridiniidae and Erniettidae, the group Petalonamae includes *Arumberia* Glaessner et Walter, *Baicalina* Sokolov, *Namalia* Germs, *Nazepia* Germs, *Zimniella* Fedonkin, *Archangelia* Fedonkin, *Inkrylovia* Fedonkin, *Podolimirus* Fedonkin. The state of the art of these organisms, appreciably differing in morphology, does not allow us to formally place them into taxa of higher rank. At the same time, we have no reason either to regard them as plants, sponges or archaecyatheans, or to assume that the level of their organization was higher than in the Coelenterata. Therefore, some workers propose to consider them as of uncertain affinities in the Coelenterata (Glaessner and Walter 1975). For example, *Archangelia valdaica* Fedonkin is a true, bilaterally symmetrical animal. Its ovoid body is separated by a segmented median zone with the segments decreasing in length unilaterally. Lateral parts demonstrate gentle transverse crenulation with the number of crenations equalling the number of segments of the median zone. From modern and fossil material it is still impossible to infer forms close in morphology, however, the metamerous structure of the bilaterally symmetrical organism may suggest a rather high organization.

The other species, *Inkrylovia lata* Fedonkin, is also bilaterally symmetrical. The bag-shaped body of this large organism is divided into uniform, longitudinal segments. Considerable distortions and folding suggest that the body of *Inkrylovia* was extremely elastic and thin-walled. Flat, symmetrical segments detach from the median furrow and, not varying in width, bend towards one side. The length of the segments of each specimen is constant, although it varies in different specimens as a function of their dimensions. On one of the casts, the median furrow ends with a triangular depression, possibly equating to the oral cavity. In another specimen, from the end of large *Inkrylovia* extends a small leaf-like body, similar to the former in the character of segmentation. The leaflike, segmented body may well be the impression of the juvenile, not yet separated from the parent individual. One specimen exhibits a slight expansion of the terminal part of *Inkrylovia*, which can be regarded either as an adhesion talon, or as an attachment disc, or the like. If new finds support the above suggestions, this would indicate that *Inkrylovia* were bag-shaped, bilaterally symmetrical, uniformly segmented organisms which possessed an oral cavity; they may be characterized by the attached mode of life and budding.

In general, the bag-shaped forms are not unusual among known Vendian organisms. Nesov has recently described a new genus of such fossils. *Yarnemia ascidiformis* Nessov (Chistyakov et al. 1984). The form was found together with some other Vendian fossils in silty sediments in the middle Onega River.

Yarnemia is described as representing "organisms possessing a bag-shaped body, variable in shape, and dense, slightly folding covers or an integument. A projection with a probable three- or four-valvate cavity (oral siphon) slightly elevates at one side; close to it there is another small projection with a probable cavity (anal siphon)... The surface of the body is smooth, about 2–3 cm in diameter" (Chistyakov et al. 1984, p. 18).

Nesov places *Yarnemia* into the class Ascidiacea of Tunicata (Chordata). *Inkrylovia*, described above, is also interpreted as representing tunicates. According to Nesov, the segments of *Inkrylovia* may well be transverse muscular bands being more effective, as compared to ascidians, in passage of water through the pharynx. Some representatives of the class Salpae possess eight or nine such bands. The median furrow of *Inkrylovia* is assumed to have corresponded to the position of the endostyle in the pharynx; the endostyle is an organ for food catchment in filter-feeding organisms of the primitive Chordata.

As for *Yarnemia*, in addition to the above features suggesting, according to Nesov, a close similarity to the Ascidiacea, there are also longitudinally striated, stolonlike projections, in some cases linking two specimens (probably parent and daughter individuals). One of the indirect arguments for a comparison of the above bag-shaped organisms with tunicates is a higher (by one order of magnitude) vanadium content in the enclosing rocks, as compared to other rocks. A considerable amount of vanadium compounds is present in the blood of many living ascidians. However, it is unclear whether this vanadium content was sufficient to increase the amount of vanadium in the rock and whether or not vanadium compounds could be lost as a result of lithogenesis and further alteration of enclosing rocks.

The above cited work is one of the first which raises the question relevant to a possible existence of chordates in the Precambrian. Problematics which were compared to tunicates occur in the Early Cambrian (Walcott 1911) and Late Silurian, and relatively highly organized chordates, jawless vertebrates (Agnata), existed as early as the Lower and Middle Ordovician (Ritchie and Gilbert-Tomlinson 1977). Probably, the existence of conodonts in the fossil record since the Middle Cambrian has to be taken into account. The known fossil animals, conodont carriers, probably belong to chordates (Melton and Scott 1973). All the above makes the existence of chordates in the Vendian highly probable, therefore, some forms of the Precambrian Metazoa are to be observed from a different point of view.

Representatives of the family Pteridiniidae Richter are very abundant in the Vendian. They were recorded from the Nama Group, Namibia; Precambrian Pound Quartzite, South Australia; southeastern White Sea area; Dniester Basin of Podolia, Ukraine; northwestern Canada (Mackenzie Mountains); North Carolina, USA. Quite different statements about the nature of these organisms were made: they were assigned to Gorgonaria (Richter 1955); to Ctenophora (Gürich 1930a,b, 1933; Wells and Hills 1956); to Pennatularia (Glaessner 1959; Glaessner and Wade 1966); to a hypothetical type Petalonamae (Pflug 1970a,b, 1972a,b) The concept of the Petalonamae as a separate branch of the Precambrian evolution of organic life, transitional between the Plantae and Animalia, was subjected to sharp criticism by Glaessner and Walter (1975). In this work the authors revise the ideas on *Pteridinium* as a representative of the Pennatularia and suggest the allocation of the genus, together with *Ernietta* Pflug, *Namalia* Germs, *Nasepia* Germs, *Baicalina*

Sokolov and *Arumberia* Glaessner et Walter as of uncertain affinities among the Coelenterata.

In the territory of the USSR, *Pteridinium* are abundant in the Vendian deposits of the Letny Coast, White Sea area; rare occurrences are known from the Zimny Coast, White Sea area, and from the lower Valdai Superseries of the Dniester Basin of Podolia Ukraine (Fedonkin 1981a; Velikanov et al. 1983). In 1977, in the Syuzma River area, letny Coast, Onega Peninsula, one specimen of *Pteridinium* was found lying vertically, relative to the bedding plane. Probably, this is the lifetime position of the animal. No traces of basal attachments have been found, however, the fixed position of one end of *Pteridinium* is suggested by concentric furrows of a trace fossil *Suzmites volutatus* Fedonkin, occurring in one and the same specimen with *Pteridinium*, which laid this trail (Fedonkin 1976a).

All the above facts suggest that *Pteridinium* were attached organisms, resembling the present marine pennate forms Pennatularia in outward appearance, but such a taxonomy of *Pteridinium* may raise serious objections. Unlike marine pennate forms, the body of *Pteridinium* is undivided, it bears no polyps, and the segments themselves are hardly polyps. At the same time, the suggestion of a higher organization of *Pteridinium* as compared to the Coelenterata (Glaessner and Walter 1975) appears to be doubtful.

The most abundant material on *Pteridinium* was obtained from the Nama Group, Namibia (Pflug 1970a,b, 1971a, 1973). In the state of preservation the African material resembles that recorded from the southeastern White Sea area and South Australia. The African specimens often show three half-leaflike elements (petaloids) extending from the axial line. One of the Autralian specimens also demonstrates three petaloids arising from the axis; two of these, very closely spaced, are separated by a thin (1 mm) layer of quartzite, and the third occupies a more or less symmetrical position on the other side of the axial line (Jenkins and Gehling 1978). The three-pennate shape of *Pteridinium*, known also from the White Sea area, as well as other morphological features, suggest a separate group of organisms under the Petalonama.

Bag-shaped *Arumberia banksi* Glaessner et Walter, which are rather abundant in the Upper Precambrian of Central Australia, have been recently recorded from the Vendian of the Central Urals. The walls of these cup-shaped organisms bear elongate, subparallel elastic fibrils which perform a supporting function, and their shape slightly resembles primitive thin-walled, bowl-shaped sponges, such as Early Paleozoic Leptomitidae (Glaessner and Walter 1975).

Seilacher (1984) has recently stated that Petalonamae and many other Precambrian organisms could well be segmented forms resembling an inflatable mattress in build. Such a construction enabled them to sustain a sufficiently high internal water pressure (hydrostatic skeleton). The organisms fed by intake of organic matter, soluble in water, through the walls of the body. In fact, the mode of feeding of many Petalonamae is still uncertain. Thin, near-axial structures, which could be interpreted as pouch-like organs, possibly serving the intake of food particles, are known in *Pteridinium nenoxa*. Charniidae could use numerous slits between lateral outgrowths for food intake. The absence of obvious polyps on the lateral outgrowths of *Charnia* and some other similar forms makes us regard these organisms either as extremely integrated colonies of polyps, with indiscernible individuals, or as certain solitary forms, i.e. individuals. In any case these forms deserve thorough study.

Apart from the above colonial forms, these are known bush-, fan-, spindle- and crest-shaped colonies, some of these being rather large (see Fig. 13). The forms are long known from the Vendian Avalon faunal assemblage, Newfoundland, but they have not been described in detail yet (Anderson and Misra 1968; Anderson and Conway Morris 1982).

As a whole, we have reasons to suggest the existence of much more abundant problematic forms in the Vendian.

Some generalizations have emerged from the above brief discussion of major Vendian animal groups. Multicellular organisms of the Vendian biota, immediately preceding the Cambrian biota, are characterized by:

1. Lack of sclerotized skeletal tissues (elements) in most invertebrates.
2. High diversity of life forms.
3. Significant differentiation at the level of taxa of high rank and insignificant differentiation at a specific level, suggesting the pre-Vendian (though unlikely long-term) stage of metazoan history.
4. Sharp prevalence of coelenterates (partly due to their relatively large dimensions and preferential preservation of abundand sedentary forms).
5. Presence of all ecological groups: stationary and vagile-benthon, nekton, plankton (Fig. 14) and pleuston.
6. Preferential occurrence of benthon and other groups in normal marine, shallow water environments.
7. Similarity of Vendian faunal assemblages in various parts of the world as a result of the absence of strong ecological barriers during Vendian post-glacial transgression (Table 1).
8. Presence among Vendian Metazoa of adults resembling, in body plan, larval stages of organisms known from later geological periods.
9. Gigantism of the most primitive solitary organisms and a decrease in body size as the organization of animals becomes higher. The same trend in the decrease of abundance and specific diversity probably due to the preservation of the largest fossil forms. This tendency may be ascribed to the history of appearance and development of the main metazoan phyla.
10. Presence in the different Vendian metazoan groups of characteristic types of symmetry and architectonics, unknown or extremely rare in the Phanerozoic, for instance, the concentric structure of cyclomerous coelenterates or three-fold radial symmetry in some representatives of this group.
11. First traits of an organic skeleton, and the development of a mineralized skeleton in some groups at the close of Vendian.
12. Absence among the known Vendian soft-bodied animals of obvious ancestors of minute skeletal organisms, which appear at the level of the Tommotian, Lower Cambrian.
13. Low activity of vagile predators and necrophages.
14. Short trophic chains.
15. Low specific weight of infauna in benthic communities of the Vendian sea with sharp predominance of sestonophages and detritophages.

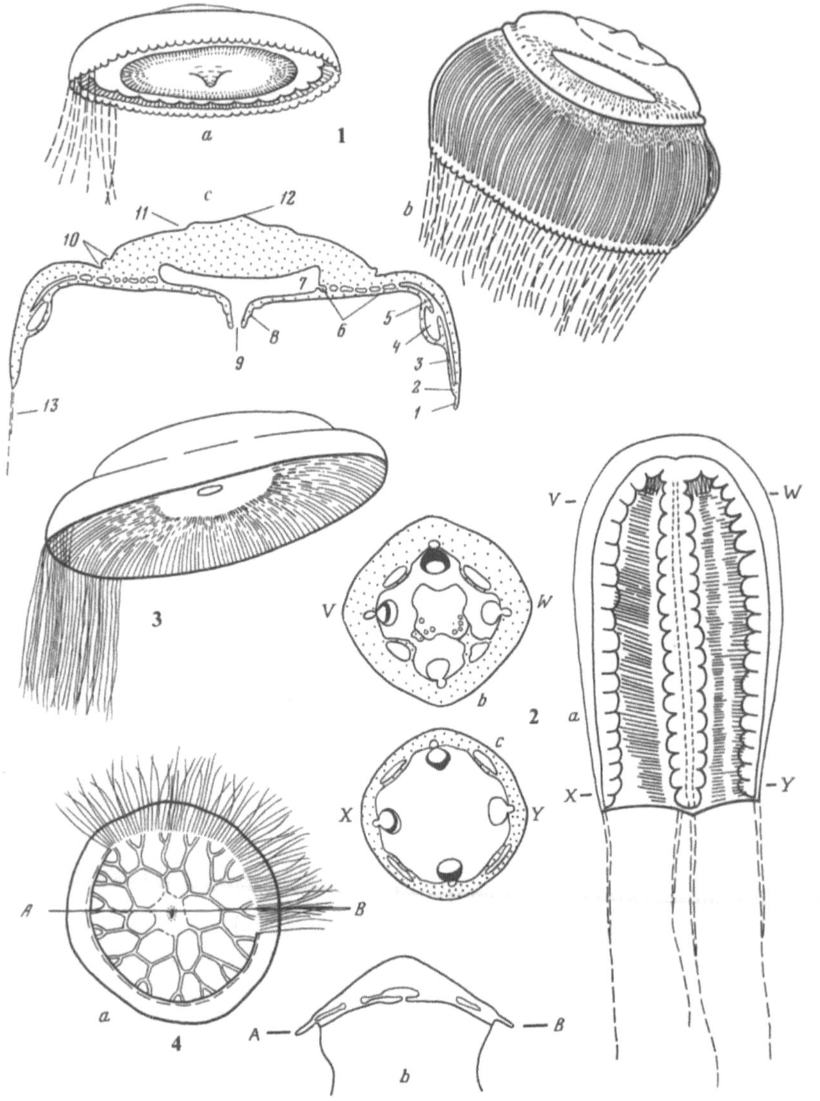

Fig. 14. Some typical medusae of Ediacaran biota according to Wade's (1972a) interpretation. **1***a,b,c* *Brachina delicata* Wade; *a* oral side; *b* lateral view; *c* cross-section along axis **2***a,b,c* *Kimberella quadrata* Glaessner et Wade; *a* lateral view; *b,c* cross-sections perpendicular to axis; **3** *Ediacaria flindersi* Sprigg; **4***a,b* *Rugoconites enigmaticus* Glaessner et Wade *a* oral side; *b* cross-section along axis

Table 1. Geographic distribution of remains of Vendian non-skeletal organisms[a]

	1	2	3	4	5	6	7	8	9
Nemiana simplex Palij		x	x	x					
Beltanelliformis brunsae Menner	x	x							
Bronicella podolica Sokolov			x						
Sekwia excentrica Hofmann									x
Nimbia occlusa Fedonkin									
N. dniesteri Fedonkin		x	x						
Tirasiana disciformis Palij		x							
T. coniformis Palij		x							
Cyclomedusa davidi Sprigg		x	x		x		x		
C. radiata Sprigg		x	x		x				
C. plana Glaessner et Wade		x	x		x				
C. gigantea					x				
C. serebrina Palij			x						
C. minuta Fedonkin	x								
C. delicata Fedonkin	x								
Ediacaria flindersi Sprigg		x	x	x	x				x
Beltanella gilesi Sprigg					x				
Medusinites asteroides (Sprigg)	x	x	x		x				x
Medusinites patellaris Sokolov			x						
Planomedusites grandis Sokolov			x						
Kullingia concentrica Føyn et Glaessner				p					x
Kaisalia mensae Fedonkin		x							
Ovatoscutum concentricum Glaessner et Wade		x		x	x				
Chondroplon bilobarum Wade		x			x				
Eoporpita medua Wade		x	x		x				
Mawsonites spriggi Glaessner et Wade					x				
Protodipleurosoma wardsi Sprigg					x				
P. rugulosum Fedonkin		x	x						
Hallidaya brueri Wade									x
Lorensinites rarus Glaessner et Wade					x				
Pseudorhizostomites howchini Sprigg		x	x		x				
Rugoconites enigmaticus Glaessner et Wade					x				
R. tenuirugosus Wade					x				
Hiemalora stellaris Fedonkin		x	x		x				
Pomoria corolliformis Fedonkin		x							
Bonata septata Fedonkin		x							
Armillifera parva Fedonkin		x							
Evmiaksia aksionovi Fedonkin		x							
Irridinites multiradiatus Fedonkin			x						
Veprina undosa Fedonkin		x							
Elasenia aseevae Fedonkin			x						
Tribrachidium heraldicum Glaessner		x	x		x				
Scinnera brooksi Wade									x
Albumares brunsae Fedonkin	x								
Anfesta stankovskii Fedonkin		x							
Staurinidia crucicula Fedonkin		x							
Conomedusites lobacus Glaessner et Wade		x	x		x				
Persimedusites chahgazensis Hahn et Pflug									x
Ichnusina cocozzi Debrenne et Naud									x
Glaessnerina longa (Glaessner et Wade)				p	x			p	
G. grandis (Glaessner et Wade)									x
Arborea arborea (Glaessner)					x				
Charniodiscus concentricus Ford		p					x		

Table 1 *(Continued)*

	1	2	3	4	5	6	7	8	9
C. planus Sokolov			x						
C. oppositus Jenkins et Gehling		x	x		x				
C. longus Glaessner et Wade					x				
Phyllozoon hanseni Jenkins et Gehling					x				
Charnia masoni Ford		x		x			x	x	x
Pteridinium simplex (Gürich)						x			
Preridinium nenoxa Keller	x	x	x		x				x
Rangea schneiderhoehni Gürich						x			
Nasepia a!tae Germs						x			
Baicalina sessilis Sokolov									x
Arumberia banksi Glaessner et Walter									x
Namalia villiersiensis Germs						x			
Ernietta plateaensis Pflug						x			
Inkrylovia lata Fedonkin	x								x
Archangelia valdaica Fedonkin	x								
Podolimirus mirus Fedonkin			x						
Valdainia plumosa Fedonkin			x						
Ramellina pennata Fedonkin		x							
Vaveliksia velikanovi Fedonkin			x						
Zolotytsia biserialis Fedonkin		x							
Vaizitsinia sophia Sokolov et Fedonkin		x							
Lomosovis malus Fedonkin			x						
Dickinsonia costata Sprigg	x	x	x		x				
D. elongata Glaessner et Wade		x			x				
D. tenuis Glaessner et Wade		x	x		x				
D. lissa Wade		x			x				
D. brachina Wade									x
Platypholinia pholiata Fedonkin		x							
Vladimissa missarzhevskii Fedonkin		x							
Palaeoplatoda segmentata Fedonkin	x								
Praecambridium sigillum Glaessner et Wade					x				
Parvancorina minchami Glaessner		x			x				
Velancorina martina Pflug						x			
Vendomia menneri Keller	x								
Vendia sokolovi Keller									x
Pseudovendia charnwoodensis Boynton et Ford							x		
Onega stepanovi Fedonkin	x								
Spriggina floundersi Glaessner					x				
S. borealis Fedonkin	x								
Marywadea ovata Glaessner					x				
Protechiurus edmondi Glaessner						x			
Bomakellia kelleri Fedonkin	x								
Mialsemia semichatovi Fedonkin		x							

[a]Numbers denote: *1* Letny Coast, White Sea; *2* Zimny Coast, White Sea; *3* Dniester River Basin, Podolia, Ukraine; *4* Khorbusuonka River Basin, Olenek Uplift; *5* Flinders Ranges, South Australia; *6* Namibia; *7* Charnwood Forest, England; *8* Newfoundland; *9* other locations; *x* presence; *p* forms tentatively assigned to this species.

Chapter 2

Systematic Description of Vendian Metazoa

1 Phylum Coelenterata

1.1 Class Cyclozoa

Genus *Nemiana* Paliy 1976

Type Species. N. simplex Paliy 1976.

Diagnosis. Impressions of sedentary organisms in the form of a positive hyporelief or negative epirelief. Casts are smooth-surfaced, convex discs, arranged in groups or (rarely) isolated.

Species Composition. Monotypic genus.

Nemiana simplex Paliy 1976 (Plate 5, 3)

Synonymy. See Paliy (1976, p. 70).

Description. Casts, as a rule, round, convex and smooth. Most irregularities (pits, rugae, folds, etc.) are traces of soft body deformations during burial. However, the finest concentric grooves, sometimes observed in the largest specimens, most likely point to the presence of annular muscles in *Nemiana*. In cases, when it is possible to separate the cast from the overlying rocks, the oral part of the organism can be observed, particularly the mouth, preserved as a small, rounded depression on the relatively flat, upper surface of a discoidal body (see Fig. 2A). *Nemiana* casts, commonly buried in situ, occur mainly as groups from several to hundreds of specimens, sometimes covering the entire bedding plane on an area of many metres. Due to close contact, the casts become deformed, sometimes acquiring a polygonal shape. Some specimens exhibit stages of budding or simple longitudinal division.

Dimensions (in mm). Diameter, 2–6; height, 1–10; mouth diameter, about one-third of that of the disc.

Distribution. Vendian, Valdai Series, Zimny Coast of the White Sea; the Mogilev Formation (Yampol Beds), the Yaryshev Formation (Bernashevka Beds) and the

Nagoryany Formation (Dzhurzhevka Beds) of Podolia; the Khatyspyt Formation of the Olenek Uplift (Khorbusuonka River).

Material. Nine well-preserved casts from Bed 11 of the Valdai Series section, Zimny Coast of the White Sea, Medvezhy Creek (between the Veprevsky and Zimnegorsky lighthouses); hundreds of specimens from the Mogilev, Yaryshev and Nagoryany Formations of Podolia and abundant material from the Khatyspyt Formation of the Olenek Uplift.

Genus *Beltanelliformis* Menner (in Keller et al. 1974)

Type Species. B. brunsae Menner in Keller et al. (1974).

Diagnosis. Rounded impressions with an outlined central, almost entirely smooth portion, framed around the periphery by a delicate exumbrella zone. Specimens contain no traces of gonads or an oral aperture. The new genus resembles *Beltanella* in the differentiation of central and peripheral parts, with a concentric folding observed on the latter. It differs from *Beltanella* in the absence of gonads and oral aperture traces and a flat, even, concave, but not inflated central portion; its size does not exceed 1.5 cm (Keller et al. 1974).

Species Composition. Monotypic genus.

Beltanelliformis brunsae Menner (in Keller et al. 1974; Plate 4, *2,3*; Plate 5, *2*)

Synonymy. See Fedonkin (1981a, p. 58).

Description. Casts are convex or slightly flattened (positive hyporelief); plan outline is circular or slightly deformed with fine concentric rugae, mainly around the periphery, with a smooth, almost flat, central part. Rugae are not regularly concentric, but sometimes fuse and branch. They are not always preserved, being rare (if any) in small forms less than 5 mm across. Large specimens to 20 mm across sometimes have small depresssions near the centre (Plate 5, *2*). This peculiar feature can be of great taxonomic significance, and we conditionally assign specimens with such a depression to B. brunsae until additional data are obtained.

Distribution. Valdai Series of the northern Russian Platform.

Material. Over 200 specimens were reported from the Valdai Series in the Onega Peninsula, 5 km above the mouth of the Syuzma River.

Genus *Cyclomedusa* Sprigg 1947

Type Species. C. davidi Sprigg 1947.

Diagnosis. Configuration is circular, the disc surface has several or numerous concentric grooves, slightly separating elevated and protruding areas (rugae). The pattern of the latter points to an initially conic shape of the central part, and in some species, of the entire body. Many specimens display thin, straight, radial grooves, interpreted as gastrodermal canals. The oral side of these frequently occurring organisms is quite rarely preserved. Most likely, they were attached to the substrate by aboral cones.

Species Composition. Seven species of the genus *Cyclomedusa* have been described: *C. davidi* Sprigg 1947; *C. radiata* Sprigg 1947; *C. plana* Glaessner et Wade 1966; *C. gigantea* Glaessner et Wade 1966; *C. serebrina* Paliy 1966; *C. minuta* Fedonkin 1981.

Cyclomedusa cf.davidi Sprigg 1947 (Plate 1, *3*)

Synonymy. See Fedonkin (1981, p. 58).

Description. The central tubercle is commonly low (positive hyporelief). Spacings between concentric grooves are sometimes inconsistent around the periphery and range within 0.5–4 mm. Diameter of the umbrella is 10–50 mm. The shelf, separating the inner and outer zones, has a more distinct relief in large specimens. No radial grooves are observed.

Distribution. Vendian, Valdai Series, Zimny Coast of the White Sea; the Cherny Kamen Formation of the Sylvitsa Series of the Central Urals; the Pound Quartzite of South Australia; the Conception Series, southeastern Newfoundland.

Material. Twelve specimens recorded from the Valdai Series on the Zimny Coast of the White Sea.

Cyclomedusa cf.plana Glaessner et Wade 1966 (Plate 2, *3* and *5*; Plate 5, *7*)

Description. Broad flat disc, sometimes divided by irregular radial grooves, the feature, characteristic of large specimens over 30 mm across. Small specimens commonly have a smooth umbrella. Central concentric zone on the aboral side looks like a tubercle, surrounded by concentric rugae; on the oral side, it recalls a stack of saucers, slightly shifted (Plate 5, *7*).

Dimensions (in mm). Diameter of the umbrella, 15–45 to 200–300; that of the central, concentric, rugate zone is commonly about one-third that of the umbrella.

Distribution. Vendian, Valdai Series, Zimny Coast of the White Sea; the Mogilev and Yaryshev Formations in the Dniester Basin of Podolia.

Cyclomedusa radiata Sprigg 1947 (Plate 1, *7*)

Synonymy. See Fedonkin (1981, p. 59).

Description. Flat, central dome separated by a relatively deep groove from a relatively narrow outer zone with numerous, fine radial grooves; some are also visible in the central portion. Central dome has fine concentric grooves.

Dimensions (in mm). Diameter of umbrella, 36; diameter of central dome, 24; radial grooves of the outer zone spaced at 1.5 to fractions of a millimetre.

Distribution. Vendian, Valdai Series, Zimny Coast of the White Sea; the Pound Quartzite of South Australia.

Material. One specimen from Bed 11 of the section of the Valdai Series at the Medvezhy Creek on the Zimny Coast of the White Sea (between the Veprevsky and Zimnegorsky lighthouses).

Cyclomedusa minuta Fedonkin 1981

Cyclomedusa minuta Fedonkin: Fedonkin 1981, p. 59, Plate IV, Fig. 2. Plate 20, Fig. 8.

Description. A small medusa, in the form of a flat cast with a slightly submerged central tubercle, is surrounded by regular concentric ridges of similar width.

Dimensions (in mm). Diameter of umbrella, 7; the width of central tubercle, 1; width of concentric ridges, 0.4–0.6; number of concentric ridges, 6.

Discussion. Among the described species of the genus *Cyclomedusa*, structurally this form is most similar to *C. davidi* Sprigg (Wade 1972b), differing in the more regular concentric pattern, small size and flat shape. *C. minuta* differs from the regular concentric *C. radiata* Sprigg (Wade 1972b) through the absence of fine radial grooves and a much smaller size. The described form differs from *C. plana* Glaessner and Wade in the absence of a smooth and broad outer body zone.

Distribution. Valdai Series (Vendian); Onega Peninsula.

Material. One well-preserved specimen and several poorly preserved specimens, probably belonging to this species, have been found in a clay sequence of the Valdai Series at the Syuzma River, 5.6 km above its mouth (Onega Peninsula).

Cyclomedusa delicata Fedonkin 1981 (Plate 1, *4*)

Cyclomedusa delicata Fedonkin: Fedonkin 1981, pp. 59–60.

Description. A large, ovate, flat body with a conical central part and numerous concentric ridges, which cover regularly the entire body and are separated by the finest furrows. Concentric ridges in the central part of the body are slightly narrower than in the broad and flat peripheral part, where the width of ridges actually does not change away from the centre, though it varies. Rare, broader ridges characterize the central part of the body, sometimes forming concentric shelves.

Dimensions of the Largest Specimen (No. 4464/335) (in mm): Diameter of umbrella, 116; width of fine concentric ridges, 0.5–1.2; width of broad ridges, up to 2.5. The number of ridges can reach 40 or more.

Discussion. The described form is morphologically similar to *Cyclomedusa davidi* Sprigg, differing in the greater number of more narrow ridges.

Distribution. Valdai Series, Onega Peninsula.

Material. Two well-preserved specimens found in a clay sequence of the Valdai Series at the Syuzma River, 11 km above its mouth (Onega Peninsula).

Genus *Tirasiana* Paliy 1976

Type Species. T. disciformis Paliy 1976.

Diagnosis. Impressions of sessile organisms in the form of a positive hyporelief and negative epirelief. Casts are convex to conic; the surface has a concentric ornamentation: shelves and ridges. Arranged in groups of several specimens or isolated (Paliy 1976).

Species Composition. Two species, i.e. *T. disciformis* Paliy and *T. coniformis* Paliy (Paliy 1976).

Tirasiana disciformis Paliy (Plate 29, *1*)

Synonymy. See Fedonkin (1981a, p. 57).

Description. Moderately convex casts in the form of medium-sized discs on the lower bedding planes. The main, slightly flattened, disc is superimposed by the second, smaller and relatively more convex disc with a small, round tubercle in the centre, probably corresponding to a fine, short stalk by which *Tirasiana* attached to the ground. The margin of the main disc surface, extending into the rock, is curved abruptly inward. This is confirmed by the fact that, unlike the lower, aboral body surface, which is easily separated from the underlying rock, the central tubercle is often cemented in the underlying layer; when a cast is extracted from the enclosing rock, it has a fresher split as compared to the surrounding surface of the aboral side. Lateral budding and simple longitudinal division are quite typical.

Remarks. There are reasons to suggest that, at least some specimens described as *T. disciformis,* should be identified with *Ediacaria flindersi* Sprigg (Jenkins 1981). This problem needs special study.

Dimensions (mm). Diameter of the main disc, 10–27; diameter of the second disc, 5–17; height of cast, 1–5.

Distribution. Vendian, Valdai Series, Zimny Coast of the White Sea; the Yaryshev Formation, Bernashevka Beds, central Dniester area; Cherny Kamen Formation of the Sylvitsa Series (Central Urals).

Material. Abundant material from the Yaryshev Formation of Podolia, the lower Valdai Series on the Zimny Coast of the White Sea, the Khatyspyt Formation of the Olenek Uplift.

Genus *Paliella* Fedonkin 1980b

Type Species. P. patelliformis Fedonkin 1980b.

Diagnosis. Flat and circular body composed of the central tubercle with a flattened apex and a broad, smooth, outer zone, separated by a narrow annular groove. Tentacles are absent. Radial elements (commonly irregular, isolated and paired grooves) observed only in the largest specimens.

Species Composition. Monotypic genus.

Discussion. Paliella is similar in its structure to *Cyclomedusa,* differing in the absence of a concentric groove over the entire umbrella or in the central part, as in *C. plana* Glaessner et Wade. Small forms of *Paliella* resemble *Medusinites asteroides* Sprigg, differing in the absence of distinct radial grooves, intersecting the outer zone of the latter.

Paliella patelliformis Fedonkin 1980b (Plate 3, *3,4* and *9*; Plate 10, *5*)

Paliella patelliformis Fedonkin: Fedonkin 1980b, pp. 10–11, Plate II, Figs. 1–3; Fedonkin 1981a, pp. 62–63, Plate XXXI, Figs. 2, 3; Plate XXXII, Figs. 1. 2.

Description. The largest specimens are noted for a gently undulating, marginal part of the umbrella or its lateral deformations, as well as deep and short radial grooves, diverging from the margin of the umbrella, or fine, isolated and paired grooves, extending from the margin of the umbrella to the central annular groove. The inner portion, surrounded by an annular groove, is devoid of radial furrows, being relatively flat in large specimens and protruding as a flat-tipped tubercle in smaller ones. Small juvenile specimens provide more inflated casts than larger ones. Specimens occur in groups, or in pairs, where individuals are closely pressed to each other and rimmed by a common narrow zone. Such an occurrence indicates that the reproduction of *Paliella* proceeded as a simple, longitudinal division. However, there are clear indications of multiple division (two types: linear and clustered) and a stolonal type of asexual reproduction, when a long stemlike stolon extends from the central part, sometimes with a daughter individual on the end.

Dimensions (in mm). Diameter of umbrella, 5–45; diameter of central zone, 2–12; fine, paired grooves spaced at 0.5.

Distribution. Vendian, Valdai Series, Zimny Coast of the White Sea, the Mogilev Formation (Lomozov Beds) of the Dniester Basin in Podolia.

Material. About 20 well-preserved specimens from the lower part of the Valdal Series section on the Zimny Coast of the White Sea, 5 km north of Zimnegorsky lighthouse. Several dozens of specimens from the Lomozov Beds of the Mogilev Formation of Podolia (Novodnestrovskaya Hydroelectric Power Station).

Genus *Ediacaria* Sprigg 1947

Type Species. E. flindersi Sprigg.

Diagnosis. Exumbrella surface demonstrates central disc and outer ring. A sharp annular groove, separating the disc from outer ring, can correspond to the margin of the gastral cavity. Radial grooves are mainly confined to the outer ring. On the subumbrella surface there is a round, central mouth without appendages. One specimen apparently bears numerous long and fine peripheral tentacles.

Species Composition. Monotypic genus.

Ediacaria flindersi Sprigg 1947 (Plate 1, *1,2* and *5*; Plate 2, *4*; Plate 7, *5*)

Synonymy. See Wade (1972, pp. 218–219).

Description. Representatives of this species in many associations are the most abundant fossils, which may be attributed to their greater dimensions as compared to those of many other Precambrian medusoid organisms. With rare exceptions, the aboral side or exumbrella (positive hyporelief) is commonly preserved. As a rule, these are smooth, flat casts; however, forms with a distinct relief are also frequent, which, in addition to the central disc and outer ring in the centre of the disc, also exhibit a relief tubercle, sometimes surrounded by several concentric rugae caused by post-mortem body deformations. The central disc can also be surrounded by concentric rugae; however, more often the disc and outer ring are separated by a distinct low projection or groove. The radius of the disc is two-thirds to three-fourths that of the total radius. The radius of the central tubercle is one-fourth to one-fifth that of the central disc. Sometimes, on the outer ring, frequent, very fine and straight grooves, radiating towards the periphery, are visible. They are much better preserved on the oral surface of the body, where they actually cover the entire outer ring. The diameter of the oral aperture is close to that of the central tubercle or slightly smaller. Fine, short tentacles, making up the marginal crown, are more frequently observed from the oral side. A peculiar feature of this species is the combination (or alternation?) of several types of asexual reproduction: budding, longitudinal division and stolonal reproduction, often marked in a fossil state.

Dimensions (in mm). Most common are specimens with a diameter of 40–50; however, there are forms of 450 and more across.

Distribution. Vendian, Valdai Series, Zimny Coast of the White Sea; the Mogilev Formation and Yaryshev Formation of the Dniester Basin of Podolia; the Khatyspyt Formation of the Olenek Uplift.

Material. Several specimens from Bed 1 of the Ust Pinega Formation section (Valdai Series) on the Zimny Coast of the White Sea; abundant material from the Lomozov Beds, the Mogilev Formation (Valdai Series) in the Dniester Basin of Podolia; dozens of forms from the upper Khatyspyt Formation at the Khorbusuonka River (Olenek Uplift).

Genus *Nimbia* Fedonkin 1980b

Type Species. N. occlusa Fedonkin 1980b.

Diagnosis. Flat, discoid organisms preserved as annular or ovate structures at the base of siltstone bands (positive hyporelief). The peripheral part of the disc is preserved in the form of a relief, marginal ridge. An extensive field inside the ring is commonly smooth, rarely with a small tubercle in the centre.

Species Composition. Two species: *N. occlusa* and *N. dniesteri.*

Nimbia occlusa Fedonkin 1980b (Plate 3, *1,6* and *7*)

Nimbia occlusa Fedonkin: Fedonkin 1980b, p. 8, Plate I, Figs. 1,2; Fedonkin, 1981, pp. 60–61, Plate XXXII, Figs. 3, 6.

Description. The ring is circular to ovate. The radius of the central smooth zone is three to five times the width of the outer ridge. In a low relief, the ridge is gently sloping and smooth; in a higher relief, in some cases the ridge has regular, but weak constrictions, less common are transversal furrows. In the centre of a smooth zone, a small, gently sloping tubercle with an acute apex is sometimes preserved.

Dimensions (in mm). Diameter of the ring, 4–15; width of the ridge, 3–5; ridge constrictions spaced at 1.5–4.

Distribution. Vendian, Valdai Series, Zimny Coast of the White Sea; the Mogilev Formation (Lomozov and Yampol Beds), the Dniester Basin of Podolia.

Material. About 15 specimens of differing degrees of preservation from the lower unit (Bed 1) of the Valdai Series section on the Zimny Coast of the White Sea, 5 km north of Zimnegorsky lighthouse. Six specimens from the Mogilev Formation, the Dniester Basin of Podolia.

Nimbia dniesteri Fedonkin 1983b (Plate 3, *2,8*)

Holotype. PIN, No. 3994/384-A. Valdai Series (Vendian), Mogilev Formation, Lomozov Beds exposed on the right bank of the Dniester River, near Novodnestrovskaya Hydroelectric Power Station.

Description. Small discoid organisms with a flat and smooth central part and a clearly protruding marginal ridge with a trapeziform cross-section. The oral aperture is small; it is a shallow, rounded depression in the centre of the disc. Rare, fine and short tentacles, not always preserved, diverge from the marginal ridge. Longitudinal division of the organisms is fairly common.

Dimensions (in mm). Diameter, 10–30; width of marginal ridge, 3–6; diameter of oral aperture, 2–3; length of tentacles, 5–6, with width up to 1.

Discussion. Unlike *Nimbia occlusa* Fedonkin (Fedonkin 1980b, 1981), which has a smooth or segmented marginal ridge with a rounded section, this species exhibits a low-relief ridge of a relatively greater width with a trapeziform cross-section.

Distribution. Vendian, Valdai Series, the Mogilev Formation, the Lomosov Beds, the Dniester Basin of Podolia.

Material. Five casts and several well-preserved counterparts from the Lomosov Beds of the Mogilev Formation, exposed near Novodnestrovskaya Hydroelectric Power Station.

Genus *Protodipleurosoma* Sprigg 1949

Protodipleurosoma Sprigg 1949, p. 79; Treatise ... 1956, p. F79.

Type Species. P. wardi Sprigg 1949.

Diagnosis. Rounded impression of exumbrella, broad outer ring (velum) covered by narrow concentric grooves; mid-field large, circular, with an ovoidal, irregularly compressed central area (stomach?), divided into lobes, and irregular radial grooves (radial canals?), with non-dichotomous branching; primary grooves deep, reaching the submarginal groove (radial canal?) giving rise to shorter furrows: branching near the base.

Species Composition. Protodipleurosoma wardi Sprigg, Upper Precambrian of South Australia; *B. rugulosum* Fedonkin, Upper Precambrian (Vendian) of the northern Russian Platform.

Remarks. Protodipleurosoma genus was described by R. Sprigg (1949) on the basis of very scarce material. To this fact we may partly attribute the placement of the type species *P. wardi* into the synonymy of *Ediacaria flindersi* Sprigg (Glaessner and Wade 1966). More extensive material of much better preservation from the Valdai Series of the Zimny Coast shows that *Protodipleurosoma* is a genus which stands apart from the Precambrian coelenterates.

Protodipleurosoma rugulosum Fedonkin, 1980b (Plate 5, *4,5* and *6*; Fig. 15)

Protodipleurosoma rugulosum Fedonkin: Fedonkin 1980b, pp. 13–14, Plate II, Figs. 4, 5; Fedonkin 1981a, pp. 64–65, Plate XXXI, Figs. 4–6.

Description. The median groove, commonly not strictly in the middle of an ovate impression, divides the central disc into two unequal, elongated lobes. Minor furrows and rugae, radiating from the median groove, thin towards the periphery of the central disc, not affecting the annular zone. The latter is well-preserved in large specimens, strongly reduced (if at all) in smaller ones. On the holotype, semicircular processes are seen around the body periphery, which may correspond to reproductive organs or budding daughter individuals. As a rule, one lobe of the central disc has more radial rugae than the other.

Dimensions (in mm). Diameter of body, 15–80; diameter of central disc, 15–60; width of radial rugae, 0.5–3.

Discussion. New species differs from *Protodipleurosoma wardi* in the greater number of rugae, relatively larger size and lateral processes in the largest specimens.

Distribution. Vendian, Valdai Series, Zimny Coast of the White Sea; the Mogilev Formation in the Dniester Basin of Podolia.

Material. About 20 well-preserved casts and fragments.

Fig. 15. *Protodipleurosoma rugulosum* Fedonkin. Drawing from a photograph; processes along the umbrella margin may be equated to gemmating daughter individuals or reproductive organs

Family Chondroplidae Wade 1971

Pneumatophores are chambered, bilaterally symmetric, with a narrow axial zone; chambers narrow from one axis end to the other, leaving a "cut" between chamber ends on the left and right lobes or in the area where chambers have the smallest width, or on either ends.

Genus *Chondroplon* Wade 1971

Type Species. C. bilobatum Wade 1971.

Diagnosis. Large bilobate pneumatophore; distinct axial line in the form of a ventral groove and, possibly an obtuse dorsal keel. Initial chamber is large and elongated. Successive, earliest chambers are annular; later ones do not merge, exhibiting a peripheral cut in the broader portion. The latest chambers also form a cut at the opposite end of the body where they are least wide and display a transversal arrangement. Outer ends of the chambers form an undulating margin. Height of chambers exceeds their width. Chambers are separated by deep sutures. Large radial plicae on the holotype surface seem to be occasional.

Species Composition. Monotype genus.

Chondroplon sp. (Plate 6, *4*)

Description. Fragment of a small individual (incomplete lobe of pneumatophore) preserved in a positive hyporelief. The ventral groove is poorly expressed. Initial chamber is not preserved. Over 30 undulatory curved chambers, thin near the ventral groove and widening distally. Approaching the initial chamber, other chambers become shorter, and their bend, steeper.

Dimensions (in mm). Maximum width of pneumatophore lobe, 22; visible length of lobe, 27; spacing between axial lines of chambers in their distal and proximal portions, about 1.0 and 0.5 respectively.

Distribution. Vendian, Valdai Series, Zimny Coast of the White Sea.

Material. One fragment from Bed 11 of the Valdal Series section (Vendian), exposed on the right slope of the Medvezhy Creek, Zimny Coast of the White Sea.

Genus *Ovatoscutum* Glaessner and Wade 1966

Type Species. Ovatoscutum concentricum Glaessner et Wade 1966.

Diagnosis. Scutiform pneumatophore with a weakly expressed axis. Small initial chamber, concentrically surrounded by narrow elongated, curved chambers, which form a poorly expressed triangular cut near the peripheral contact, where chambers have the smallest width. This cut runs along the body axis, not reaching the centre. Chambers are separated by deep, narrow furrows.

Species Composition. Monotypic genus.

Ovatoscutum concentricum Glaessner and Wade 1966 (Plate 6, *1–3*)

Ovatoscutum concentricum: Glaessner and Wade 1966, pp. 612–613, Plate 97, Fig. 8; Wade 1971, p. 188, Plate 6, Figs. 5, 6; Glaessner 1979, pp. A91-A92, Fig. 5; Fedonkin 1981b, pp. 98–99.

Description. Crumpling plicae and rolled-up margins of pneumatophore indicate that it was soft and flexible. Relief chambers with a rounded trapeziform cross-section are separated by deep grooves of a semicircular section. Width of groove is close to that of chambers. The opposite side of the pneumatophore, judging by a rolled-up margin on one of the specimens, had a slightly different appearance: chambers were much broader, flat, with a less distinct relief and separated by narrow, shallow grooves. The cut area was possibly sclerotized to a much lesser degree than the main portion of the pneumatophore, bearing chambers, and, therefore, more often plicated or not preserved. This cut is clearly visible only on one of the five specimens recorded from the Vendian of the Russian Platform. The maximum and minimum number of chambers is 20 and 12 respectively. The last chamber around the pneumatophore periphery is followed by a narrow and smooth rim, widest at the end of the axis opposite the cut.

Dimensions (in mm). Length, 65; width, 60; thickness of chambers, 1.5; width of rim, 5.

Distribution. Vendian, the Ust Pinega Formation of the Valdai Series on the Zimny Coast of the White Sea; the Pound Quarzite, the Flinders Ranges area, South Australia.

Material. Five well-preserved specimens from Bed 1 of the Valdai Series section (Ust Pinega Formation) on the Zimny Coast of the White Sea.

Family Incertae

Genus *Eoporita* Wade 1972a

Type Species. E. medusa Wade 1972a.

Diagnosis. Rounded or ovate outline. Oral side bears two groups of clavate polypides. Polypides on the outer ring are almost equal in length and form several concentric series. Polypides of the inner zone are arranged in series, their size decreasing towards the centre, marked by a large, central cone. Polypides are overlain by remains of a thin pneumatophore with numerous narrow, concentric, annular chambers. Aboral side of this discoid body demonstrates fine radial grooves.

Species composition. Monotypic genus.

Eoporita medusa Wade 1972a (Plate 6, *5*)

Eoporita medusa: Wade 1972a, pp. 198–204; Plate 40, Figs. 1–6, Plate 41, Fig. 6a, b, Text Fig. 2; Glaessner 1979b, p. A92, Fig. 6; Fedonkin 1981b, p. 99.

Description. In a positive hyporelief, the central cone (gastrozooid) and the surrounding zone of polypides (gonozooids) are preserved as the most protruding portion on the oral side of the body relative to a comparatively flat zone on the outer ring of polypides (dactylozooids). Polypides display a slight distal widening. A narrow ridge, surrounding a small ovate area in the centre of the cast, indicates that the gastrozooid had a thin wall. Gonozooids of the inner series are very short, thin, with an inclined downward orientation. Larger and longer gonozooids of the outer series have a radial orientation. Dactylozooids, even in the inner series, are much thicker and longer than gonozooids and have a radial orientation. Polypides of the inner and outer zones are very closely spaced without any intervals in the concentric series. The number of dactylozooid series is no less than six to seven, that of the gonozooid series is about the same. The aboral side is smooth, concave in the centre (possibly deformed during burial), in places barely discernible, fine, short, arched furrows with a concentric orientation, which can correspond to contacts of annular pneumatophore chambers. If so, the width of the annular chambers approximately equals the thickness of the polypoids of the corresponding series.

Dimensions (in mm). Diameter of discoid body enclosing pneumatophore, 65; diameter of gastrozooid opening, 9; diameter of the inner zone of polypides, 35; diameter of organism, measured at the margins of the polypides of the outer series, up to 85; maximum thickness of dactylozooids, 2.5; minimum thickness of gonozooids, 0.5.

Distribution. Pound Quartzite of the Wilpena Series (South Australia); Valdai Series on the Zimny Coast of the White Sea.

Material. One well-preserved cast from the talus of the exposed upper Valdai Series, Zimny Coast of the White Sea.

Genus *Kaisalia* Fedonkin 1984

Type Species. Kaisalia mensae Fedonkin 1984.

Diagnosis. Large discoid organism with a rounded outline. Aboral side suggests a relatively concentric structure, evidenced by the presence of regular, flat, concentric rings embracing the entire disc. Rings are separated by fine grooves. Two relief concentric ridges, one in the central zone and the other at a distance of two-thirds of the radius from the centre, point to differentiation of the internal structure. Radial elements are absent.

Generic Composition. Monotypic genus.

Discussion. Among Vendian coelenterates, many forms display a concentric structure; however, the genera *Cyclomedusa* Sprigg (Sprigg 1947) and *Kullingia* Glaessner (Føyn and Glaessner 1979) are most similar to the new genus. Unlike *Cyclomedusa*, the new genus has a dense and regular concentric structure. It differs from concentrically regular *Kullingia*, viz. in the latter, the concentric structure is marked not by broad flat rings, but by more closely spaced annular ribs, surrounding a flat conical central portion, which accounts for one-sixth of the diameter of the umbrella.

Kaisalia mensae Fedonkin 1984 (Plate 6, *6*)

Holotype. PIN, 4482/284, Ust Pinega Formation of the Valdai Series (Vendian), Zimny Coast of the White Sea.

Description. A large, flat disc is divided by relief ridges (in places resembling projections) into three concentric zones: inner, middle and peripheral. The inner portion accounts for about one-fifth of the disc diameter and is bounded by one or a pair of neighbouring ridges. The middle, broader zone accounts for about one-half of the disc radius. This zone displays distinct, regular, concentric rings of similar width. The middle zone has seven rings, the outer one being conjugated with a broad ridge, sometimes grading into a shelf, bounding the middle zone. The inner and middle zones make up a relatively elevated portion of the aboral side; the organism has the maximum thickness in this part of the disc. The outer zone, flat and thin, is composed of six rings. The thickness gradually decreases towards the margin: thus, for instance, the width of the outer ring is about two-thirds that of the second ring in this zone. Concentric rings in the middle and outer zones are commonly divided by fine furrows. Locally, these annular grooves are connected by narrow ridges, arranged rigorously along the groove, as a rule, from the inner margin of a ring, facing the centre. This may point to the presence of relatively dense and narrow concentric features, closely adjoining grooves and even coinciding with the latter in plan outline, but located in the inner portions of the body or on its oral side.

Dimensions of Holotype (in mm). Radius of disc, up to 80; radius of the central zone, about 30; width of the middle zone, about 30; width of the outer zone, 28. Maximum and minimum width of rings, 6 and 3 respectively.

Distribution. Ust Pinega Formation of the Valdai Series (Vendian), Zimny Coast of the White Sea.

Material. Holotype preserved as a flat cast at the base of siltstone Bed 1, Ust Pinega Formation, between the Medvezhy and Elovy creeks, 5 km farther north, Zimny Coast of the White Sea. Several fragments have also been found there.

1.2 Class Inordozoa

Genus *Irridinitus* Fedonkin 1983b

Type Species. I. multiradiatus Fedonkin 1983b.

Diagnosis. Small, rounded casts of the basal part of the polyp, preserved in a positive hyporelief; low relief. Central part is commonly caved in. Numerous radial ridges on rugae radiate from the centre.

Species Composition. Monotypic genus.

Discussion. Shape of the basal part of the polyp described resembles that of *Nemiana* (Paliy 1976); however, the essential difference is the presence of numerous radial structures detaching from the centre to the periphery in representatives of the new genus.

Irridinitus multiradiatus Fedonkin 1983b (Plate 8, *5, 6*)

Holotype. PIN, No. 3994/524; Valdai Series (Vendian), the Mogilev Formation, Lomozov Beds, the right bank of the Dniester River, Novodnestrovskaya Hydroelectric Power Station.

Description. Radial elements on casts of *I. multiradiatus* appear differently, depending on the degree of preservation. Most common are fine rugae, irregularly detaching from the centre to the periphery. There are deeper rugae, cutting the margin of a body, and very fine ones. Specimens with radial rugae, similar to that illustrated in Plate 8, *5*, are possibly casts, which display characteristic features of the polyp basal surface. Therefore, it is difficult to say to what extent the above rugae reflect the number and arrangement order of inner radial structures. Casts, upon which structure of the inner part of these polyps is impressed (Plate 8, *6*), appear different. These casts have a relatively low relief and are preserved on the top of a bed. From a small central depression, which corresponds to an oral aperture, numerous unbranched radial ridges detach towards the periphery (canals of a gastrovascular system?) and reach a narrow marginal ridge. Most likely, the marginal ridge corresponds to an annular canal of a gastrovascular system. The number of radial canals can reach 40 and more.

Dimensions (in mm). Diameter of polyp, up to 20; diameter of oral aperture, 2–3; thickness of radial canals, 0.5; width of annular canal, up to 1.5.

Distribution. Vendian, Valdai Series, the Mogilev Formation in the Dniester Basin of Podolia, the Ust Pinega Formation in the southeastern White Sea area.

Material. Over ten casts and impressions of different degrees of preservation from the Lomozov Beds of the Mogilev Formation, exposed near the Novodnestrovskaya Hydroelectric Power Station. Several specimens from borehole cores and natural outcrops of the Ust Pinega Formation (Valdai Series), southeastern White Sea area.

Genus *Veprina* Fedonkin 1980b

Type Species. V. undosa Fedonkin 1980b.

Diagnosis. The large umbrella, clearly divided into two broad zones, is separated by a gently dipping, smooth, annular depression. The outer zone is characterized by a sharper relief and regular radial ribs, separated by relatively narrow furrows. The inner zone, close in width, has a similar, but a less regular ribbing and extends around an elongated depression in the centre.

Species Composition. Monotypic genus.

Discussion. By the ribbing of its umbrella *Veprina* resembles *Peytonia* Walcott from the Middle Cambrian of Canada or *Actinophyllum* Phillips from the Devonian of England (Treatise. . . 1956), differing in having two concentric zones, numerous ribs and, apparently, a more elastic body.

Veprina undosa Fedonkin 1980 (Plate 4, *6*)

Veprina undosa: Fedonkin 1980b, pp. 9–10, Plate I, Fig. 6; 1981, pp. 61–62, Plate XXXI, Fig. 1.

Description. Inner, ribbed zone extends smoothly around a narrow elongated depression only from one end, whereas from the opposite end of the depression, the inner zone becomes narrower and rests against (or grades into) an outer zone, crossing an annular groove separating these zones. In a positive hyporelief, in addition to broad ribs, narrow and long ridges, particularly abundant in the inner zone, are also clearly seen on the cast. They parallel radial ribs or cut the latter obliquely, often joining or intersecting each other. These narrow ridges are either casts of rugae, or those of tentacles, which become shorter and more numerous in the centre, while they are longer and sparse in the outer zone. An elongated central depression, which imparts a bilateral character to the body, possibly corresponds to the oral aperture.

Dimensions (in mm). Diameter of body, up to 60; width of the outer ribbed zone, 10–15; width of the inner ribbed zone, 6–11; width of the central depression, 2–4; width of the ribs in the outer zone, 1.5–3; width of the fine ridges, 0.3–0.5; length, up to 15.

Distribution. Vendian, Valdai Series; Zimmy Coast of the White Sea.

Material. One well-preserved specimen (cast and incomplete impression) and several fragments, which possibly belong to this species, from Bed 1 of the Valdai Series section, Zimny Coast of the White Sea.

Genus *Armillifera* Fedonkin 1980b

Type Species. Armillifera parva Fedonkin 1980b.

Diagnosis. Small, ovate body clearly divided into two zones, the relatively narrow "fimbriate" zone bears numerous short and fine tentacles of similar length. The inner zone, smooth and caved-in in the centre (negative hyporelief), bears relief projections around the periphery, which in plan outline look like a comma, whose head faces the centre, and the "tail" extends along the contact between the inner and outer zones.

Species Composition. Monotypic genus.

Armillifera parva Fedonkin 1980b (Plate 7, *3*)

Armillifera parva: Fedonkin 1980b, pp. 14–15, Plate II, Fig. 7; 1981, pp. 65-66, Plate XXXII, Fig. 4.

Description. The ovate shape of the impression available to the author may be attributed to a lateral compression of an originally round body, which is confirmed by the different dimensions and shape of shelves in the inner zone. The shape of these arched shelves resembles flat, ovate, stacking bowls with a strong lateral displacement, so that each lower bowl is only partly overlapped by the upper one. Tentacles are not preserved over the entire outer zone, though the zone itself is clearly visible around the entire body periphery. Inner and outer zones are separated by a narrow annular groove. Outer termination of the impression is distinct and preserved in the form of a shelf or fine furrow.

Dimensions (in mm). Minimum diameter of body, 8; maximum, 11; width of the outer zone (length of tentacles), 1.5; thicknes of tentacles, 0.25; maximum width of shelves in the inner zone, 1.0.

Distribution. Vendian, Valdai Series, Arkhangelsk District, Zimny Coast of the White Sea.

Material. Holotype, Bed II of the Valdai Series section at the Medvezhy creek, 5 km north of Zimnegorsky lighthouse (Zimny Coast of the White Sea).

Genus *Elasenia* Fedonkin 1983b

Type Species. E. aseevae Fedonkin 1983b.

Diagnosis. Small discoid organism, preserved in a positive hyporelief as a semispherical cast, whose inner zone is taken up by a relatively large central disc; the outer annular zone contains several minor, round tubercles adjoining the central disc.

Species Composition. Monotypic genus.

Discussion. New genus slightly resembles *Medusinites* Glaessner et Wade 1966; however, the outer annular zone of the latter bears no tubercles, but only rare radial grooves.

Elasenia aseevae Fedonkin 1983b (Plate 8, *7,8*)

Holotype. PIN, No. 3994/338; Valdai Series (Vendian), Mogilev Formation, Lomozov Beds, the right bank of the Dniester River, Novodnestrovskaya Hydroelectric Power Station.

Description. Central disc, possibly corresponding to gastral cavity, is relatively flat; its diameter is about one-third of that of the holotype body. Seven rounded tubercles of the outer annular zone, adjoining the central disc, possibly belong to the gonads; however, the distribution around the circumference is not quite regular: some of them are pressed to each other, the others are spaced at small intervals. On some tubercles, an apical depression is seen, Margins of the body, judging by their deformation, were fine and delicate. An oral portion is unknown.

Dimensions of Holotype (in mm). Diameter of body, 11; diameter of the central disc, 4; diameter of tubercles in the outer zone, 2.

Distribution. Vendian, Valdai Series, the Mogilev Formation, Lomozov Beds; the Dniester Basin of Podolia.

Material. Holotype and one larger specimen, possibly belonging to this species, from the Lomozov Beds of the Mogilev Formation, exposed near Novodnestrovskaya Hydroelectric Power Station.

Family Bonatiidae Fedonkin, fam.nov.

Description. Flat discoid organisms. Broad peripheral zone is smooth or covered by irregular concentric rugae; no radial elements are observed in this zone. Relatively more relief in the inner zone displays wedge-shaped or ovate elements, radially oriented around a small central depression. Symmetry is of some high order.

Genus *Bonata* Fedonkin 1980b

Type Species. B. septata Fedonkin 1980.

Diagnosis. Round or ovate casts, made up of two concentric zones. Outer annular zone is relatively broad and flat with rare concentric rugae. Inner zone, with a more distinct relief and relatively elevated (positive hyporelief) is made up of numerous radial, wedge-shaped lobes or petals with rounded distal and acute proximal ends, not reaching the centre.

Generic Composition. Monotypic genus.

Discussion. Central portion of *Bonata* resembles *Peytonia nathorsti* Walcott from the Middle Cambrian of British Columbia (Treatise . . . , 1956), differing from the latter in the absence of tetraradiate symmetry and the presence of a broad outer zone.

Bonata septata Fedonkin 1980b (Plate 8, *4*)

Bonata septata: Fedonkin 1980b, pp. 12–13, Plate I, Figs. 8, 9; 1981a, pp. 63–64, Plate XXX, Figs. 5–7.

Description. Wedge-shaped lobes of central zone commonly closely adjoin each other. Sometimes, in the case of body deformation, lobes are clustered together and compressed in one part of the central zone, whereas in the other one they are separated by shallow radial grooves, which are wide near the centre and narrow distally. On the floor of these grooves, there are sometimes small ridges, not reaching the periphery of the central zone. Despite the fact that wedgelike lobes have distinct terminations, the entire central zone was apparently a single, undivided crimped disc with a small oral aperture in the centre. Crumpling plicae and rolled-up body margins suggest a rather elastic and delicate, particularly at the margins, outer annular zone. In some cases, the outer zone is not preserved. In juvenile specimens, the number of wedgelike lobes is commonly much less than in adult ones, and the outer zone is relatively narrow.

Dimensions (in mm). Diameter of the body, up to 40; diameter of the central zone, 8–20; length of wedgelike lobes, up to 5 and more; minimum and maximum number of lobes, 11 and 19 respectively.

Distribution. Vendian, Valdai Series, Zimny Coast of the White Sea.

Material. Four casts of different preservation from Bed 1, the Valdai Series section (Vendian), Zimny Coast of the White Sea.

Family Hiemaloriidae Fedonkin, fam.nov.

Description. Medusoid, bowl-shaped organisms with a convex, semispherical aboral side; oral side is slightly concave with a small, rounded oral aperture in the middle. Dense marginal tentacles point to a radial symmetry of a high order. Other radial elements (oral aperture, canals of gastrovascular system, etc.) are not observed.

***Genus Hiemalora* Fedonkin 1982**

Pinegia Fedonkin: Fedonkin 1980b, p. 9; 1981a, p. 61.

Type Species. H. stellaris Fedonkin (= *Pinegia stellaris* Fedonkin 1980b).

Diagnosis. Bowl-shaped body with a smooth or concentrically rugate aboral part an numerous tentacles, radiating from the outer margin of body. Preserved in a negative epirelief, less frequently, in a positive hyporelief; round oral aperture. Manubrium is absent.

Species Composition. Monotypic genus.

Discussion. Plan outline of *Hiemalora* complies with morphology of isolated, lower hydrozoan polyps; however, the absence of fine structural characters prevents a more detailed comparison. A morphologically similar form was described by R. Sprigg as medusa under the name *Medusina filamentis* from the Precambrian Pound Quartzite of South Australia (Sprigg 1949). Later, this name was introduced into synonymy of *Pseudorhizostomites howchini* Sprigg (Glaessner and Wade 1966).

Hiemalora stellaris Fedonkin 1982 (Plate 7, *1,4,6*)

Pinegia stellaris Fedonkin: Fedonkin 1980b, p. 9, Plate I, Figs. 3–5; 1981a, p. 61, Plate XXX, Figs. 1–3.

Description. Casts, as a rule, with a distinct relief, since the central part is considerably sunken relative to the peripheral part of the body, bearing tentacles. Absolutely flat impressions are less frequent. Aboral side is smooth, flat or convex; rarely, concentric rugae are observed. Differences in structure of the basal portion in different specimens may be attributed to a different preservation degree and body deformation after burial. Tentacles are short and fine, sometimes bulbous; their number in different specimens varies from 15–20 to 40 and more.

Dimensions (in mm). Diameter (without tentacles), 3–40; thickness of tentacles, 0.3–2; their length, 3–15; cast height, up to 4.

Distribution. Vendian, Valdai Series, Zimny Coast of the White Sea; the Mogilev Formation of the Valdai Series in the Dniester Basin of Podolia; the Khatyspyt Formation of the Olenek Uplift.

Material. Over 20 impressions and casts from Bed 1 of the Valdai Series section, Zimny Coast of the White Sea; over ten specimens from the Lomosov and Yampol Beds of the Mogilev Formation in the Dniester Basin of Podolia; 12 specimens from the middle and upper Khatyspyt Formation of the Olenek Uplift (Khorbusuonka River).

Genus *Evmiaksia* Fedonkin 1984

Type Species. Evmiaksia aksionovi Fedonkin 1984.

Diagnosis. Medusae, preserved as discoid, flat impressions, within which two broad concentric zones are discernible. In the inner zone, around a relatively elevated central tubercle, there are numerous ovate structures with a radially oriented long axis. The outer zone is relatively smooth, separated by a narrow ridge from the inner zone.

Species Composition. Monotypic species.

Discussion. The new genus bears a slight resemblance to *Elasenia*, differing in the shape of ovate structures and their arrangement in the inner zone, a broad outer zone and presence of a central tubercle.

Evmiaksia aksionovi Fedonkin 1984 (Plate 20, *2*)

Holotype. PIN, No. 3993/657; Vendian, Valdai Series, Ust Pinega Formation, Zimny Coast of the White Sea.

Description. Broad, flattened central tubercle, delimited by a narrow ridge. Ovate structures in the inner zone, possibly corresponding to gonads, are elongated, and their proximal ends are at some distance from the central tubercle, whereas the distal ends adjoin a narrow ridge, delimiting the inner and outer zones. The outer zone is relatively broad and flat, without traces of radial structures, but it has numerous irregular, concentric rugae. Margin is smooth. No tentacles are observed.

Dimensions of Holotype (in mm). Diameter of umbrella, 56; diameter of outer zone, 24; central tubercle, 10; length of ovate structures, about 5; their width, up to 2; width of ridge delimiting inner and outer zones, 1.

Distribution. Vendian, Valdai Series, the Ust Pinega Formation, Zimny Coast of the White Sea.

Material. Holotype in the form of a well-preserved cast and impression in Bed 1 of the Ust Pinega Formation between the Medvezhy and Elovy creeks, Zimny Coast of the White Sea.

1.3 Class Trilobozoa

Family Albumaresidae Fedonkin fam.nov.

Description. Small medusoid organisms; umbrella is discoid or lobed with or without marginal tentacles. Gastrovascular system is in the form of branching canals. This family is noted for triradiate symmetry.

The family comprises three Precambrian genera of scyphozozoans: *Skinnera* Wade 1969; *Albumares* Fedonkin, in Keller and Fedonkin 1976; *Anfesta* Fedonkin 1984.

Genus *Albumares* Fedonkin (in Keller and Fedonkin 1976)

Type Species. A. brunsae Fedonkin in Keller and Fedonkin 1976.

Diagnosis. Small medusae with a flattened umbrella of a trilobate structure. Three ridges detach from the centre of the umbrella, narrowing radially and dividing the lobes of the umbrella. From the centre of each of the three lobes, three canals of the gastrovascular system detach; they dichotomize four times towards the margin of the umbrella. No traces of an annular canal are observed. Numerous fine and short tentacles are arranged along the margin of the umbrella.

Species Composition. Monotypic genus.

Albumares brunsae Fedonkin 1976 (Plate 9, *1,4*)

Albumares brunsae Fedonkin: Keller and Fedonkin 1976, p. 38, Plate I, Figs. 1, 2; Fedonkin 1981a, p. 60, Plate IV, Figs. 1, 6.

Description. Small medusae with a semispherical, flattened umbrella, characterized by triradiate symmetry. Three oral lobes, preserved as ridges, extend from the centre of the umbrella. They begin with small thickenings '(gonads), thinning towards the periphery of the umbrella, and terminating in small tubercles, possibly corresponding to rhopalia near its margin. In each of the three sectors of a medusa, divided by oral lobes, three canals of the gastrovascular system extend from the central part; they dichotomize four times towards the margin of the umbrella. There are no traces of an annular canal. There are numerous, fine, short tentacles, more than 100, along the margin of the umbrella disc.

Dimensions (in mm). Diameter of umbrella, 13; length of oral lobes, 5; thickness of tentacles along the margin of the umbrella, 0.15.

Distribution. Vendian, Valdai Series of the Onega Peninsula.

Material. Three well-preserved specimens from sandy strata of the Valdai Series section at the Syuzma River, 5 km above the mouth (Onega Peninsula).

Genus *Anfesta* Fedonkin 1984

Type Species. Anfesta stankovskii Fedonkin 1984.

Diagnosis. Three elongated ridges radiate from the centre at an angle of 120°; they have rounded distal and proximal ends, probably corresponding to gonads. The umbrella is cut into narrow radial lobes, their number divisible by three. Distal ends of the lobes are bifurcate.

Discussion. Structurally, the new genus is very similar to the other two genera of Vendian medusae: *Skinnera* Wade 1969 and *Albumares* Fedonkin in Keller and Fedonkin 1976. These medusae also have a common order of radial symmetry, not typical of coelenterates. The new genus mainly differs from *Skinnera* in shape and arrangement of gonads, character of the marginal zone of the umbrella and presence of lobes. It differs from *Albumares* in a discoid, but not trilobate shape of the umbrella and presence of radial lobes.

Anfesta stankovskii Fedonkin 1984 (Plate 9, *2*)

Holotype. PIN, No. 3993/260, lower unit of the Ust Pinega Formation of the Valdai Series (Vendian), Zimny Coast of the White Sea between the Medvezhy and Elovy creeks.

Description. Length of sausagelike ridges, apparently corresponding to gonads, is approximately two-thirds of the umbrella radius; therefore, distal ends of the gonads do not reach the margin of the umbrella. In interradii, limited by the gonads, there are narrow, long, slightly distally widening lobes. Only lobes located at the same distance from gonads, limiting this interradius, exhibit a strictly radial orientation. Other lobes are oriented towards the gonads with their proximal ends and start above the latter. Thus, gonads occupy a relatively lower position on the subumbrella than lobes. The distal ends of the lobes bifurcate twice: they are first divided by a single, relatively long groove, and then each lobe is cut into two by a relatively narrower and shorter groove. The number of grooves is about 18, since one lobe is located over each gonad, the others being confined to interradii. Sometimes, fine tentacles are preserved, their number is difficult to determine.

Dimensions (in mm). Diameter of umbrella, up to 18; length of gonads, up to 5; their width, up to 1.3.

Distribution. Ust Pinega Formation of the Valdai Series (Vendian), Zimny Coast of the White Sea.

Material. Two well-preserved specimens from Bed 1 of the Ust Pinega Formation, the Valdai Series (Vendian) between the Medvezhy and Elovy creeks, 5 km north of Zimnegorsky lighthouse.

Family Incertae

Genus *Tribrachidium* Glaessner 1959

Type Species. Tribrachidium heraldicum Glaessner 1959.

Diagnosis. Discoid organism, slightly convex, with a relatively steep peripheral boundary. One side (oral?) has three elevated arms, which radiate at the same angle, curving clockwise, and which join the periphery of the disc, where they become thinner. Occasionally, a small Y-shaped furrow is seen in the centre between the arms. From each arm on its concave side, a small elevated area ("bulla") diverges. At about 0.7 of the length, distal parts of the arms bear short, strong tentacles on the outer side and on the acute side of each arm. Between arms sometimes there are fine, long, straight or slightly curved setiform structures, extending from the convex part of each arm across an acute portion of a neighbouring arm. The opposite side (aboral?) of the disc displays only several concentric grooves.

Generic Composition. Monotypic genus.

Tribrachidium heraldicum Glaessner 1959 (Plate 9, *3,5,6*)

Synonymy. See Fedonkin (1981a, p. 72).

Description. Small discoid fossils. In a negative epirelief three arms radiate from the central part. They curve smoothly counterclockwise, and then parallel peripheral termination occurs, tapering towards the ends. Distal portions of the arms, for about two-thirds of their length, bear numerous short tentacles (about 20 on each arm), forming a fringe along the outer margin of the body. Each of the three areas, between branches, ramifies, starting from a convex part of the arm near the centre and almost touching with its end the concave portion of the neighbouring arm at a point separated by about one-fourth of its length from the distal end.

Cast of upper, oral surface (positive epirelief), not very common in a fossil state, displays the following structural features. The upper side is scutiform and slightly convex. In the middle portion, there are three arched depressions, oriented towards the centre of the disc with their narrow ends. The outer sides of the arched depressions are relatively broader and deeper due to small, rounded, terminal widenings. The disc surface is speckled with fine radial grooves and ridges entering the arched depressions. Such a structural pattern of the oral side enables the interpretation of a *Tribrachidium* as a more primitive organism than it was previously assumed (Glaessner and Wade 1966; Fedonkin 1981a). Arched depressions with a terminal widening may belong to oral apertures, though, possibly, only terminal widening served for introducing food. Fine radial grooves, entering the arched depressions, may be interpreted as microrelief, favouring transportation of nutrient particles from the entire umbrella area towards the oral apertures by means of ciliated epithelium. A further analysis of *Tribrachidium* morphology from the same viewpoint shows that "arms", bearing short "tentacles" in their distal parts, may and even should be regarded as a distribution system, structurally similar to canals of the gastrovascular system. If this interpretation proves reasonable, *Tribrachidium* should be assigned to coelenterates rather than to some other group of Metazoa. Indirect evidence is also provided by the third-order symmetry of this organism, a rather common feature in Vendian medusae (*Albumares* Fedonkin, *Skinnera* Wade), but not characteristic of other groups of Vendian and living Metazoa. The above facts provide some grounds for assigning *Tribrachidium* to coelenterates of Class Trilobozoa.

Dimensions (in mm). Diameter of the smallest and largest specimens, 7 and 15 respectively.

Distribution. Vendian, Valdai Series, Zimny Coast of the White Sea; the Mogilev Formation in the Dniester Basin of Podolia; the Pound Quartzite of South Australia.

Material. Five impressions from Bed 1 and two impressions from Bed II of the Ust Pinega Formation, the Valdai Series, Zimny Coast of the White Sea between Veprevsky and Zimnegorsky lighthouses. Four impressions from the Lomozov Beds of the Mogilev Formation (Valdai Series), exposed near Novodnestrovskaya Hydroelectric Power Station, the Dniester Basin of Podolia. Two impressions from clay strata of the Ust Pinega Formation at the Lower Syuzma River (right bank, 5.5 km above the mouth), Letny Coast of the White Sea.

1.4 Class Conulata Moore and Harrington 1956

Order Conulariida Miller and Gurley 1896; Suborder Conchopeltina Moore and Harrington 1956.

Family Conchopeltidae Moore and Harrington 1956

Genus *Conomedusites* Glaessner and Wade 1966

Type Species. Conomedusites lobatus Glaessner and Wade 1966.

Diagnosis. The theca, forming a low cone with concentric rugae, is divided by four deep radial grooves. Thus, the peripheral boundary is tetralobate. Each lobe can be further divided by small grooves. Along the peripheral boundary, sometimes there is a crown of rather thick tentacles.

Species Composition. Monotypic genus.

Conomedusites cf. lobatus Glaessner and Wade 1966 (Plate 8, *1*)

Description. A large specimen, preserved in a positive hyporelief, is characterized by a relatively smooth, central, conical portion of a circular-square outline. Along the margin of the theca, a distinct ridge is observed. The theca radius in plan outline is approximately one-half that of a body. Four grooves, cutting the broad outer ring, do not extend onto the theca. Margins of the body point to its tetralobate structure, minor undulations of the margin are also observed. Tentacles are not preserved. Absence of concentric rugae and deep radial grooves in the thecal area suggest that not the outer (lower) thecal surface, but rather its inner surface or the body surface of the organism is impressed in the central part.

Dimensions (in mm). Diameter of body, up to 50; diameter of the theca, up to 22; width of ridge surrounding theca, up to 2.

Distribution. Representatives of the genus *Conomedusites* are known from the Precambrian Pound Quartzite of South Australia (Glaessner and Wade 1966). The specimen herein described is from the Lomozov Beds of the Mogilev Formation exposed near Novodnestrovskaya Hydroelectric Power Station (Dniester Basin of Podolia).

Material. One well-preserved specimen from the Lomozov Beds, Mogilev Formation (Vendian), the right bank of the Dniester River near Novodnestrovskaya Hydroelectric Power Station. Several small specimens from Bed I, the Zimnegorsky section of the Ust Pinega Formation.

1.5 Class Scyphozoa (?)

Family Incertae

Genus *Stauridinia* Fedonkin, gen.nov.

Type Species. Stauridinia crucicula Fedonkin, sp.nov.

Diagnosis. Small medusae, characterized by a simple gastrovascular system, consist of four radial canals, starting from a small gastral cavity in the centre of the umbrella. Cordate, swollen, distal portions of the radial canals apparently correspond to gonads. Marginal tentacles, fine and rare, are not always preserved.

Discussion. Such organisms are as yet unknown among Precambrian medusae. Structurally, *Stauridinia* strongly resemble the living hydroid medusa *Obelia* (Leptomedusa). This similarity shows up in the distribution and number of radial canals and gonads, as well as in the absence of velum, rudimentary in *Obelia* (Ivanov et al. 1981). The difference is that *Obelia*, though much smaller, has an annular canal, which, however, might well be not preserved in *Stauridinia*; *Obelia* has also an oral stem and numerous marginal tentacles.

Generic Composition. Monotypic genus.

Stauridinia crucicula Fedonkin, sp.nov. (Plate 4, *1,5*)

Holotype. PIN, No. 3993/392 A, Ust Pinega Formation of the Valdai Series (Vendian), Zimny Coast of the White Sea.

Description. Medusae preserved as rather deep impressions on the basal part of the interbeds (negative hyporelief), indicating the height of the umbrella and the considerable thickness of the mesogloea. The umbrella impression displays narrow relief, radial ridges, corresponding to the four radial canals of the gastrovascular system. The distal ends of the ridges grade into broad, cordate and even more relief swollen portions, probably corresponding to gonads. The margin of the umbrella is very thin, irregular and, possibly devoid of an annular canal. Fine and short marginal tentacles are rarely preserved.

Dimensions (in mm). Diameter of umbrella, 6–10; minimum width of radial canals, up to 1; maximum width of gonads, up to 3; length of marginal tentacles, 4–5.

Distribution. Ust Pinega Formation of the Valdai Series (Vendian), Zimny Coast of the White Sea.

Material. Five specimens of varying degrees of preservation from Bed 5 of the Ust Pinega Formation, the Valdai Series (Vendian) near Medvezhy Creek, 5 km north of Zimnegorsky lighthouse (Zimny Coast of the White Sea).

Family Pomoriidae Fedonkin fam.nov.

Description. Small medusae, characterized by the presence of two crowns of tentacles: one along the umbrella margin, and the other one, surrounding the manubrium. The shape of the latter points to the presence of a sixth-order radial symmetry axis; however, it was impossible to determine the multiplicity of the number of tentacles. Fine tentacles of the labral crown radiate in clusters.

Genus *Pomoria* Fedonkin 1980b

Type Species. *P. corolliformis* Fedonkin 1980b

Diagnosis. Small medusae with two crowns of closely spaced tentacles: labral and marginal. Labral crown embraces strongly protruding manubrium with a small, round, oral aperture. Aboral side is unknown.

Species Composition. Monotypic genus.

Discussion. The author does not know of any forms with a similar morphology among fossils or living medusae. Structurally, the subumbrella of *Pomoria* slightly resembles the oral end of actinias and anemones, though a double crown of tentacles is known not only in Actiniaria (Anthozoa), but also in Hydrozoa, particularly isolated polyps, viz. Branchiocerianthus, Syncoryne, and Corymorpha (Anon. 1937).

Pomoria corolliformis Fedonkin 1980b (Plate 7, *2*)

Pomoria corolliformis: Fedonkin 1980b, p. 11, Plate I, Fig. 7; 1981a, p. 63, Plate XXX, Fig. 4.

Description. Tentacles of labral crown forming small clusters (two to four), not going beyond the umbrella margin. The manubrium is apparently rather long. Lateral, semicircular processes are seen in its transversal section, their number, possibly equalling six. On the holotype, two (and, probably, more) long, bent processes radiate from the manubrium, their thickness is two to three times that of the tentacles. They are probably reproductive organs.

Dimensions (in mm). Diameter of umbrella (without marginal tentacles), 16–20; length of labral tentacles, 4–6; length of marginal tentacles, up to 10; width of labral tentacles, 0.3–0.4; width of marginal tentacles, up to 0.7–0.8; thickness of processes, up to 1.5; their length, 10–12; length of manubrium, no less than 5–6; diameter of oral aperture, 2.

Distribution. Vendian, Valdai Series, Zimny Coast of the White Sea.

Material. Two well-preserved specimens from Bed 11 of the Valdai Series section (Vendian), Zimny Coast of the White Sea.

2 Phylum Plathelminthes (?)

Genus *Platypholinia* Fedonkin gen.nov.

Type Species. *Platypholinia pholiata* Fedonkin, sp.nov.

Diagnosis. Flat, leaf-shaped organisms with a narrowed and smoothly acute anterior end and smoothly rounded posterior end of the body. Mid-ridge, lost in the middle part of the body, starts from the anterior end. On either side of the ridge and subparallel to it are finer ridges and furrows. There are no traces of an anus or reproductive organs.

Discussion. Anterior-posterior orientation, leaf-shaped form of the flat body and simple structure make the new genus similar to some turbellarians, particularly rhabdocoelan planarians; some forms are fairly large. However, the lack of evidence on the structure of reproductive organs in *Platypholinia* makes a more detailed comparison impossible. It should be noted that a terminal mouth is not a characteristic feature of turbellarians: most representatives of this class have a ventral mouth. Nevertheless, we have no grounds for assigning the organization higher than that of planarians, to the new genus; therefore, we place *Platypholinia* into the type Plathelminthes with a certain degree of assurance.

Species Composition. Monotypic genus.

Platypholinia pholiata Fedonkin sp.nov. (Plate 19, *5,6*)

Holotype. PIN, No. 3993/195; Valdai Series (Vendian), Redkino Horizon, Bed 10 of the Ust Pinega Formation, Zimny Coast of the White Sea.

Description. Surface of the leaf-shaped body is absolutely flat and smooth, except for the axial median zone, where linear structures are located. These structures extend from a smoothly acute anterior end and are gradually lost in the middle portion of the body or its posterior third. A small depression in the extreme portion of the anterior end of the body points to the presence of a terminal mouth. Near this point a long narrow mid-ridge starts, apparently corresponding to a simple pharynx or a straight, unbranched rectum. Long, narrow rugae on either side of the mid-ridge may correspond to nerve trunks or, more likely, longitudinal muscles, which drew the anterior end of the body in by capture of food particles.

Dimension (in mm). Length of the largest specimen, about 60; width, up to 28; width of mid-ridge, 1.5–2.

Distribution. Vendian, Valdai Series (Redkino Horizon), Ust Pinega Formation, Zimny Coast of the White Sea.

Material. Two complete impressions and one well-preserved fragment from Bed 1 of the Ust Pinega Formation (Valdai Series), Zimny Coast of the White Sea, 5 km north of Zimnegorsky lighthouse.

Genus *Vladimissa* Fedonkin gen.nov.

Type Species. *V.missarzhevskii* Fedonkin, sp.nov.

Diagnosis. The flat, leaf-shaped body is broad in the middle and acute at the anterior and posterior ends; fine-scalloped edging. In the central part of the body, two rows of small longitudinal tubercles are seen.

Discussion. Structurally, representatives of the new genus are similar to that of the now living rhabdocoelan turbellarians Mesostoma (Ivanov 1981, p. 305, Fig. 268). These large rhabdocoellids have two rows of large static eggs on each side of the intestine. Unfortunately, the structure of the digestive and other physiological systems of *Vladimissa* is still unknown. However, the available data enable the assignment of *Vladimissa* to turbellarians with a considerable degree of assurance.

Species Composition. Monotypic genus.

Vladimissa missarzhevskii Fedonkin, sp.nov. (Plate 19, *7*)

Holotype. PIN, No. 3993/204. Valdai Series (Vendian), Redkino Horizon, Bed 11 of the Ust Pinega Formation, Zimny Coast of the White Sea.

Description. Judging by the cast, the leaf-shaped body had a considerable thickness (several millimetres). The body surface is smooth. Anterior-posterior orientation is determined by a marginal notch, directed obliquely to the plane of bilateral symmetry. In the middle part of the body, two rows of rounded tubercles of similar size are closely spaced. Each row with a longitudinal orientation comprises four (possibly more) tubercles, probably corresponding to reproductive organs, particularly seed sacs or egg capsules. Central portion is strongly curved, possibly by post-mortem alteration. No oral aperture is observed.

Dimensions (in mm). Length, about 45; width, 32; width of tubercles, about 1.5; (maximum) spacing between neighbouring notches on the body margin in its central part, up to 2. Notches become smaller towards the anterior and posterior parts of the body.

Distribution. Valdai Series (Vendian), Redkino Horizon, Ust Pinega Formation, Zimny Coast of the White Sea.

Material. One cast with a well-preserved marginal part and central zone from Bed 11 of the Ust Pinega Formation, Zimny Coast of the White Sea, 5 km north of the Zimnegorsky lighthouse.

3 Phylum Proarticulata

3.1 Class Dipleurozoa

Family Dickinsoniidae

Genus *Palaeoplatoda* Fedonkin 1979

Type Species. Palaeoplatoda segmentata Fedonkin.

Diagnosis. Flattened, leaf-shaped body, with an undulating deformation around the periphery and narrowed from the sides in the middle. Ventral(?) side of the body is covered with fine, peripheral ridges, diverging from the median zone and slightly curved in the same direction.

Species Composition. Monotypic genus.

Palaeoplatoda segmentata Fedonkin 1979. (Plate 16, *1,3,6*)

Description. Lateral portions of an elongated body, devoid of appendages, commonly convex. Segmentation in the form of fine, transverse ridges clearly observed on one (ventral?) side of the body, whereas on the other side (dorsal?) segmentation is weak (if any). Irregular, undulating, deformed body margins indicative of its elasticity during the lifetime of organisms.

Dimensions (in mm). Length of body, over 70; width of body, 30; width of mid-ridge, 2; length of segments, 0.6.

Distribution. Valdai Series, Onega Peninsula.

Material. Twelve specimens of different preservation from the lower sandy strata of the Valdai Series at the Syuzma River, 5 km above the mouth, Onega Peninsula.

Genus *Dickinsonia* Sprigg 1947; *Dickinsonia costata* Sprigg 1947 (Plate 16, *2*; Plate 17, *1,2,5,6,7*; Plate 18, *1,2,3*)

Synonymy. See Fedonkin (1981a, pp. 69–70).

Description. Flat, ovoidal, bilaterally symmetric body, with a distinct segmentation, its length approaching the width. On some impressions, a longitudinal mid-groove is seen. The body is divided into narrow and long segments, slightly widening towards the outer ends. The closer to the anterior or posterior ends of the body the segments lie, the greater their curvature towards the corresponding end and the smaller the angle between them and the mid-line. On one of impressions, two ridges are seen on each side of the mid-groove, where segments start. The smallest specimens are round.

Dimensions (in mm). Length of the smallest complete impression of *D. costata*, 10; width, 9; number of segments, 26. The largest impression is, unfortunately, incomplete; its width is 50; visible length, 40; visible number of segments, 30. Fragments suggest the existence of larger specimens.

Distribution. Vendian, Valdai Series: Ust Pinega Formation of the Onega Peninsula (Letny Coast) and Zimny Coast of the White Sea; the Mogilev Formation of Podolia; Pound Quartzite of South Australia.

Material. Eleven well-preserved specimens from the Valdai Series section at the Syuzma River (Onega Peninsula) and from the Zimny Coast of the White Sea between Veprevsky and Zimnegorsky lighthouses.

Dickinsonia cf. tenuis Glaessner and Wade (Plate 16, *3*)

Description. Fragment of a relatively small, flat body whereby the length is two-fold greater than the width. A distinct mid-ridge is slightly lowered relative to the lateral segmented portions of the body, gradually thinning towards the posterior end. Fine segments (25 cm^{-1} body length) diverge from deep grooves on each side on the mid-ridge. The description of the specimen is based on an impression on the lower surface of a sandstone bed.

Dimensions (in mm). Visible length of body, 42; width of body, judging by the complete right half of the body, did not exceed 20; length of segments measured along the mid-line, 0.3.

Occurrence. Vendian, Valdai Series, Zimny Coast of the White Sea.

Material. One well-preserved, incomplete impression from Bed 11 of the Valdai Series section, Zimny Coast of the White Sea at the Medvezhy Creek.

Dickinsonia lissa Wade 1972 (Plate 16, *5,7*; Plate 17, *3*)

Dickinsonia lissa Wade: Wade 1972b, Plate 6, Fig. 6; Plate 7, Figs. 1,4.

Description. Elongated, leaf-shaped body with deformed posterior portion. Very fine segments at an acute angle (50–60°) to mid-line are inclined towards the posterior portion of the body. Mid-ridge is strongly elevated and does not thin towards the posterior end.

Dimensions (in mm). Length of body, 21; width in the central part of the body, 7; width of the mid-ridge, 0.4; number of segments per 1 cm body length exceeds 40.

Distribution. Vendian, Valdai Series, Zimny Coast of the White Sea, Pound Quartzite of South Australia.

Material. Five well-preserved specimens from Bed 11 of the Valdai Series section at the Medvezhy Creek, Zimny Coast of the White Sea, between Veprevsky and Zimnegorsky lighthouses.

3.2 Class Vendiamorpha

Family Vendomiidae Keller

Genus *Onega* Fedonkin in Keller and Fedonkin 1976; *Onega stepanovi* Fedonkin in Keller and Fedonkin 1976 (Plate 19, *1,2,4,8,9*)

Onega stepanovi Fedonkin: Keller and Fedonkin 1976, p. 42, Plate I. Figs. 3,6; Paliy et al. 1979, p. 68, Plate LVIII, Fig. 2; Fedonkin 1980b, Plate I, Fig 6; 1981a, p. 71, Plate XIII, Figs. 5,6.

Description. Small organisms with a distinctly outlined, flat, ovate body; in the centre, a relief protruding from the segmented zone, slightly shifted towards the narrower posterior end of the body. The structure of the zone is the following: a non-segmented crescent area along the longitudinal axis of the body is followed by five paired, longitudinal lobes with a normal orientation to the body axis and divided by a deep and broad axial groove; outer ends of the lobes are slightly turned up towards the broad end of the body. Length of lobes gradually decreases towards the narrow end of the body.

Dimensions (in mm). Length of holotype body, 6; width of body, 3.8; width of the relief-segmented zone, 1.9; length of the relief-segmented zone, 2.8; width of lobes, 0.3; of eight impressions from the lower surface of the clay sandstone bed, two (one of them is the holotype) are similar in dimensions and structure. The third impression, 4 in length and about 2.5 in width, has only three segments, poorly discernible because of poor preservation. Maximum length of body, 7.5.

Distribution. Valdai Series, Onega Peninsula.

Material. Eight well-preserved impressions from the lower sandy strata of the Ust Pinega Formation (Valdai Series) at the Syuzma River, 5 km above the mouth (Onega Peninsula).

Genus *Vendomia* Keller in Keller and Fedonkin 1976; *Vendomia menneri* Keller in Keller and Fedonkin 1976 (Plate 21, *7*)

Vendomia menneri Keller: Keller and Fedonkin 1976, p. 43, Plate I, Fig. 4; Paliy et al. 1979, p. 69, Plate LVII, Fig. 5; Fedonkin 1980b, Plate II, Fig. 2; 1981a, p. 72, Plate XIII, Fig. 2.

Description. Small, ovoid body; more than one-third is occupied by a broad semicircular cephalon with two symmetrical depressions, possibly corresponding to the eyes. Mid-ridge begins in the cephalis; paired elongated segments, diverge from the mid-ridge and become smaller towards the posterior part of the body. In the same direction, the angle between segments and mid-ridge also decreases from 85° to 45°.

Dimensions (in mm). Length of body, 4; width of body, 3; width of mid-ridge, 0.2; number of segments, 6.

Distribution. Valdai Series, Onega Peninsula.

Material. One well-preserved, complete impression and two small impressions, possibly belonging to this species.

Genus *Vendia* Keller 1969

Type Species. Vendia sokolovi Keller, Redkino Horizon (Vendian) of the Russian Platform.

Diagnosis. An egg-shaped impression. In the anterior part, there is a crescent-divided field, which bears a slight resemblance to the cephalon of trilobite. Its crescent-shaped margins are posteriorily oblique, merging with the generally ovate outline of the impression. From the inner side, the anterior field is adjoined by a mid-ridge, from which lateral processes diverge. Along the mid-ridge, extends a poorly discernible depression, strongly narrowed where lateral processes attach.

Remarks. Systematic position of the genus *Vendia* is unclear. Some authors interpreted it as trilobitelike organisms; however, this form lacks bilateral symmetry: lateral processes are not arranged opposite each other, but are alternately arranged. Other workers (A. Yu. Rozanov) believed that *Vendia* was a medusa, lying on the side; and still others assigned it to some ancient, unknown coelenterates.

Vendia sokolovi Keller 1969 (Plate 21, *1*)

1963. Trilobitelike organism. In: Upper Precambrian . . . , Plate 18, Fig. 11.

1969. Vendia sokolovi: Keller, p. 175, Plate XVI, Fig. 5.

Holotype. Specimen, illustrated in Plate 18, Fig. 11 (Upper Precambrian . . . 1963).

Description. A small representative of the genus *Vendia*, 14 mm long and 8 mm wide. The anterior field is smooth, without ornamentation. The width of the mid-ridge reaches 1 mm; the length of diverging processes, 3–4 mm; their width,

0.8 mm. A small depression with lateral processes extends along the mid-ridge. The posterior portion of the organism is not fully preserved. Apparently, the central ridge ends abruptly; its lateral processes in the posterior part could be much longer.

Material. One impression.

Distribution. Arkhangelsk District, Yarensk borehole, depth 1552 m, Redkino Horizon, Ust Pinega Formation.

4 Phylum Arthropoda

4.1 Class Paratrilobita

Genus *Bomakellia* Fedonkin, gen.nov.

Type Species. *B. kelleri* Fedonkin sp.nov.

Diagnosis. Large organism with a clearly discernible crescentiform cephalis and an elongated body bearing numerous lateral appendages. A broad cephalis in its axial area is taken up by a convex, ovate, axial lobe. Lateral portions of the cephalis and the posterior part of the cephalic axial lobe have a less distinct relief. The posterior part of the cephalic axial lobe has a rounded process and two notches on both sides. The junction of the body with the cephalis follows this sinuous boundary. The body, relatively narrow at the front, gradually widens towards the middle, narrowing appreciably towards the posterior end. The margins of the relatively flat dorsal side display two rows of tubercles of similar diameter. Closely spaced lateral appendages are ovate with an ornamented bipinnate surface. The size of the appendages increases towards the posterior end of the body.

Discussion. The similarity between the genera *Bomakellia* and *Mialsemia* Fedonkin gen.nov. is determined by the presence of a characteristic system of lateral appendages with a pinnate ornamentation and a non-segmented surface of the elongated body. The difference consists in the varying proportions: *Mialsemia*, despite a narrow elongated body, is generally ovate, owing to the character of appendage growth. *Bomakellia* has a large cephalis.

Species Composition. Monotypic genus.

Bomakellia kelleri Fedonkin, sp.nov. (Plate 21, *6*)

Holotype. PIN, No. 3992/401; Vendian, lower sequence of the Valdai Series section at the right bank of the Syuzma River, 5 km above the mouth, Letny Coast of the White Sea.

Description. Relief cephalis with steeply inclined frontal and lateral sides. The large ovate axial lobe is probably separated from the lateral lobes by a narrow groove (clearly seen on the left and less clearly on the right side of the cephalis). At the front, a narrow frontal furrow separates the axial lobe from a narrow frontal field, grading into wider lateral lobes. Short narrow ridges, directed forward, extend along the anterior margin of the cephalis. Looking to the right and backwards from the frontal field, one can see five ridges, directed from the axial lobe towards the cephalic margin on the lateral lobe. These ridges resemble surface undulations of the lateral lobe with the smallest wave in front, and the greatest, behind. The poorly preserved left part of the cast prevents a conclusion as to whether this is an initial undulation of the lateral lobes, or whether it is an artefact.

The width of the posterior process is at least one-fourth that of the cephalis. Generally, the cephalis is clearly separated by a sinuous groove from the thorax. There is no distinct segmentation on the dorsal side. In the anterior half, two possible segments are seen on the right, holding an intermediate position relative to the lateral appendages, which is generally not typical of articulates. Therefore, these two elements were possibly formed in the process of burial due to mechanical damage of the right half of the dorsal side.

The presence of two rows of tubercles on the dorsal side margins and pleuralike appendages can point to a metameric structure of the thorax. However, this metamerism did not apparently affect the shield in the axial part of the thorax as, for instance, in trilobites. It is still unclear how the number of dorsal tubercles correlates with that of the appendages: in front, on the right, one appendage corresponds to a tubercle; however, in the middle portion, two tubercles correspond to one appendage. On the right, 11 appendages are preserved; on the left, 7 incomplete ones. The size of the appendages increases from the anterior to posterior ends more than two-fold, the angle between the axial line of appendages and the lateral side of the thorax becomes more acute, and the appendages themselves are slightly curved in the same direction. Ornamentation is most distinct on the largest appendages: from mid-ridge, straight near the anterior appendages and curved near the posterior ones, narrower short ridges diverge from both sides.

The posterior part of the thorax is smoothly rounded, definitely owing to some longitudinally elongated linear structures, whose fragments are preserved on the cast, relatively lower than the surface of the dorsal side.

The systematic position of *B. kelleri* is not quite clear. The cephalis strongly resembles the cephalon of trilobites, and its axial lobe can correspond to a large smooth glabella. Lateral appendages are similar to pleura, whereas the thorax is not divided into axial rings, as in trilobites. A bipinnate relief of pleuralike appendages resembles pre-epipodites and epipodites of some crustaceans. The appendages could well perform the function of breathing or movement (swimming), if they were mobile. *B. kelleri* may belong to a special group of Metazoa, which represents the Vendian phase of Articulata radiation. New data will probably enable one to precisely place them among the major types of this group.

Dimensions (in mm). Visible length of holotype, about 95; length of thorax, about 80; maximum width of thorax without appendages, 22; width of cephalis, about 30; length of appendages, from 7 to 18; width, from 4 to 8; diameter of dorsal tubercles, 1–2.

Distribution. Vendian, Redkino Horizon (Valdai Series), Ust Pinega Formation of the southeastern White Sea area.

Material. New species described from a single specimen, preserved as a cast and a counterpart. Left side and anterior part of the cast are not completely preserved. Besides, along the body, left of the symmetry plane, there is a narrow furrow, cutting the cephalis and grading into a narrow ridge in the posterior portion of the body. This linear structure may be attributed to mechanical damage of the non-lithified sediment. The specimen is from the lower unit of the Ust Pinega Formation (Valdai Series, Vendian), exposed on the right bank of the Syuzma River, 5 km above the mouth, Onega Peninsula, Letny Coast of the White Sea.

Genus *Mialsemia* Fedonkin, gen.nov.

Type Species. M.semichatovi Fedonkin, sp.nov.

Diagnosis. Ovate organism; easily discernible, relatively narrow fusiform thorax, with elongated, flat appendages on both sides and rounded, distal ends. The thorax is smooth and non-segmented. One end (posterior?) tapers abruptly; the narrow mid-ridge is bounded by two short furrows. Two short, narrow ridges radiate from the smooth, acute end of the thorax. The other end of the thorax (anterior?) is smoothly rounded. The cephalis is crescent-shaped. The largest appendages in the central portion of the body gradually decrease towards both ends of the elongated thorax. Along the junction line of the appendages and thorax, there are closely spaced tubercles on the lateral parts of the latter. There are also appendages with a bipinnate ornamentation in the form of a mid-ridge and diverging, alternating, short transverse ridges. The outer ends of the appendages are limited by a narrow, smooth zone.

Discussion. Absence of distinct segmentation of the thorax and the presence of pleuralike appendages with a characteristic bipinnate ornamentation makes the new genus somewhat similar to *Bomakellia* Fedonkin, gen.nov.; however, differences in proportions and shape, and growth of appendages allow the recognition of a new genus.

Composition. Monotypic genus.

Mialsemia semichatovi Fedonkin, sp.nov. (Plate 21, *2,3*)

Holotype. PIN of the USSR Ac.Sci., No. 3993/401.

Description. A unique specimen represented by a well-preserved cast and an incomplete counterpart. A slight lateral deformation shows that the animal was rather soft or elastic. A cigar-shaped thorax without traces of segmentation, except for lateral rows of weakly expressed tubercles in zones, where appendages join the thorax. These tubercles are well-preserved only on the right. At the level of the junction point of the second (from the posterior end) appendage, the thorax tapers abruptly. There is a narrow mid-ridge, bounded by narrow, longitudinal grooves. Two short, narrow, linear structures, resembling furcae, diverge from the acute end of the body. The organism had seven pairs of appendages. The proximal ends of the lateral appendages, as well as their outer distal ends, are smoothly rounded. These flat, pleuralike appendages display a marked distal widening. The appendages are arranged opposite each other. Each has a distinct limit, a double narrow longitudinal ridge, extending from the junction point and almost reaching the distal end (possibly, a result of poor preservation), and short, simple transverse, alternating ridges, which diverge from a longitudinal ridge. On the cast, in a more fully preserved part, flat pinnate appendages seem to be lying on a rather thick, flat, semiround support, its outer margin sometimes extending beyond the distal ends of the appendages. A lateral limit, which extends beyond the distal ends of the appendages, is clearly seen as a narrow, smooth zone in the front portion on the left, deformed part of the cast.

The systematic position of *Mialsemia* is unclear. The cephalon is relatively small, crescent-shaped and very poorly preserved; therefore, no judgement can be made on the details of its structure. It is not quite clear which side of the organism the fossil demonstrates: dorsal or ventral. No other appendages, besides lateral pinnate ones, are observed. These appendages are very similar to pre-epipodites and epipodites of crustaceans; like *Bomakellia*, they were petal-shaped and undivided; their bipinnate shape is due to a relief mid-ridge, and transverse ridges radiate from it.

Family Sprigginidae Glaessner

Genus *Spriggina* Glaessner 1958; *Spriggina borealis* Fedonkin 1979 (Plate 21, *5*)

Spriggina borealis: Paliy et al. 1979, pp. 67, 68, Plate LXI, Fig. 1; Fedonkin 1980b, Plate I, Fig. 1; 1981a, p. 71, Plate X, Fig. 5.

Description. Horseshoe-shaped protostomium, its width slightly greater than that of the segmented body in the broadest, middle portion. Segments of the middle portion are longest, broadest, and slightly turned up towards the caudal end. Posterior segments are short and straight. A shallow mid-groove is sometimes narrowed due to small tubercles, giving rise to segments.

Dimensions (in mm). Length with prostomium, 55; width of segmented part of the body, up to 20; maximum length of segment, 2.5; width of mid-groove, up to 1.5; visible number of segments, 19.

Discussion. The structure of the cephalon and segmentation resembles that of *Spriggina floundersi* Glaessner (Glaessner 1958; Glaessner and Wade 1966), differing from it in having a much smaller number, shape of segments, and relatively larger body dimensions.

Distribution. Valdai Series (Vendian), Onega Peninsula.

Material. One well-preserved specimen and several fragments, possibly, belonging to this species, from sandy strata of the Valdai Series section at the Syuzma River, 5 km above the mouth, Onega Peninsula.

Colonial Forms, Petalonams, and Other Problematica

Genus *Pseudorhizostomites* Sprigg 1949

Type Species. Pseudorhizostomites howchini Sprigg 1949.

Diagnosis. Grooves, radiating from centre, branching and thinning towards the periphery. A distinct peripheral boundary is absent. A canal, filled with undifferentiated sediments, extends upward from the centre through them.

Composition. Monotypic genus.

Pseudorhizostomites howchini Sprigg 1949 (Plate 10, *2,3*)

Synonymy: See Fedonkin (1981a, p. 68).

Description. Fine furrows, radiating from the centre and irregularly branched towards the periphery, become smaller and disappear. The central portion is commonly strongly sunken (negative hyporelief). No lateral terminations of the structures are observed.

Dimensions (in mm). Width of radial structures, 9–10 to 50–60; height of vertical canal, 10–12.

Distribution. Ust Pinega Formation of the Valdai Series (Vendian), Zimny Coast of the White Sea; the Mogilev Formation of Podolia; Pound Quartzite of South Australia.

Material. About 20 specimens from the Valdai Series section, Zimny Coast of the White Sea between Veprevsky and Zimnegorsky lighthouses. Several specimens from the Lomozov Beds, the Mogilev Formation, exposed near Novodnestrovskaya Hydroelectric Power Station (Dniester Basin of Podolia).

5 Phylum Petalonamae Pflug 1972a

Family Charniidae Glaessner 1979b

Description. Single, elongated, leaf-shaped bodies ("petaloids"), often with a narrow stem and widened discoid base, with a mid-groove or other median structure. Secondary branches form convex parallel structures between the primary branches.

The family comprises three Precambrian genera: *Charnia* Ford; *Charnodiscus* Ford and *Glaessnerina* Germs.

Genus *Charnia* Ford 1958

Type Species. *C. masoni* Ford 1958.

Diagnosis. Narrow petaloids with a sinusoidal mid-line and distinct primary grooves, forming acute angles with corresponding secondary grooves and branches; hence, an almost transverse position on the petaloid.

Composition. Monotypic genus.

Charnia masoni Ford 1958 (Plate 12, *4*; Plate 13, *2,3,4*)

Description. Fragments of large frondlike bodies. Irregular, zigzag mid-groove. Primary branches at 30° to 50° to body axis. Secondary branches are divided by deep grooves. Their maximum number on one primary branch reaches 15. Each secondary branch bears fine grooves (8–10) parallel to the petaloid axial line.

Dimensions (in mm). Width of frondlike colony, up to 80; visible length, up to 170; width of secondary branches in the axial portion of the colony, up to 6; length, up to 13; grooves on secondary branches spaced at 1.

Distribution. Vendian, Valdai Series, Zimny Coast of the White Sea, Khatyspyt Formation of the Olenek Uplift, Charnwood Series in the English Midlands.

Material. Fifteen well-preserved specimens from Bed 1 of the Valdai Series section, Zimny Coast of the White Sea between Veprevsky and Zimnegorsky lighthouses. Five specimens from the Khatyspyt Formation of the Olenek Uplift (Khorbusuonka River).

Family Pteridiniidae Richter 1955

Description. Leaf-shaped, elongated structures ("petaloids") with a coarse bilateral symmetry, possibly complexly and elastically deformed in the process of burial, with a clear mid-groove, displaying a row of small rhombic elements ("commissures"); more distinct transverse, sharply incised grooves and convex branches; sometimes an almost smooth marginal zone is preserved.

Genus *Pteridinium* Gürich 1933

Synonymy: See Glaessner (1979b, p. A99).

Pteridinium nenoxa Keller 1974 (Plate 11, *1–4*)

Synonymy. See Fedonkin (1981a, pp. 66–67).

Description. Leaf-shaped, bilaterally symmetric segmented forms with fine, intersegmental ribs curved in one direction and converging in the lateral parts. Segments, divided by ribs, are mainly of two types: sharply angular and rounded (and in large forms, flattened) in a vertical section parallel to the mid-line. No complete impressions have been found; however, fragments, apparently belonging to this species, suggest that the entire body was metamerized. Arrangement of segments is not opposite, but alternating. Commissural tubercles are not always preserved. Some specimens bear small tubercles on each segment near the mid-line. One specimen, tentatively assigned to *Pteridinium nenoxa*, has bifurcations in the form of paired claws on the distal edges of the relatively short segments (Fedonkin 1981a, Plate VII, Fig. 4).

Dimensions (in mm). Visible length of impressions, up to 200; width, up to 80; length of segments, measured along the mid-line, 4–9; visible number of segments, about 25.

Remarks. The above form is identical to *Pteridinium cf. simplex* described from the Pound Quartzite of South Australia (Glaessner and Wade 1966); however, it is quite different from *Pteridinium simplex* (Gürich), Nama Group of southwestern Africa, described in detail by R. Richter (1955) and, particularly, by H. Pflug (1970a). They differ in the junction type of adjacent segments and paired symmetrical segments with a mid-line, the shape of the segments and their number, the absence of a smooth zone (zona levis), and a structurally complex mid-zone with clearly expressed commissures and commissural pores, as well as traces of commissural and petaloid canals in specimens from the Valdai Series. The described well-preserved fossils prevent one from attributing most of the above differences between *Pteridinium* from the Valdai Series and Nama Group exclusively to taphonomic or preservation differences. However, it is difficult to determine the degree of difference between fossils from the Nama Group and those from the Valdai Series without a careful comparison (it is likely that such differences are of a superspecies or higher rank).

Distribution. Vendian, Valdai Series, Onega Peninsula (Letny Coast) and Zimny Coast of the White Sea; Pound Quartzite, South Australia; Mogilev Formation, the Dniester Basin of Podolia.

Material. Over 50 specimens of varying degrees of preservation from the lower sandy strata of the Valdai Series section at the Syuzma River, 5 km above the mouth

(Onega Peninsula). Three specimens from the Valdai Series, the Zimny Coast of the White Sea near Zimnegorsky lighthouse. Two impressions from the Lomozov Beds of the Mogilev Formation, the Dniester Basin of Podolia.

Genus *Inkrylovia* Fedonkin 1979

Type Species. Inkrylovia lata Fedonkin 1979.

Diagnosis. Large, sacculate, bilaterally symmetric forms are regularly segmented in a longitudinal direction. Flat segments are curved to one side and divided by fine grooves, radiating from a deeper mid-groove towards the lateral sides of the body. The ends of a sacculate body slightly bifurcating. The arrangement of segments is not strictly opposite.

Species Composition. Monotypic genus.

Inkrylovia lata Fedonkin 1979 (Plate 12, *3,5*)

Inkrylovia lata Fedonkin: Paliy et al. 1979, pp. 69–70, Plate LVI, Figs. 1–4; Fedonkin 1980b, p. 29, Plate I, Fig. 5; 1981, pp. 68–69, Plate VII, Fig. 6; Plate VIII, Figs. 1–4; Plate IX, Figs. 1–4.

Description. Inkrylovia apparently had a very elastic body, as evidenced by random, gently sloping plicae, formed in the course of burial. On one of the casts (Fedonkin 1981a, Plate IX, Fig. 4), the mid-groove ends in a triangular depression, possibly corresponding to the oral aperture. However, no features of interior organs are observed on the specimens. On one specimen (Fedonkin 1981a, Plate VIII, Fig. 4) a small, leaf-shaped segmented body, similar in its segmentation type, diverges from the end of *Inkrylovia*. It may be a juvenile individual, not separated from the parent Another specimen (Fedonkin 1981a, Plate IX, Fig. 2) displays a slight widening of the terminal part of *Inkrylovia*. If these facts are confirmed by additional material, the form described can be interpreted as an attached, sacculate mollusc, reproducing by budding.

Dimensions (in mm). Width of the sacculate body, 40–60; visible length of a completely preserved specimen (Fedonkin 1981, Plate VIII, Fig. 3), about 70, but judging by fragments, could reach 100. Length of segments varies between 4.5 and 6.5, being constant in each specimen.

Distribution. Valdai Series, Onega Peninsula.

Material. Fifteen specimens from the sandy strata of the Valdai Series section at the Syuzma River, 5 km above the mouth, Onega Peninsula.

Genus *Archangelia* Fedonkin 1979

Type Species. Archangelia valdaica Fedonkin 1981a.

Diagnosis. A bilaterally symmetric organism with an ovoid body. The axial zone consists of successive, short, broad segments, forming a zigzag line in the longitudinal vertical section. Segments gradually decrease in size in one direction. Lateral parts display a gentle, transverse undulation.

Species Composition. Monotypic genus.

Archangelia valdaica Fedonkin 1979 (Plate 11, *5*)

Archangelia valdaica Fedonkin: Paliy et al. 1979, p. 69, Plate LVII, Figs. 1, 3; Fedonkin 1980b, p. 29, Plate II, Fig. 4; 1981a, pp. 72–73, Plate XII, Figs. 4, 5.

Description. In plan outline: elongated, egg-shaped body, divided by an axial zone, the ratio of its width to that of the organism is 1:5. Visible number of segments in the axial zone, 12; but with reconstructed missing part, the number unlikely to exceed 15. Length of segments (measured along the body axis) decreases quicker than their width. Number of gentle, transverse waves in the lateral parts apparently corresponds to the number of segments in the axial zone.

Dimensions (in mm). Visible length of body, 45; width, 27; length of segments in the axial zone: maximum, 4.5, minimum, 2.5; width of segments in the axial zone: maximum, 7, minimum, 5.

Distribution. Valdai Series, Onega Peninsula.

Material. One cast and counterpart fragments from sandy strata of the Valdai Series section at the Syuzma River, 5 km above the mouth, Onega Peninsula.

Genus *Podolimirus* Fedonkin 1983b

Type Species. Podolimirus mirus Fedonkin 1983b.

Diagnosis. Large, bilaterally symmetrical organism, composed of paired, arched segments, wide in the proximal part and narrowing distally. Segments, diverging at 80–85° from the symmetry plane, curve abruptly, becoming subparallel to the above plane in their distal part. Size of segments decreases successively in the same direction, i.e. towards the growth zone. The ends of the last segments in this zone become aligned, forming a sort of "comb".

Species Composition. Monotypic genus.

Discussion. The genus described is similar to *Pteridinium* Gürich 1933, by its segmentation and, probably, both genera can be placed into the family Pteridiniidae Richter 1955. The new genus differs in much greater dimensions and proportions of the segments, and a structurally different growth zone.

Podolimirus mirus Fedonkin 1983b (Plate 10, *4*)

Holotype. PIN, No. 3994/524; Valdai Series (Vendian), Mogilev Formation, Lomozov Beds, the right bank of the Dniester River, Novodnestrovskaya Hydroelectric Power Station.

Description. On the holotype (impression), nine segments preserved in the right part; and six in the left one. Smooth, flattened segments with narrow marginal ridges. Their ends are probably irregularly bifurcated. Arrangement of segments is not strictly opposite, with considerable displacement. Judging by an irregular decrease in the width of certain segments from the proximal to distal ends, segments were relatively mobile and could be partly overlapped, at least in their distal parts. However, one should bear in mind possible disturbances of continuity of the body after the death of an organism and in the course of burial. Distal parts of the last 10–12 segments are subparallel, and their ends are aligned, forming a pectinate edge. The first segment on the right is much wider than successive ones, particularly in its proximal part, where a relatively small trapeziform area with a distinct relief is located. Unlike the second and subsequent segments, the front part at the beginning of the first segment is not rounded, but rounded-angular.

Dimensions (in mm). Length of body, 170; visible width, 100; width of the first proximal segment, 22; width of marginal ridges, to 1.5.

Distribution. Vendian, Valdai Series, Mogilev Formation, Lomozov Beds, the Dniester Basin of Podolia.

Material. Holotype and two impressions (fragments) from the Lomozov Beds of the Mogilev Formation exposed near Novodnestrovskaya Hydroelectric Power Station.

Genus *Valdainia* Fedonkin 1983b

Type Species. Valdainia plumosa Fedonkin 1983b.

Diagnosis. Bipinnate, segmented organism. Short, broad body. Rounded-zigzag mid-line. Broad, smooth segments are gently curved towards the growth zone; distinctly alternating segments. Proximal ends of the segments are rounded; distal ends taper smoothly and obliquely. Segments are divided by a pair of grooves, with a long narrow ridge in-between. The end of the body is bifurcated in the growth zone.

Discussion. The genus described is similar to *Pteridinium* Gürich 1933 in the general structural outline, but differs in proportions, shape of segments and the presence of a ridge between neighbouring segments diverging to the sides; the finest segments in the growth zone form the bifurcated end of the body.

Species Composition. Monotypic genus.

Valdainia plumosa Fedonkin 1983b (Plate 10, *1*)

Holotype. PIN, No. 3994/276. Valdai Series (Vendian), Mogilev Formation, Lomozov Beds, the right bank of the Dniester River, Novodnestrovskaya Hydroelectric Power Station.

Description. Narrowing of segments from the proximal to distal part begins at some distance from the mid-line, mainly due to a steeper curvature of the segment margin, which faces the side, opposite the growth zone. The distal ends of the largest segments are thin and taper obliquely. According to holotype, the segments are separated by a relatively narrow zone, consisting of two grooves with a narrow ridge in-between. This intersegmental ridge begins near the mid-line and extends towards the distal ends of the adjacent (one side) segments. The width of the intersegmental ridges is variable: the broader the ridge, the narrower the segment. Besides, from the side of the growth zone of the organism, segments are separated by a much shallower groove from the intersegmental ridges than from the opposite side. These specific features suggest that intersegmental ridges are parts of the segments, or rather rigidly conjugated structures, located on the side of the growth zone of the organism. These parts or structures were apparently impressed in the sediment in those parts of the body, where the segments are to a certain extent deformed. The largest segments extend at an almost right angle from the body axis, but approaching the growth zone, the angle between the segments and the body axis decreases; in the growth zone it equals 30–40°.

Dimensions. (in mm). Visible length of the holotype, 70; maximum width, 82; maximum width of segments, 12, minimum, 4; width of intersegmental ridges, 1-3.5.

Distribution. Vendian, Valdai Series, Mogilev Formation, Lomosov Beds, the Dniester Basin of Podolia.

Material. Two casts and two-well preserved counterparts from the Lomosov Beds of the Mogilev Formation, exposed near Novodnestrovskaya Hydroelectric Power Station.

Family Incertae

Genus *Ramellina* Fedonkin 1980b

Type Species. Ramellina pennata Fedonkin 1980b.

Diagnosis. Small bipinnate forms, preserved in a positive hyporelief as an axial ridge (main polyp?) and extending on both sides of it at a right angle as a shorter ridge (secondary polyps or lateral processes, bearing secondary polyps). Lateral ridges have an alternating arrangement, their width being close to that of the axial ridge.

Species Composition. Monotypic genus.

Ramellina pennata Fedonkin 1980b (Plate 14, *1*)

Description. Secondary polyps (or lateral processes) have sligthly widened distal ends; being of maximum length in the middle, and minimum at the ends of the colony; the colony is leaf-shaped. One end of the colony is gently rounded; the other one, gently tapering. On the acute end, secondary polyps (or lateral growths) are thinner, and the angle between them and the axial polyp is more acute than in the middle part of the colony. This suggests a growth zone of the colony from the side of the gently tapering end. Lateral terminations in the form of a projection or marginal groove are rather distinct.

Dimensions. (in mm). Length of axial polyp, 20–35; width, 1–2; length of secondary polyps (or lateral processes) in the middle part of the colony, 4,5–6; their width, 1–1,5; number, about 20.

Distribution. Vendian, Valdai Series, Zimny Coast of the White Sea.

Material. Two well-preserved casts and incomplete counterparts from the lower part of the Valdai Series section, Zimny Coast of the White Sea.

Genus *Vaveliksia* Fedonkin 1983b

Type Species. Vavelikia velikanovi Fedonkin 1983b.

Diagnosis. Problematic organisms, preserved as elongated casts and impressions with two distinct parts: a relatively small, regular disc, which may imply a discoid attachment of the organism, and an ellipsoidal part, adjoining the disc by the long axis end and, apparently, corresponding to a soft ampullar sac. On the other end of ellipsoidal part, opposite to the disc, no smooth closure of the ellipse is observed; short linear elements follow the long axis, forming a pectinate margin.

Species Composition. Monotypic genus.

Discussion. Obconical body with a discoid attachment at the base is characteristic of many groups of sessile Metazoa. The primitive morphology of the genus described, first of all, implies a comparison with the most simple coelenterates, particularly some solitary hydroid polyps. It is difficult to accomplish a more detailed comparison until more detailed data on the order of symmetry and other structural features of these organisms are available.

Vaveliksia velikanovi Fedonkin 1983b (Plate 12, *1*)

Holotype. PIN, No. 3994/581. Valdai Series (Vendian), Mogilev Formation, Lomozov Beds, the right bank of the Dniester River, Novodnestrovskaya Hydroelectric Power Station.

Description. Length of ampullar sac, 3–4; its width is 1.5 times that of the diameter of the discoid attachment area. The sac is joined to the disc by a short stalk, not always observed in fossils, if overlain by fixing of the disc during crushing of the organism in the course of burial. An oral end, opposite the disc, bears a crown of short tentacles, surrounding the oral aperture. The surface of the casts of the ampullar sac is smooth or tubercular, possibly depending on whether the inner or outer surface of the sac is impressed in the sediment.

Dimensions (in mm). Length, 30–80; maximum width of sac, up to 35; diameter of discoid attachment area, 8–20.

Distribution. Vendian, Valdai Series, Mogilev and Yaryshev formations, the Dniester Basin of Podolia.

Material. Several casts of differing degrees of preservation from the Lomozov Beds, the Mogilev Formation, exposed near Novodnestrovskaya Hydroelectric Power Station; and two casts from the Bernashevka Beds, the Yaryshev Formation, exposed near Ozarintsy Village.

Genus *Zolotytsia* Fedonkin 1981

Type Species. Zolotytsia biserialis Fedonkin 1981a.

Diagnosis. Two rows of ovate bodies (polyps or zooids) on both sides of the mid-groove. In each of the rows, ovate bodies are divided by deep grooves, narrower than the mid-groove. The length of the ovate bodies, measured along the mid-groove, regularly decreases in the same direction in both rows.

Composition. Monotypic genus.

Zolotytsia biserialis Fedonkin 1981a (Plate 12, *6*; Plate 14, *2*)

Holotype. GIN, No. 4482/158; Valdai Series, Zimny Coast of the White Sea.

Description. Arrangement of ovate bodies on both sides of the mid-line cannot be regarded as symmetrical. Holotype also demonstrates a difference between the size of the rounded bodies in the right and left rows. This discrepancy may be attributed to the extension of the right row and compression of the left row of rounded bodies during left-sided bending of the entire organism. However, it is impossible to prove the initially symmetrical arrangement and correspondence between the size of the bodies in both rows from the available data.

Dimensions of the Holotype (in mm). Visible length, about 70; width, 17; length of ovate bodies, 3–6; their visible number in one row, 9.

Distribution. Vendian, Valdai Series: Zimny Coast of the White Sea, the Dniester Basin of Podolia.

Material. Two incomplete specimens from Bed 1 of the Valdai Series section, Zimny Coast of the White Sea between the Medvezhy and Elovy creeks, and one specimen, possibly belonging to this species, from the talus in the same locality. One poorly preserved fragments from the Bernashevka Beds, the Mogilev Formation, on the left bank of the Dniester River, above Bernashevka Village near the dam of the hydroelectric power station.

Genus *Vaizitsina* Sokolov and Fedonkin 1985

Type Species. V. sophia Sokolov and Fedonkin 1985; Vendian, Valdai Series; Zimny Coast of the White Sea, Vaizitsa Village.

Diagnosis. Small pinnate colonies. Bilaterally, two rows of lateral processes diverge from the main polyp. Basal part of the colony resembles a short stem, ending in an ellipsoid thickening. A characteristic, distinct, lateral termination of the colony.

Composition. Monotypic genus.

Discussion. The form of the colony is very typical of modern Pennatularia (Octocorallia); however, there are no adequate reasons for assigning *Vaizitsinia* to this group. Among Precambrian organisms, the new genus is very similar to *Charniodiscus* Ford with one exception, the latter (in addition to a larger size) has a discoidal attachment in the basal part of the main polyp.

Vaizitsinia sophia Sokolov and Fedonkin 1985 (Plate 14, *3,4,6*)

Holotype. PIN, No. 3993/1305, Arkhangelsk district, Zimny Coast of the White Sea, Vaizitsa Village, borehole 202, depth 328 m; Vendian, Valdai Series.

Description. Bipinnate, leaf-shaped colony with a monopodial type of branching. The main polyp is relatively broad at the base, gradually thinning upwards. The basal part of the main polyp forms a short stem, ending in an elongated, egg-shaped thickening, which could serve for boring into muddy ground. The lateral sides of the main polyp bear dense, irregular, lateral processes (probably, daughter polyps), their visible number in one row exceeding 22. Short and thick at the base, these lateral processes reach their maximum length in the middle part of the colony; and their size decreases upwards. The distal ends of the lateral processes have a rounded termination; their width in this part is slightly greater than in the proximal one. In some area, a slight transverse segmentation is preserved as regular hollows. The lateral termination of the colony is distinct and preserved as a broad, flat ridge with a slight downward widening. At the level of the distal end of the eighth (from the bottom) lateral process, a thin "train" extends from both sides of the bounding ridge, widening downwards. It suggests the existence of a film, which partly covers the lower processes and forms symmetrical cavities in the lower part of the colony.

Dimensions (in mm). Visible length of colony, 70; maximum width, up to 37. Greatest thickness of lateral processes, up to 3.5; length, up to 15. Length of the stem of colony without lateral processes, 18.

Distribution. Valdai Series (Vendian), Zimny Coast of the White Sea, Arkhangelsk District.

Material. Two fairly preserved, incomplete specimens.

Genus *Lomozovis* Fedonkin 1983b

Type Species. Lomozovis malus Fedonkin 1983b.

Diagnosis. Large, dendritic, colonial organisms. The basal part is a high, relatively broad obconical stem with a small discoid basal attachment. Relatively fine, tubular processes detach from the upper part of the stem and end in a crown of fine, long setae. Such processes can also diverge from the lateral sides of the basal stem; they often display a dichotomous branching. The surface of the organism is flat.

Discussion. Structure and size of the representatives of the new genus make one look for analogies, first of all, among relatively primitive organisms, such as sponges or coelenterates; however, the lack of adequate material makes a detailed comparison difficult.

Species Composition. Monotypic genus.

Lomozovis malus Fedonkin 1983b (Plate 15, *2*)

Holotype. PIN, No. 3994/418. Valdai Series (Vendian), Mogilev Formation, Lomozov Beds, the right bank of the Dniester River, Novodnestrovskaya Hydroelectric Power Station.

Description. The upper part of the basal stem bears two semicircular processes, either continuing as long, tubular growths or as shorter brushlike terminations. Below, the stem has two or three processes of different length. These processes are commonly rather thick proximally; they then become narrow, and instead of branching, become slightly thicker. The surface of the processes and basal stem are smooth, though the processes are mainly covered by fine and long longitudinal rugae. The basal stem in one case bears distinct transverse rugae: fine double grooves, almost straight, cut the stem from margin to margin. Fine occasional grooves, curvatures and traces of contortion, crumpling processes, as well as the basal layer seem to be fairly soft and flexible. The ends of the processes and offshoots are commonly brushlike.

Dimensions (in mm). Length of basal stem, 70–80; minimum width, 15–20; maximum width, 25–40; length of processes, 60–100; their thickness, 6–11; length of racemose terminations, 10–30.

Distribution. Vendian, Valdai Series, Mogilev Formation, Lomozov Beds, the Dniester Basin of Podolia.

Material. Three specimens with preserved basal portion and processes, several fragments on one plate of fine-grained sandstone from the Lomozov Beds (Mogilev Formation), exposed near Novodnestrovskaya Hydroelectric Power Station.

Chapter 3

Vendian Metazoa from the Urals

Precambrian fossils and trace fossils in ancient strata of the Ural fold area were discovered in the Central Urals, and later, in the South Urals. These regions comprise a number of localities, where non-skeletal metazoans are confined to certain stratigraphic horizons. Fauna of each region has its own specific features. Impressions of the central Uralian fauna are noted for a distinct relief, the presence of large, often giant specimens, in association with grey psammitic deposits, where trace fossils are quite rare.

South Uralian biota contains small forms with a weak relief, confined to grey siltstone, often yielding trace fossils.

1 Central Urals

Impressions of molluscs as isolated specimens were first reported in 1972 from the Kosva Basin in the Urals (Becker 1977). Later, the collection was replenished and presently contains over 100 specimens. Besides the Kosva River, the Ediacaran fauna is also recorded in an other locality, near the Koiva River, some 100 km from the first locality. The age of impressions from both localities is defined by their belonging to the Sylvitsa Series. However, at the Kosva River, they are confined to the Cherny Kamen Formation; and at the Koiva River, to the Ust Sylvitsa sediments. Thus, in the Sylvitsa Series composite section, two levels of the Ediacaran fauna development are recorded. The fauna of each level is rather diverse and, judging by preliminary data (possibly, due to the lack of data), contains no common forms. Therefore, the description of fauna from each level is given separately.

2 Fauna from the Cherny Kamen Sediments

On the Kosva River, impressions of non-skeletal organisms are recorded in the upper part of the Shirokovskoe Reservoir, sligthly lower than Nyara Village. There, the steep right bank of the reservoir exposes greenish-grey and grey, flaggy sandstone and siltstone talus. Fine-grained sandstone is often micaceous and thin-bedded. Downstream of the Kosva River, sandstone and siltstone are replaced by numerous rock fragments of greenish-grey and claret-coloured mudstone. The sandstone surface is covered by ligulate hieroglyphs and rill marks. In sandstone and siltstone, Liesegang rings and "curled" layering are observed, pointing to slumping.

The impressions, doubtlessly, come from the Cherny Kamen sediments of the Sylvitsa Series. Moreover, the position of deposits in the section points to their apparent association with the lower half of the Cherny Kamen Formation.

Metazoa occur on the surface of sandy plates as rounded relief impressions. They are represented only by medusoids, imparting a peculiar character to the Kosva locality; therefore, it stands apart among a number of other regions, where such fauna is developed. With a cursory acquaintance, one can observe certain features in common between some specimens and the Ediacaran fauna. A more detailed examination shows that a direct identification of the Kosva and Ediacaran forms is difficult because of strong differences in structure and size. Thus, for instance, the Kosva forms bear no distinct radial ribbing, which is highly typical of the Ediacaran medusoids, representatives of the genus *Cyclomedusa* (Sprigg 1947; Wade 1972a). Though this feature is negative and, thus, can be defined more precisely in the future, we cannot help taking it into account. The genetic nature of radial ribbing in Precambrian medusoids is not yet clear. Presumably (Wade 1972a), the radial structure resembles gastrodermal canals of younger forms. According to other statements, radial ornamentation is not an impression of living structures, but rather a result of posthumous fracturing of medusoid bodies. Therefore, it is difficult to account for the absence of radial ribbing in Uralian forms, which were not immortal. In this context, it is also difficult to interpret a distinct radial ornamentation in some Ediacaran forms, particularly *Cyclomedusa radiata* Sprigg.

At present, it is difficult to discriminate between the Uralian and Ediacaran medusoids. M. Glaessner, who examined the author's collection, stated (oral communication) that forms, similar to the Uralian ones, are also knowm among the Ediacaran fauna, but have not as yet been examined in detail.

At the same time, the Uralian specimens bear a certain resemblance to the remains described by V.M. Paliy as representatives of the genus *Tirasiana* in 1975 and 1977 from the Vendian of the Dniester Basin.

Genus *Tirasiana* Paliy 1976; *Tirasiana disciformis* Paliy (Plate 29, *1*)

Description. Isolated casts are slightly elevated above the bedding surface and are formed by discs of different sizes inserted into each other. Boundaries of the disc, particularly the outer one, are fairly distinct. A rounded tubercle does not always occupy the central position, being sometimes displaced towards the margin. The central disc radius approaches one-half of that of the entire cast radius.

Dimensions (in mm)		
	1st specimen	2nd specimen
Diameter of the outer disc	22	32
Diameter of the inner disc	10	18
Diameter of the rounded tubercle	2	8
Height of the cast	2	3

Discussion. It differs from other representatives of the genus *Tirasiana* in its simple structure, resulting from the presence of two discs. Unlike *Ediacaria flindersi* Sprigg, it is characterized by the total absence of radial ribbing and giant forms, so typical of an Ediacaran medusoid (Glaessner and Wade 1966).

Distribution. Perm district, Shirokovskoe Reservoir, Nyar Village, the Sylvitsa Series, the Cherny Kamen Formation.

Material. Two well-preserved specimens: No. 1/11406, CGM, sample 1/74.

Tirasiana concentralis Becker, sp. nov. (Plate 29, *3*)

Etymology. Latin "concentralis", concentric.

Holotype. CGM, No. 1/11406, sample 232/72; Perm district, Shirokovskoe Reservoir, Nyar Village; the Sylvitsa Series, the Cherny Kamen Formation.

Diagnosis. A relief-rounded cast on the bedding surface with a concentric ornamentation and a tubercle in the centre.

Description. A relief-ellipsoidal cast, composed of four annular ridges inserted into each other. The ridges have a gradually decreasing diameter with a tubercle in the central, most convex part of the ellipse. The second and fourth ridges have the most distinct relief; on the surface, the second ridge has a rounded groove with a weak relief, which is locally absent. As a whole, the cast is a flattened cone with four shelves of different relief.

Dimensions (in mm)	Holotype Maximum	Minimum	2nd specimen Maximum
Diameter of the first outer ridge	24	20	50

Dimensions (in mm)	Holotype Maximum	Minimum	2nd specimen Maximum
Diameter of the second ridge	16	14	38
Diameter of the third ridge	11	9	34
Diameter of the fourth ridge	6	5	25
Diameter of the central tubercle	3	2.5	2.5
Height of the cast from the surface	5–4.5	–	3

As follows from the above data, the size of *Tirasiana concentralis* ranges widely (20–50 mm).

Discussion. It differs from *Tirasiana coniformis* Paliy in the more complex ornamentation, displaying four (instead of three) ridges and a less convex cone. It is easy to discriminate it from *T. disciformis* on the basis of the simple structure with two concentric ridges. As compared to *Cyclomedusa davidi* Sprigg, the form described has fewer annular grooves and a more relief surface instead of radial ribbing.

Distribution. Cherny Kamen Formation of the Sylvitsa Series of the Central Urals.

Material. Two well-preserved specimens.

Tirasiana cf. coniformis Paliy (Plate 29, *2*)

The collection contains one relief, although it is not an entirely preserved impression, on the surface of the grey micaceous siltstone (No. 1/11406, CGM; sample 2/72). It is ellipsoid and has three annular ridges of different relief inserted into each other with a convex central tubercle. The diameter of the first (outer) ridge, 18 mm; second, 11 mm; third, 7 mm; central tubercle, 3 mm; height of relief, 2–3 mm.

Examination of V.M. Paliy's collection shows that this impression may be similar to forms which he described as *T. coniformis.* The difference consists in a smaller size (for holotypes, 28, 21, 13 and 2 respectively) and a lower relief in the holotype (6 mm). *T. coniformis* Paliy has been reported from the Cherny Kamen Formation, the Sylvitsa Series of the Central Urals and from the Yaryshev Formation (Bernashevka Beds) of the Middle Dniester area.

Tirasiana cocarda Becker, sp.nov. (Plate 29, *4*)

Named after "cockade"

Holotype. CGM, sample No. 1/11406, sample 2/1972; Perm district, Shirokovskoe Reservoir, Nyar Village, Sylvitsa Series, Cherny Kamen Formation.

Diagnosis. A single relief convex cast on the bedding surface with concentric rugae and a tubercle in the centre.

Description. A relief convex cast, resembling an ellipsoid and consisting of five annular ridges inserted into each other with a gradually decreasing diameter. The first two outer ridges are distinct, but not ubiquitously; particularly poorly discernible in the left-hand portion of the figure. The other three inner ridges are continuously traceable ubiquitously. A rounded tubercle is seen in the most convex, central part of the ellipse. The third ridge has the most distinct relief, which extends above the surface formed by the first two outer ridges. The other peculiar feature of the third and second ridges are the fine longitudinal rugae surrounding these elevations. The second ridge also bears three rugae, and the fourth one appears locally. The number of rugae on the third ridge reaches seven; however, only two of them are traced around the entire perimeter of the ridge, the others are sometimes lost. Unlike the other ridges, the third one has three radial cuts, dividing it into three different segments. In general, the cast is a flattened cone with five shelves of different relief.

Dimensions (in mm)	Holotype	
	Maximum	Minimum
Diameter of the first outer ridge	60	52
Diameter of the second ridge	47	39
Diameter of the third ridge	34	30
Diameter of the fourth ridge	20	16
Diameter of the fifth ridge	8	6
Diameter of the central tubercle	1.5	1.5
Height of the cast from the surface	6	5

Discussion. It differs from other representatives of the genus in size, a stronger convexity and a more intricate ornamentation: five annular ridges and fine longitudinal rugae.

It differs from the majority of representatives of the genus *Cyclomedusa* in having almost no radial ribs and in morphology.

Its outward appearance resembles the holotype of *Cyclomedusa annulata* (Sprigg); however, a considerable difference in size and certain morphological features prevents the identification of these forms.

Distribution. See Holotype.

Material. One well-preserved specimen.

Tirasiana from the Kosva sediments inhabited an extensive shallow, slightly freshened basin. Freshening has been confirmed by geochemical studies of the Sylvitsa rocks and, particularly, by the Sr/Ba ratio, which falls within 0.2–0.6. In this respect, they resemble some representatives of modern scyphozoans, inhabiting desalted water of fjords and estuaries. The Cherny Kamen medusoids inhabited highly agitated waters, as evidenced by the different structural features of enclosing rocks.

Metazoa from the Cherny Kamen sediments from the Kosva locality are most similar to the Vendian non-skeletal fauna of the Dniester area (Paliy 1976). Common forms are: *Tirasiana disciformis* Paliy, *T.cf. coniformis* Paliy, *Nemiana simplex* Paliy.

In the Urals, new forms, unknown from the Vendian of Podolia (*Tirasiana concentralis* Beck., *T. cocarda* Beck.) have been found.

Non-skeletal fauna of the Cherny Kamen Horizon has also been reported from some other localities of the Central Urals, which are being investigated at present.

3 Fauna of the Ust Sylvitsa Sediments

On the Koiva River, fossils were recorded near the mouth of the right tributary to the Chusovaya River, west of the main localities of the Sylvitsa sediments, marked by Devonian linear folds. The Sylvitsa Series crops out over a relatively small area and is mainly composed of the youngest Ust Sylvitsa rocks, resting on the erosion surface of various Devonian strata and bounded by extensive folds and faults. For a long time, these deposits were placed into the Lower Devonian, due to the famous occurrence of 'psilophyte flora" and an impression of the core of "*Murchissonia cf. demidoffi* Vern." No other occurrences have been recorded, and their organic nature seems questionable.

The Ust Sylvitsa Formation is composed of brownish-grey, medium-grained sandstone, greenish-grey siltstone, variegated claret-coloured and greenish-grey silty pelite. Terrigenous rocks exhibit a distinct rhythmic pattern. Structurally, the rocks are highly diverse. They are marked by symmetric and asymmetric ripple marks, rill marks, locally resembling psilophyte remains, raindrops, desiccation fissures and abundant hieroglyphs, yielding fossils similar to brachiopod cores. Slump folds are less common. Siltstone and silty pelite are characterized by peculiar lustrous surfaces.

The casts, described below, were taken from bedrock exposures on the right bank of the Koiva River near its mouth. They are almost identical to the forms reported from the Amadeus Basin of Central Australia.

Genus *Arumberia* Glaessner and Walter 1975

Diagnosis. A series of predominantly rectilinear ribs, radiating from a gently sloping elevation and separated by shallow grooves.

Species Composition. One species, i.e. *A. banksi* Glaessner et Walter. Its impressions are known from the Arumbera Sandstone, the northern Amadeus Basin (Central Australia), Kuibis and Schwarzrand Formations of the Nama Group (South Africa) and the Ust Sylvitsa Formation of the Central Urals.

Similar forms are recorded in the Longmindian on the British Isles, Signall Hill and Masgravetown groups in the Avalon Peninsula, and in some other sections (Bland 1984).

Arumberia banksii Glaessner and Walter 1975 (Plate 29, *5a-c*)

Description. Three casts with radial ribbing on sandstone surface. On each of them, predominantly rectilinear ribs are separated by shallow grooves. The casts display differing degrees of preservation and structural features; therefore, they are described separately.

On the first cast, 8 × 7 cm, the central part of the sample (Plate 29, 5a) has a distinct elevation with an area of 1.5 × 2 cm and 1.5 × 2 mm height, from which the ribs, separated by parallel grooves, radiate semicircularly. The ribbing is fairly regular, and rarely bifurcates. The heigth of the ribs does not exceed 1-1.5 mm, and their length varies between 1 and 4.5 cm. The width of the ribs is 2–5 mm. In some cases, the width of the ribs increases away from the central elevation. The relief of the ribs also decreases in the same direction. The profile of ribbing is commonly convex, less frequently, flat. Grooves, separating the ribs, are much narrower than the latter (see below for measurements). All grooves clearly flatten towards the central rise and, in rare cases, pinch out away from the apex. Apart from numerous longitudinal groves, rare, short, oblique grooves, dividing the ribs into separate segments, are observed.

On the second cast, 7 × 5 cm, due to the preservation degree, one can see an elevation. A distinct parallel relief ribbing is about 5 cm long. In a preserved adjacent layer, ribbing of a similar trend is also visible (Plate 29, *5b*).

On the third cast, 10 × 5 cm, only part of the elevation is preserved. Rectilinear relief ribbing is fragmentary. Adjacent layers exhibit a distinct ribbing (Plate 29, *5c*) of two different directions. The angle between the two different directions of ribbing reaches 90°. Glaessner and Walter (1975) believe that this type of ribbing reflects an impression of the distal edge of the forms discussed.

Dimensions (in mm)	Width		Length	
	Ribs	Grooves	Ribs	Grooves
First cast	2–5	1–2	7–45	7–5
Second cast	1–4	1–3	25–45	15–40
Third cast	1–4	1–2	20–30	10–30

On some casts, or in the immediate vicinity, we find small (2–3 mm) spheroidal formations, described elsewhere (Bland 1984). However, they have not been recorded in most impressions, and in the Uralian form casts have peculiar organic films.

Discussion. As compared to the forms described (Glaessner and Walter 1975), the Uralian specimen seems to be the smallest, but its grooves are slightly wider.

Remarks. The reconstructions available are not simple. Some of them illustrate *Arumberia banksii* as cup-shaped sessile organisms 5–20 cm high. Their inner and outer walls are complicated by longitudinal ribbing. They might well belong to coelenterates of an uncertain systematic position.

The others state that these are impressions of a colony with numerous tubular elements, forming a single cluster, "creeping" above the sediment. The difficulties in diagnosis and interpretation of the structure of *Arumberia* are sometimes attributed to its abiogenic nature. However, a narrow stratigraphic range, a distinct attachment trace and a close association with organic films do not quite confirm this statement.

Material. Over 20 well-preserved casts and fragments, CGM, No. 1/11406, sample 264b/1969. Perm district, Ust Koiva Village, Sylvitsa Series, Ust Sylvitsa Formation.

4 South Urals

In 1986–87, in the basin of the left tributaries of the Belaya River on the western slope of the South Urals we found several localities of Ediacaran fauna. The Ryauzyak locality is the best known. Ediacaran fauna of the South Urals, similar to that of the Central Urals, is dominated by medusoid forms. However, they are smaller in size and have a weaker relief. This may be attributed to the fact that they occupy a different ecological niche, associated with the emplacement of redbeds. Among the forms inhabiting this niche, the following have been identified: *Pseudorhizostomites howchini* Sprigg, *Paliella patelliformis* Fed., *Medusites* sp., *Protodipleurosoma* sp. nov. and forms similar to *Kullingia* sp.

The Vendian of the South Urals also contains such trace fossils as: *Neonereites uniserialis* Sei, *Palaeopasoichnus delicatus* Paliy, as well as some other, less frequent forms.

Genus *Protodipleurosoma* Sprigg 1949

Type Species. Protodipleurosoma wardi Sprigg 1949.

Diagnosis. The diagnosis is based on the similarity to living Dipleurosoma with radial canals (Sprigg 1949). The exumbrella impression is circular; the outer ring (velum?) is concentrically striated and wide; the mid-field is large and circular, with an ovoid, unevenly constricted, lobate central area (stomach?) and irregularly developed radial grooves (radial canals?) branched non-dichotomously; the primary grooves are strong, reaching the submarginal groove (ring canal?), giving rise to shorter grooves by branching near the base (Treatise 1956).

Species Composition. Protodipleurosoma wardi Sprigg, *P. rugolasum* Fed. and *P. paulus* sp. nov.

Protodipleurosoma paulus Becker, sp. nov. (Plate 89, *1*)

Etymology. Latin "paulus", small.

Holotype. CGM, No. 1/11406, sample 176/86. Gornaya (Mountain) Bashkiria, Ryauzyak River. Vendian, Asha Series, Basa Formation.

Description. A well-delimited, rounded impression with a clearly discernible outer ring around the periphery. The inner disc is divided by a central groove into two approximately equal parts. On one side, the central groove also cuts the outer ring, which is slightly deformed at the intersection point. Small grooves radiating from the central groove are flattened and do not reach into outer ring zone. The degree of preservation of the outer zone does not depend on the size of impressions: in the smallest forms it is distinct, and in larger ones it is sometimes reduced.

Diameter (mm). Outer ring, 8, 20; inner disc, 5.

Discussion. Protodipleurosoma paulus differs from known representatives of this genus in its smaller size, absence of lateral processes in the outer zone, branching elements of radial canals and their weaker development in the inner zone.

Material. Three impressions and fragments.

Distribution. Vendian, Asha Series, Basa Formation of the South Urals, Ryauzyak River.

Genus *Pseudorhizostomites* Sprigg 1949; *Pseudorhizostomites howchini* Sprigg (Plate 89, *2*)

Description. Convex impression without distinct lateral terminations. A row of branched relief grooves, flattening around the periphery of the impression and imparting an irregular pustular outline to its surface, radiate from a lower slitlike central part.

Dimensions (mm). Length, 12; width, 8.

Material. Well-preserved specimen.

Distribution. Vendian, Asha Series, Basa Formation of Gornaya Bashkiria, Ryauzyak Basin; Mogilev Formation of Podolia; Ust Pinega Formation, Zimny Coast of the White Sea, Pound Quartzite of South Australia.

Genus *Paliella* Fedonkin 1980

Type Species. Paliella patelliformis Fedonkin.

Species Composition. Monotypic genus.

Paliella patelliformis Fed. (Plate 89, *4*)

Description. Two zones, outer and inner, are clearly discernible within a rounded and more or less flattened impression. The outer zone contains radial grooves of variable intensity, from near-relief to barely visible or almost obscured. The inner zone, clearly separated by an annular ridge, has no radial ribbing. Some impressions have a rounded elevation in the central part.

Dimensions (mm). Outer zone, 14–16; inner zone, 5–6; central elevation, 1–2.

Discussion. In general outline and size, *Paliella* is similar to *Medusites asteroides* Sprigg; however, it is flattened and lacks an epimarginal groove. The form described differs from *Tirasiana disciformis* Paliy in having radial ribbing.

Material. Six impressions from the Basa Formation of the Asha Series on the Ryauzyak River (South Urals) and three specimens from the Cherny Kamen Formation of the Sylvitsa Series (Central Urals).

Distribution. Vendian of the South Urals (Asha Series, Basa Formation), Central Urals (Sylvitsa Series, Cherny Kamen Formation), Zimny Coast of the White Sea (Valdai Series) and Dniester area (Mogilev Formation, Lomozov Beds).

Genus *Medusinites* Glaessner and Wade 1966

Type Species. Medusinites asteroides (Sprigg).

Medusinites sp. (Plate 89, *3*)

Description. Two zones, outer and central, are visible within a flattened, conical impression. Conical surface of the outer zone is cut by deep, radial grooves. The outer zone is slightly deformed around the periphery, giving rise to a weak concentric, marginal groove. The central zone is flattened and surrounded by a weak relief ridge.

Diameters (mm). Central zone, 5; outer zone, 25. Ratio of diameters, 1:5.

Discussion. In a number of important, specific features the form described is similar to *Medusinites patellaris* Sok.
It is more conical, smaller and has a deformed outer zone.

Material. Three impressions and their fragments.

Distribution. Vendian, Asha Series, Basa Formation, the Ryauzyak Basin of the South Urals.

Chapter 4

Paleoichnology of Vendian Metazoa

Trace fossils are one of the few transit faunal groups at the Precambrian/Cambrian boundary. This fact makes us pay particular attention to this faunal group, despite the low stratigraphic importance of trace fossils. On the other hand, the enormous heuristic value of trace fossils for paleontology and paleoecology is well known, especially for the understanding of the earliest stages of faunal evolution.

Trace fossils allow passing a judgement on the amount and diversity of animals within a biotope, density and general dynamics of populations, trophic phenomena and links. Paleoichnology provides information on the movement modes of animals, specific features of their behaviour and responses to diverse ecological factors. It is often possible to obtain information on the level of physical, physiological, and even the nervous organization of an animal, and in some cases, its taxonomic affinity. As to the Precambrian, most likely, it is the trace fossils that provide tools for the determination of the time of appearance of the first benthic Metazoa in the geological record and contribute to the history of habitation of the seafloor.

The establishment of the most ancient movement tracks is a highly important, but difficult task: the oldest Precambrian trace fossils are extremely rare, morphologically fuzzy, and quite frequently give us reason to doubt their biological origin. One can even grow more sceptical because of dubious descriptions of numerous pseudofossils reported as remains or trace fossils of Metazoa from the Precambrian strata, or erros in dating of enclosing rocks. Under the assumption of relatively poor knowledge of the pre-Vendian deposits (from the viewpoint of paleoichnology), one can still state that the overwhelming majority of Precambrian trace fossils have been described from the Vendian strata. Presumably, the epoch of the global Laplandian or Varanger Ice Ages, i.e. when a shelf zone was extremely limited in life forms, and was confined mainly to the pelagic region, was succeeded by an epoch of great transgressions and a time of giant epiplatform seas and extensive shelf zones. It would be logical to assume that intense land erosion during this epoch could give rise to extremely extensive littorals. One of the most important taxonomic and population outbreaks in the history of faunal evolution on our planet took place under these unique conditions. Paleontologically, this outbreak is documented by abundant remains of non-skeletal fauna and trace fossils.

Vendian was apparently a peculiar stage in a long-term process of the appearance of organisms on the Precambrian ocean floor.

During Vendian, the abundance and diversity of trace fossils progressively increased, particularly in shallow-water facies (Glaessner 1969; Banks 1970; Sokolov 1972b; Germs 1973; Cloud et al. 1976; Fedonkin 1976a, 1977, 1979, 1980a, 1981a,b; Paliy 1976; Paliy et al. 1979; Crimes and Germas 1982, etc.).

Trace fossils, known from the Vendian deposits (570–650 m.y.), point to a high organization of vagile benthos. In sedimentary strata, these organisms moved owing to peristaltic changes in the shape of the entire body (in the same way as the living nemertines, annelids, sipunculids, etc.), passing through their digestive canal a residue, rich in organic matter. This is confirmed by such trace fossils as Planolites and Neonereites. In the subsurface layer and on the sediment surface, Vendian Metazoa moved by peristaltic waves, passing through the ventral part of body, like living planarians, chitons and gastropods (e.g. trace fossils such as Nenoxites, Bilinichnus and Aulichnites), less frequently by means of wavelike curvings of the body (Cochilichnus). Imprints of crawling trails by means of the motion of parapodias are extremely rare.

A peculiar feature of Vendian ichnocoenoses is the significant share of tracks with repeated symmetry elements. These traces were apparently left by repetition of similar movements by mud-eaters for the optimum capture of sediment. Some organisms, moving forward, consumed a broad area of detritus-rich sediment by transverse movements with the anterior end of their body. Others methodically ate the sediment away, alternating with abrupt right and left turns, etc. Such imprints are widespread in Vendian deposits of different regions and point to the increasingly complicated behaviour of mud-eaters in the process of alimentation.

Of utmost interest are those Vendian deposits, which along with trace fossils, also contain imprints of non-skeletal Metazoa. Judging by trace fossils, vagile benthos is generally characterized by a higher level of organization (possibly annelids and mollusc), whereby morphological (and taxonomic) diversity is much lower. If the interpretation of trace fossils is correct, the differences may be attributed to a peculiarly directed and relatively fast pre-Vendian evolution of vagile benthos as compared to that of pelagic and sedentary forms.

Typical Vendian ichnocoenoses can be exemplified by trace fossil assemblages from the Valdai Series of the southeastern White Sea area and Dniester Basin of Podolia (Table 2), associated with imprints of non-skeletal fauna. Let us dwell on the most characteristic forms of trace fossils.

The studies accomplished in the last few years show that Pascichnia (grazing trails) is the most abundant and diverse group in the Vendian ichnology. Trace fossils of this group are commonly characterized by complex movement patterns, systematically repeated movements of the same type and an intense biological processing of the sediment, essentially horizontally. As a rule, animals do not cross their own trail.

This group includes trace fossils of the genus *Ilmenites bacillaris* Fedonkin, i.e. densely meandering burrows with a smooth lower surface, and *Aulichnites*, preserved on the top of sandstone beds in the form of two gently sloping ridges, separated by a groove; the latter also exhibit an intense reworking of the sediment. Closely meandering trails, whose pattern resembles that of Helminthorhaphe, occur as well. Judging by trace fossil morphology, it appears that they were produced by systematically different organisms, which, however, used the same means of motion (peristaltic) and similar behavioural programs (meandering). *Bilinichnus simplex* Fedonkin et Palij, in the form of two parallel narrow grooves, resembles trails of small gastropods, whose lateral parts of the foot are slightly deeper sunken into the sediment than its central part, i.e. it is a trail of peristaltic crawling, when the contraction waves pass along the ventral side of the body.

Nenoxites curvus Fedonkin are meandering ribbonlike trails with transverse rugae. The shape of a trail shows that an organism was moving by peristaltic waves passing through the ventral part of the body (like living Turbellaria, Loricata, and Gastropoda); and its track bears a great resemblance to those of modern gastropods, which eat food from the substrate surface. Transverse rugae are a reflection of the work of the mud-collecting apparatus rather than peristaltic contractions of the ventral side. Slightly curved trails occur together with sinusoidal ones, exhibiting closely adjoining meanders. This points to a more complicated behaviour of organisms, which were aiming at the optimum use of the sediment, saturated in organic matter, upon which they were feeding.

Palaeopascichnus delicatus Palij is a succession of narrowly arched grooves, closely adjoining each other and apparently reflecting the systematic repetition of movements of a crawling organism across the direction of its motion. As a result, the organism embraces a broader area of the sediment than in case of a simple forward movement. It was probably a peristaltic motion, as in the case of *Nenoxites* or *Planolites.*

Palaeopacsichnus sinuosus Fedonkin displays a more complicated behaviour of the organisms as compared to *P. delicatus:* along with movements across the direction of motion, the track is also sinusoidal and, thus, the utilization of the sediment becomes optimal.

Harlaniella podolica Sokolov, preserved in a positive hyporelief as obliquely hatched ridges, most likely belong to feeding traces of wormlike organisms.

Feeding tracks and trails (Fodinichia) are probably the second largest and diverse group of Vendian trace fossils.

Planolithes Nicholson are subhorizontal dwelling burrows of a circular section with smooth and rugate walls. Such trails could be left by mud-eating annelidomorphic organisms, moving within a silty bed by peristaltic waves, involving the entire body.

Neonereites uniserialis Seilacher are preserved as round pellets and are trails of rather highly organized soil-eaters, which had a wormlike body, a developed digestive system, and which moved in the sediment by peristaltic waves, involving the entire body. There are numerous areas of the substrate surface, which were subject to biological processing and repleted with fecal pellets.

Neonereites biserialis Seilacher differs in having a paired arrangement of small fecal pellets, oriented obliquely to the directed motion of an organism. Such an arrangement of pellets may be accounted for by the fact that they were slightly displaced when the organism pushed itself with its posterior end. On one side of some pellets we can see fine dents, probably formed when an organism pushed itself, while moving through the sediment. Intersection of bedding planes by trails of *N. biserialis* (as well as *N. uniserialis*) indicates that these organisms moved below the sediment surface, whereas chains of relatively large nodular fecal pellets (*N. renarius* Fedonkin) were left on the sediment surface.

The proportion of dwelling burrows (group Domichnia) in Vendian ichnocoenoses is generally low and, as in subsequent geological periods, they are confined to extremely shallow shelf zones. Thus, in the upper unit (Bed 11) of the Valdai Series section, Zimny Coast of the White Sea, we found inclined, fine dwelling burrows of *Skolithos declinatus* Fedonkin sp.nov., which gradually thin downwards. The walls of the dwelling burrows could well be lined with mucus. The population density of the wormlike organisms, inhabiting these dwelling burrows, was rather high: up to 20–30 individuals per square decimetre. The occurrence of other ichnofossils in this unit indicates that in Vendian the littoral region was very densely populated.

As to Bergaueria Prantl, the Vendian yields low and not very large specimens as compared, for example, to Bergaueria major Palij, described from the Khmelnitsk Formation of the Baltic Series (Palij 1976). Sometimes, there are trails of a horizontal movement of polyps, leaving trails of the Bergaueria type. In this case, one can observe a series of arched ridges, as if inserted into each other, with the concave side facing the cast of the basal part of the polyp. Such forms were recorded from the Lomozov Beds of the Mogilev Formation, the Dniester Basin of Podolia, and seem to occur also in the Upper Precambrian of northwestern Canada (Hofman and Aitken 1979).

Group Cubichnia (resting tracks) is very poorly represented in the Vendian inchnocoenoses, though usually it is this group that gives the best information on the morphology of organisms. *Vendichnus vendicus* Fedonkin, a trail of a resting organism in the form of two winglike dents, separated by the mid-ridge, is described from the Valdai Series at Syuzma River. Short transverse grooves, observed on both halves of the trails, possibly appeared due to limb motions of an organism. *Vendichnus* is oriented parallel to linear current marks, which reflects the response of organisms to a stream of water (rheotaxis), typical of most of the living, bilaterally symmetric invertebrates in an aqueous environment. Common features between the trail described and those ascribed to trilobites, particularly *Ixalichnus* Callison (Häntzchel 1975), suggest that a producer of *Vendichnus* was a bilaterally symmetrical nektonic organism, which actively swam, rather than crawled on the floor, because no morphologically similar crawling trails have been found.

In Vendian ichnocoenoses, the overwhelming majority of trace fossils cannot be correlated taxonomically with the mollusc known from imprints. This is due to the fact that the trails were mainly left by mobile benthic fauna, whereas most non-skeletal Metazoa, known from impressions, were nektonic or sedentary forms, except for *Suzmites volutatus*. These trails could have been left by *Pteridinium*. A peculiar Vendian feature is a shallow burrowing of organisms, which moved either on the sediment surface, or just below it, since Vendian trails are mainly subhorizontal.

Vendian ichnocoenoses are dominated by trace fossils with repeated elements of symmetry. Of utmost interest among them are trails of soil-eaters, left by repeated movements (behaviour on the principle of maximum coverage of the sediment surface during movement). Among trace fossils of this type are *Harlaniella, Palaeopascichnus, Nenoxites, Vimenites, Aulichnites,* etc.

The predominance of trace fossils of the group Pascichnia in Vendian ichnocoenoses implies that most of vagile benthos were of microphage affinity, i.e. organisms feeding on small nutrient particles, which they collected from soil.

The occurrence of common forms of trace fossils in the strata, containing identical or similar molluscs, permits one to use them for a distant correlation of the Vendian deposits. Of course, this primarily concerns the terrigenous facies, since Vendian carbonate facies have not been adequately studied in terms of paleontology.

A striking change in the character of ichnocoenoses is observed at the Vendian/Lower Cambrian boundary, in particular at the level of the maximum abundance of sabellitids on the Russian Platform (Rovno Horizon) and in the Nemakit-Daldyn Horizon of Siberia. Generally, this change shows a higher degree of biological reworking of sediments. Many new, earlier unknown taxa appear in Vendian: Treptichnus, Phycodes, Gyrolithus, Didymaulichus, large flat-spiral forms, etc. Many forms are large and fairly deep; morphologically complicated feeding trails with repeated elements of symmetry are widespread. Along with subhorizontal and shallow (as in Precambrian) forms, there also appear "two"- and "three -dimensional" systems, which are not characteristic of Vendian ichnocoenoses, particularly Spreiten burrows and vertical spiral feeding trails.

A much more intense colonization of the seafloor, observed in the Precambrian-Cambrian transition zone, was possibly caused by the influence of certain biotic and (or) abiotic factors. Presumably, interspecific competition and predation became more intense under conditions of a rapid increase in the density of the benthic population. Apparently, penetration into the depth of sediment became possible due to better aeration, owing to a higher oxygen content in the atmosphere, and a more active utilization of fossils by diverse reducers (more active as compared to Vendian), and in many cases it precluded a possible hydrogen sulphide pollution of the water and sediment bottom layers, thus, favouring a more active colonization of ecological niches in this zone.

Beginning with the Atdabanian (Lower Cambrian) and further, the arthropods began holding a leading position in the ichnocoenoses of offshore areas; though, sometimes, tracks of rest and trails of unknown arthropods on the floor surface occur in the Tommotian deposits.

The similarity between Lower Cambrian ichnocoenoses in different parts of the world is remarkable. Dozens of forms, which occur almost ubiquitously, beginning with the lowermost Cambrian, could be named. Their specific character and differences from Vendian trace fossils (ichnofossils) make this group a useful marker of the Precambrian/Cambrian boundary (Table 2), at which the most important events in the fate of marine benthos at one of the most significant phases in organic evolution took place.

Table 2. Stratigraphic distribution of trace fossils in Upper Vendian and Lower Cambrian of East European Platform[a]

	Vendian				Cambrian	
	Volyn Series	Valdai Series		Baltic Series		
	1	2	3	4	5	6
Harlaniella podolica Sokolov			—			
Palaeopascichnus delicatus Palij		—	—			
P. sinuosus Fedonkin		—				
Cochlichnus Hitchock		—	—	—	—	—
Neonereites uniserialis Seilacher		—	—	—	—	—
N. biserialis Seilacher		—	—	—	—	—
N. renarius Fedonkin		—				
Nenoxites curvus Fedonkin		—				
Vendicus vendicus Fedonkin		—				
Vimenites bacillaris Fedonkin		—				
Intrites punktatus Fedonkin		—				
Bilinichnus simplex Fedonkin et Palij		—	—	—	—	—
Didymaulichnus tirasensis Palij				—		
Bergaueria Palij		—	—	—	—	—
Bergaueria major Palij				—		
Gordia Emmons		—	—	—	—	—
Skolithos Haldemann		—				
S. linearis Haldemann					?	—
Diplocraterion parallelum Torell				?	—	—
Rhizocorallium Zenker					—	—
Teichichnus rectus Seilacher				—	—	—
Treptichnus bifurcus Miller				—	—	—
Treptichnus triplex Palij				—		
Phycodes pedum Seilacher				—	—	—
Aulichnites Fenton et Fenton		—			?	—
Planolites Nicholson		—	—	—	—	—
Gyrolithes polonicus Fedonkin				—	—	—
Rusophycus Hall					—	—
Cruziana d'Orbigny						—

[a] Numbers denote horizons: *1* Drevlyany; *2* Redkino; *3* Kotlin; *4* Rovno; *5* Lontova; *6* Talsy.

Chapter 5

Vendotaenids – Vendian Metaphytes

In life evolution on earth, the advent of megascopic plants showing a tissuelike structure goes back to pre-Vendian time, i.e. earlier than 680–700 m.y. ago. For example, megascopic non-mineralized, ribbonlike remains (0.1–2 mm wide) studied on rocks were reported from deposits with an isotopic age of 1300 m.y. from the Belt Supergroup, North America. First they were described as Helminthoidichites (Walcott 1914) and later redescribed as Proterotaenia, Lanceoforma and Grypania (Walter et al. 1976). "Beltlike" forms, up to 6 mm wide, of presumed algal origin were reported from the pre-Nama Quartzite (approximately Upper Riphean), Southwest Africa (Glaessner 1963). Unfortunately, plant tissue of the fossils is fully replaced by iron oxides. Moreover, ribbonlike fossils, proposed vendotaenids, were isolated by maceration from rocks of the Vtoroi Kamen (Derevnya) Formation (approximately Upper Riphean), Turukhansk Uplift on Siberian Platform (collection of Dragunov). On the Siberian Platform, ribbonlike fossils (0.3–0.7 mm wide) were also isolated from roughly coeval deposits of the Lakhanda Formation (Upper Riphean) with an isotopic age of 870–900 m.y. (collection of T.N. German). However, these are regionally scattered occurrences. Only starting from the Vendian at 580–600 m.y. do ribbonlike macrophytes (megaphytes) occur in abundance over large areas.

Non-mineralized, megascopic, ribbonlike fossils are very common in the Vendian of the East European Platform. At present, they are well known from Poland, the Leningrad District, the Baltic Sea area, Byelorussia, central regions of Russia, the Ukraine and Moldavia. Abundant occurrences were recorded also from coeval deposits of the Siberian Platform, but they have not yet been adequately studied. All these non-mineralized macrophytes were placed into the formal group, Vendotaenids Gnilovskaya – vendotaenids (Gnilovskaya 1971, and others).

At present, the Vendian macrophytes are interpreted as the class Vendophyceae Gnil., 1986 containing two orders: Eoholyniales (bushy forms) and Vendotaeniales (ribbonlike forms). For details, the reader is referred to "Vendotaenids of the East European Platform" (1988).

Vendotaenids are associated, as a rule, with extremely abundant, also non-mineralized, organic, shapeless brown films measuring 100 × 100 mm. Since the end of the past century they are known as "laminaritic" films according to Eichwald (1854), who was the first to study them; he assigned them to algae and described them under the name *Laminarites antiquissimus*. However, the name cannot be applied to the Precambrian fossils because the genus *Laminarites* Sternberg, originating from the Mesozoic, belongs to higher plants. In addition, the nature of

Precambrian organic films remains uncertain. As yet, we also could not establish their external appearance or determine any anatomical structure. Apparently, we are in agreement with those who do not consider these organic films or at least most of them as fossils, but rather as products of reworking of some plant remains, viz. sapropelite.

Detailed study shows that vendotaenids are a fairly diverse group. On morphological grounds they can be subdivided into at least two subgroups: ribbonlike, occasionally bifurcating forms, viz, the genera *Vendotaenia, Dvinia, Aataenia, Tyrasotaenia;* and cordlike, intensely branching forms, e.g. the genus *Eoholynia*. The genus *Fusosquamula*, recognized by E.A. Aseeva (1976, p. 57, Plate XX, Figs. 1–2) from the Vendian of the Dniester Basin of Podolia, Ukraine, may also be placed into the first subgroup. These fairly short (about 10 mm long) organisms with a sharp thickening in the middle part and tapering ends resemble the pre-Vendian genus *Lanceoforma* Walter, Oehler, Oehler. Somewhat apart stands the genus *Pilitela*, forming bundles of fine fibres totalling 1.5 mm in width (Aseeva 1976, p. 57–58, Plate XX, Figs. 3, 5).

The genus *Vendotaenia*, which gave the name to the entire vendotaenid group is the best studied (Gnilovskaya 1971, 1975, 1979b). In a rock, vendotaenids look like elastic, dark brown ribbons fully covering the bedding planes (Plates 30, 31). They are up to 150 mm long and have a constant width in a single specimen, but it varies from 0.5 to 3 mm in different specimens, their thickness equalling some fractions of a millimetre. Ribbons occasionally branch into two ribbons, almost equal to the original one in length (Plate 30, *3*; Plate 33, *3*); there is a specimen with a ribbon branching into four daughter ribbons (Plate 30, *1*). The substance of *Vendotaenia* gives the following chemical composition: 41.3% C, 5.1% H, 0.5% N, S + 0–53.1%. A high content of trace metals Fe, Co, V and Cu in vendotaenid thallus may imply an autoheterotrophic metabolism. A high concentration of porphyrins suggests the presence of chlorophyll in the original substance and confirms the plant nature of vendotaenids. The microscopic study of vendotaenid fragments extracted from rocks by solution in hydrofluoric acid proves to be very useful. Vendotaenid ribbons have a tissuelike structure lacking any evidence of a vascular system indicative of their algal nature. The thallus seems to be two-layered, and multiseriate cells form longitudinal rows along the thallus. The rows are fairly well visible under the light microscope and form the fibrous surface of the thallus (Plate 33, *1–2*), while individual cells, seen under an anoptral and phase microscope, are about 3 × 5 μm in size. In *Vendotaenia* we find sporangia arranged in longitudinal rows on the thallus surface. Two differing types of sporangia structure and their arrangement in rows can be distinguished.

In one case (Plate 36, *2*) these are longitudinal rows extending along the entire length of the preserved thallus fragments. There are two to four rows, two stretching along the thalus margins. Sporangia are very densely spaced within the rows. In adjacent rows they lie at about the same level. They are 100–135 μm in size, round or somewhat elongated in shape. Each sporangium apparently consists of many small spores(?) 10–20 μm in size.

In the other case (Plate 32, *1–4*) sporangia form short longitudinal rows in the middle of the thallus fragments, commonly two rows, three to five sporangia in each. In adjacent rows, sporangia exhibit a staggered arrangement, they are about

60–100 μm in size. The sporangia are spherical in shape. Their material is divided into four to eight parts. One can assume that these are multicellular sporangia, oogonia, containing large egg cells. It is noteworthy that forms interpreted here as sporangia are similar to those assigned to microphytoplankton (Sphaeromorphida in the sense of Timofeev).

The above suggests that *Vendotaenia* is one of the oldest metaphytes. Metaphytes, macrophytes with a tissuelike (tissue) thallus structure, originated in the course of the long-term evolution of unicellular, planktonic, microscopic algae. Modern paleoalgologists (Schopf et al. 1973) believe that major lines of the evolution were the advent of sexual reproduction in ancient eukaryotes, regular alternation of sexual (haploid) and asexual (diploid) generations in a life cycle and, finally, the leading role of asexual (diploid) generation in a life cycle. *Vendotaenia* seem to be a diploid generation of metaphytic algae. The general habit of the thallus, the nature of sporangia and the type of metabolism suggest that *Vendotaenia* are brown algae.

A brief description of all known vendotaenids is given below.

Genus *Vendotaenia* Gnilovskaya 1971 (Plates 30, *1–5*; 31, *1–6*; 32, *1–4*; 33, *1–3*, 36, *2*)

Vendotaenia: Gnilovskaya 1971, p. 105–106; 1974, p. 59; 1975, p. 954; 1976, p. 10–12; 1979, p. 40.

Laminarites (part., quoad specimen, hoc loco): Timofeev 1966, p. 16–17, Plate I, Fig. 1, Plate II, Fig. 1a.

Type Species. V. antiqua Gnilovskaya 1971; Leningrad District: Vendian, Kotlin Horizon.

Diagnosis. Thallus is ribbonlike, rarely branching; the surface is longitudinal-fibrous, 150 mm long, 0.25–3.5 mm wide. Thallus is multicellular, multilayered; multiseriate, longitudinal rows of cells (3–5 μm in size) are clearly visible. Sporangia are spherical, arranged in long or short longitudinal rows, 50–90 μm in size.

Species Composition. One species, *V. antiqua* Gnilovskaya 1971, represented by four morphological varieties, is easily distinguishable in rock in which the width of ribbons is described. The ribbon width in each group ranges as follows: *V. antiqua* forma prima, 0.25–0.5 mm; *V. antiqua* forma secunda. 0.5–1.0 mm; *V. antiqua* forma tertia, 1.0–1.5–2.0 mm; *V. antiqua* forma quarta, 2.0–3.5 mm. Photomicrographs show no differences in thallus structure.

Geological and Geographic Distribution. Vendian, Kotlin Horizon; East European and Siberian Platforms.

Genus *Dvinia* Gnilovskaya 1979b (Plate 35, *5*)

Dvinia: Gnilovskaya, 1979b, p. 171–172.

Type Species. D. fibrosa Gnilovskaya 1979b; Byelorussia, Estonia; Vendian-Cambrian, Baltic Series, Rovno and Lontova Horizons.

Diagnosis. Thallus is flattened-tubular, 0.2–0.3 mm wide; the surface relief is roughly longitudinal-fibrous. Thallus is multicellular and multilayered. Cells are arranged in longitudinal or oblique-longitudinal rows. Cells are subsquare, ca. 5 μm. Sporangia are unknown.

Species Composition. See type species.

Discussion. In those cases, when in *Dvinia* only a fragment of a single wall is preserved, its longitudinal-fibrous texture resembles that of *Vendotaenia*. However, even a single wall of *Dvinia* is much denser than the multilayered thallus of *Vendotaenia*, and the longitudinal structure of *Dvinia* is much more pronounced and does not exhibit such a regular structure as that of *Vendotaenia*. In addition, reproductive organs in *Dvinia* have not been found as yet, which precludes the understanding of its nature in general.

Geological and Geographical Distribution. Vendian-Cambrian. Rovno, Nemakit-Daldyn, Lontova Horizons; East European and Siberian Platforms.

Genus *Aataenia* Gnilovskaya 1976 (Plate 35, *1*)

Aataenia: Gnilovskaya 1976, p. 11–12.

Type Species. A. reticularis Gnilovskaya 1976; northeastern Estonia; Kohtla-Järve area, borehole A-165; depth interval, 138–140 m; Vendian, Kotlin Horizon.

Diagnosis. Thallus has a distinct cellular structure, multiseriate, up to 12 cellular rows. Cells are subrectangular and lie at the same level in adjacent rows, 6 × 4 μm. Reproductive organs have not been found.

Species Composition. Only one species is described.

Genus *Tyrasotaenia* Gnilovskaya 1971 (Plate 34, *1–4*)

Tyrasotaenia: Gnilovskaya 1971, p. 106–107; 1974, p. 59; 1976, p. 11–12; 1979, p. 41.

Diagnosis. Thallus is ribbonlike or flattened-tubular, the surface is smooth 0.3–0.5 mm wide. Thallus is thin, transparent, readily folding or dense, not folding. Thallus is multicellular, cells commonly uniformly arranged, about 3 μm in size. Sporangia are unknown.

Species Composition. T. podolica Gnil. 1971; Vendian, upper Kotlin and Rovno Horizons; East European Platform. *T. tungusica* Gnil., 1979; Vendian, Rovno Horizon of the East European Platform and Nemakit-Daldyn Horizon of the Siberian Platform.

Genus *Eoholynia* Gnilovskaya 1975 (Plate 35, *2–4, 6*)

Eoholynia: Gnilovskaya 1975, p. 954–955; 1979, p. 42.

Type Species. E. mosquensis Gnilovskaya 1975; Moscow syneclise; Vendian, Redkino Horizon.

Diagnosis. Thallus is cordlike and intensely branching. Small bush is about 10 mm, the attachment is high, central, near the base, apparently a bundle of three to five cordlike major branches, dividing monopodially, branches are 100–150 μm wide. Side branches dichotomize; they are 50–80 μm wide and taper to 20 μm at the ends. Thallus is multicellular; cells are about 3 μm in size. The entire thallus surface is repleted with numerous spherical bodies interpreted here as sporangia. They are sessile and occur mainly on the surface of major and side branches, less commonly on the tops of terminal branches; they are 50–100 μm in size. It is noteworthy to mention the preservation of abundant, much smaller (15–20 μm) spherical bodies on sapropelic films near the thallus, except for sporangia of the above described appearance. These could well be spores spilled out of sporangia. The general appearance of the thallus and the arrangement of sporangia on it suggest that *Eoholynia* belongs to brown or red algae. The branching pattern of the thallus implies a central attachment of the plant and a benthic mode of life.

Species Composition. See type species.

Thus, vendotaenids, or at least most of them, are macroscopic, multicellular algae, some of them are true benthos and therefore, can be assigned to the earliest or one of the earliest metaphytes.

At present, vendotaenids show the following distribution in the Vendian succession of the East European Platform.

The most ancient vendotaenids of the East European Platform are related to the lower half of the Valdai Superseries. Small bushy vendotaenids with a cordlike thallus belonging to the genus *Eoholynia* were reported from the Redkino Series, Moscow syneclise. They were recorded from the Soligalich-1, Soligalich-7, Galich 1, Roslyatino-1, Bobrovskaya boreholes. So far *Eoholynia* is known only from the Nizhniy Valdai Redkino Series and not from the Verkhniy (Upper) Formation.

In addition, peculiar organisms, probably fungi, *Caudina* Gnilovskaya (Dorogobuzh and Vorobievo boreholes) were recorded from the Redkino Series,

Moscow syneclise. The Redkino Series yields abundant remains of colonial algae, *Orbisiana* Sokolov (Plate 33, *7*) forming chains and nets of angular cells differing in size in various colonies. Also numerous are *Leiothrichoides* Hermann covering the organic films.

Starting from the upper half of the Valdai Superseries, *Vendotaenia* shows abundant distribution in the Povarovo Series, Moscow syneclise; the Lyublin Series, Poland, the Kotlin Horizon, Baltic Sea area; the Kotlin Horizon, Byelorussia; the Nagoryany and Kanilovka Formations, Ukraine; the Avdarmin Series, Moldavia, within the Kaushany, Sokolets and Feropontievo Formations.

The most complete and well-known data are available on the Dniester Basin of the Ukraine and Moldavia. In this section, vendotaenids show their first appearance in the Dzhurzhevka Beds of the Nagoryany Formation of the Ukraine, viz. Kremenitsy village, at the mouth of the Lyadova River (author's data); the villages of Loevtsy, Minkovtsy, Lyadova, Kalyus, Maly Bereg (Ishchenko and Korenchuk 1979), and from Moldavia: the right side of the Dniester River, station 220 on the single-track railway (author's data). Dark brown ribbons of *Vendotaenia* occur in abundance on the bedding planes. Ribbons of differing width occur together, commonly folded, twisted and are not preserved unfolded, i.e. fairly thin and elastic. The most abundant is *V. antiqua* forma tertia, less common is *V. antiqua* forma prima; *V. antiqua* forma quarta and forma secunda are scarce. Phytocenoses studied by A.A. Ishchenko yield abundant *V. antiqua* forma secunda and sparce forma prima, the ratio of forma tertia to forma quarta is the same as that shown by the author. Micrographs of ribbon fragments show dense, unfolded thallus. However, the anatomy of *Vendotaenia* from the Dzhurzhevka Beds must be studied in more detail.

The overlying Kalyus Beds, Kaushany Formation, Moldavia (Bolotnino-1 borehole, depth interval, 755–784 m, author's data) contain abundant plant remains evenly distributed throughout the section. The most numerous are *V. antiqua* forma quarta, preserved as dense, glossy black, very wide and short ribbons (up to 3.5 mm wide and 30 mm long) lying freely on the bedding planes. Another characteristic feature of the Kalyus flora is narrower and shorter black ribbons covering the entire bedding plane, their width is similar to that of *V. antiqua* forma secunda-tertia. They produce an impression of branching, though it may be due to overlapping of ribbons. These short ribbons might well have broadened towards one end, most likely to the upper one. *V. antiqua* forma tertia are often folded into distinct and long, longitudinal folds. Micrographs show a dense, dark brown relief thallus with a fine-granular surface. A tissuelike structure is easily discernible and very fine. In the Ukraine, the Kalyus Vendotaenia, studied by Ishchenko (Ishchenko and Korenchuk 1979) (villages of Minkovtsy, Maly Bereg, Loevtsy, Pilipy) occur at two levels, separated by a sequence without *Vendotaenia,* at the base and top of the succession. The lower unit is dominated by forma tertia and scarce forma secunda, while the upper unit contains forma quarta and secunda with a predominance of forma tertia.

The overlying Salkuts and Lunguts Beds, Kaushany Formation, Moldavia, contain *V. antiqua* forma secunda and forma tertia.

A mass abundance of well-preserved *V. antiqua* was recorded from the overlying Sokolets Formations, Avdarmin Series, Moldavia (Bolotino-1 borehole section, depth interval, 738–641 m was studied in detail) and from the coeval Sokolets Beds, Kanilovka Series, Ukraine, which may be approximately equated to the Danilovo, Zharnov and Kryshany Formations, whose vendotaenids were studied by Ishchenko (1982). *V. antiqua* is represented mainly by forma tertia with ribbons about 1 mm, rarely 2 mm wide. *V. antiqua* forma secunda is less common, though also abundant, its ribbons cover the entire bedding plane. As a rule, these varieties occur on different bedding planes and very rarely occur together. Narrow *Vendotaenia antiqua* forma prima are very scarce. Phytocenoses studied by Ishchenko occasionally contain forma quarta and the genus *Kanilovia* recognized by Ishchenko. Under the microscope we can see fairly thin, light brown, elastic thalli, showing a distinct tissuelike structure, best visible along the margins. Less common are very thin, light yellow thalli with a poorly discernible tissuelike structure.

The above lying Ferapontievo Formation, Moldavia (Bolotino-1, depth interval, 462–527 m) is divided by means of vendotaenid flora into the lower sequence, about 26 m thick, containing *Vendotaenia*, and the thin, about 10 m, upper sequence yielding *Tyrasotaenia*. They are separated by a 32-m-thick sequence containing no vendotaenids. The lower sequence is overcrowded with well-preserved vendotaenid ribbons lying on the bedding surface unfolded along the length and width, often bent but not twisted, suggesting fairly dense thalli. They belong to morphological varieties of *V. antiqua* forma secunda and less commonly to *V. antiqua* forma tertia, which occur in association. Micrographs show thin, pale yellow thalli with a distinct tissuelike structure. It is interesting that by solution we isolated a fragment of thallus at the branching point (Plate 33, *3*). We found several narrow and dense thalli with a characteristic inner structure formed by crossing dark brown strands, whose function is not clear (Plate 33, *6*). There are also ribbons which lack a characteristic longitudinal-fibrillar structure. The upper sequence up to the boundary with the Cambrian is replete with well-preserved, short and narrow ribbons of *Tyrasotaenia*. On the bedding plane, they lie folded, bent and twisted, suggesting thin and elastic material. They definitely differ from the typical appearance of vendotaenids in the rock. Microscopic study shows extremely thin, pale yellow, structureless ribbons of *Tyrasotaenia podolica*, which form various folds, including thin transverse compression folds. However, some ribbons exhibit an uncertain longitudinal structure.

Moreover, we have found several ribbonlike fragments displaying interlaced, narrow and long, brown filaments (Plate 33, *4*). Filaments within a ribbon show a fairly irregular pattern, being in general longitudinally oriented, and there are no traces of a cellular structure.

In the Ukraine, the Studenets Formation, approximately corresponding to the Ferapontievo Formation, Moldavia, and crowning the Vendian section, contains abundant vendotaenids throughout the succession. They are represented mainly by the same two morphological varieties of *V. antiqua* forma tertia and forma secunda; according to Ishchenko, forma quarta and the genus *Kanilovia A.* Ishchenko are scarce. It should be noted that a continuous section, at Kitaigorod village, studied by Ishchenko, has a thin unit, several tens of centimetres, at the top of the Studenets Formation, which is generally dominated by very narrow and short ribbons of *Tyrasotaenia* on the rock, but which contains large ribbons of *Vendotaenia* and tubes of Sabelliditids as well. Now it is the only known joint occurrence of *Vendotaenia* and *Tyrasotaenia* and it should be studied in detail.

The other sections of the Valdai Superseries have not been thoroughly sampled for vendotaenids, therefore, we have only a general idea about the history of their development in other regions of the East European Platform.

Within the Moscow syneclise, the Povarovo Series yields abundant typical vendotaenids represented by *Vendotaenia antiqua* forma tertia, 1.0–1.5 mm wide. Microscopic study shows well-preserved thalli with a tissuelike structure. The thickness of the thallus and hence the intensity of colouration vary from dense, red brown ribbons to thin, almost transparent pale yellow ribbons. There are thalli (Toropets, Rovno boreholes) with preserved sporangia of two types. Sporangia form long, longitudinal rows (Plate 36, *2*). These are rare multicellular sporangia of type 2 (Plate 32, *1–4*). Apart from typical structures which may be interpreted with any assurance as sporangia, very common are thalli covered by dissipated, spotted, organic clots (Plate 33, *5*) of uncertain nature, however, they may be interpreted as reproductive organs of type 1.

Very many vendotaenid thalli from the Moscow syneclise are covered with characteristic organic structures which are epibionts of Primoflagella, once entirely covering the thalli and organic films which occurred together.

The Povarovo Series of the Moscow syneclise yields abundant sapropelic films reworked to a different degree. They form an excellent substrate, on which both small organisms and fine structural features of larger organisms are preserved, otherwise, if plant remains occur in the rock, they can be destroyed by chemical reworking. For example, first *Vendotaenia* sporangia were found on a sapropelic film. Another vendotaenid genus, *Aataenia* Gnilovskaya, with a multiseriate, ribbonlike thallus was found on organic films. Tendrils of "Helminthoidichnites" Walcott are preserved on the organic substrate. Organic films fully covered by interlaced, thin filaments of *Leiothrichoides* Hermann are extremely abundant, especially in the lower Povarovo Series.

Well-preserved, abundant *Vendotaenia* were reported from the Kotlin Series, Byelorussia. However, *Tyrasotaenia* has not been recorded as yet even from the most complete sections. *Vendotaenia* from Byelorussia are mainly *V. antiqua* forma tertia but, unlike those from the Moscow syneclise, they are 1.5–2.0 mm wide. Forma quarta is also fairly common, however, ribbons are not dense and dark brown as those from Moldavia, but pale yellow. Micrographs show pale yellow thalli with a very fine, tissuelike structure. There are also dense thalli, but most of them are very thin, pale yellow to transparent. Most vendotaenids have thalli entirely covered with peculiar epibionts of Primoflagella resembling those from the Moscow syneclise.

In the Leningrad District and in the Baltic Sea area, the Kotlin Series is replete with pale brown vendotaenid ribbons associated with organic films. *Tyrasotaenia* have not been found. *Vendotaenia* are represented by *V. antiqua* forma tertia, about 2 mm wide. Under the microscope pale yellow thalli with a distinct tissuelike structure are visible.

In Poland, in the Lyublin Series, like in coeval deposits of Moldavia and the Ukraine (in the author's opinion), two successive sequences can be distinguished, viz. the lower and upper sequences containing *Vendotaenia* and *Tyrasotaenia* respectively. Well-preserved, abundant *V. antiqua* forma secunda were recorded from the Radzin borehole, depth interval 1630.1–1640.1 m. Micrographs show fine tissuelike structure of thalli. In the Kaplonosy borehole, depth interval

1354.2–1372.0 m, rocks are overcrowded with well-preserved *V. antiqua* forma tertia. Well-preserved thalli exhibiting a distinct tissuelike structure are visible on micrographs. The upper sequence is exposed in the Radzin borehole at a depth interval of 1612.1–1612.7 m. The bedding plane is covered by short and thin small ribbons of *Tyrasotaenia*, 0.37–0.50 mm wide. Two types of ribbons are distinguished under the microscope: various pale yellow structureless and folded *T. podolica*, like those recorded from the Upper Vendian successions of Moldavia and the Ukraine, and dense, dark brown, almost unfolded *T. tungusica*. The same holds true for the Kaplonosy borehole, where numerous ribbons of *Tyrasotaenia*, represented by thin, pale yellow, elastic *T. podolica* and dense, dark brown *T. tungusica*, were recorded from the depth interval 1340.2–1340.7 m. However, in both sections the sequence yielding *Vendotaenia* is separated by a break from that containing *Tyrasotaenia*.

The Rovno (Baltic) vendotaenids so far have been studied only from sections of Moldavia (Chok-Maidan borehole); Byelorussia (Drissa borehole, depth interval 450–459 m; Brest-17 borehole, depth interval 811–814m); Estonia (Narva borehole, depth interval 106–107 m); Moscow syneclise (Gavrilov Yam, depth interval 1908–1860 m); and the Siberian Platform, Platonovskaya Formation in the Sukhaya Tunguska River Basin, and Nemakit-Daldyn Formation in the Kotuikan River Basin, Anabar Uplift. Vendotaenids are represented by the genus *Tyrasotaenia, T. tungusica* and scarce *T. podolica*, and by the genus *Dvinia* in the lowermost Platonovskaya Formation.

Data available (Moldavia, Chok-Maidan-379, Vendian Ferapontievo Formation, Cambrian Tigech Formation) suggest that near the Precambrian-Cambrian boundary the predominant species *T. podolica* gradually gives way to *T. tungusica*, in a 10-m unit they occur together. In the Chok-Maidan borehole, in a lithologically homogeneous black shale sequence, the Precambrian-Cambrian boundary was drawn from the mass appearance of sabelliditids of the genus *Sokoloviina* Kirjanov at a depth interval of 1237–1242 m. Near this boundary, the distribution of *Tyrasotaenia* is as follows: the depth interval of 1245–1248 m is dominated by *T. podolica* and occasional *T. tungusica*; at the depth interval of 1243–1246 m *T. tungusica* starts dominating over *T. podolica*; at 1235–1237 m *T. tungusica* far surpasses *T. podolica*; at 1232–1235 m *T. podolica* disappears, here we find only *T. tungusica* and first sabelliditids. The picture slightly differs in the Precambrian-Cambrian boundary beds of Byelorussia (Stradech-17 borehole, Breast Depression, and Drissa borehole), where *Vendotaenia*, overcrowding the rocks up to the top of the Vendian, give way to *Tyrasotaenia* and *Dvinia* in the Vendian Rovno sediments (earlier the Precambrian-Cambrian boundary was placed at the base of the Rovno Horizon). It should be noted that this change apparently takes place in a continuous section, in a lithologically monotonous succession of the Stradech-17 borehole. For example, the depth interval 814–817 m still contains *Vendotaenia*, while they are absent from the depth interval 808.5–814 m, yielding abundant *Tyrasotaenia* (*T. podolica*) and *Dvinia*, as well as numerous fragments of Sabelliditida. In this section, the boundary is drawn at a depth of 814 m (Bessonova et al. 1979), thus, the first appearance of the genera *Tyrasotaenia* and *Dvinia* marks the disappearance of *Vendotaenia*. A similar drastic change is observed in the Drissa borehole, which may be attributed to a small break near the boundary.

In the Platonovskaya Formation, Sukhaya Tunguska River Basin, from the equivalents of the Nemakit-Daldyn Horizon, well-preserved *T. tungusica*, together with abundant sabelliditids, were reported. Fragments of *Dvinia* associated with abundant Sabelliditida were recorded from the Nemakit-Daldyn Formation, on the right side of the Kotuikan River, 5.5 km above the place where it falls into the Kotui River. On the Siberian Platform, the Nemakit-Daldyn Horizon is assigned to the Upper Cambrian, while on the East European Platform deposits, containing very similar organic remains, were placed at the base of the Cambrian. Therefore, the age of *T. tungusica* and *Dvinia* depends on the definition of the Precambrian-Cambrian boundary.

In conclusion, it may be said that the use of vendotaenids in the Precambrian stratigraphy allows us to recognize three successive floras, namely:

1. Redkino flora, *Eoholynia* Gn., *Orbisiana* Sokolov, *Caudina* Gn., tendrils of *Helminthoidichtes* Walcott, *Leiothrichoides* Hermann type.
2. Kotlin flora, *Vendotaenia* Gn., *Kanilovia A.* Istch., *Aataenia* Gn., *Leiothrichoides* Hermann, *Sarmenta* Gn., *Primoflagella* Gn.
3. Rovno flora, *Tyrasotaenia podolica* Gn., *T. tungusica* Gn., *Dvinia* Gn.

Chapter 6

Vendian Actinomycetes and Organisms of Uncertain Systematic Position

In 1970, when the anatomy of *Vendotaenia* was studied under the amplitude light microscope we found organisms described in 1979 as *Primoflagella* Gnilovskaya. These organisms, whose parts measure microns and fractions of a micron, cover in great abundance, reaching thousands, the thallus surface of *Vendotaenia* and associated sapropelic films. The latter and occurrences of *Vendotaenia* thalli without *Primoflagella* eliminate the statement that *Primoflagella* is an anatomic element of *Vendotaenia*. Now it is obvious that *Primoflagella* is an organism morphologically independent of *Vendotaenia*, which are epibiont with respect to it. The geological and geographical distribution of *Primoflagella* coincides with those of *Vendotaenia*; the fact that they are very common and abundant throughout the entire Valdai Superseries of the East European Platform precludes contamination from younger deposits.

For a long time, *Primoflagella* was studied only under the amplitude light microscope and, as a result, two morphological elements were recognized and tentatively named "head" and "tail" parts. The head part is elongated-ellipsoidal but varies greatly in the shape and position of a pale yellow swell on the upper end, averaging $3 \times 10\,\mu m$ in size. The head part extends more or less drastically into the tail part. The pale yellow tail part is less than $0.5\,\mu m$ wide and about $50\,\mu m$ long. The tail parts of *Primoflagella* braid the surface of the substrate, possibly burrowing into it. It remained, however, absolutely unclear what these "head" and "tail" parts are, and whether they are extended and connected in some way with each other. The study under the amplitude microscope could not provide an answer to the question. The same structure of *Primoflagella*, without any additional details, was observed under the scanning electron microscope. In 1979, this organism was described as *Primoflagella*. The question of its nature remained open.

Tangibly new results were obtained from the study under the anoptral, interference, phase-contrast light and transmission electron microscopes. Such an experience is of principal importance because following tradition, the micropaleontologists, who study Precambrian organisms, use mainly the amplitude microscope which by itself prevents the solution of many problems.

Application of the above techniques shows that tail parts of *Primoflagella* have a projection and are interconnected through a common system of fine branching filaments (several fractions of a micron wide) into which the tail parts pass. Fossil organisms with such a structure are not known as yet. To make interpretation of the nature of *Primoflagella* conclusive it was repeatedly and widely discussed among microbiologists. Despite some incompleteness and uncertainty of data, bacteriologists, whose advice we asked (Zavarzin and Kuznetsov, Institute of Microbiology, Moscow; and Kalakutsky, Institute of Physiology and Biochemistry of Microorganisms, Moscow), with assurance put *Primoflagella* under Actinomycetes. Based on modern concepts, actinomycetes are prokaryote mycelium organisms, bacteria, many of these exhibit antibiotic activity. The ancient origin of actinomycetes has been stated long ago, however, only recently we obtained reliable evidence that this group had been highly organized as early as 600 m.y. ago. The conclusion on the actinomycete nature of *Primoflagella* is consistent with the evidence on the relative paucity of the community of microorganisms, which could be represented by a single species of *Primoflagella*, a fact which may be attributed to its antibiotic activity.

Thus, *Primoflagella* consists of a vegetative part represented by the mycelium, whose finest hyphas braid the substrate like a web, and a generative part, represented by sporangiferous hyphas ("tail" parts) detaching from the mycelium and sporangia ("head" parts) sitting on sporangiophores. Mycelium hyphas are often broken into small fragments slightly separated from each other. However, their general distribution on the substrate shows with assurance that the mycelium once consisted of fairly long, branching hyphas.

Mycelium hyphas and partly sporangiophores and sporangia of *Primoflagella* are sunken into a transparent, layered film covering discontinuously the surface of *Vendotaenia*. The film is visible under the interference and phase-dark-field microscopes, as well as under the transmission electron microscope. In reflected light, one can see mineral particles. This film seems to be a fossilized mucus excreted either by *Vendotaenia* or it is the product of the activity of *Primoflagella*.

We could study the internal structure of *Primoflagella* sporangia only using the interference and transmission electron microscopes. Under the interference microscope, we can see that most sporangia have a distinct outline and smooth surface, they are homogeneous and, as a rule, exhibit no internal structure. However, some sporangia display a certain internal structure: based on the degree of visibility of the structure the latter may be classified as follows: (1) the sporangia surface exhibits longitudinal, wavy, fine, lighter lines bounding parallel rows (about ten on one side of the sporangium); (2) fine particles densely packed, arranged linearly one below the other, are seen within the rows; (3) particles begin to move as if beside each other, arranging more freely; the outline of sporangia becomes indistinct and botryoidal in shape; (4) particles are ellipsoidal, slightly bent; and (5) sporangia appear as if spilled out. Particles are individual and paired, the latter being parallel to the long axes, whose concave and convex parts are adjacent. The above features of the internal structure may be attributed to the successive degrees of sporangia maturity. Apart from individual and paired particles, we discern accumulations of particles ten-fold the size of the sporangium. By analogy with recent actinomycetes, the particles in such clusters can be stuck together by mucus excreted by the actinomycetes. The nature of the particles is not yet clear. Morever, we cannot even be sure that individual particles, their pairs and large accumulations are the same bodies we see in sporangia. The size of the particles is beyond the resolution of optical techniques, therefore, some other approaches should be used to answer these questions. Nevertheless, we believe that the described particles inside the sporangia are spores. Such recent actinomycetes as *Ampullariella* Couch and *Pilimelia* Kane, whose spores within the sporangium are aligned in parallel, vertical chains, exhibit a similar structure. The validity of such a sporangium structure in *Primoflagella* was confirmed by V.D. Kuznetsov.

However, the only evidence provided by transmission electron microscopy is inconsistent with the above statements. Fairly large, rounded bodies, aligned in a row inside the sporangium, are visible on the longitudinal section along it. If they were interpreted as spores, then the sporangium structure will resemble that of another actinomycetes genus, *Kitasatoa* Matsumae et Hata. Only further studies and the application of new techniques will help to elucidate the sporangium structure.

In conclusion, we may say that the actinomycete nature of *Primoflagella* has been proved, though some structural details remain uncertain, and the potentialities of morphological and biochemical methods have not yet been exhausted.

Hence, there is a necessity to introduce important changes into the earlier generic diagnosis of *Primoflagella*.

Genus *Primoflagella* Gnilovskaya 1979b (Plate 37, *1–6*; 38, *1, 3–6*)

Primoflagella: Gnilovskaya 1979b, p. 44, Plate XLVII, Figs. 5,6; p. 45, Text Fig. 1.

Type Species. P. speciosa Gnilovskaya; the Vendian, the Kotlin Formation; the East European Platform.

Diagnosis. The vegetative part of the organism consists of the filamentous mycelium, densely braid in the substrate; the reproductive part contains sporangia ("head" part of the organism) and sporangiferous hyphas ("tail" part) branching from the mycelium. The mycelium is tubular, very thin, 0.1–0.3 μm in diameter, of variable thickness with short thickenings two-fold thicker than common ones. The mycelium branching is distinct, monopodial, sometimes sharply differing in intensity even within small sites. Sporangiferous hyphas are of variable length, locally five times as long as sporangia. Sporangia are elongated-ellipsoidal (about 3 × 10 μm), variable in shape and arrangement of swells at the upper end of the sporangium, viz. symmetrical, asymmetrical, almost unswollen. These differences may also be attributed to the position of sporangia on the substrate. Arrangement of spores in the sporangium has not been reliably established.

Species Composition. See Type Species.

Discussion. Interference micrographs show longitudinal rows composed of densely packed, tiny particles (some fractions of a micron) one on top of the other in some sporangia. The particles can be interpreted as spores arranged in linear, parallel, vertical rows. Among recent actinomycetes, *Pilimelia* Kane and *Ampullariella* Couch exhibit a similar structure. However, the size of particles in *Primoflagella* sporangia is very small, therefore the validity of observations can be checked only by applying other techniques. Furthermore, the above proposed pattern of the structure is inconsistent with data obtained by means of the transmission electron microscopy, viz. spores in sporangia are aligned in rows like peas in a pod. Such a structure is also known in recent actinomycetes, for example, in *Kitasatoa* Matsumae et Hata.

Distribution. The Valdai Superseries, the East European Platform. Apart from representatives of the above described genus *Primoflagella*, other peculiar microorganisms were reported from the Vendian of the East European Platform. In 1979, two new genera morphologically somewhat resembling *Primoflagella*, in the meaning at that time were described. Their characteristic but not quite understandable structure and the absence of close analogies among both fossil and recent organisms prevented a reliable interpretation of their nature. The study of these microorganisms is at an early stage and at present we have no evidence contributing to present knowledge. Therefore, the description of the genera *Sarmenta* and *Caudina* is the same as that of 1979.

Genus *Sarmenta* Gnilovskaya 1979b

Sarmenta: Gnilovskaya 1979b, p. 42–43, Plate XLVII, Fig. 7.

Type Species. S. capitula Gnilovskaya; the Vendian, the Povarovo Series; the Moscow syneclise, the Vorobievo borehole.

Diagnosis. The organism is tubular, about 0.5 mm long and about 8 μm wide. The tube gradually tapers to the upper end, which is sharply and globularly widened and bent downward. Attachment to the substrate is apparently by means of short, thin processes.

Species Composition. See Type Species.

Remarks. The organisms were apparently epibionts. In our material, they occurred on the sapropelic film. However, they might well be the free ends of a filamentous colony or processes of a single organism, and the like.

Distribution. The Vendian, the Povarovo Series, the Moscow syneclise, the Vorobievo borehole.

Sarmenta capitula Gnilovskaya 1979b (Plate 36, *4*)

Holotype. IGGD, prep. No. Vor. 1029-1; the Moscow syneclise, the Vorobievo borehole, depth interval 1029–1033 m; the Vendian, the Povarovo Series.

Description. The organisms are almost tubular and occur on ovate sapropelic film (0.5 × 1.5 mm). They seem to be oriented normal to the substrate; their length (0.1 to 0.7 mm) may be attributed to different stages of ontogeny. The width at the base is 10–12 μm, decreasing towards the end to 2.5 μm. The very end of the tube globularly widens to 10 μm. In high (mature?) forms it is bent downward, while in low (young?) forms it is bent tightly down to the tube, the widened portion being somewhat flattened on both sides. They are dark brown, except for the tapering upper part in high forms, which are pale yellow. We found several high specimens with a broken (fallen off?) globular termination, one specimen had a globular termination, appearing cracked. The organisms were apparently attached to the substrate by means of short and thin processes (about 5 μm long and several fractions of a micron wide). Each organism could have about ten processes, but they are rarely visible (preserved?).

Remarks. Processes visible under the microscope seem to be too small to be able to attach (keep normal to the substrate?) such a fairly large organism. The processes could have some projection not yet found.

Distribution. As for the genus.

Material. About 50 specimens, 13 of these are very well preserved.

Genus *Caudina* Gnilovskaya 1979b

Caudina: Gnilovskaya 1979, p. 43, Plate XLVII, Figs. 1–4.

Type Species. C. cauda Gnilovskaya; the Vendian, the Redkino Series, the Moscow syneclise.

Diagnosis. The organism is elongated-ellipsoidal, smoothly tapering towards the lower end, 3–4 mm long. The upper third of the body is more intensely coloured and is commonly bent down. The entire organism is covered by a closely fitting, transparent sheath, freely drooping from the lower end of the body.

Species Composition. See Type Species.

Distribution. The Vendian, the Redkino Series; the Moscow syneclise, the Vorobievo and Dorogobuzh boreholes.

Caudina cauda Gnilovskaya 1979b (Plate 36, *1a,b, 3, 5*)

Holotype. IGGD, prep. No. Vor. 813–1; the Moscow syneclise, the Dorogobuzh borehole, depth 813 m; the Vendian, the Redkino Series.

Description. The organism is elongated-ellipsoidal, 3–4 mm long, 0.5–0.6 mm wide. The lower end smoothly tapers to 0.3 mm. The upper end is 0.7 to 1.0 mm long in different specimens, which accounts for one-third of the total length, commonly bent down. In most specimens, it is bent at an angle of 180°, thus closely fitting the main body or even lying on it. In the first case, the bend is smooth, while in the latter case it is twisted, whereby here the upper part could break off. One specimen has the top bent aside at an angle of about 90°. The entire organism is brown, the upper end is commonly darker, i.e. dark brown or even black. Sometimes, only the middle longitudinal part (about 0.4 mm wide) is coloured. The lower end has a darker rim.

The entire organism is covered by a tightly attached, thin sheath, which hangs freely beyond the lower end for about one-fourth the entire length of the organism. Because of the fineness of the sheath we are not certain whether the lower end is complete or torn. Only one specimen shows a distinct rupture of the sheath and the bundle of fine filaments, of which it probably consists. In some cases we can see that the sheath has been folded into thin, transverse folds. However, with the material available we cannot conclude that the sheath was always a part of the organism or that at some stage of evolution it passed through the rupture to the outside.

Distribution. As for the genus.

Material. About ten well-preserved specimens.

Chapter 7

Vendian Microfossils

1 Introduction

This chapter consists of six parts. The first part presents a brief account of the studies of microfossils from the East European Platform, complemented by the analysis of their assemblages from separate areas. Vendian microfossil assemblages have some specific features, enabling one to distinguish them from the Riphean and Lower Cambrian forms. The next two parts contain a detailed characterization of Vendian assemblages from the Arkhangelsk District and the Urals; the fourth part describes filamentous algae in the Volyn-Podolia area. The data on Vendian microfossils of the Siberian Platform are fairly scarce. The sixth part provides a description of silicified Vendian algae (probably containing silicified equivalents of calcareous algae from the southern Siberian Platform and Altai-Sayan area); the fifth part deals with acritarchs of the Yudoma Complex. The latter needs comment. Among acritarchs from the Yudoma Complex, representatives of the genus *Baltisphaeridium* are described; they are characteristic of the Lower Cambrian of the East European Platform. Occurrences of these forms have repeatedly cast doubt on the age of enclosing rocks (Volkova et al. 1980; Treshchetenkova et al. 1982). These acritarchs were reported from below salt deposits, penetrated by a number of wells within the Nepa-Botuoba Anteclise (Pyatiletov 1980b; Rudavskaya 1980). The deposits were commonly correlated with the Moty Formation and assigned to the Upper Precambrian (Yudomian). Occurrences of Lower Cambrian acritarchs, according to some workers (Volkova et al. 1980), can confirm the Lower Cambrian age of the enclosing rocks; whereas others (Pyatiletov 1980b) believe that they appeared on the Siberian Platform in Precambrian (Yudomian) time. However, such an assumption is inconceivable. Acanthomorphic acritarchs of the genus *Baltisphaeridium* appear in beds containing trilobites known not only from the East European Platform, Kazakhstan (Korolev and Ogurtsova 1982), the Yunan Province of China (Xing 1982), Scandinavia (Vidal 1981a,b) and northwestern Canada (Walton 1980). It is unlikely that the southern Siberian Platform was a specific province where these forms appeared earlier. The assignment of deposits containing acritarchs of the genus *Baltispaeridium* to Yudomian has not been proved.

2 Acritarchs and Other Plant Microfossils of the East European Platform

The study of Vendian acritarchs on the East European Platform began only 30 years ago. They were first studied by S.N. Naumova (1951) from the Asha Formation, the western slope of the southern Urals and a little later by B.V. Timofeev (1952) from the oldest deposits of Moldavia, which equate to the Ushitsa Formation of Podolia. The first investigators considered them as spores of terrestrial plants. Therefore, at the beginning of the 1950s, the Soviet publications stated that the most ancient deposits, as well as younger ones, contain abundant spores of terrestrial plants. However, at the end of this decade, B.V. Timofeev (1959) showed that many Precambrian and Lower Paleozoic microfossils belong to hystricosphaeres, a problematic group of organisms, whose representatives were long ago reported from the Lower Paleozoic deposits in Western publications. This represented major progress in the studies of the oldest microfossils.

At the beginning of the 1960s, owing to the works of several Western scientists (Downie et al. 1963), it was established that typical Mesozoic hystricosphaere genera belong to dinoflagellates. The other microfossils (mainly Paleozoic) were named "acritarchs".

Since that time, many Soviet workers placed the oldest microfossils into the acritarch group. However, B.V. Timofeev had not acknowledged the acritarch group. He proposed his own classification for the oldest fossils (Timofeev 1966) and named them microphytofossils. Starting from the 1960s, two names, i.e. "acritarchs" and "microphytofossils" and, hence, two classifications have been used in the Soviet literature for the Precambrian and Lower Paleozoic microfossils. It should also be noted that some foreign (Loeblich 1970; Cramer and Diez 1977) and Soviet (Kiryanov 1978) investigators do not apply suprageneric ranks to describe acritarchs, regarding them as unwarranted, and give genera and species in an alphabetical order.

Recently, English scientists (Diver and Peat 1979) proposed a new classification for the oldest microfossils with an organic wall. They placed them into two groups. The former, group Acritarcha, whose range they have limited, and the newly created group Cryptarcha. The latter comprises filamentous and colonial forms of microfossils, which were left beyond the limits of the acritarch group, as well as spheromorphic forms (Sphaeromorphitae) from the group Acritarcha. However, this classification has not yet gained proper recognition among specialists, though it seems quite convenient, because practically all microfossils, both extracted by maceration from Precambrian and Lower Paleozoic rocks and studied in thin sections, will fall into the groups Acritarcha and Cryptarcha. In the present chapter, the group Acritarcha is regarded in its former volume.

The main works, pertinent to the study of Vendian acritarchs, were accomplished during the last 2 decades. These studies were carried out by Naumova (1960, 1968, 1974) and Timofeev (1966–1977). Naumova (1968) recognized eight acritarch assemblages from the Vendian of the Russian Platform. However, these assemblages were distinguished from isolated sections and have not been confirmed. Timofeev (1966) distinguished assemblages, characteristic of the Gdov and Laminarite Beds of the Baltic area, and from the Asha Series of the Southern Urals.

From the Vendian of the Ukraine (Timofeev 1973a) he recorded three phytozones based on microphytofossils.

During this period, a large team of scientists studied Vendian acritarchs: Andreeva (1960, 1966, 1973a,b); Shepeleva (1963, 1969, 1973, 1974); the author (Volkova 1964, 1968, 1973); Aseeva (1971, 1974, 1976); Chepikova (1971a,b, 1973; Lagutenkova and Chepikova 1976); Golub (1974, 1979; Balashova et al. 1973), Paškevičiene (1976, 1980; Gnilovskaya et al. 1979); Piskun (Bessonova et al. 1979) and Jankauskas (1980). The above authors gave a rather detailed micropaleophytological description of the Upper Vendian deposits (Redkino and Kotlin Horizons) of the Baltic area, the Leningrad District, the Moscow Syneclise, Volyn-Podolia area, the Urals and Cis-Ural area.

Before we pass on to the description of Vendian microfossil assemblages, it is necessary to dwell briefly upon microfossils from the Pachelma Series, underlying Vendian deposits. The papers by Shepeleva and Timofeev (Shepeleva and Timofeev 1963; Shepeleva 1967, 1974) are rather informative. According to their data, the Vedenyapino, Vorona and Krasnoe Ozero Formations, the Pachelma Series, contain *Kildinella hyperboreica* Tim., *K. sinica* Tim., *Leiosphaeridia aperta* (Schep.), *L. gigantea* (Schep.), *L. minor* (Schep.), *L. pelucida* (Schep.), *L. warsanofiewae* (Naum.), *Brochopsophosphaera minima* (Schep.) and *Oscillatorites wernadskii* (Schep.). All these species, except for a local taxon, Brochopsophosphaera minima, were reported from the Vendian, some of them (*L. aperta, L. minor, L. pelucida*) being quite abundant.

Microfossils of the Laplandian Horizon (sections of Laplandian and Volyn type) have not been adequately studied. Acritarchs from sections of the Laplandian type are known only from the Churochnaya Formation on the western slope of the northern Urals. There, Golub (Borovko and Golub 1966) found single specimens of *Kildinella sinica*, widely reported from the Riphean and Vendian of the USSR (Timofeev 1966).

Acritarchs from sections of the Volyn type were studied by Aseeva from the Grushka (Berestovets) Formation, the Volyn Series. They are represented by two species: *Leiosphaeridia aperta* and *L. parva* Aseeva. The former occurs in the Proterozoic, Upper Precambrian and Cambrian of Eurasia (Aseeva 1976); and the latter, in Vendian of the Ukraine and Byelorussia, Lower Cambrian (Baltic Series) of the Ukraine, Byelorussia, Baltic area (Aseeva 1976; Paškevičiene 1980). In addition, Aseeva and Piskun (Bessonova and Chumakov 1979) also reported on *Protosphaeridium aff. parvulum* Tim. (= *Leiosphaeridia* minor) and *Oscillatorites* sp. from the Svisloch Formation, the Volyn Series, the Orsha Depression; Shepeleva (1974) recorded very small *Leiominiscula* and individual *Orygmatosphaeridium rubiginosum* Andr. The first three taxa occur in the Riphean, Vendian and Lower Cambrian of the East European Platform; the last one is known from the Vendian and Lower Cambrian (Naumova 1960; Shepeleva 1967, 1974). Thus, the Lower Vendian deposits yield seven microfossil species of an extensive stratigraphic range.

Acritarchs of post-Laplandian Vendian time are best known from the western East European Platform; to a lesser degree, from the western slope of the Urals and Cis-Ural area. Analyzed below are Upper Vendian acritarch assemblages from three areas: (1) the northwestern and central East European Platform (Baltic area,

Byelorussia, Leningrad District, Moscow Syneclise); (2) the southwestern East European Platform (Volyn-Podolia area); (3) the western slope of the Urals and cis-Ural area.

Upper Vendian deposits in the northwestern and central East European Platform are represented by the Valdai Series. In the last few years, the most important studies of acritarchs from the Valdai Series in this region were accomplished by Shepeleva (1974); Paškevičiene (1976, 1980; Gnilovskaya et al. 1979); Golub (1979), Ragozina (Sect. 3) and the author Volkova 1964, 1968, 1973). All the above workers distinguish two acritarch assemblages, viz., Redkino and Kotlin, the Valdai Series.

According to Shepeleva, the Redkino assemblage, which she distinguished in the Redkino Formation of the Moscow Syneclise, contains the following acritarchs: *Orygmatosphaeridium rubiginosum*; and the filamentous forms: *Oscillatorites wernardskii.*

Paškevičiene studied acritarchs of the Redkino sediments (Redkino Subseries) from the northern Baltic area. She found small, thin-walled acritarchs, *Leiosphaeridia culta* (Andr.), *L. minor*; acritarchs with a thick wall (*L. undulata* Tim.), and *Leiosphaeridia* sp., *Orygmatosphaeridium rubiginosum* from sediments corresponding to the Redkino Horizon; among the filamentous forms, *Leiothrichoides typicus* Hermann and *Oscillatorites wernadskii* were determined. Golub reported on problematic microfossils, assigned to a special group *Rudnjana* from the Vendian of the Orsha Depression, these deposits are equivalents of the Redkino and Gdov Formations.

Acritarchs of the Redkino Horizon were studied by the author from the Redkino Formation of the Moscow Syneclise (Redkino borehole, 1732.0–1636.0 m; Volkova 1964 and this Vol.). The following acritarchs are most common in the Redkino assemblage: *Bavlinella faveolata* Schep. (Plate 41, 8,9), small leiosphaerids with a thin wall (*Leiosphaeridia minor,*[1] Plate 39, *1–4, 8–9*) and dark yellow to red-brown opaque forms of Orygmatosphaeridium rubiginosum (Plate 40, *6–9*). Larger, thin-walled leiosphaerids (*L. pelucida*; Plate 39, *5–7*) and leiosphaerids with a small thickened wall (*L. aperta*, Plate 40, *1–3,5*) are less common. The author (Volkova 1974b) still believes that *Bavlinella faveolata* forms are secondary structures, which appeared from organic walls modified by pyrite framboids. It should be also noted that *Orygmatosphaeridium rubiginosum* were not mentioned in the first publication of the author (Volkova 1964), since they were interpreted as strongly modified plant remains, whose original morphology cannot be reconstructed. However, the majority of workers disagree with this statement; therefore, we just note their occurrence in the Redkino acritarch assemblage.

Filamentous forms of the Redkino Horizon are represented by numerous narrow (10–15 μm), non-septate fragments (Plate 41, *1,2*), rare *Oscillatorites* sp. (Plate 41, *3*) with a preserved cellular structure and individual fragments of relatively broad (40–90 μm) filaments with easily discernible cellular frameworks

[1]Species names of leiosphaerids were approved by the Colloquium on the Revision of Systematics of Sphaeromorphic Acritarchs held on September 15–26, 1980, in Irkutsk.

(Plate 41, *7*; filament, about 90 μm wide; individual cells, about 10 μm long; *10*, filament about 40 μm wide; individual cells, about 35 μm long). Besides, the Redkino assemblage contains *Rudnjana* (Plate 41, *4–6*).

Thus, the Redkino Horizon in the northwestern and central parts of the platform is noted for abundant, small, thin-walled leiosphaerids, *Leiosphaeridia minor*, dense opaque forms of *Orygmatosphaeridium rubiginosum* and fragments of narrow filamentous, non-septate forms. *Rudnjana* and the genus *Oscillatorites* are common.

A much more diverse microfossil assemblage was distinguished by Ragozina and Sivertseva (Sect. 3) from the Zimnie Gory Beds, the Ust Pinega Formation, the Zimny Coast of the White Sea. They assign these microfossils to the Redkino Horizon. The assemblage contains representatives of *Leiosphaeridia, Trachysphaeridium, Stictosphaeridium, Orygmatosphaeridium, Bavlinella, Favososphaeridium, Stratimorphis, Synsphaeridium* and *Symplassosphaeridium*. Filamentous forms are represented by the species *Leiotrichoides, Polytrichoides, Tortunema, Oscillatoriopsis, Oscillatorites* and *Volyniella*, fragments of non-septate forms and the group *Rudnjana*. It contains all forms, typical of the Redkino Horizon in the northwestern and central parts of the platform. A high diversity may be due to a peculiar environment.

Acritarchs of the Kotlin assemblage were studied from the Kotlin Formation of the Moscow Syneclise by Shepeleva (1974) and the author (Volkova 1964); at the type locality of the Kotlin Formation (northern Baltic area and Leningrad District), by Paškevičiene (1980; Gnilovskaya et al. 1979) and the author (Volkova 1968, 1973).

According to Shepeleva, the Kotlin assemblage is dominated by acritarchs with a foamy surface sculpture, belonging to the genera *Asperatopsophosphaera* and *Bavlinella faveolata*, and filamentous forms of *Oscillatorites wernadskii*. The associated forms are thin-walled acritarchs, *Leiosphaeridia minor, L. pelucida*, and filaments with a rhomboid surface sculpture.

According to Paškevičiene, the Kotlin assemblage (Gdov and Kotlin Formations) at the type locality contains acritarchs, represented by seven species of the genus *Leiosphaeridia* with a thin or thickened wall: *Leiosphaeridia pelucida, L. culta, L. effusa* (Schep.), *L. aperta, L. infriata* (Andr,), *L. parva* and *Leiosphaeridia* sp. 1.

Filamentous forms are represented by *Leiotrichoides typicus, Oscillatorites wernadskii* and trichomes with a characteristic rhomboid striation. Equivalents of the Kotlin Formation in southwestern Byelorussia, in addition to the above species, also yield abundant *Leiosphaeridia parva*, spirallike forms of the species *Volyniella*, and single *Leiovalia* sp.

Acritarchs, filamentous forms and other microfossils from the Kotlin Horizon in Latvia, Estonia, Leningrad District and the Moscow Syneclise, studied by the author, are illustrated in Plates 42–49. Acritarchs are dominated by leiosphaerids. They are represented by numerous, thin-walled forms, viz., *L. minor* (Plate 42, *1*), *L. pelucida* (Plate 42, *3,4*; Plate 44, *10, 13*), *L. gigantea* (Schep.) (Plate 43, *6*), and forms with a thickened wall, viz., *L. aperta* (Plate 42, *2,5–7,9,10*; Plate 44, *12*), *L. effusa* (Plate 42, *8,13–15*; Plate 43, *1–5*). Walls with a median rupture, *L. bicrura* Jank. (Plate 42, *11,12*), are sometimes recorded; they were first reported from the

Kotlin Horizon, the Leningrad District and Latvia. Prior to this, such walls on the East European Platform occurred only starting from the Rovno Horizon. Small acritarchs with processes, i.e. *Micrhystridium* sp. 1 (Plate 48, *5,6*) and *M. tornatum* Volk. (Plate 48, *7,8*), are quite rare. The former species is known only from the Leningrad District, the latter one, from the Redkino borehole. There are few *Orygmatosphaeridium rubiginosum* (Plate 44, *11*) and *Bavlinella faveolata* (Plate 48, *9,11*). Filamentous forms are dominated by fragments of smooth, non-septate filaments, shown in Plate 45 (*1-8*) and Plate 47 (*6-9*). Fragments of such filaments, tapering at the ends, are less common (Plate 48, *1-3*). Filaments with a preserved cellular structure, i.e. *Oscillatorites* div. sp. (Plate 46, *1-5,8*) and filaments with annular thickenings (Plate 46, *6,7,9*) are fairly scarce. Fragments of filamentous forms with longitudinal thickenings, creating a rhombic pattern by intersection (Plate 47, *1-5*) and microfossils of group *Rudnjana* (Plate 49, *1-11*) are quite common. The latter were reported from the Kotlin Horizon of the area for the first time. They have been already recorded in the Kotlin Formation of Latvia, Leningrad District, and Moscow Syneclise. *Tortunema* sp. (Plate 48, *4, 10*) and filament bundles (Plate 48, *12*) are rare. We should note also some occurrences of small spores with a tetrad scar (Plate 44, *1-9*), recorded only from the Kotlin Formation of Latvia and Leningrad District.

Thus, the Kotlin Horizon of the northwestern and central parts of the platform is noted for abundant leiosphaerids with a thin and thickened wall. [2]They are represented by nine species: *Leiosphaeridia aperta, L. bicrura, L. culta, L. effusa, L. gigantea, L. infriata, L. minor, L. pelucida* and *Leiosphaeridia* sp. 1 (Paškevičiene 1980). All the species survived in the Cambrian. Among the filamentous forms typical are smooth, non-septate filaments, filaments with a rhomboid pattern, different species of *Oscillatorites*, and *Rudnjana*. Rare specimens of small acritarchs with processes also appear, namely *Micrhystridium tornatum* and *Micrhystridium* sp. 1, and spores of *Ambiguaspora parvula* Volk.

Microfossils from the southwestern East European Platform (Volyn-Podolia area) were recently studied by Aseeva (1976; Aseeva et al. 1979; Velikanov: see Vol II of this book). She distinguished six microfossil assemblages in the Vendian of the Volyn-Podolia area. The first one was recorded from the Grushka Formation, the Drevlyane Horizon; the second, third and fourth ones were recognized respectively, in the Mogilev, Yaryshev and Nagoryany Formations of Podolia, or the Chartoryjsk, Roznich and Kolki Formations of Volyn, and were assigned to the Redkino Horizon. The fifth and sixth assemblages are known from the Kanilov Series (Kotlin Horizon): the fifth one, from the Danilov and Zharnovka Formations; the sixth one, from the Krushanovka and Studenitsa Formations.

It should be noted that Timofeev (1973a) provides a different micropaleophytological characterization of Vendian formations in the Volyn-Podolia area. He placed the Nagoryany Formation of the Redkino Horizon and the Kanilov Formation of the Kotlin Horizon into a single phytozone (Podoliella and Ven-

[2]Forms of the genus *Asperatopsophosphaera* with a foamy surface sculpture, are poorly preserved representatives of the genus Leiosphaeridia, since the foaminess of their walls is not primary and cannot be regarded as morphological character.

dotaenides). This may be attributed to a different approach to the interpretation of Vendian acritarch morphology.

Below, the assemblages, corresponding to the Redkino and Kotlin Horizons, are discussed together and based on Aseeva's data, since only two assemblages are known from other parts of the East European Platform. In the Redkino Horizon of the Volyn-Podolia area, Aseeva discovered the following acritarchs: *Leiospaeridia aperta, L. asapha* (Tim.), *L. effusa, L. gigantea, L. jacutica* (Tim.), *L. laccata* (Tim.), *L. minor, L. pelucida, L. pruniformis* Aseeva, *Favososphaeridium* sp., *Orygmatosphaeridium induratum* Aseeva, *O. rubiginosum, Polycavita bullata* (Andr.), *P. concentrica* Aseeva, *Stictosphaeridium sinapticuliferum* (Tim.), *Trachysphaeridium bavlense* (Schep.), *T. magnum* (Schep.), *T. partiale* (Schep.), *Trematosphaeridium holtedahlii* and *Stratimorphis plana* Aseeva. The Redkino acritarchs are illustrated in Plate 55, *1,2,5–16* (collection by Aseeva). It should be noted that among the above 20 species, representatives of the genus *Trachysphaeridium* are poorly preserved forms of the genus *Leiosphaeridia*; four species are local, and the remaining ones occur in the Upper Precambrian and Cambrian of the USSR. Acritarchs of the Redkino Horizon in the Volyn-Podolia area are more diverse as compared to those from the northwestern and central parts of the platform, where the Redkino assemblage comprises only eight species. The following species are common to both regions: *Leiosphaeridia aperta, L. minor, L. pelucida*, and *Orygmatosphaeridium rubiginosum*. Filamentous forms of the Redkino Horizon of the Volyn-Podolia area are also rather diverse (Plate 56, *1–13*; collection by Aseeva). These are forms of wide stratigraphic and geographic distribution (*Arctacellularia* sp., *Leiotrichoides* sp., *Paleolyngbia catenata* Hermann, *Polytrichoides* sp., *Rugosoopsis* sp., *Trachytrichoides ovalis* Hermann and *Oscillatorites wernadskii*). Local species include *Bicuspidata fusiformis* Aseeva, *Circumiella mogilevica* Aseeva, *C. podolica* Aseeva, *Ljadovia perforata* Aseeva, *Striatella coriacea* Aseeva, *Taenitrichoides jaryschevicus* Aseeva, *Tubulosa corrugata* Aseeva, *Volyniella canilovica* Aseeva and *V. valdaica* Aseeva, and taxa, presently known only from the Vendian of the East European Platform (group *Rudnjana*), and smooth, non-septate fragments of filaments (Aseeva 1976, Plate XVII, Figs. 4,9).

Therefore, acritarchs and filamentous forms of the Redkino assemblage in the Volyn-Podolia area, as compared to those from the northwestern and central parts of the platform, exhibit a much higher diversity of species due to the presence of many species, unknown outside the region. The latter may be explained by peculiar local environments, and also by the fact that recognition of many local species has not been well substantiated. However, all the forms typical of the Redkino Horizon in the northwestern and central parts of the platform, are present there, i.e. *Leiosphaeridia minor, Orygmatospharidium rubiginosum, Rudnjana*, representatives of *Oscillatorites* genus and fragments of filamentous, non-septate forms.

The composition of acritarchs of the Kotlin Horizon in the Volyn-Podolia area is not so diverse. The following forms were defined there: *Leiosphaeridia effusa, L. parva* Aseeva, *L. undulata* Tim., *L. volynica* Tim., *Orygmatosphaeridium* sp., *Podoliella irregulare* Tim., *Podoliella* sp., *Stictosphaeridium implexum* Tim. and *Trachysphaeridium magnum* (Schep.). Among them, local taxa, described by Timofeev and Aseeva from the Upper Vendian of the Volyn-Podolia area, are species of the genera *Podoliella, Leiosphaeridia undulata* and *L. volynica*. The species *L. parva* is known from the Lower Vendian of the region, and from the

Upper Vendian of Byelorussia, Lower Cambrian of the Ukraine, Byelorussia, and Baltic area (Paškevičiene 1980). The species *Strictosphaeridium implexum* is known from the Riphean, Vendian and Cambrian of different regions of the USSR, and from China, Poland and Spitsbergen (Timofeev 1966). The genus *Trachysphaeridium* was distinguished on the material of poor preservation. Only the species *L. effusa* is known also from the Kotlin assemblage of the northwestern and central parts of the platform.

Filamentous forms are represented by local species: *Bicuspidata fusiformis* Aseeva, *Flagellis tenuis* Aseeva, *Studenicia bacotica* Aseeva, *Volyniella canilovica* Aseeva, and the forms, known from the Vendian of the northwestern and central parts of the platform: *Oscillatorites wernadskii*, smooth, non-septate filaments and filaments with a rhomboid pattern (Aseeva 1976; Plate 17, *10*), group *Rudnjana* (Plate LVI, Figs. 9,10; Aseeva's collection).

Thus, despite a peculiar character of the Kotlin microfossil assemblage in the Volyn-Podolia area, it comprises a number of typical Kotlin species from the northwestern and central parts: *Leiosphaeridia effusa*, the genus *Oscillatorites, Rudnjana*, filaments with a rhomboid pattern and fragments of smooth, non-septate filaments.

Upper Vendian microfossils on the western slope of the Urals and in the Cis-Urals area are described in detail by Shepeleva (1963, 1966); Golub (1974; Balashova et al. 1973) and Jankauskas (1980).

Shepeleva first described several species, now widely used to characterize the Vendian of the other regions, from the Lower and Upper Bavly Series of the Volga-Ural area (Shepeleva regards the upper Bavly Series as an equivalent to the Valdai Series of the Leningrad District). They are: *Leiosphaeridia aperta, L. effusa, L. gigantea, L. minor* and *L. pelucida*. Besides, she has found the genera *Brochopsophosphaera* and *Asperatopsophosphaera* (which are poorly preserved leiosphaerids), and *Bavlinella faveolata* and *Oscillatorites wernadskii* from the Upper Bavly sediments.

Golub established three acritarch assemblages in the Kairovo and Shkapovo Formations, the Upper Bavly Series of the Cis-Urals near the town of Perm. The first assemblage comes from the lower Kairovo Formation correlative with the Redkino deposits. It contains *Leiosphaeridia minor, L. pelucida, Bavlinella faveolata* and *Oscillatorites wernadskii*. It should be noted that opaque brown forms of the genus *Orygmatosphaeridium*, common in the Redkino Horizon of the western parts of the East European Platform, are not presented here. The second and third assemblages occur in the upper Kairovo Formation and in the Shkapova Formation. They are correlated with the Gdov and the Laminarite Beds of the western part of the platform. Both assemblages have a similar composition and contain: *Leiosphaeridia minor, L. pelucida, L. warsanofiewae*, species of *Asperatopsophosphaera, Bavlinella faveolata, Kildinella sinica, K. hyperboreica, Oscillatorites wernadskii*, etc. Generally, according to Shepeleva's and Golub's evidence, the Upper Vendian assemblages of the Cis-Urals are similar in composition to that of the Redkino and Kotlin assemblages of the northwestern and central parts of the platform.

Jankauskas (1980; see also Sect. 4) reported on interesting occurrences of the Upper Vendian microfossils from the Urals and Cis-Urals. The Zigan Formation on the western slope of the South Urals, at the level of the Kotlin Horizon, yields abundant fossils of the group *Rudnjana* and acritarchs, i.e. *Kildinella, Leiosphaeridia* and *Bavlinella faveolata.* Of utmost interest are occurrences of *Rudnjana,* typical of the Vendian of western regions of the East European Platform.

The second locality of the Upper Vendian microfossils was discovered by Jankauskas in the Bashkirian of the Cis-Urals, the Baikibashevo (Lower Kairovo) Formation at the level of the lower Redkino Horizon (Sergeevskaya-800 borehole, 2942.4–2946.4 m). He reported on a peculiar microfossil assemblage, containing many Riphean species, and those described until now only from the given locality. The genus *Toromorpha* (= *Circumiella*) is the only Vendian taxon, known from the Redkino Horizon of Volyn-Podolia. A peculiar type of this assemblage may be accounted for by the facies environment. Probably, the deposits, occurring at the stratigraphic level of the Baikibashevo Formation, are older and are absent from the western regions of the East European Platform.

Vendian acritarchs of the Siberian Platform have been not adequately studied. However, brief information, available from the literature, enables the conclusion that Vendian acritarch assemblages of the Siberian and East European platforms are similar in composition. On the Siberian Platform, Vendian acritarchs were studied by Faizulina, Treshchetenkova, Rudavskaya and Pyatiletov (Volkova et al. 1980). They were reported from the lower Moty Formation (below the Markovo Horizon), from the lower Kursov and Borulakh Formations of the Nepa-Botuoba zone, as well as from the carbonate (upper) part of the Kamovsky Series and the Kuyumba Formation of the Kuyumba field. The deposits contain many acritarch species, widespread in the Vendian of the East European Platform: *Leiosphaeridia pelucida, L. gigantea, L. minor, Orygmatosphaeridium* sp., *Trachysphaeridium bavlensum*, etc. The Moty Formation, yielding Lower Cambrian (*Baltisphaeridium* genus, etc.) acritarchs, typical of the East European Platform, was assigned to the Lower Cambrian.

Acritarchs of the Baltic Series differ markedly from the Vendian forms. They have been adequately studied from both the western and central parts of the East European Platform, were the Baltic Series on the basis of body fossils and acritarchs is divided into the Rovno and Lontova Horizons. For detailed characterization of acritarch assemblages from these horizons, the reader is referred to recent publications (Volkova et al. 1979; Paškevičience 1980). The lower Rovno acritarch assemblage is similar to the Vendian assemblage in having numerous representatives of the genus *Leiosphaeridia* (Plate 50, *1–9, 11;* Plate 51, *5,6, 8–21*). Besides, as in the Vendian, acritarchs with processes are represented here by *Micrhystridium tornatum* (Plate 51, *1–4, 7*); however, their number abruptly increases. The Rovno assemblage clearly differs from the Vendian assemblage in the appearance of nine new species, among which *Leiosphaeridia dehisca* Pašk. (Plate 51, *13, 14, 16–19*), *Teophipolla lacerata* Kirjan., *Ceratophyton vernicosum* Kirjan. (Plate 52, *6*) and *C. duplicum* Pašk. (Plate 52, *5*) are very abundant throughout the entire study area. The other species, viz., *Disnella forcipata* Pašk., *Retisphaeridium densum* Pašk., *Leiosphaeridia pylomifera* Pašk., *Leiovalia striatella* Pašk. and *Navifusa* sp. are less common and have, as yet, been recorded only from the Baltic area (Paškevičiene 1980).

Filamentous bodies of the Rovno Horizon are represented by forms surviving from the Valdai Series. They are: *Leiothrichoides typicus* (Plate 52, *1*), fragments of smooth, non-septate filaments (Plate 52, *2,8,9*), fragments of filaments with annular thickenings (Plate 52, *3,4*) and *Oscillatorites* sp. (Plate 52, *11*). Tubes with annular thickenings (Plate 52, *10*) first appear at this level.

The Lontova Horizon is marked here by the first appearance of acritarchs with a zone, i.e. *Granomarginata prima* Naum. (Plate 53, *4,5*), *G. squamacea* Volk. (Plate 53, *1,2*), *Leiomarginata simplex* Naum. (Plate 53, *8*), and those with porecanals, viz. *Tasmanites tenellus* Volk. (Plate 54, *8,9*). Besides, *Dictyotidium birvetense* Pašk., *Leiosphaeridia cerebriformis* Volk., *Brestovia annulata* Pašk., *Pulvinosphaeridium antiquum* Pašk. (Plate 53, *11*), forms of type A (Plate 54, *2*), and forms of type B (Plate 53, *7*) appear at this level, their occurrence is confined to the Baltic area. Leiosphaerids are abundant here (Plate 54, *1,3–5*); and some forms with a granular surface are encountered (*Leiosphaeridia* sp. 1, Plate 53, *9,10*). Filamentous bodies contain forms known from the Valdai Series and the Rovno Horizon, viz. fragments of non-septate filaments (Plate 54, *6,7*), fragments with annular thickenings (Paškevičiene 1980; Plate 51, *5*), *Leiothrichoides typicus* (Paškevičiene 1980; Plate 51, *6*) and *Oscillatorites wernadskii* (Paškevičiene 1980; Plate 51, *3*).

Acritarchs of the Rovno Horizon are known only from the East European Platform; those of the Lontova Horizon occur in Siberia (Ogurstova 1975a) and Kirghizia (Ogurtsova 1975b) as well.

The following conclusions can be drawn:

1. Rare specimens of microfossils, represented by six species (Table 3), occur in the Lower (Laplandian) Vendian of the East European Platform. Four species continue from the Riphean, and two species (*Orygmatosphaeridium rubiginosum* and *Leiosphaeridia parva*) appear in sections of the Volyn type. However, the appearance of these species cannot be considered compelling evidence for placement of the lower boundary. The nature of *O. rubiginosum* is doubtful, and recognition of species from the genus *Leiosphaeridia* as yet has no solid footing.
2. On the East European Platform, the post-Laplandian Vendian yields abundant microfossils. The Redkino and Kotlin Horizons are dominated by leiosphaerids (*Leiosphaeridia aperta, L. pelucida* and *L. minor*) and filamentous forms (*Oscillatorites div.* sp., *Leiothrichoides typicus* and fragments of smooth, non-septate filaments) from the Pachelma Series (see Table 3). These horizons are characterized by the group *Rudnjana,* whose range is limited only by this stratum. As noted above, large leiosphaerids with a thickened wall (*L. effusa*) are known to appear from the Redkino Horizon, and hence cannot be regarded as typical forms. The Kotlin microfossil assemblage, unlike the Redkino assemblage, contains filaments with a rhomboid pattern (as yet found only from the Kotlin Formation throughout the entire platform). Some sections of the western and central parts of the platform are marked by the rare specimens of leiosphaerids with a median rupture (*Leiosphaeridium bicrura*), small acritarchs with processes (*Micrhystridium tornatum, Micrhystridium* sp. 1), and small spores with a tetrad scar (*Ambiguaspora parvula*).
3. The lower boundary of the Rovno Horizon is drawn by the appearance in the section of characteristic leiosphaerids (*L. dehisca*), conelike walls of Ceratophyton, ovate *Teophipolia* and by an abrupt increase in the number of *Micrhystridium tornatum.*

Table 3. Vertical range of characteristic Vendian and Lower Cambrian microfossil species on the East European Platform[a]

Riphean	Vendian					Cambrian	Species
Pachelma Series	Laplandian		Redkino	Kotlin	Rovno	Lontova	
	Laplandian type	Volyn type					
●							*Brochopsophosphaera minima* Schep.
●			x	x		x	*Kildinella hyperboreica* Tim.
●	●		x	x			*Kildinella sinica* Tim.
●		●	●	●	●	●	*Leiosphaeridia aperta* Schep.
●			●	●	●	●	*Leiosphaeridia pelucida* Schep.
●		●	●	●	●	●	*Leiosphaeridia minor* Schep.
x			x	x			*Leiosphaeridia gigantea* Schep.
●		●	●	●	●	●	*Oscillatorites* div. sp.
●				●			*Leiosphaeridia warsanofiewae* Naum.
○			●	●	●	●	*Leiotrichoides typicus* Herm.
○			●	●	●	●	Fragment of non-septate, smooth filament
		●		●	●	●	*Leiosphaeridia parva* As.
		●	●	●	●		*Orygmatosphaeridium rubiginosum* Andr.
			●	●			*Rudnjana*
			●	●	●	●	*Leiosphaeridia effusa* Schep.
				●			Filament fragment with a rhomboid pattern
				x			*Ambiguaspora pervula* Volk.
				x			*Micrhystridium* sp. 1
				x	●	●	*Leiosphaeridia bicrura* Jank.
				x	●	●	Micrhystridium tornatum Volk.
					●	●	*Leiosphaeridia dehisca* Pašk.
					●	●	*Ceratophyton vernicosum* Kirjan.
					●	●	*Ceratophyton duplicum* Pašk.
					●		*Teophipolia lacerata* Kirjan.
						●	*Granomarginata squamacea* Volk.
						●	*Granomarginata prima* Naum.
						●	*Leiomarginata simplex* Naum.
						●	*Tasmanites tenellus* Volk.
						x	*Pulvinosphaeridium antiquum* Pašk.

[a] Symbols: ●, common occurrences; x, rare occurrences; ○, occurrences reported from other regions.

4. The upper boundary of the Vendian is easily established by the appearance in the section of acritarchs with a zone, characteristic of the Lontova Horizon of the Baltic Series (Granomarginata, Leiomarginata), porecanals (*Tasmanites tenellus*) and peculiar *Pulvinosphaeridium antiquum*.

3 Microfossils of the Valdai Series in the Northwestern Arkhangelsk District

The northwestern Arkhangelsk District is underlain by deeply metamorphosed rocks of the crystalline basement and thick Riphean and Vendian rocks of the sedimentary cover. Vendian deposits (up to 1 km thick) rest unconformably on the erosion surface of rocks, assigned to the Middle and Upper Riphean, or directly on the basement rocks. According to the correlation chart accepted by the Interdepartmental Stratigraphic Committee of the USSR in 1974, these deposits are assigned to the Valdai Series and divided into the Redkino and Kotlin Horizons. The Redkino Horizon comprises the Pletnev and Ust Pinega Formations; the Kotlin Horizon includes the Mezen and Padun Formations (former Lyubim and Reshma Formations).

Deposits assigned to the Redkino Horizon are made up mainly of grey and greenish-grey clay-mudstone rocks and subordinate siltstone and sandstone. The Kotlin deposits are made up predominantly of red sandstone and siltstone. The discovery of unique fauna of most ancient non-skeletal organisms, Metazoa, in the Vendian (Keller and Fedonkin 1976; Fedonkin 1978; Sokolov 1977a, 1978), abundant microfossils and the presence of tuffaceous marker horizons made it possible to improve the detailed stratigraphic chart for the Upper Precambrian of the northwestern Arkhangelsk District (Stankovsky and Verichev 1981 and others).

Samples for micropaleophytological analysis were taken from boreholes and natural exposures by Stankovsky and Verichev (Arkhangelsk PGO), Sivertseva (Leningrad State University) and Ragozina (Paleontological Institute, Moscow). Vendian deposits of the northern Arkhangelsk District yield abundant microfossils of plant origin; as a result, four assemblages: Lower Redkino, Redkino, Upper Redkino and Kotlin, were distinguished. Their study shows that microfossils are mainly confined to clay-silty facies; their changes were traced throughout the entire Valdai Series (Upper Vendian) (see Table 4).

The lowermost horizons of the Valdai Series, the Tamitsa and Lyamtsa Beds (stratigraphic subdivision, proposed by a team of geologists working in Arkhangelsk PGO) (Stankovsky et al. 1981), yield a rather poor assemblage of microfossils, known as the Lower Redkino assemblage.

3.1 Assemblage I (Plate 57) (Lower Redkino)

The Lyamtsa and Tamitsa Beds (boreholes 41,45,11,1, outcrops near Lyamtsa Village) are characterized predominantly by microfossils of the genus *Leiosphaeridia* Eis. *Leiosphaeridia laccata* (Tim.) As., *L. aperta* (Schep.) and *L. asapha*

Table 4. Stratigraphic distribution of plant microfossils in the Vendian of the northwestern Arkhangelsk district[a]

Vendian								
Redkino Horizon							Kotlin Horizon	
Tamitsa Beds	Lyamitsa Beds	Arkhangelsk Beds	Verkhovsk Beds	Syuzma Beds	Vaizitsa Beds	Zimnie Gory Beds		Microfossils
1	2	3	4	5	6	7	8	9
x	x	x	x	x				*Kildinella sinica* Tim.
x	x	x			x	x		*K. hyperboreica* Tim.
o	o	o		o		o	o	*Leiosphaeridia minor* Schep.
o	o	o		o		o	o	*L. aperta* Schep.
		o	o	o	o	o	o	*L. pelucida* Schep.
						o	o	*L. culta* Andr.
o	o	o	o	o				*L. laccata* Aseeva
		o	o	o		o		*L. bituminosa* Tim.
				o	o	o	o	*L. effusa* Schep.
				o		o		*L. giganthea* Schep.
x	x	x	x					*Protosphaeridium densum* Tim.
o	o	o	o	o		o		*Trachysphaeridium laminaritum* Tim.
						o	o	*T. magnum* Schep.
	o	o	o	o	o	o	o	*T. bavlense* Schep.
		o		o		o		*T. partialum* Schep.
		o				o		*Stictosphaeridium pectinale* Tim.
		o				o	o	*S. sinapticuliferum* Tim.
o	o	o				o		*S. implexum* Tim.
x	x	x	x	x		x		*Favososphaeridium favosum* Tim.
						x		*Stratimorphis plana* As.
		o	o	o	o	x	o	*Orygmatosphaeridium rubiginosum* Andr.
x	x	x				x		*Podoliella regulare* Tim.
				x	x	x		*Bavlinella faveolata* Schep.
		o	o	o			o	*S. sorediforme* Tim.
		x	x			x		*Synsphaeridium bullatum* Andr.
		o				o	o	*S. conglutinatum* Tim.
			o			o		*Symplassosphaeridium* sp.
x	x	x	x			x	x	*Nucellosphaeridium* sp.
		x	x					*Pterospermopsimorpha insolita* Tim.
	x	x				x		*Volyniella valdaica* (Schep.) As.
						x	x	*V.* sp.
				x		x		*Leiominuscula minuta* Naum.
		x	x			x		*Gloeocapsomorpha* sp.
				o		o	o	*Leiotrichoides typicus* Herm.
						x		*L.* sp.
						o		*Polytrichoides lineatus* Herm.
						o		*Tortunema sibirica* Herm.
						o		*Oscillatorites wernadskii* Schep.
						o		*Oscillatoriopsis magna* Tynni et Bonner
						o	o	*O. constricta* T.et B.
							o	*O. rhomboidalis* sp. nov.
						o		*O. funiformis* sp. nov.
						x		*Palaeolyngbia* sp.
		x	x	x	x	o		*Rudnjana* Golub

[a] Symbols· o abundant occurrences: x, rare occurrences.

(Tim.). *Kildinella sinica* Tim., *K. hyperboreica* Tim., *Protosphaeridium densum* Tim., *Stictosphaeridium implexum* Tim., *Trachysphaeridium laminaritum* Tim., *Favososphaeridium* sp., *Podoliella regulare* Tim., *Nucellosphaeridium* sp. and *Pterospermopsimorpha insolita* Tim. are scarce.

3.2 Assemblage II (Plates 57, 58) (Redkino)

1. *Arkhangelsk Beds* (boreholes 1,11,179,202). Its composition is more diverse. Microfossils belong mainly to two genera: *Leiosphaeridia* Eis. and *Trachysphaeridium* Tim. The following species were distinguished: *Leiosphaeridia pelucida* (Schep.), *L. laccata* (Tim.) As., *L. aperta* (Schep.), *L. bituminosa* Tim., *L. minor* (Schep.) *Trachysphaeridium laminaritum* Tim., *Tr. bavlense* (Schep.), *Tr. partialum* (Schep.) and *Tr.* sp. Besides, there are *Stictosphaeridium implexum* Tim., *St. pectinale* Tim., *St. sinapticuliferum* Tim., *Protosphaeridium densum* Tim., *Orygmatosphaeridium rubiginosum* Andr., *Favososphaeridium favosum* Tim., *Nucellosphaeridium* sp., *Pterospermopsimorpha insolita* Tim., *Podoliella regulare* Tim., *Synsphaeridium conglutinatum* Tim., *S. sorediforme* Tim., *S. bullatum* Tim., *Volyniella valdaica* As., *Gloecapsomorpha* sp. and *Pterospermopsimorpha binata* Tim. The Arkhangelsk Beds are marked by the first appearance of problematica of the group *Rudnjana* Golub. and *Volyniella valdaica* (Schep.) As. They were repeatedly reported from the Redkino deposits of the East European Platform. Brown and red-brown films of organic matter are also fairly common. They are specially abundant in the Ust Pinega borehole (from a depth of 698 m, borehole No. 1).

2. *Verkhovsk Beds* (Boreholes 1,202). These deposits yield scarce microfossils, mainly of the genus *Trachysphaeridium: Tr. bavlense* (Schep.), *Tr. laminaritum* Tim. and *Tr. sp.* Also present are: *Leiosphaeridia pelucida* (Schep.), *L. bituminosa* Tim., *Kildinella sinica* Tim., *Protosphaeridium densum* Tim., *Favososphaeridium favosum* Tim., *Nucellosphaeridium* sp., *Pterospermopsimorpha* sp., *Orygmatosphaeridium rubiginosum* Andr., *Synsphaeridium bullatum* Tim., *S. sorediforme* Tim., *Gloecapsomorpha* sp. and *Volyniella valdaica* (Schep.) As. The fairly poor composition of these microfossils may be explained by a high ash content, which affects the preservation of fossils.

3. *Syuzma Beds* (Boreholes 247,202,219,1,279, and outcrops near Syuzma Village). Microfossils of the genus *Leiosphaeridia* Eis. are predominant. The following species were distinguished: *Leiosphaeridia aperta* (Schep.), *L. effusa* (Schep.), *L. pelucida* (Schep.), *L. bituminosa* (Tim.), *L. minor* (Schep.), *L. laccata* (Tim.) As., *L. gigantea* (Schep.) and *Leiominuscula prima* (Naum.). Less abundant are: *Trachysphaeridium laminaritum* (Tim.), *Tr. bavlense* (Schep.), Tr. *partialum* (Schep.) and *Tr.* sp. The abundance of *Orygmatosphaeridium rubiginosum* (Andr.), *O. sp.* and *Stratimorphis plana* As. microfossils slightly increases. *Bavlinella faveolata* (Schep.) and abundant dark brown and red-brown films of organic matter were also recorded. *Kildinella sinica* (Tim.), *Favososphaeridium favosum* (Tim.), *Synsphaeridium sorediforme* (Tim.) and *Symplassosphaeridium* sp. occur sporadically. Filamentous algae are represented by a few smooth forms, i.e. *Leiothrichoides typicus* Herm. and *Leiothrichoides* sp.

4. *Vaizitsa Beds* (Boreholes 1,202,219,247,235,8). Microfossils, recorded from these beds, are similar in composition to the Syuzma microfossils. The following species were distinguished: *Leiosphaeridia pelucida* (Schep.), *L. effusa* (Schep.), *Trachysphaeridium bavlense* (Schep.), *Tr.* sp., *Orygmatosphaeridium rubiginosum* (Andr.) and *Bavlinella faveolata* (Schep.).

3.3 Assemblage III (Plate 59) (Upper Redkino)

Zimnie Gory Beds (boreholes 1, 202, 219, 247, 235 and exposures on the Zimny Coast). The assemblage is dominated by microfossils of the genera *Leiosphaeridia* Eis., *Trachysphaeridium* Tim. and *Stictosphaeridium* Tim. There appear rather large forms, 200–300 μm; less common are even larger specimens. The following species were distinguished: *Leiosphaeridia pelucida* (Schep.), *L. effusa* (Schep.), *L. minor* (Schep.), *L. culta* (Andr.), *L. bituminosa* (Tim.), *L. aperta* (Schep.), *L. gigantea* (Schep.), *Trachyspaeridium laminaritum* (Tim.), *Tr. bavlense* (Schep.), *Tr. partialum* (Schep.), *Tr. magnum* (Tim.), *Tr. laminaritum* (Tim.), *Stictosphaeridium sinapticuliferum* (Tim.), *St. pectinale* (Tim.) and *St. implexum* (Tim.). Less common are: *Favososphaeridium favosum* (Tim.), *Stratimorphis plana* (As.), *Orygmatosphaeridium rubiginosum* (Andr.), *Bavlinella faveolata* (Schep.), *Synsphaeridium conglutinatum* (Tim.), *Podoliella regulare* (Tim.), *Leiominuscula* sp., *Symplassosphaeridium* sp. and *Gloeocapsomorpha* sp. Occurrences of microfossils with a denser wall, placed into *Kildinella hyperboreica* (Tim.) and *K. sinica* (Tim.) are occasional. Of special note are spiralomorphic microfossils, *Volyniella valdaica* (Schep.) As. and *V.* sp., which become more abundant than in the underlying Arkhangelsk Beds. These forms have been repeatedly recorded and are characteristic of the Upper Vendian of the East European Platform. The assemblage comprises abundant filamentous algae and plant films. Among them are: *Leiothrichoides typicus* Herm., *L. gracilis* Pjat., *L.* sp., *Polytrichoides lineatus* Herm., *Tortunema sibirica* Herm., *Oscillatoriopsis magna* Tynni et Donner, *O. constricta* Tynni et Donner, *O.* sp., *Oscillatoriopsis wernadskii* Schep. and microfossils of the group *Rudnjana.*

3.4 Assemblage IV (Plate 60) (Kotlin)

The Padun and Mezen Formations (boreholes 247, 235, 271).

The Kotlin deposits are dominated by microfossils of the genus *Leiosphaeridia* Eis. The following species were distinguished: *Leiosphaeridia minor* (Schep.), *L. aperta* (Schep.), *L. effusa* (Schep.), *L. culta* (Andr.), *L. pelucida* (Schep.) and *L. gigantea* (Schep.). Leiosphaerids often form clusters. Second in abundance are: *Trachysphaeridium bavlense* (Schep.), *Tr. magnum* (Tim.), *Stictosphaeridium sinapticuliferum* (Tim.), *Synsphaeridium conglutinatum* (Tim.), *S. sorediforme* (Tim.) and *Nucellosphaeridium* sp. No microfossils with a dense wall of the genus *Kildinella* (Tim.) were recorded. There are abundant fine filamentous algae, viz. *Leiothrichoides typicus* Herm. and *L.* sp. and filaments with a rhomboid pattern distinguished as a new species *Oscillatoriopsis rhomboidales* sp. nov. The Kotlin microfossil assemblage is noted for the predominance of smooth leiosphaerids, abundant fine filamentous algae and the appearance of trichomes with a rhomboid pattern.

These deposits are almost devoid of coarse, tubelike forms, red-brown films and associated microfossils of the genus *Orygmatosphaeridium* Tim.

As a result of micropaleophytological studies of the Upper Vendian deposits in the northern Arkhangelsk District, four microfossil assemblages have been distinguished; they were assigned to the Redkino and Kotlin Horizons. These assemblages enable the subdivision of the Redkino Horizon into three differing composition levels: Lower Redkino, Redkino and Upper Redkino. Microfossils of the Redkino Horizon are very diverse. They comprise abundant filamentous algae, smooth, with a shagreen and meshed surface and cellular walls, numerous organic films and associated Orygmata-like microfossils. Sphaeromorphic microfossils are dominated by thin-walled forms with a diverse sculpture. The Kotlin microfossil assemblage is characterized by a fairly monotonous composition. It is dominated by smooth leiosphaerids, often forming clusters, and it yields very few microfossils of the genus *Orygmatosphaeridium* Tim. There are abundant, fine filamentous algae and organic films. We should note the appearance of peculiar filamentous algae with a rhomboid surface structure.

The Vendian is noted for the predominance of thin-walled microfossils, mainly represented by the genera *Leiosphaeridia* Eis., *Trachysphaeridium* Tim. and *Stictosphaeridium* Tim. The assemblages, occurring in deposits and containing tuffaceous material, are replete with vesicles of the genus *Orygmatosphaeridium* Tim. Filamentous microfossils are abundant and diverse, and represented by non-septate and septate forms of *Leiotrichoides* Herm., *Oscillatorites* Schep. and *Oscillatoriopsis* Schopf; there are narrow (10–20 μm) and rather broad (40 to 80–100 μm) varieties. Filamentous forms include numerous problematic fragments from the group *Rudnjana* Golub. The above assemblages change throughout the section. The lower Redkino Horizon (Tamitsa and Lymatsa Beds) are dominated by thick-walled forms of the genus *Leiosphaeridia* Eis., and less common are *Kildinella* Tim.; there occur also *Stictosphaeridium* Tim., *Favososphaeridium* Tim., *Trachysphaeridium*, etc. In the middle of the horizon (Arkhangelsk, Verkhovsk, Syuzma and Vaizitsa Beds), the species diversity of microfossils increases. The composition of the assemblage is as follows: eight microfossil species of the genus *Leiosphaeridia* Eis., four species of the genus *Trachysphaeridium* Tim., three species of the genus *Stictosphaeridium* Tim., *Orygmatosphaeridium* Tim., abundant *Synsphaeridium* Tim., *Symplassosphaeridium* Tim., *Nucellosphaeridium* Tim., *Pterospermopsimorpha* Tim. and a few filamentous forms and characteristic microfossils of the genus *Volyniella* (Schep.) As. The upper part (Zimnie Gory Beds) shows an abrupt increase in the number of filamentous forms and thin-walled bodies of the genera *Leiosphaeridia* Eis. and *Trachysphaeridium* Tim. There were reported also *Orygmatosphaeridium* Tim., *Bavlinella* Schep., *Nucellosphaeridium* Tim., *Synsphaeridium* Tim., *Symplassosphaeridium* Tim., *Volyniella* As., *Stictosphaeridium* Tim., *Favososphaeridium* Tim., etc. Filamentous forms (genera *Leiothrichoides* Herm., *Oscillatoriopsis* Schopf, *Oscillatorites* Schep. and *Palaeolyngbia* Schopf) are abundant and diverse.

Plant remains assemblages, distinguished in the Redkino Horizon from the northern Arkhangelsk District, can be roughly correlated with the Redkino assemblages of the southwestern East European Platform (Aseeva 1976, 1981), Moscow Syneclise, and Pachelma Trough (Volkova 1964; Shepeleva 1974), the Mezen Depression (Rudavskaya et al. 1977) and the Cis-Urals area at the town of Perm (Golub 1974).

The Kotlin microfossil assemblage of the northern Arkhangelsk District is correlated with those of the Kotlin Horizon distinguished by Volkova (1962, 1964, 1968, 1973) in the Kotlin Formation of Estonia, Leningrad District, Moscow Syneclise, the Ushitsa Formation of the Dniester Basin, by Shepeleva (1974) from the Kotlin deposits (Povarovo Formation) of the Moscow Syneclise, by Aseeva (1976) from the Kanilov Formation of Volyn-Podolia, by Rudavskaya et al. (1977) from the Mezen Depression, and by Paškevičiene (1980) from the Kotlin deposits of the Baltic area, Byelorussia, and Leningrad District.

Below, two new species of filamentous microfossils from the Valdai deposits of northwestern Arkhangelsk District are described.

Oscillatoriopsis Schopf 1968; *Oscillatoriopsis funiformis* Ragoniza, sp. nov. (Plate 59, *2,2a,3*)

Etymology. Latin "funiformis", cordlike.

Holotype. PIN, slide 10/6; northern Arkhangelsk District, Zimny Coast, Upper Vendian, Redkino Horizon, Zimnie Gory Beds.

Description. Single, slightly curved trichomes with a distinct cellular structure. Shagreen or fine-meshed surface. A characteristic feature is the presence of annular thickenings or strands, commonly of darker colouration. Length of filaments is 250–300 μm; their width, 20–70 μm; width of strands, 4–6 μm; annular thickenings vary from 8–10 to 20–50 μm in size. Height of different cells is 4–6 μm.

Discussion. Differs from other species of the genus in having strands and annular thickenings.

Distribution. Northern Arkhangelsk district, Zimny Coast of the White Sea; Upper Vendian, Redkino Horizon.

Material. Ten specimens.

Oscillatoriopsis rhomboidalis Sivertseva, sp. nov. (Plate 60, *4,6,7*)

Etymology. Latin "rhomboidalis", rhombic.

Holotype. LGU, slide 57/1; Arkhangelsk District, borehole 247 "Zimny Coast", depth 57 m; Vendian, Valdai Series, Kotlin Horizon.

Description. Individual, unbranched trichomes without traces of a cellular structure. Characteristic surface sculpture displays a dense, oblique, parallel, rhombic network pattern. Trichomes may owe their appearance to the ribs (about 1 μm wide) in the form of spirals along a filament, when a flattened, tubular trichome forms a rhombic crossing network. Areas with a rhombic striation pattern are aligned discontinuously along a filament and have no distinct boundaries with the smooth parts of the thallus. Apparently, diagonal striae belong to the inner structural elements of the trichome and its framework pattern. Trichomes of constant width along their entire length often form bundles. Trichomes are 13 to 16 μm wide; fragments are 80-90 μm long.

Remarks. Filaments of *Oscillatoriopsis rhomboidalis* often occur together with thin, smooth trichomes of the genus *Leiothrichoides* Herm., which are widespread in the Kotlin sediments.

Discussion. Differs from other representatives of the genus in having an indistinct cellular structure and rhombic pattern.

Distribution. Northern Arkhangelsk District, Upper Vendian, Kotlin Horizon.

Material. About 15 specimens.

4 Plant Microfossils of the Urals

Microphytologically, the Vendian deposits of the Urals and Cis-Urals are poorly known. The data on composition and morphology of the Vendian plant microfossils of the region were published by Chibrikova (1961 and others); Shepeleva (1962, 1963); Chepikova (1966, 1971b); Naumova (1968); Pykhova (1973) and Jankauskas (1980). All these publications give brief and preliminary information on the studies accomplished in this area.

Initially, lists of plant microfossils for different parts of this huge area included only names of morphologically simple acritarchs from the sphaeromorphid subgroup and filamentous algae, viz. *Oscillatorites wernadskii* Schep. Acritarchs were mainly assigned to the genera *Kildinella, Leiosphaeridia, Leiominuscula, Bavlinella, Orygmatosphaeridium, Brochopsophosphaera* and *Asperatopsophosphaera.* Their diagnosis was precluded by the absence of any distinct characters, hence their low stratigraphic value.

In addition, it should be noted that the Vendian deposits of the Urals and Cis-Urals are shallow-water facies, unfavourable for the preservation of acritarchs in life position and their burial. Therefore, the occurrences of plant microfossils are very rare in this region (Shepeleva 1963, p. 7).

The study of facies favourable for acritarchs in the South Urals and Cis-Urals in Bashkiria was specially accomplished by Jankauskas (1980 and others), who succeeded in finding a number of important occurrences of microphytofossils. On the left flank of the Zigan Valley, the Zigan Formation exposed a thin unit of black mudstone, overcrowded with tubular algae from the group *Rudnjana.*

They were first described by Golub (1979) from the Smolensk and Gdov Formations of the Orsha Depression (western East European Platform). They have not been recorded at higher and lower stratigraphic levels. This enabled one to finally prove the Vendian age of the Asha Series.

Among *Rudnjana* fossils, occurring at the Zigan River, most abundant are *Omalophyma gracilis* Golub (Plate 62, *4,5,7,8*). *O. angusta* Golub and *O. ovalis* Golub are less frequent. At this locality, *Rudnjana* are associated with such acritarchs, as *Kildinella* and *Leiosphaeridia*.

The other locality crops out on the right bank of the Mendym River, the middle Zigan Formation. It contains abundant *Bavlinella faveolata* Schep., widely known from the Vendian of the East European Platform.

In the Cis-Urals of Bashkiria, mass accumulation of plant microfossils comes from the section of Sergaeevskaya-800 borehole, with a depth interval of 2942.4–2946.4 m (see Plates 61 and 62). Plant microfossils occur in a 1-cm thick black mudstone layer; they are extremely diverse and abundant. The stratigraphic interval above corresponds to the middle Baikibashevo (= Lower Kairovo) Formation; in this section, it is 62 m thick (depth interval 2922–2984 m). This formation rests with an abrupt unconformity on deeply eroded Upper Riphean deposits. According to a unified scheme, the Baikibashevo Formation has an age equivalent to the Pletenev Formation of the Moscow Syneclise, Liozno Formation of the Orsha and Brest depressions, Yashyunai Formation on the northwestern slope of the Mazury-Byelorussian Anteclise, and Mogilev Formation of the southwestern slope of the Ukrainian Shield, hence its placement at the level of the lower Redkino Horizon. However, many other workers assign the Liozno Formation and its age equivalents to the upper part of the much older Drevlyane Horizon. Therefore, we cannot precisely place the Sergeevskaya locality in the unified scheme. Doubtlessly, it is one of the oldest deposits in the Vendian section.

The Sergeevskaya microphytofossil assemblage comprises mainly acritarchs, including structurally complex forms, and to a lesser degree, filamentous algae. Its characteristic feature is the joint occurrence of abundant relict forms from older Riphean deposits and the predominant amount of common Vendian species. The Sergeevskaya microbiota can be regarded as transitional between Riphean and Vendian ones.

Riphean relics contain the following acritarchs: *Kildinella hyperboreica* Tim., *K. sinica* Tim. (18% of the total number of microfossils), *Nucellosphaeridium minutum* Tim. (4%), *Pterospermopsimorpha* sp. (0.5%) (Plate 61, *6*), and individual filamentous algae: *Arctacelluluria ellipsoidea* Herm. (Plate 61, *1,2*), *A. aff. doliiformis* Herm. (Plate 61, *5*), *Tortunema eniseica* Herm., *Polytrichoides lineatus* Herm. (Plate 61, *3*), *Leiothrichoides typicus* Herm., *Ostiana microcystis* Herm., *Glenobotrydion* (?) sp. (Plate 62, *1,2a,b*) and *Cholorogloaeopsis* (?) sp. (Plate 62, *3*).

Vendian species are represented by abundant and diverse *Leiosphaeridia* (20%). They differ from the species *Kildinella* in having a thin delicate wall and intricate crumpling plicae. Such forms are scarce in Riphean deposits.

Among leiosphaerids, forms with a wall callus are distinguished (*Leiosphaeridia incrassatula* Jank., Plate 61, *7,8,10,11*). Thick-walled, spherical vesicles make up abundant accumulations (*Synsphaeridium, Symplassosphaeridium* genera and large, thick-walled acritarchs, *Trachysphaeridium*).

This locality is marked by interesting occurrences of annular algae, Toromorpha (Plate 61, *4*), also known from the Redkino Horizon of the East European

Platform. Perhaps, the spiny acritarch, assigned to the new species of the genus *Baltisphaeridium*, should be regarded as a unique occurrence. This is *B. perrarum* Jank. (Plate 61, *9*), a thick-walled, spherical vesicle up to 20–30 μm across, bearing short, rare spines with slightly thickened distal ends.

Large, structurally coarse acritarchs, flattened elliptical in shape, composed of separate polygonal plates, *Satka granulosa* Jank. (Plate 62, *9* and *10*) are also rather abundant. The above listed forms can be considered as index forms for the stratigraphic level of the Baikibashevo Formation.

All in all, Vendian microphytofossils of the Urals enabled the recognition of over 20 well substantiated species. However, such a faunal characteristic of the Ural Vendian is far from being complete. Further studies should add depth and detail to current knowledge.

5 Vendian Filamentous Algae

The data provided by the authors, who reported on the occurrences of filamentous forms from the Upper Precambrian sections of the East European and Siberian platforms, point to an extensive regional development of non-mineralized microorganisms with a filamentous structure.[3] However, most authors have placed major emphasis on the study and description of sphaeromorphids (acritarchs); therefore, most of filamentous microfossils are poorly known.

The first description of the occurrences of blue-green filamentous algae was given by Shepeleva. In 1960, she described two species from the Lower Cambrian of the Leningrad District: *Oscillatorites wernadskii* Shepeleva 1960 and *Nodularites maslovi* Shepeleva 1960. The first species should be assigned to the *Botuobia* Pyatiletov 1979.

The earliest appearance of filamentous forms in the Vendian was recorded from the Yartsevo Formation (Volyn Series), the central Russian Platform (Shepeleva 1974, p. 18). The Sergeevskaya Formation in the Cis-Urals of Bashkiria contains microbiota, which can be correlated with the early Redkino level (Jankauskas 1980). The above author has discovered there a mixed acritarch assemblage dominated by Vendian forms. Filamentous microfossils are represented by Riphean species: *Acrtacellularia ellipsoida* Herm., *A. aff. doliiformis* Herm., *Tortunema eniseica* Herm., *Polyrichoides lineatus* Herm. and *Leiothrichoides typicus* Herm.

Filamentous forms were reported from three microfossil assemblages, distinguished in the course of studies of the Upper Bavly deposits in the Cis-Urals at Perm (Golub 1974, pp. 24–26). They span the interval: Redkino deposits (Rudnjana borehole)-Gdov and Laminarite Beds.

[3]The term "filament" in the description of fossil material does not refer to the sense of the trichome in a vesicle. These are mainly fragments of a filamentous-trichomous structure, mainly without cellular walls; in the latter case, they are empty vesicles of filamentous algae and bacteria.

Filamentous microfossils are known from microfossil assemblage IV of the Upper Precambrian of the Arkhangelsk District. This assemblage is correlated with that of the Redkino Formation, reported from many localities of the Pachelma Trough and Moscow Syneclise (Sivertseva and Stankovsky 1979, p. 158).

Many filamentous microorganisms are known from the Valdai Series (Vendian) of the Baltic area (Paškevičiene 1980, p. 18) and from the Vendian/Cambrian boundary beds (upper Vendian) of the Brest Depression, Stradech-17k borehole, Byelorussia (Bessonova et al. 1979, p. 167), and also from the boundary beds on the Russian Platform and Laminarite Beds in Estonia (Vakhrameev and Volkova 1968, p. 126).

On the Siberian Platform, filamentous forms are described among microfossils of the Yudoma Complex, South Yakutia (Pyatiletov 1979, 1980a).

The study shows that filamentous microorganisms are an integral part of the Vendian microbiota; however, they have as yet been reported only from isolated localities. The most complete Vendian reference section is the one in the Dniester Basin of Podolia, marked by abundant filamentous forms (Timofeev 1973a,b; Aseeva 1974, 1976). The author has studied the material from this section in an ascending order: from the Grushka Formation to the Kanilov Formation inclusive.

Filamentous microfossils are known to occur almost throughout the entire section and contribute to the paleontological characterization of beds and formations, along with other faunal groups.

The first appearance of a few filamentous algae was recorded from the Lomozov Beds. They occur as individual specimens and in clusters, which form turfs with a longitudinal alignment of long, multicellular, unbranched filaments. Their length in fragments attains 1400 μm; and they are mainly 5 μm wide, cells being as long as wide.

Filamentous organisms become more abundant from the base of the Lyadova Beds. They contain numerous fragments of straight and slightly curved individual, unbranched filaments with a non-cellular structure and a rough surface. Fragments, exhibiting a cellular structure, are numerous. Filaments are 5–10 μm wide. The most distinct are thalli about 100 μm wide or slightly greater. They display a distinct, narrow, transverse fragmentation and a narrow outer vesicle (genus *Contextuopsis* As., 1983) (Plate 74, *1–2*). These deposits also contain characteristic Vendian coiled algae of the genus *Volyniella* Schep. et As. (Plate 74, *3–7*).

The lower Lyadova Beds are separated from the upper part by a redbed unit. Just above the redbeds, the base of the upper Lyadova Beds yields peculiar microorganisms of the genus *Podolina* Herm., noted for an inconsistent shape (Timofeev et al. 1976). As yet, it is the only known locality in the section of the Dniester Basin of Podolia.

The Bernashevka Beds contain fine filaments 20–25 μm wide (*Leiothrichoides gracilis* Pjatiletov, Plate 75, *6*), filaments with a hardly discernible septation, assigned to *Botuobia wernadskii* Shepeleva 1960. Less common are non-septate filaments 5–10 to 12.5 μm wide, i.e. *Leiothrichoides typicus* Herm. There are numerous filaments with a rough, sometimes cellular surface structure, genus *Lyadovia* gen. nov. (Plate 75, *5*). The base of the Bernashevka Beds is marked by fossils, exhibiting a peculiar morphology. These are vesicles of the genus *Octaedryxium* Rudavskaya 1973.

The Zinkov Beds are overcrowded with filamentous forms. These are mainly fine, narrow, strongly folded filaments 10–12.5 μm wide. These beds also yield abundant *Contextuopsis* gen. nov. The beds are marked by the first appearance of filaments with characteristic, irregularly arranged, fine inclusions (Plate 75, *2,3,4*). Inclusions range from single grains to tiny rounded accumulations, which leave isodiametric meshes on the filament surface (Plate 75, *1,9,10*). Filaments of this type were placed into a new genus, *Zinkovioides* gen. nov., having two species (*Z. inclusis* sp. nov., *Z. perforata* sp. nov.).

The composition of filamentous organisms from the Dzhurzhevka Beds is similar to that of the Zinkov Beds; they are dominated by filaments of the genus *Zinkovioides* gen. nov.

Filamentous microfossils are extremely abundant, but morphologically uniform in the Kalyus Beds. There are filaments with a rough surface structure, *Ljadovia* gen. nov., 20–25 μm wide and thinner ones, 5–12.5 μm with a smooth surface structure (*Leiothrichoides typicus* Herm.).

Filamentous forms from the Sokolets and Komarovo Beds are noted for fragmentation of filaments into short portions. The density of the filament wall and its width varies within a fragment (Plate 74, *9*). Very common are fragments with a thickened central portion, tapering towards the ends. Fossils of the Kanilov Formation are darker in colour (brown) than those of the overlying beds.

A peculiar feature of Vendian filamentous microfossils is the predominance of individual (along the sections) and simple (in accumulations), unbranched filaments with a non-cellular structure, of variable width and differing in surface structure, as well as of filaments with an *Oscillatoria*-like structure. Neither complete organisms nor terminal cells have as yet been found among filamentous microfossils, despite their mass abundance in the section. Hence, they differ from morphologically more diverse Riphean filamentous algae, where numerous complete trichomes and fragments with a preserved structure of terminal cells were reported (Hermann 1981). Most likely, Vendian filamentous forms were fairly long; they display a tendency towards coiling into spirals and rings. A pale yellow colour of the filaments reflects their thin-walled structure.

The published species of filamentous microfossils and those reported from the section of the Dniester Basin of Podolia are marked by a reference, showing the place where the holotype is kept, and regarding their distribution; the description of new occurrences of Vendian filamentous algae from the Dniester Basin of Podolia is given. The collection No. 1973 is kept at the Laboratory of Biostratigraphy and Paleobiology at the Institute of Precambrian Geology and Geochronology (IGGD) of the USSR Academy of Sciences in Leningrad.

Genus *Volyniella* Schepeleva et Aseeva 1974; *Volyniella valdaica* Schepeleva et Aseeva 1974; (Plate 74, *3,6*)

Holotype. IGN AN UkrSSR, No. 242/1; Podolia; Gusyatin Village, borehole 11,673, depth 574 m; Yaryshev Formation, Zinkov Beds.

Distribution. Moscow Syneclise, Vendian, Yartsevo Formation; Volyn-Podolia, the upper Yampol Formation, Lyadova, the Zinkov Beds of the Yaryshev Formation.

Volyniella jampolica Aseeva 1974 (Plate 74, *5*)

Holotype. IGN AN Ukr.SSR, No. 15/2; Podolia, Kaskady Village, borehole 14,274, depth 123.9–124.0 m; the upper Yampol Formation.

Distribution. Podolia, Vendian, the upper Yampol Formation, the Lyadova Beds of the Yaryshev Formation.

Genus *Leiothrichoides* Hermann 1974 emend. 1979; *Leiothrichoides typicus* Herm., 1974 emend. 1979 (Plate 74, *8*)

Holotype. IGGD, No. 49/27; Krasnoyarsk Province, Turukhan District, Miroedikha River, Upper Riphean, the Miroedikha Formation.

Distribution. Upper Riphean of the Siberian Platform; Vendian-Lower Cambrian of the East European Platform.

Leiothrichoides gracilis Pyatiletov 1980a (Plate 75, *6*)

Holotype. IGiG SO AN SSSR, No. 803; South Yakutia; Yudomian, the Botuoba Formation, borehole 750, depth 1786.4 m.

Distribution. South Yakutia, Yudomian, Upper Vilyuchan, Bochugunor (undifferentiated) and Botuoba Formations, Dniester Basin of Podolia, Vendian, the Ushitsa Series.

Genus *Siphonophycus* Schopf 1968; *Siphonophycus kestron* Schopf 1968 (Plate 75, *7*)

Holotype. Harvard University, No. 58469; Central Australia; Late Precambrian, the Bitter Springs Formation.

Distribution. Central Australia, Late Precambrian, the Bitter Springs Formation; the Dniester Basin of Podolia, Vendian, the Lyadova Beds.

Genus *Botuobia* Pyatiletov 1979 emend. Kolosov 1984; *Botuobia wernadskii* (Shepeleva 1960) emend. Kolosov 1984 (Plate 74, *12*)

Holotype. VNIGNI, No. 8/5122; Leningrad District; Lower Cambrian (Gdov Beds, blue clay), borehole Smerdovitsa 3, depth 2.115–2.381 m.

Distribution. East European and Siberian platforms; Riphean-Vendian-Lower Cambrian.

Genus *Striatella* Aseeva 1983a; *Striatella coriaceae* Aseeva 1983a; *Contextuopsis zonata* Hermann 1985 (Plate 74, *1,2*-holotype)

Holotype. IGN AN Ukr SSR, No. 1796/2; Podolia, Karpovka Village, Mogilev-Podolia Series; the Yaryshev Formation, Lyadova Beds.

Distribution. The Dniester Basin of Podolia, Vendian, the Lyadova and Zinkov beds.

Genus *Zinkovioides* Hermann, gen. nov.

Genus named after the Zinkov Beds.

Type Species. L. inclusis sp. nov.; the Dniester Basin of Podolia; Vendian, the Yaryshev Formation.

Diagnosis. Individual, unbranched, straight and strongly curved filamentous thalli 5–12.5 μm wide with tiny, irregularly arranged, granular inclusions and isodiametric perforations, i.e. meshes.

Species Composition. Two species, i.e. *Z. inclusis* sp. nov., *L. perforata* sp. nov.

Zinkovioides inclusis Hermann, sp. nov. (Plate 75, *3*-holotype; *2,4*)

Etymology. Latin "inclusio", inclusion.

Holotype. IGGD, No. 1083/3; the Dniester Basin of Podolia; Vendian, the Yaryshev Formation, the Zinkov Beds.

Description. Fragments of individual, unbranched, filamentous thalli, straight to strongly curved and plicated, 5 to 12.5 μm wide with abundant, fine inclusions as single grains and their irregularly arranged clusters. Size of accumulations do not exceed 5 μm. Some portions of filaments without inclusions are preserved. Filaments are pale yellow.

Distribution. Vendian, the Dniester Basin of Podolia, the Zinkov and Dzhurzhevka Beds.

Material. Abundant occurrences of different degrees of preservation, the Zinkov Beds, La-Izvaara exposure.

Zinkovioides perforata Herman, sp. nov. (Plate 75, *1*-holotype, *9,10*)

Holotype. IGGD, No. 1082/1; the Dniester Basin of Podolia; Vendian, the Yaryshev Formation, the Zinkov Beds.

Description. Fragments of individual, unbranched, straight and curved, intensely plicated filamentous thalli 5–12.5 μm wide. Numerous closely spaced, rounded and angular meshes variable in size are seen on the thalli surface. Their maximum size does not exceed 5 μm. Filaments are pale yellow.

Distribution. The Dniester area, Vendian, the Zinkov and Dzhurzhevka Beds.

Material. Abundant occurrences of different degrees of preservation. The Zinkov Beds, La-Izvaara exposure.

Genus *Ljadovia* Hermann, gen. nov.

Genus named after Lyadova Beds.

Type Species. L. exasperata sp. nov., the Dniester area of Podolia; Vendian; the Yaryshev Formation, the Lyadova Beds.

Diagnosis. Individual, unbranched, straight and curved, thin-walled filaments with a non-cellular structure and rough surface. Filaments are 20–62.5 μm wide.

Species Composition. See Type Species.

Discussion. Similar to filaments of *Leiothrichoides gracilis* Pyatiletov 1980a; however, it has an extremely rough, nearly corroded surface and much wider filaments.

Ljadovia exasperata Hermann, sp. nov. (Plate 75, *5,11*)

Etymology. Latin "exasperatus", rough.

Holotype. IGGD, No. 718/1; the Dniester area of Podolia; Vendian, the Yaryshev Formation, the Lyadova Beds.

Description. Individual, unbranched, straight and curved, thin-walled filaments with a non-skeletal structure and an extremely rough, almost corroded surface. Filaments are 20–62.5 μm wide; up to 500 μm long in fragments. Filaments are pale yellow.

Distribution. The Dniester Basin of Podolia, Vendian, the Lyadova, Bernashevka, Zinkov and Kalyus Beds.

Material. Numerous occurrences of different degrees of preservation. Borshchov-Yar exposure, the Lyadova Beds.

6 Acritarchs of the Yudoma Complex

Group Acritarcha Evitt 1963

Subgroup Acanthomorphitae Downie et al. 1963

Genus *Baltisphaeridium* Eisenack 1958 emend. Downie et al. 1963; *Baltisphaeridium cerinum* (?) Volkova 1968[4] (Plate 65, *4*)

Synonymy. See in Volkova et al. (1979, p. 8).

Description. A rounded-ovate, dark, dense vesicle. Light, short (2–3 μm), numerous spines. Base of spines is slightly thickened. Vesicles are 25–28 μm in diameter.

Discussion. Differs from *B. ciliosum* Volk. in the length of spines.

Distribution. Lower Cambrian of the East European and Siberian Platforms.

Material. Individual specimens.

Distribution. East Siberia, the middle Lena River, Kazarka borehole-4, depth 2997–3000 m; the Moty Formation.

Baltisphaeridium ciliosum Volkova 1969[5] (Plate 65, *2,3*)

Synonymy. See Volkova et al. (1979, p. 8).

Description. Flattened vesicle with crumpling plicae with a circular-ovate outline. Numerous processes with a widened base. Ends of some processes are funnel-shaped, frequently broken. Processes are lighter than the central body. Diameter of a vesicle without processes is 28–37 μm, with processes, 37–47 μm.

[4] The material was submitted by Rudavskaya (VNIGRI, Leningrad)
[5] See reference, p. 380.

Discussion. Differs from *B. cerinum* Volk. in length of spines.

Distribution. Lower Cambrian of the East European and Siberian Platforms.

Material. Six specimens.

Distribution. East Siberia, the middle Lena River; Nepa borehole-1, depth 2368–2421 m; the Moty Formation, the Middle-Upper Moty Member.

Baltisphaeridium primarium Jankauskas 1979[6] (Plate 63, *1–4*)

Baltisphaeridium primarium: Volkova et al. 1979, p. 12, Plate III, Figs. 8,9.

Description. Vesicles are circular in outline, thin, with a smooth surface, sometimes bearing very narrow, short, sinuous plicae. The vesicle has several (six to ten) spiny processes. Length of processes, 40–60 μm; width at the base, 12–20 μm. Diameter of vesicles, 80–125 μm; they are pale yellow and yellow in colour.

Variability. Observed in the number of processes and vesicle density.

Discussion. Differs from other species of this genus in having characteristic spiny processes.

Distribution. Lower Cambrian (Lyukati Horizon) of the East European Platform; Yudomian of the Siberian Platform.

Material. Six well- and fairly well-preserved specimens.

Occurrence. Siberian Platform, Nepa-Botuoba area, Byuk No. 715 borehole, depth 1968.8 m; the upper Kursov Formation.

Baltisphaeridium (?) strigosum Jankauskas 1976[6] (Plate 63, *7–9*; Plate 64, *1–3*)

Synonymy. See Volkova et al. (1979, p. 12).

Description. Vesicles circular or ovate in outline. Vesicle surface is smooth, sometimes bears short, narrow, sinuous plicae and is densely covered by sinuous hairlike processes, slightly thickened at the base; their length, 8–17 μm. Vesicle diameter, 70–150 μm; pale yellow to brown in colour.

Variability. Observed in the number of processes, their size and vesicle density.

---

[6]The material submitted by Pyatiletov (IGiG SO AN SSSR, Novosibirsk).

Discussion. Differs from other species of *Baltisphaeridium* in the character of processes.

Remarks. The forms discovered are much larger in size as compared to microfossils of this species described by Jankauskas.

Distribution. Lower Cambrian (Lyukati, Vergale and Rausvene Horizons) and Middle Cambrian (Kibartai Horizon) of the East European Platform; Yudomian of the Siberian Platform.

Material. Abundant, well- and fairly well-preserved specimens.

Occurrence. Siberian Platform, the middle Lena River, the Moty Formation, Middle Member, Nepa-Botuoba area, the Kursov Formation (Nepa No. 1 borehole, depth 2925–2944 m; Middle Botuoba No. 4 borehole, depth 1,924–1930 m; Peledui No. 750 borehole, depth 1,836; Byuk No. 715 borehole, depth 1962–1971 m).

Micrhystridium sp. (Plate 65, *13,14*)

Description. Spherical, thick, dense, often flattened vesicles, circular or circular-ovate in outline with or without short narrow plicae. Surface covered by short, acute, thin, sinuous processes. They vary in number and are 2–4 μm long. Vesicles are 8–20 μm in diameter, dark grey to brown in colour.

Material. Few well-preserved and fairly well-preserved specimens.

Distribution. East Sayan, Mana Trough, Anastasjina, Ungut, Leiba and Krol Formations; East Siberia, the middle Lena River, the Moty Formation.

Subgroup Prismatomorphytae Downie et al. 1963

Genus *Octoedryxium* Rudavskaja 1973; Octoedryxiumtruncatum Rudavskaja 1973; (Plate 66, *5–7,9,10*)

Octoedryxium truncatum: Rudavskaya 1973, p. 7, Plate I, Fig. 1; 1973, p. 19, Figs. 1–16.

Quadratimorpha ordinata: Sin and Lin 1973, p. 60, p. 4; Figs. 10–12.

Octoedryxium truncatum: Vidal 1976, p. 22, Fig. 1 A-P.

Description. Octahedral vesicles, with rough, granular and tubercular surface sculpture. Ribs are 0.1–1 μm thick and more in large forms. Sometimes ribs are lenticular, which is observed mainly in large forms. Size varies extensively from 1-3 to 20-45 μm. Well-preserved vesicles show apertures at the octahedral apices.

Variability. Considerable, due to size of forms, surface sculpture and thickness of the vesicle.

Discussion. Differs from *O. simmetricum* Tim. in dimensions of the vesicle.

Distribution. East Siberia, Upper Riphean, Yudomian and Cambrian; China, Upper Sinian; Sweden, upper Visingco Formation.

Material. Over 50 specimens.

Occurrence. East Siberia, the middle Lena River, Moty Formation, Lower Member.

Genus *Polyedryxium* Deunff 1954 emend. Deunff 1971; *Polyedryxium pritulai* Rudavskaja 1971 (Plate 64, *6,7*)

Polyedryxium pritulai: Rudavskaya 1971, p. 98, Plate I, Figs. 1-4.

Description. Vesicles are parallelipiped in shape, sometimes close to a cube; hollow walls are thin, glassy. Faces (or one of them are concave inside towards the central cavity, which is commonly rectangular. In poorly preserved vesicles the shape of the central cavity is not seen. A few vesicles have low membrane ridges and processes at the junction of the ribs, at angles to the parallelipiped. Parallelipiped sides are 10-30 μm long. Vesicles locally form accumulations.

Variability. Considerable, in size and shape of vesicles and in the presence or absence of membrane ridges; however, it cannot be reliably stated due to poor preservation of vesicles.

Discussion and Remarks. Differs from *P. neftelenicum* Rud. species in the parallelipiped shape of the vesicle. Poor preservation of vesicles allows specimens to be assigned to a single species, despite their differences.

Distribution. Yudomian of central East Siberia.

Material. Over 50 specimens.

Occurrence. East Siberia, the middle Lena River, the lower Nepa Formation (Morkovo borehole No. 42, depth 2675-2649; 2975-2963; Chastin borehole No. 2, depth 2402-2392); the Borulakh Formation (Middle Vilyuchan borehole No. 609, depth 2442 m).

Polyedryxium neftelenicum Rudavskaja 1971 (Plate 64, *4,9*; Plate 66, *4,8,11*)

Polyedryxium neftelenicum: Rudavskaya 1971, pp. 99, Plate I, Figs. 6–8.

Description. Prismatic, hexahedral, sometimes pentahedral, hollow vesicles with thin, glassy walls. Lateral faces are commonly slightly convex. Bases (or one of them) are convex inside towards the central cavity, which repeats the outlines of a prism. In poorly preserved vesicles, the forms of the inner cavity are not seen. Ribs are smooth, without membrane ridges. There are no apertures. The sides of the prism are 15–35 μm long. Accumulations of forms were recorded.

Variability. Considerable in outlines of prisms. However, it cannot be reliably stated because of poor preservation.

Discussion and Remarks. Differs from *P. pritulai* species in the prismatic shape of the vesicle. Because of poor preservation, pentahedral and octahedral vesicles are assigned to one species.

Distribution. Yudomian of the central East Siberia.

Material. Over 20 specimens.

Occurrence. East Siberia, the middle Lena River, the lower Nepa Formation, the Bezymyanny Horizon (Markovo, borehole No. 42, depth 2675-2649 m, borehole No. 46, depth 2975-2963.5 m; Chastin borehole No. 2, depth 2402-2398 m.

Subgroup Tasmanititae (Sommer) Staplin et al. 1965

Genus *Paracrassosphaera* Rudavskaja in Faizulina and Trestschetenkova 1979; *Paracrassosphaera dedalea* Rudavskaja in Faizulina and Trestschetenkova 1979 (Plate 65, *15*)

Paracrassosphaera dedalea: Faizulina and Treshchetenkova 1979, p. 165, Plate XXXVII, Fig. 12.

Description. Spherical vesicles are 6–32 μm across; wall thickness, 2.5–4 μm, sometimes up to 10 μm. Wall canals are slightly conical. Surface sculpture is sinuously looped. Large, not very thick-walled forms display a distinct network sculpture.

Discussion. Differs from *P. actinomorpha* Trest. in wall thickness and length of wall canals.

Distribution. Yudomian-Lower Cambrian of the Siberian Platform.

Material. Abundant occurrences (about 100 specimens).

Occurrence. East Siberia, the middle Lena River, Moty Formation (Markovo, borehole No. 18, depth 2808–2817 m; borehole No. 50, depth 2614–2624 m; Kazarki, borehole No. 1, depth 2725–2729 m; Srednebotuobinskaya borehole No. 4, depth 1817–1915 m); the Usolka Formation (Srednebotuobinskaya borehole No. 6, Yaraktinskaya borehole No. 10).

Subgroup Sphaeromorphitae Downie et al. 1963

Genus *Bavlinella* Shepeleva 1963; *Bavlinella faveolata* Shepeleva 1963 (Plate 65, *19,20*)

Bavlinella faveolata: Shepeleva 1963, p. 16, Plate II, Fig. 10.

Description. Vesicles have a circular or circular-ovate outline with a wavy, outer margin. Fine-meshed vesicle surface is 10–25 μm in diameter; pale yellow to brownish in colour.

Discussion. Differs from *B. faceta* Schep. in smaller meshes.

Distribution. Upper Riphean-Lower Cambrian of the East European Platform; Yudomian-Lower Cambrian (less common) of the Siberian Platform.

Material. Abundant, well- and fairly well-preserved specimens.

Occurrence. Siberian Platform, the middle Lena River, the Moty Formation; Yudoma-Maya Trough, the Yudoma Series.

Genus *Granomarginata* Naumova 1960; *Granomarginata prima* Naumova 1960 (Plate 65, *16–18*)

Synonymy. See Volkova et al. (1979, p. 19).

Description. Thick vesicles, circular or circular-ovate in outline. Granular or shagreen surface. Wall thickness is 1–2 μm, outer margin slightly undulating. Vesicle diameter is 10–27 μm; pale to dark brown in colour.

Discussion. Differs from *G. squamacea* Volk. in narrower rim.

Distribution. Lower Cambrian (Lontova and Lyukati Formations) of the East European Platform; Yudomian-Lower Cambrian of the Siberian Platform.

Material. Abundant, well- and fairly well-preserved specimens.

Occurrence. Siberian Platform, Yudoma-Maya Trough, the Yudoma Series; Olenek Uplift, the Kessyuse Formation; Irkutsk Amphitheater, the Moty Formation; Nepa-Botuoba area, Boguchunor and upper Kursov Formations; Baikal area, the Ushakovka and Kurtun Formations; Patom Highland, the Tinnovka Formation, the western East Sayan, Mana Trough, the Anastasjina and Zhistyk Formations.

Genus *Leiomarginata* Naumova 1960; *Leiomarginata simplex* Naumova 1960 (Plate 65, *9–12*)

Synonymy. See Volkova et al. (1979, p. 20).

Description. Vesicles are circular in outline with a smooth, outer margin. Vesicle surface is smooth or granular. Wall thickness is 0.5–2 μm; vesicle diameter 8–25 μm; dark grey to brown in colour.

Variability. In vesicle size and wall thickness.

Discussion. Differs from *Granomarginata prima*. Naum. in having a smooth outer margin.

Distribution. Lower Cambrian of the East European Platform (Lontova and Lyukati Formations), Yudomian-Lower Cambrian of the Siberian Platform.

Material. Abundant, well- and fairly well-preserved specimens.

Occurrence. Siberian Platform, Yudoma-Maya Trough, the Yudoma Series; Anabar Rise, the Nemakit-Daldyn Formation; Olenek Uplift, the Kessyune Formation; the western East Sayan, Mana Trough, the Anastasjina, Ungut and Leiba Formations.

Leiosphaeridia div. sp. (Plate 66, *1–3*; Plate 67, *1–11*)

Description. Vesicles are circular or circular-ovate in outline, thin or thick. Vesicle surface is smooth, granular, shagreen with broad crescentiform, sometimes sinuous and narrow plicae. Occur as individual forms and in accumulations. Vesicles are 10 to 400–500 μm in diameter.

Material. Abundant, well- and fairly well-preserved specimens.

Distributions. Siberian Platform, Yudoma-Maya Trough, Yudoma Series; Anabar Massif, the Nemakit-Daldyn Formation; Nepa-Botuoba area, the Borulakh, Bochugunor, Kursov Formations; the middle Lena River, the Nepa, Moty Formations.

Subgroup Netromorphitae Downie et al. 1963

Genus *Leiovalia* Eisenack 1965; *Leiovalia tenera* Kiryanov 1974 (Plate 63, *5,6*)

Synonymy. See Volkova et al. (1979, p. 24).

Description. Vesicles are very thin, ovate in outline, with a smooth surface, bearing narrow sinuous plicae. Vesicles are 110–130 μm long, 35–45 μm wide. They are pale yellow in colour.

Variability. In the size of the vesicle and arrangement of plicae.

Discussion. Differs from the above species of this genus in having a very thin vesicle.

Distribution. Lower-Middle Cambrian of the East European Platform; Yudomian of the Siberian Platform.

Material. Five well-preserved specimens.

Occurrence. Siberian Platform, South Yakutia, the Kursov Formation (Byuk borehole No. 715, depth 1965–1987 m).

Filamentous Algae

Genus *Botuobia* Pyatiletov 1979; *Botuobia vermiculata* Pyatiletov 1979 (Plate 64, *5,10*)

Botuobia vermiculata: Pyatiletov 1979, p. 715, Plate I, Fig. 2.

Description. Individual, symmetrical trichomes, straight or slightly curved with conical, rounded edges of similar shape. In the central part of the trichome, cells are slightly longer than at its margin. Septae between cells are slightly compressed, hence a somewhat undulating outer margin of the entire trichome. Pale yellow to yellowish-brown in colour. Complete specimens are 80–660 μm long (holotype, 360 μm); 20–26 μm wide (holotype, 25 μm).

Variability. In size and colour of the trichomes.

Discussion. Trichomes of the genus exhibit some similarity to the trichomes of the genus *Tortunema* Hermann, particularly *T. sibirica* (Timofeev et al. 1976); however, the latter differ in having acute terminations of trichomes and non-compressed septae between cells.

Distribution. Yudomian of the Siberian Platform.

Material. A great number of diverse fragments and five well-preserved specimens.

Occurrence. Siberian Platform, South Yakutia, the Kursov, Borulakh and the Bochugunor formations (Peledui borehole No. 750, depth 1786–1794.5 m; Byuk borehole No. 715, depth 1931–2012 m).

Genus *Leiothrichoides* Hermann 1974 emend. 1978; *Leiothrichoides typicus* Hermann 1974 (Plate 65, *1*)

Leiothrichoides typicus: Hermann 1974, p. 7, Plate VI, Figs. 1,2.

Description. Smooth trichomes without visible traces of cellular structure, equally wide lengthwise. Individual or interwoven multiple trichomes, 8 to 16 μm wide.

Variability. In width and colour of trichomes.

Distribution. Upper Riphean of the Siberian Platform and the South Urals; Yudomian of the Siberian Platform.

Material. Abundant, well-preserved specimens.

Occurrence. Siberian Platform, South Yakutia, the Borulakh, Bochugunor and Kursov Formations (Peledui borehole No. 750, depth 1786–1794 m; Byuk borehole No. 715, depth 1931–2012 m).

Leiothrichoides gracilis Pyatiletov 1980a (Plate 65, *7,8*)

Leiothrichoides gracilis: Pyatiletov 1980a, Plate IV, Figs. 4,5.

Description. Individual trichomes, equally wide lengthwise, without any visible traces of a cellular structure. Very thin, straight to slightly curved trichomes with a shagreen surface. Trichomes are up to 26 μm wide, pale yellow to yellow in colour.

Discussion. Differs from *Leiothrichoides typicus* Hermann in having a thinner vesicle and a shagreen surface.

Distribution. Yudomian of the Siberian Platform.

Material. Numerous, well-preserved specimens.

Occurrence. Siberian Platform, South Yakutia, the Kursov, Bochugunor and Iktekh Formations (Peledui borehole No. 750, depth 1786–1794.5 m; Byuk borehole No. 715, depth 1931–2022 m).

Genus *Palaeolyngbya* Schopf 1968; *Palaeolyngbya catenata* Hermann 1974 (Plate 65, *6*)

Palaeolyngbya catenata: Hermann 1974, p. 8, Plate IV, Fig. 5.

Description. Individual, slightly curved cellular trichomes; cells are 4–5 μm long; the trichome is equally wide lengthwise, i.e. 18–23 μm. Septa between cells are slightly compressed. Trichomes are yellowish-brown to brown in colour.

Distribution. Upper Riphean-Yudomian of the Siberian Platform.

Material. Individual, fairly well-preserved specimens.

Occurrence. Siberian Platform, South Yakutia, the Kursov Formation (Peledui No. 750 borehole, depth 1794–1794.5 m).

Genus *Peleduica* Pyatiletov 1980a; *Peleduica doliiformica* Pyatiletov 1980a (Plate 65, *5*)

Peleduica doliiformica: Pyatiletov 1980a, Plate IV, Fig. 10.

Description. Individual, symmetrical, as a rule, slightly curved trichomes with conical, rounded edges, 16–20 μm wide. Length of cells is 2–3 μm, or one-seventh to one-tenth the width. Septa between cells are not compressed. Five to eight cells commonly form sections 16 to 18 μm. Septa between sections are compressed, hence the barrel shape. Trichomes are brown to dark brown in colour.

Distribution. Yudomian of the Siberian Platform.

Material. A few well- and fairly well-preserved specimens.

Occurrence. Siberian Platform, South Yakutia, the Kursov Formation (Peledui No. 750 borehole, depth 1794–1794.5 m).

7 Silicified Vendian Algae of the Southern Siberian Platform and Altai-Sayan Area

Type Cyanophyta Sachs 1874; Class Hormogonophyceae (Geitler) Elenkin 1934

Order Oscillatoriates Elenkin 1934; Family Oscillatoriaceae (Kirchn.) Elenkin 1934

Genus *Obruchevella* Reitlinger 1948

Obruchevella: Reitlinger 1948, p. 78; 1959, p. 21; Kolosov 1977, p. 73.

Type Species. Obruchevella delicata Reitl.; Siberian Platform; the Sinyaya River, a tributary to the Lena River, Lower Cambrian, the Lena stage, the Sina Formation.

Diagnosis. Calcified and non-calcareous trichomes form a regular spiral, comprising 8 to 13 whorls. Whorls are slightly extended and closely adjoin each other. A longitudinal section demonstrates a circular trichome outline, whose diameter equals the trichome width. If a spiral is curved in different planes, part of it remains undissected. A spiral may be cylindrical, i.e. with equal width of the whorls lengthwise; sometimes it is fusiform, when the width of whorls gradually and symmetrically decreases towards the ends. Some specimens at 600 times magnification show a transverse striation, which may be interpreted as traces of cellular septae.

Discussion. Differs from the genus *Girvanella* Nich. et Ether. in having spirallike trichomes.

Species Composition. Obruchevella delicata Reitlinger 1948; *O. parva* Reitlinger 1959; *O. sibirica* Reitlinger 1959; *O. ditissima* Schipitzyn et Yakshin.

Distribution. Yudomian-Ordovician of the Siberian Platform; Upper Riphean of the Altai-Sayan fold area.

Obruchevella parva Reitlinger 1959 (Plate 71, *1-3*)

Obruchevella parva: Reitlinger 1959, p. 21, Plate VI, Figs. 1,2; Yakshin, Luchinina 1981, Plate X, Figs. 1-3.

Description. Silicified, spirally coiled trichomes with closely adjoining whorls. Width of spiral coils is 35–43 μm; width of the trichome, 9.3–10.1 μm; width of the trichome vesicle, 0.4–0.6 μm; length of spiral, 100–160 μm. If well-preserved , the axial filament is seen on which the entire spiral coils; diameter of the axial filament, 12–14 μm; diameter of the axial zone, 14–20 μm.

Discussion. Similar in shape to *Obruchevella delicata* Reitl. Differs from it in smaller dimensions of all morphological elements.

Distribution. Yudomian of the Siberian Platform.

Material and Occurrence. Numerous specimens of different degrees of preservation. The Lena River, near the Nokhtuisk Village, lower third of the Tinnovka Formation, sample YaM-1/4.

Genus *Oscillatoriopsis* Schopf 1968

Oscillatoriopsis: Schopf 1968, p. 665.

Type Species. Oscillatoriopsis obtusa, Schopf 1968; Central Australia, Late Precambrian, Bitter Springs Formation.

Diagnosis. Individual and clustered trichomes, 5–6 to 50–60 μm across with a distinct cellular structure and the width consistent throughout the entire length. Cells along the height (along the filament) are, as a rule, much smaller than the width of a trichome; and the height to width ratio is characterized by 1/3 to 1/12 values. In the rock, trichomes are both rectilinear and curved as a loop, interwoven, sometimes forming accumulations in the form of "balls". A trichome is often compressed near the transverse septa. The main features of the species are the shape and width of the trichome, and the shape and relation of the linear cell dimensions.

Discussion. A great similarity of morphological elements with modern algae of the genus *Oscillatoria* Vauch. is recorded.

Species Composition. Oscollatoriopsis obtusa Schopf 1968; *O. breviconvexa* Schopf 1971; *O. nochtuica* Yakshin 1980; *O. tomica* Yakshin 1980.

Remarks. In the present work, the extreme limits of the linear dimensions of morphological elements of algae, belonging to the genus *Oscillatoriopsis* Schopf, are changed, since it apparently should include most fossil algae, which are morphologically similar to modern *Oscillatoria* Vauch.

Oscillatoriopsis nochtuica Yakshin 1980 (Plate 72, *3a,b*)

Oscillatoriopsis nochtuica: Yakshin 1980, Plate XI, Figs. 3a,b.

Description. Trichomes are knit into small bundles of three to six trichomes, each forming a sort of loop in the rock. Visible length of trichomes attains 300 μm; their width, 23–27 μm; cell height, 4–5 μm; length to width ratio, 1:7. Thickness of the trichome wall is 1.5–2 μm; the septum between cells, 0.6–0.8 μm. Trichome near each intercellular septum is distinctly compressed.

Discussion. Differs from the type specimen in having a much greater width of the trichome, smaller height to width ratio of the cell and a compressed trichome.

Distribution. Siberian Platform, the Yudoma Complex (Vendian).

Material and Occurrence. In three thin sections, over ten specimens of different degrees of preservation. The Lena River, near Nokhtuisk Village, the Tinnovka Formation, sample YaM-1 (Yakshin's collection).

Genus *Catenuata* Yakshin 1980

Catenuata: Yakshin 1980.

Type Species. Catenuata solida, Yakshin 1980; Siberian Platform, the Lena River, at Nokhtuisk Village, Vendian, the Tinnovka Formation.

Diagnosis. Trichomes differ in length, are straight and curved, with a distinct cellular structure. Width of trichomes is 10 to 70 μ; the cell height to width ratio, about unity. Cells are cylindrical and barrel-shaped. Trichomes are commonly compressed.

Discussion. Morphologically, similar to *Oscillatoriopsis* Schopf 1968; however, the dimensions are much greater.

Species Composition. Catenuata solida Yakshin 1980; *C. implicata* Yakshin 1980; *C. articulosa* Yakshin 1980.

Catenuata solida Yakshin 1980 (Plate 71, *4*; Plate 72, *2*)

Catenuata solida: Yakshin 1980, Plate XI, Fig. 2; Plate III, Fig. 3; Vendian, Tinnovka Formation.

Description. Trichomes are straight or slightly curved; visible length of fragments, 200–250 μm; width of trichomes, 28–35 μm; thickness of vesicle, 2.5–3 μm; cell height, 18–40 μm; thickness of the septum between cells, 1–1.5 μm; cell height to width ratio, about 1. Sometimes trichomes display a slight compression. Apical part of trichomes is acute.

Discussion. Morphologically similar to modern algae of the genus *Oscillatoria* Vauch.

Distribution. Vendian of the Siberian Platform.

Material and Occurrence. Two thin sections, which contain up to ten well-preserved specimens. The Lena River in the vicinity of Nokhtuisk Village, the Yudoma Complex, sample YaM-1 (Yakshin's collection).

Catenuata implicata Yakshin 1980 (Plate 72, *5*)

Catenuata implicata: Yakshin 1980, Plate XI, Fig. 5.

Description. Trichomes are commonly fancifully curved, and occur in the rock as single filaments. Length of trichomes is up to 200–250 μm; width, 11–16 μm; thickness of trichome vesicle, 1–2 μm, height of cell, 6–12 μm; the height to width ratio, 1:2 to 1; thickness of septum between cells, 0.6–0.9 μm. Trichome is commonly compressed in some places. Apical part of the trichome tapers slightly, forming a direct angle in projection.

Discussion. Differs from the type species in the smaller size of morphological elements and curved trichomes.

Distribution. Vendian of the Siberian Platform.

Material and Occurrence. One thin section, several well-preserved specimens. The Lena River at Nokhtuisk Village, the Yudoma Complex. Sample YaM-1 (Yakshin's collection).

Genus *Fistularia* Yakshin 1980

Fistularia: Yakshin 1980.

Type Species. Fistularia volubila Yakshin; Siberian Platform, the Lena River, at Nokhtuisk Village, Vendian, the Tinnovka Formation.

Diagnosis. Tubular forms, gently curved, with a thin transparent vesicle and a dense nucleus of dark brown organic matter with longitudinal fibres. Thickness of tubes, 50 to 150 μm.

Discussion. A tubular shape, fairly similar to the trichomes of blue-green algae; however, it differs from the known fossil forms in having a larger size.

Species Composition. See Type Species.

Fistularia volubila Yakshin 1980 (Plate 73, *1–4*)

Description. Tubes are, as a rule, individual, up to 2–3 mm long, straight or curved. Tubes are 60 to 150 μm thick, mainly 70–90μm; vesicle transparent, brown-yellow, 5–8 μm thick (sometimes 10 –12 μm). Inside the tube, there is a stem, 25 to 40 μm thick, of a dense dark brown substance, seen in places in transmitted light, when a longitudinal, fibrous microstructure can be observed. The stem is ruptured. The stem is inside the tube near the vesicle or adjacent to it and is coiled inside around a very gently sloping spiral.

Chapter 8

Calcareous Algae from the Nemakit-Daldyn Horizon

Calcareous algae, known also as carbonate- or lime-secreting algae, occur in mass abundance starting from Nemakit-Daldyn time. One can assume that they appeared before the Vendian. Their probable occurrences were recorded from the Riphean (Vologdin 1962; Spjeldnaes 1963; Kolosov 1975; and others). From the end of the Vendian, calcareous algae had become rock-forming organisms and major bioherm-builders. In 1969, Voronova (Voronova and Missarzhevsky 1969) for the first time described an assemblage consisting of only four genera of calcareous algae from the Nemakit-Daldyn Horizon, northwestern slope of the Anabar Massif. Later (Voronova and Radionova 1976; Meshkova et al. 1976, 1979; Kolosov 1977; Luchinina et al. 1978; Shishkin and Stepanova 1978; Akulcheva et al. 1981; and others), the assemblage of this stratigraphic level was completed and specified using material from other regions as well. At present, the following nine genera are recorded from the horizon: *Renalcis, Korilophyton, Girvanella, Protoortonella, Botomaella, Gemma, Subtrifloria, Botominella.*

Now the distribution of calcareous algae in the Upper Vendian-Cambrian sections is as follows. On the Siberian Platform, in the Anabar Massif, their first appearance is recorded from numerous exposures of the Nemakit-Daldyn Horizon. *Renalcis polymorphus* (Masl.), *R. gelatinosus* Korde, *Korilophyton inopinatum* (Vor.), *Botomaella sibirica* Vor. and *Protoortonella flabellata* (Vor.) have been found at the Kotui River. Farther east, at the Eriechka and Fomich Rivers, calcareous algae from the Nemakit-Daldyn Horizon are represented by *Renalcis polymorphus* (Masl.), *R. gelatinosus* Korde, *Korilophyton inopinatum* (Vor.), *Girvanella problematica* Nich. et Ether., at the Rassokha River they include *Botomaella zelenovii* Korde. *Renalcis gelatinosus* Korde, *Korilophyton inopinatum* (Vor.) and *Gemma inclusa* Luch. were recorded at the Bolshaya Kuonamka River. In this area, calcareous algae form bioherm beds, 6–12 m thick, which are good markers known as the "Koril" Member (Fig. 16). Anabaritids occur between bioherms and in their peripheral parts, both this biota of calcareous algae and that of first tubular organisms are fairly persistent in time and space (Voronova and Missarzhevsky 1969; Tommotian ... 1969; Meshkova et al. 1976, 1979; Zhuravleva et al. 1982).

Outside the Anabar Massif, the mass appearance of calcareous algae was recorded in the Igarka Basin (northwestern Siberian Platform) from the upper Sukharikha Formation which may be considered as an equivalent of the Nemakit-Daldyn Horizon, where *Renalcis polymorphus* make up bioherms. *Renalcis polymorphus* (Masl.), *Girvanella* sp. and *Proaulopora glabra* Krasnop. were also reported from the uppermost part of the formation, from the upper 15 m of the

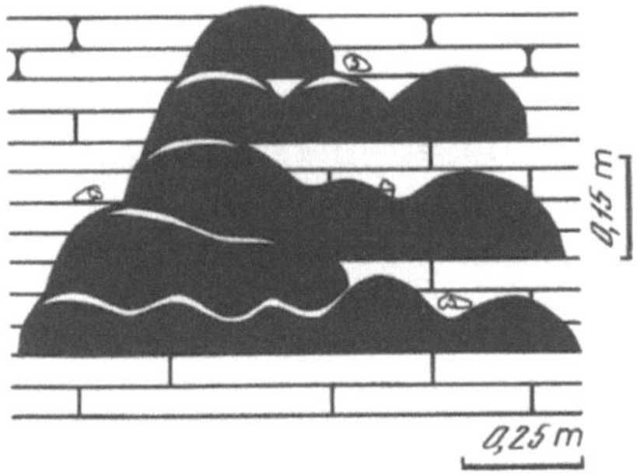

Fig. 16. Fragment of Koril Member at the B. Kuonamka River. Tubular organisms are visible between bioherms

Vendian section (Luchinina 1969; Tommotian ... 1969; Voronova and Radionova 1976).

Of importance are occurrences reported from the middle Lena River (southeastern Siberian Platform), Upper Tolba Formation, which also may be equated to the Nemakit-Daldyn Horizon. Here, *Renalcis galatinosus* Korde and *Korilophyton inopinatum* (Vor.), which build bioherms 18.8 m below the boundary with the Pestrozvet Formation, were reported from borehole No. 1 in the type area. Above occur *Renalcis gelatinosus* Korde and *Subtifloria delicata* Masl. which are also associated with the anabaritids (Zhuravleva et al. 1983).

The occurrence of *Renalcis* sp., *Girvanella* sp. is associated with problematic skeletal organisms as recorded from the Ust Yudoma Formation at the Aim River (Shishkin and Stepanova 1978). Calcareous algae *Obruchevella parva* Reitl. are known from the Nokhtuisk section (Lena River) of the Tinnaya Formation (Reitlinger 1948), the fossilized limestone there yields also varieties of the genus *Obruchevella*, however, their thallus is mineralized, has a complex chemical composition and cellular filaments (Yakshin and Luchinina 1981). The lower Nokhtuisk Formation assigned to the Vendian contains *Renalcis jacuticus* Korde and *R. polymorphus* (Masl.). In different regions of the Baikal, Sayany area and Irkutsk Amphitheater, remains of calcareous algae are also known from the Vendian and represented by the genus *Renalcis* (Khomentovsky 1976; Akulcheva et al. 1981).

The account of the distribution of calcareous algae in the Nemakit-Daldyn deposits shows that they were recorded only from the Siberian Platform and rarely from the adjacent areas. The studied algal assemblage is not diverse, but occupies the same stratigraphic position and its forms exhibit a similar composition.

The Nemakit-Daldyn calcareous algae display a high level of morphological organization. We may almost definitely state that their appearance was very sudden and the calcareous sheath, due to a number of reasons, formed in algae living as early as Riphean. The occurrence of the genus *Obruchevella* Reitl., whose spiral shape of the filaments exhibits three variations, namely, (1) with a calcareous sheath; (2) mineralized in cherts; (3) in the form of phytoleimmas, i.e. with the thallus free of any mineralization, argues for this conclusion (Pyatiletov 1979; Yakshin and Luchinina 1981). In addition, Pyatiletov's macerated material contains the genera *Glomovertella* Reitl. and *Proaulopora* Vologd., whose calcareous thalli are abundant in the Lower Paleozoic deposits (Pyatiletov et al. 1981).

Of special interest are such calcareous forms as *Korilophyton* Vor. and *Gemma* Luch., first recorded in the Nemakit-Daldyn deposits. Voronova (Voronova and Missarzhevsky 1969), who found the genus *Korilophyton* in the Nemakit-Daldyn Horizon (northwestern slope of the Anabar Massif), first considered them as the genus *Epiphyton*. However, a more detailed study showed that *Korilophyton* has short and thick filaments with poorly developed random branching, if visible (Fig. 17).

Luchinina (in Zhuravleva et al. 1982) described the genus *Gemma* Luch. also from the Nemakit-Daldyn Horizon, the northeastern slope of the Anabar Massif. The author recognizes three modifications of *Gemma*: (1) a dense, black sheath of whimsical outline; (2) a grey sheath enclosing small circular bodies; and (3) an empty, pale grey sheath with torn margins adjoining ball-shaped bodies which could be located earlier inside (Fig. 18).

As to the other calcareous algae of this age, they are very common in the overlying Paleozoic deposits and fairly well known. *Korilophyton* and *Gemma* have been recently recorded from the Tommotian of Mongolia (Precambrian-Cambrian boundary ... 1982).

We should emphasize the fairly drastic changes in stromatolite structures in Late Vendian time, when bioherm beds consisted mainly of calcareous algae, though stromatolites have not yet disappeared. Of particular interest are bioherms found at the Kotui River, 15 m below the base of the "Koril" biogenic buildups. They are made up of small 10–15 cm branching stromatolites and forms of *Renalcis polymorphus* (Masl.) (Figs. 19, 20) (Zhuravleva et al. 1982). It is the only known example, showing stromatolites and calcareous algae occurring together; later, algae acquired a leading role in building of biogenic structures. After the advent of calcareous algae, the distribution of stromatolites becomes less common. Bioherms of the so-called Koril Member are similar in structure and can be traced throughout the Anabar sections.

A calcareous sheath in ancient calcareous algae is formed in such a way that it gives rise mainly to a mould reflecting only a superficial appearance of the thallus,

Fig. 17. Diagrammatic sketches of *Korillophyton inopinatum* (Vor.) forms

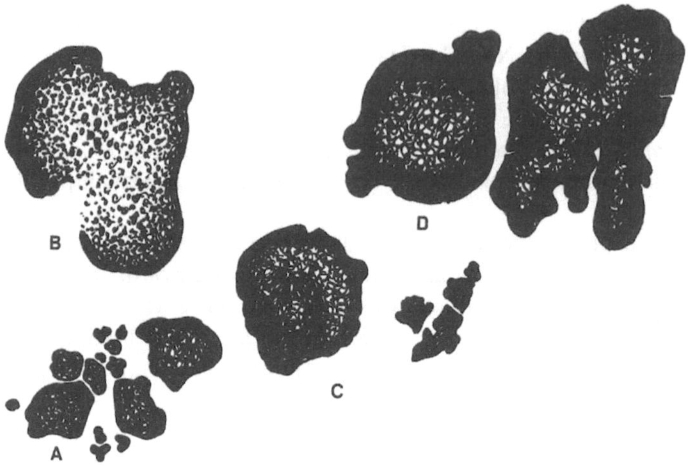

Fig. 18A-D. Diagrammatic sketches of fossil algae of the genus *Gemma* Luch. **A-B** General view of the thallus: **A** colony with closely packed, small, round bodies; **C** young colonies with homogeneous, calcareous sheath with fragments of closely spaced, round bodies inside; **D** round bodies not enclosed by a sheath, apparently an old colony

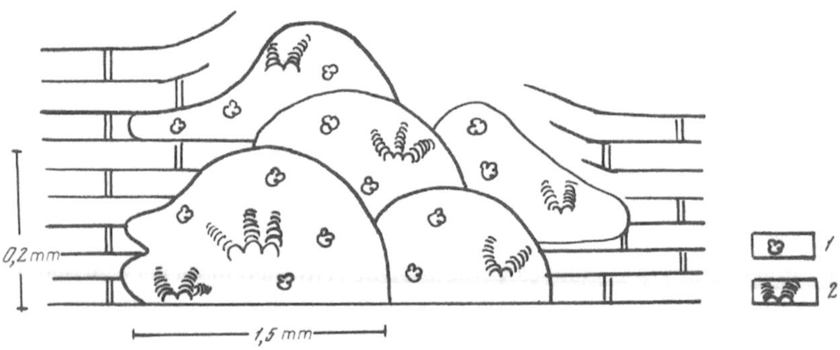

Fig. 19. Bioherms at the Kotui River consisting of branching stromatolites and calcareous algae, *Renalcis polymorphus* Masl. *1 R. polymorphus* Masl.; *2* stromatolites

Fig. 20. Calcareous sheath (sometimes torn); *Renalcis polymorphus* Masl.

therefore, it is difficult to make a detailed reconstruction of its morphological structure. Thus, any statements about a systematic position of ancient algal flora are often inconclusive and remain disputable.

Genus *Korilophyton* Voronova, in Voronova and Radionova 1976

Epiphyton (part.): Voronova and Missarzhevsky 1969, p. 208.

Korilophyton: Voronova and Radionova 1976, p. 83.

Type Species. Epiphyton inopinatum: Voronova and Missarzhevsky 1969, p. 208, Text Figs. 1/1, 2.

Diagnosis. Microscopic, calcareous thallus is apparently attached to the substrate. The thallus consists of upright, short filaments, not always distinctly branching, sometimes lateral surfaces that are knit or coalesced. Branching is scarce, irregular, random. Abundant lateral processes, whose further development leads to the formation of closely spaced clusters. Transverse sections of filaments are circular and consist entirely of dark, fine-grained carbonate (in thin sections).

Species Composition. Two species: *Korilophyton inopinatum* (Vor.), *Korilophyton angustum* Vor.

Remarks. Voronova earlier assigned the forms to the genus *Epiphyton* Born. In fact, they slightly resemble circular transverse sections of *Epiphyton* entirely composed of micrograined carbonate or longitudinal sections of forms with short branches. However, sparce, irregular branching in species of the genus *Korilophyton* is the most characteristic feature allowing recognition of the genus.

Discussion. Korilophyton is akin to some representatives of the genus *Epiphyton*, in particular, to species having fairly short, closely spaced bushes with thick, short branches. The difference is that branching in *Korilophyton* is irregular and always distinct.

Age and Distribution. The Precambrian-Cambrian, Nemakit-Daldyn Horizon and Lower Cambrian, Anabar Massif (Kotui, Kotuikan, Bolshaya Kuonamka, Eriechka, Fomich and Olenek Rivers); Lower Cambrian, Tommotian, West Mongolia.

Korilophyton inopinatum (Voronova, in Voronova and Radionova 1976) (Plate 68, *6*, Plate 70, *6–9*)

Epiphyton inopinatum: Tommotian 1969, p. 208, Fig. 1/1, 2.

Korilophyton inopinatum: Voronova and Radionova 1976, p. 84, Plate XVII, Figs 4,5,6.

Holotype. GIN, No. 3481, thin section 325, specimen 1, Eriechka River, left side, 1.5 km below the Nemakit-Daldyn River; the Nemakit-Daldyn Horizon.

Description. Thalli are sparcely branching, sometimes coalescing and forming clusters irregular in shape with short ramified filaments. Branching is irregular, dichtomous: usually they are the less common of two orders leading to the formation of squat, sometimes curved small bushes. Dimensions of colonies reach up to 3 mm. Individual branches are 45–75 μm in diameter. Branching filaments are 135–200 μm long.

Discussion. The described species differs from *Krilophyton angustum* in the smaller size of the thallus and thinner branches.

Age and Distribution. The Nemakit-Daldyn Horizon, the Precambrian-Cambrian, the Anabar Massif; the Lower Cambrian, the Tommotian of West Mongolia.

Material and Locality. Abundant, well-preserved specimens. Specimen 9–1.0/76, 10–8/72, the Kotui River; specimen 63–15/74, the Bolshaya Kuonamka River; specimen 19/9–72, the Rassokha River; 113/69, 146/69, the Olenek River; specimen M321/1, M321/20, M321/24, the Eriechka River; B79/30, B79/41, West Mongolia.

Genus *Gemma* Luchinina, in Zhuravleva et al. 1982

Gemma: Zhuravleva et al. 1982, p. 100.

Type Species. Gemma inclusa Luchinina: Zhuravleva et al. 1982, p. 100, Plate XII, Figs. 1–4.

Diagnosis. Calcified thallus, mainly shapeless, less common globular, is apparently attached to the substrate, forming lobate processes. Locally, when growing it is divided into portions with readily discernible boundaries of the junction. A young, undeveloped colony with a homogeneous, dense, nearly black sheath, in old colonies, it is distinct along the periphery, sometimes diffused. In some cases, the internal structure is reticulate or cellular, rarely dark, rounded or feebly elongated, small bodies. It is locally scattered in clusters and not enclosed in a sheath.

Discussion. It differs from the genus *Renalcis* Vologd. in the shapeless, angular outlines of the colonies, homogeneous, dense, almost black sheath, but mainly in the internal structure consisting of small spherules.

Remarks. In a general aspect of thalli, their size and internal structure, representatives of the genus *Gemma* resemble modern, blue-green algae of the genus *Chlorogloea* Wille (Gollerbach et al. 1953).

Species of *Gemma* were reported from Siberia only from the "Koril" Member at the Bolshaya Kuonamka River. In these cases, when representatives of the genus were found without a sheath in the form of small bodies in clusters, it is difficult to differentiate them from those of the genus *Korilophyton*, also known from the "Koril" Member and its equivalents.

Age and Distribution. The Precambrian-Cambrian, the Nemakit-Daldyn Horizon, the Anabar Massif, the Bolshaya Kuonamka River; the Lower Cambrian, the Tommotian of West Mongolia.

Gemma inclusa Luchinina, in Zhuravleva et al. 1982, Plate 69, *5*, Plate 70, *1-3*

Holotype. IGiG, No. 677, section 89–57/74; the Bolshaya Kuonamka River, the Anabar Massif; the Nemakit-Daldyn Horizon.

Description. Calcified, shapeless and rounded thalli with dense, nearly black sheath covering the entire colony. The outline of the sheath around the circumference is locally open and torn. Fine, circular bodies, appearing as if composing the colony, occur sometimes inside the colony.

Dimensions. The coalesced colonies attain 5 mm; diameter of the colony is 1.1-1.5 mm; bodies inside colonies, 30–60 μm.

Discussion. Differs from *G. maculosa* Vor. et Drosd. in having a dense, nearly black sheath.

Age and Distribution. The Precambrian-Cambrian, the Nemakit-Daldyn Horizon, the Anabar Massif, the Bolshaya Kuonamka River.

Material and Locality. Above 50 well-preserved specimens. Specimen 63-15/74; 89/52/74; 89–57/74, the Bolshaya Kuonamka River.

Genus *Renalcis* Vologdin 1932

Renalcis: Vologdin 1962, p. 447; Korde 1958, p. 107; 1961, p. 55; 1963, p. 51; Reitlinger 1959, p. 10; Voronova and Missarzhevsky 1969, p. 190; Voronova and Radionova 1976, p. 69; Luchinina 1969, p. 184; *Nubecularites* (pars.): Maslov 1937, p. 336.

Type Species. Renalcis granosus, Vologdin 1932, Middle Cambrian, Altai.

Diagnosis. Calcareous colonies, microscopic, mainly shapeless, rarely more or less globular or elongated. Calcified sheath covers regularly either the entire colony, or only the peripheral portion. Outline of the central portion of the inner hollow is very diverse and filled with light calcite.

Species Composition. Renalcis granosus Vologd., *R. cibus* Vologd., *R. polymorphus* Masl., *R. jacuticus* Korde, *R. pectunculus* Korde, *R. gelatinosus* Korde, *R. levis* Vologd., *R. devonicus* Johnson, *R. tubritus* Wray, *R. jaskovitchi* Luch.

Discussion. In the shape of circular colonies, *Renalcis* is similar to some representatives of the genus *Chabakovia*, in particular to the species with a thin, calcified sheath covering peripheral parts of the colonies. They differ in the irregular cluster pattern of *Renalcis.*

Age and Distribution. The Nemakit-Daldyn Horizon, the Lower and Middle Cambrian, Lower and Upper Devonian of the Siberian Platform, the Altai-Sayany area; Lower and Middle Cambrian of Central Asia, Primorie.

Renalcis granosus Vologdin 1932 (Fig. 21)

Holotype. CGM, Leningrad, No. 354, Altai, Lower Cambrian.

Description. Calcareous colonies irregular in shape with peculiar lobate and elongated processes (in thin section), locally exhibiting radiate structure (Fig. 21). The internal cavity, filled with light carbonate, is poorly visible. Colonies in general are clustered in masses of uncertain outline.

Dimensions. The coalesced colonies attain 3 mm; diameter of a colony, 600–900 μm; sheath, 500–700 μm wide; inner cavity, 0–50μm wide.

Discussion. Colony structure strongly resembles *Renalcis gelatinosus* Korde. The dimensions of the calcareous sheath of coalesced colonies in both species are very similar. They differ in having more rugged colonies than in *Renalcis gelatinosus* and in the radiating-ray structure of the sheath.

Age and Distribution. The Precambrian-Cambrian, the Nemakit-Daldyn Horizon; the Anabar Massif; the Lower Cambrian of Siberia and Far East, Central Asia, West Mongolia.

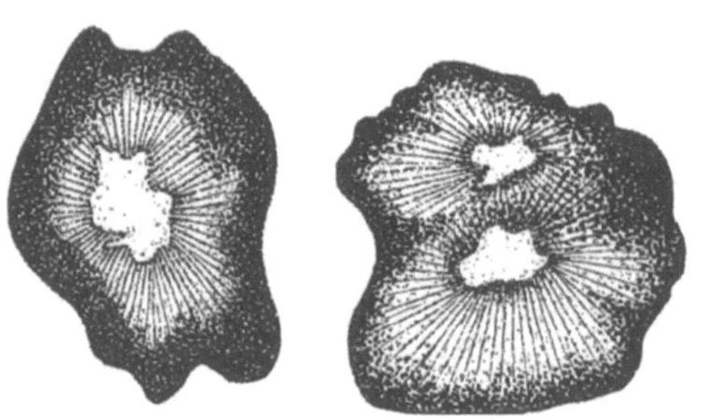

Fig. 21. Radial-ray structure of calcareous sheath: *Renalcis granosus* Vologd

Material and Occurrence. Numerous, well-preserved specimens. Sample No. 113/69, Olenek River.

Renalcis polymorphus (Maslov) 1937 (Plate 68, *1-5*, Plate 69, *1,2,4*)

Holotype. GIN, section No. 2154; Vvedenskoe Village at the Irkut River; the Cambrian.

Description. Large, calcareous colonies of irregular, whimsical outline with small, mainly circular, processes. The sheath is often torn with a very distinct, dark, outer rim. The internal cavity is indistinct, lacks any regular outline.

Dimensions. Coalesced colonies, up to 2 mm; colony diameter, 330 m; sheath width, 75 μm and that the inner cavity, 305 μm.

Discussion. The described species differs from *Renalcis gelatinosus* in having a distinct rim accentuating the sheath. The dimensions of the described species is much larger than those in all other representatives of the genus *Renalcis*. It is noteworthy that the sheath enclosing the colonies, as a rule, is torn.

Age and Distribution. The Precambrian-Cambrian, the Nemakit-Daldyn Horizon, the Anabar Massif; the Lower Cambrian of Siberia, Central Asia, and West Mongolia.

Material and Locality. Eight well-preserved specimens. Sample 8-0.5/76, 8-0.6/76 from the Kotui River; B79/68, West Mongolia; M423/1, the Kotui River.

Genus *Botomaella* Korde 1958

Botomaella: Korde 1958, p. 117; 1961, p. 54; Voronova and Radionova 1976, p. 80.

Type Species. Botomaella zelenovii Korde 1958; Siberia, the Lena River; the Lower Cambrian, the Atdabanian.

Diagnosis. Algal nodule consisting of branching tubular filaments (channels?) forming a bush. Threads in a bush are closely spaced, have a thin sheath and a wide, light internal cavity (in some species). Filaments are either straight or slightly bending; branching is rarely observed. Filaments are variable in diameter, gradually increasing towards the distal ends. Sometimes there is zonal growth at the top portion.

Remarks. Voronova compares *Botomaella* with the genus *Ortonella*. She believes that they differ slightly in the smaller dimensions of the filaments and in the larger slope by branching in *Botomaella*. *Botomaella* has no individual filament, but rather they form a bush.

Species Composition. Four species: *Botomaella zelenovii* Korde, *B. anabarica* Voronova, *B. mitis*, Voronova, *B. sibirica* Voronova.

Age and Distribution. The Precambrian-Cambrian, the Nemakit-Daldyn Horizon; the Lower Cambrian of the Siberian Platform and West Mongolia.

Botomaella zelenovii Korde 1958 (Plate 70, *5*)

Botomaella zelenovii: Korde 1958, p. 117, Plate IV, Fig. 2; 1961, p. 54, Plate 1, Fig. 7.

Holotype. PIN, No. 1297/1399; the Siberian Platform, the Botoma River; the Lower Cambrian, the Atdabanian (Kameshkov Horizon).

Description. Calcified, bushy forms made up of bundles of threads. In a bush, filaments are closely spaced. Branching of filaments is observed occasionally. A bush grows zonally. Tubular filaments are of variable diameter, gradually increasing towards the distal ends.

Dimensions. A bush is 750 μm high; the filament is 15–25 μm wide.

Discussion. The described species differs from *Botomaella sibirica* Vor. in the smaller filament width and zonal growth in the top portion.

Age and Distribution. The Precambrian-Cambrian, the Nemakit-Daldyn Horizon; the Lower Cambrian of Siberia and West Mongolia.

Material and Locality. Three specimens of medium preservation. Sample 41/72, the Rassokha River.

Genus *Protoortonella* Luch., gen. nov.

Rothpletzella: Voronova 1976, p. 82.

Type Species. Rothpletzella igarcaensis Voronova and Radionova 1976, p. 82; Lower Cambrian, Siberian Platform.

Diagnosis. Bushy colonies with adjoining filaments, occasinally branching. The transverse section of filaments is circular, outer margins are unsharp. Cavities, once occupied by filaments, consist of light, pelitomorphic carbonate, hence, the light colour of the entire bush.

Discussion. Differs from the genus *Ortonella* Garw. in having long, thin filaments and a poorly pronounced outer sheath and sparce branching. It differs from the genus *Botomaella* Korde in having much wider filaments.

Species Composition. Two species: *Protoortoonella igarcaensis*, Voronova, *P. flabellata*, Voronova.

Age and Distribution. The Vendian(?)-Lower Cambrian, the Siberian Platform.

Protoortonella flabellata (Voronova in Voronova and Radionova 1976) (Plate 70, *4*)

Rothpletzella flabellata: Voronova in Voronova and Radionova, p. 82, Plate XII, Fig. 6.

Holotype. GIN, No. 4161/55, specimen M-71-8/9; the Olenek River, the Lower Cambrian, the Tommotian.

Description. A bush consists of sparcely branching filaments, 30–40 μm wide, 20–30 μm in diameter; height of a bush, up to 1 mm.

Discussion. Differs from *P. igarcaensis* (Vor.) in the smaller size of bush units.

Age and Distribution. The Nemakit-Daldyn Horizon, the Siberian Platform, the Kotui River.

Material. Four specimens of medium preservation; 3–54.0/72.

Chapter 9

Stromatolites

1 Introduction

Stromatolites are laminar, biogenic-sedimentary structures attached to the substrate. They are mainly carbonate in composition and are formed under the influence of blue-green algae and may represent bacteria that trap fine(silty) detritus and precipitate calcium carbonate. The addition of microlaminae results in the formation of stromatolite crusts, nodules, bioherms or biostromes (Monty 1967 and others). Lamellae may be either continuous, thus composing stratiform stromatolites, or discontinuous. The latter are columnar stromatolites separated by the sediment. They have a variety of gross forms from near-horizontal to markedly convex. Columnar stromatolites differ also in the branching pattern of the columns, the structure of lateral termination and in the shape of columns.

All formal stromatolite classifications are based on the above characters. Here, we used the most widely accepted classification in which a group (formal genus) is defined by the morphology. The holotype (formal species) is determined using microstructure (Semikhatov 1962; Komar 1964, 1966 and others).

The complex nature of stromatolites made some scientists sceptical about a possible relationship between morphology and the microfabric of stromatolites and the composition of stromatolite-building algae. There existed also some doubt as to a possible evolutin of stromatolites in time and their importance for biostratigraphy. Recently, however, siliceous stromatolites of the Gunflint Formation provide evidence for a certain relation between their morphology and microstructure and the microbiota which formed them (Awramic 1976; Awramic and Semikhatov 1979). Though the data cannot be considered as conclusive, they still confirm empirical study of the Precambrian stromatolites, suggesting a change of stromatolite forms in time and the possibility to use them for recognition of several units in the Precambrian (Keller et al. 1960; Krylov and Komar 1965 and others).

The Vendian is one of such divisions characterized by a certain set of stromatolite forms. Despite the absence of stromatolites from the Vendian stratotype, they are very common in the Yudoma deposits of the Siberian Platform which are now considered as an equivalent of the Vendian in Siberia.

The Yudoma stromatolite assemblage is fairly rich and has a rather peculiar composition. It differs greatly from those of the underlying Riphean and overlying Cambrian deposits (Zhuravleva and Komar 1962 and others; Semikhatov et al. 1970; Shenfil 1975a,b and others). Korolyuk and Komar (Korolyuk 1960a,b;

Komar 1966 and others) were the first to report on the assemblage. Later, their statements were developed in a monograph on the Yudoma Complex at the type area (Semikhatov et al. 1970) and a number of papers were devoted to a greater or lesser degree to stromatolites of this age from the Aldan Shield (Krylov et al. 1969), the Patom Highland and Baikal area (Shenfil 1968; Korolyuk and Sidorov 1969, 1971; Khomentovsky et al. 1972; Dolnik and Vorontsova 1974; Dolnik 1978), the Kolyma Massif (Komar and Furdui 1969) and the Karatau (Krylov 1967). There are also descriptions of stromatolites from the Vendian successions (or their possible equivalents) of Spitsbergen (Raaben and Zabrodin 1972), North America (Cloud and Semikhatov 1969) and Australia (Preiss 1973).

Unfortunately, not all stromatolite forms from different localities have been approved at the All-Union colloquia on stromatolites (held in the USSR). Therefore, the scheme presented in Fig. 22 shows original data with respect to the degree of reliability (see legend to Fig. 22).

At present, about 30 forms of Vendian stromatolites, placed into 12 groups, are known. The most common are *Boxonia grumulosa* Kom., *B. allahjunica* Kom. et Semikh., *Paniscollenia emergens* Kom. and *Colleniella singularis* Kom. (see Fig. 22). The above forms occur mainly in the northern and eastern parts of the Siberian Platform.

Boxonia grumulosa Kom. was reported from the lower Karayutekh Formation of the Kharaulakh Mountains and lower Khatyspyt Formation, the Olenek Uplift (Komar et al. 1966 and others). Stromatolites exhibiting a similar structure were reported from the Yudoma Series of the Uchur-Maya area and Staraya Rechka Formation of the Anabar Massif; however, their morphology is inconsistent with the original diagnosis of the group *Boxonia* Korolyuk (Korolyuk 1960a; Komar 1966; Semikhatov et al. 1970; Anon 1972a; Krylov 1975). There are no data on the morphology of *Boxonia grumulosa* Kom. reported from the Lapichi Formation of the East European Platform (Zhuravleva and Chumakov 1968; Bessonova and Chumakov 1969), therefore, no definite assignment to a certain group can be made, On Spitsbergen, *Boxonia grumulosa* Kom. was recorded from the Bakaberg and Elbobreen (Raaben and Zabrodin 1972; Keller et al. 1974). Opinions differ as to the assignment to the Vendian of the above listed formations of Spitsbergen. The solution of the problem, viz. should the Bakerberg and Elbobreen be placed into the Vendian (Khomentovsky 1976) or be assigned to the upper pre-Vendian (Keller et al. 1974a; Chumakov 1978a and others), will show whether *B. grumulosa* is a Vendian form or somewhat older.

Some workers believe that *B. grumulosa* is related to the lower Yudomian of Siberia (Semikhatov et al. 1970). Recent data in general support the statement, if the correction is introduced, that stromatolites with such a microstructure occur not only in the lower formation of the Yudoma Series, but in the lower part of its upper formation as well (Shenfil et al. 1975 and others).

Boxonia allahjunica Kom. et Semikh. was recorded from the upper formation of the Yudoma Series of the Uchur-Maya area, the Porokhtakh Formation on the western slope of the Aldan Shield, the Ayankan Formation of the Baikal area and the Nokhtui Formation of the Patom Highland (Semikhatov et al. 1970; Anon 1972 and others). There is also evidence for its occurrence in the Irkutsk Formation of the

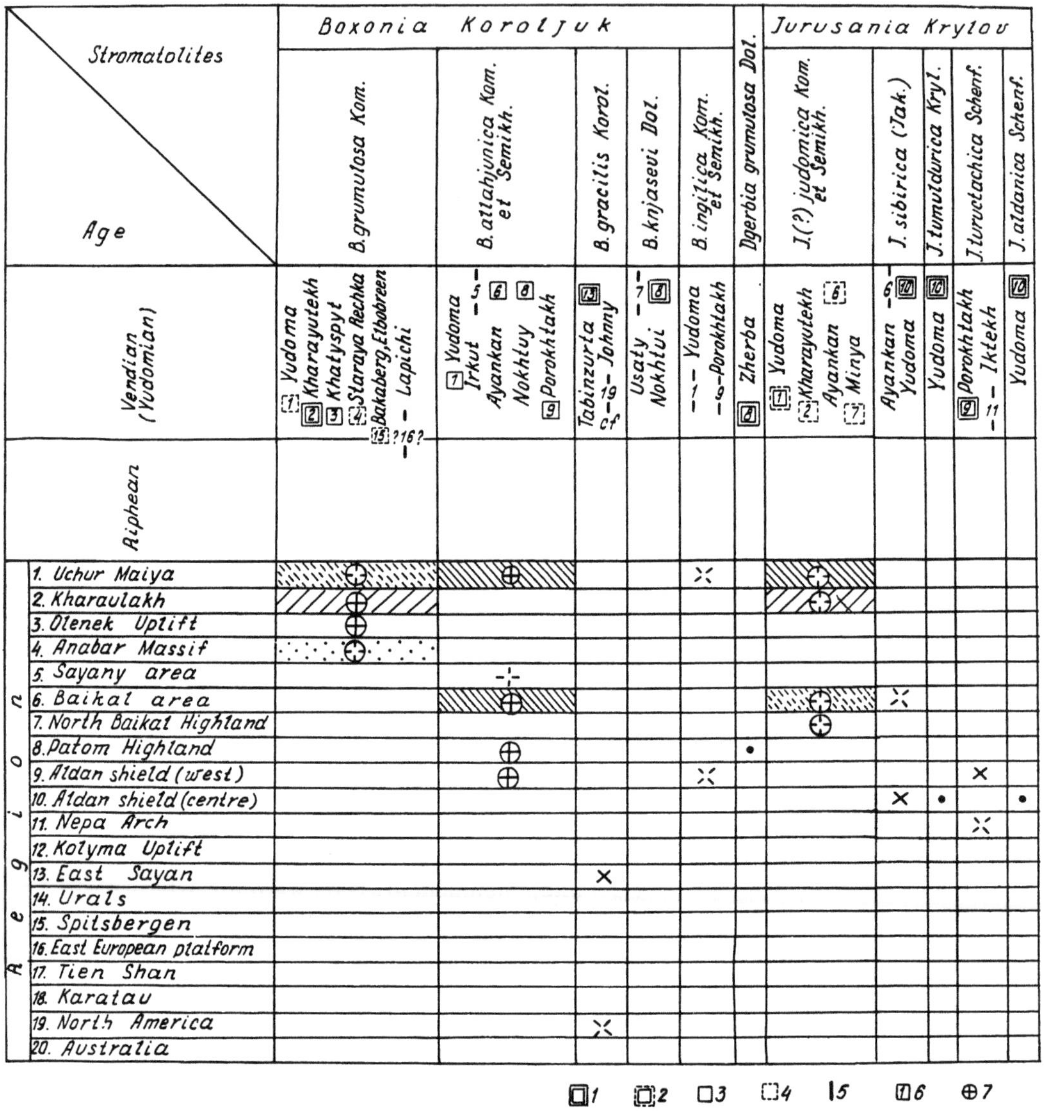

Fig. 22. Distribution of stromatolites in the Vendian deposits. *1* Holotype of stromatolites; *2* holotype with assignment to group not accepted at Colloquia; *3* determination approved by Colloquia; *4* stromatolites whose assignment to groups was either not confirmed or not approved at Colloquia, but the form is accepted; *5* stromatolites presented from published data (not approved at Colloquia); *6* Number of regions corresponding to the number at the bottom of the diagram; *7–9* interregional forms of stromatolites approved at Colloquia: *7* group and form; *8* form; *9* not discussed; *10,11* forms known from two regions: *10* holotypes and forms accepted by Colloquia; *11* published forms not discussed at Colloquia; *12* endemic forms; *13* surviving forms

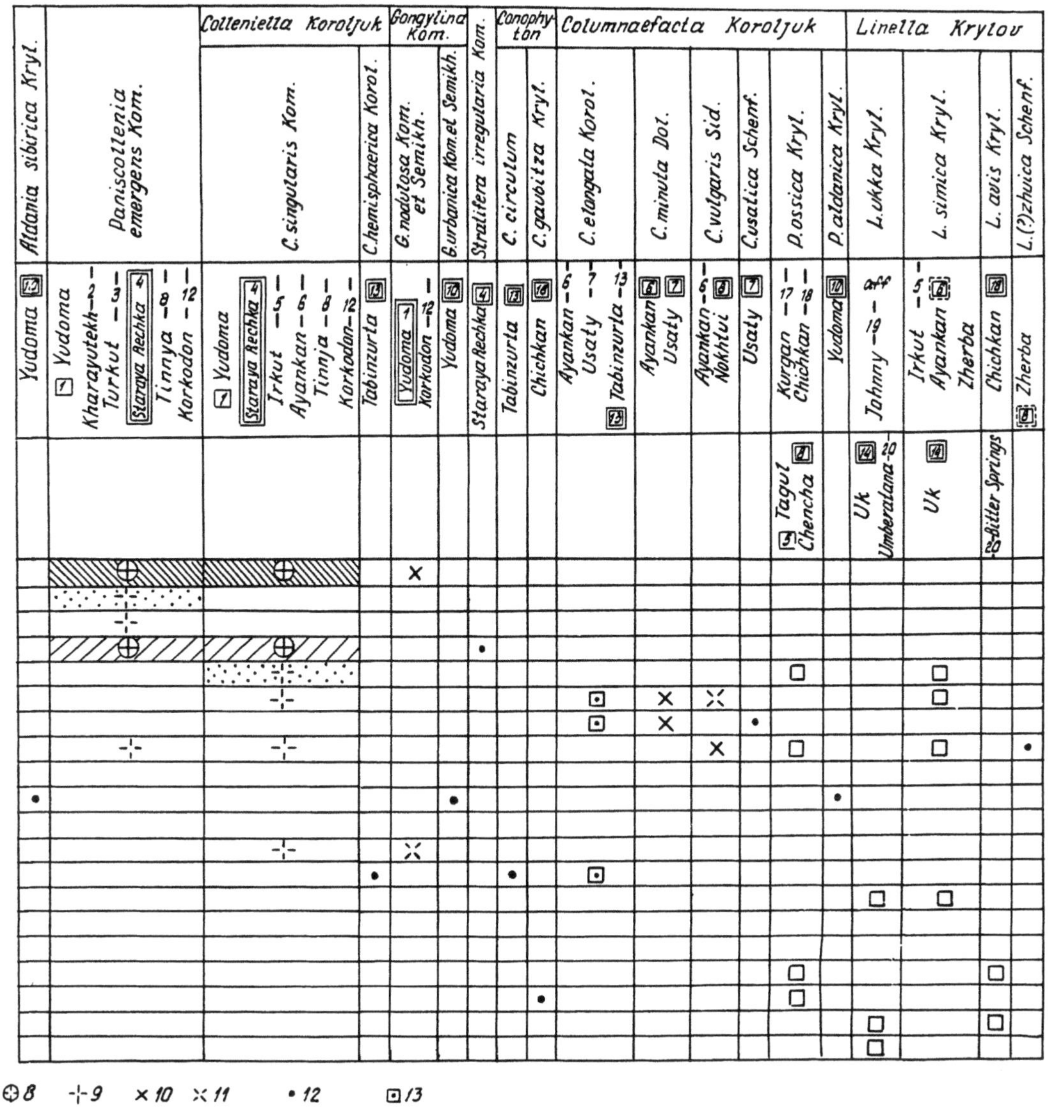

Fig. 22, part II

Sayany Mountains area (see Fig. 22). The discussed form does not appear earlier than the upper, lower half of the Yudomian and it is mainly related to the uppermost Yudomian.

Paniscollenia emergens Kom. shows a wide distribution in the Staraya Rechka Formation of the Anabar Massif (Semikhatov et al. 1970). This form was reported from the Kharayutekh Formation of the Kharaulakh Mountains, Turkut Formation, the Olenek Uplift, and Korkodon Formation, the Kolyma Massif area (Komar 1966; Komar and Furdui 1969).

Colleniella singularis Kom. was recorded from the Staraya Rechka Formation, the Anabar Massif, the Yudoma Series of the Uchur-Maya area (Komar 1966; Semikhatov et al. 1970). It was reported also from the Korkodon Formation, the Kolyma Uplift, the Irkutsk Formation, the Sayany Mountains near Irkutsk, Tinnaya Formation, the Patom Uplift and the Ayankan Formation, Baikal area. In the latter case the validity of its microstructure definition has been confirmed at the Colloquium on stromatolites (Anon 1972a), though no morphological reconstructions were presented (Komar and Furdui 1969; Korolyuk and Sidorov 1969; Anon 1972a).

There is a description of *Jurusania judomica* Kom. et Semikh. from four regions (Uchur-Maya area, Yudoma Series; Kharaulakh Mountains, Kharayutekh Formation; North Baikal area, Minka Formation; Baikal area, Ayankan Formation); however, in all the cases of assignment to a certain group, even the holotype from the Yudoma Series has not been approved by the Colloquium on stromatolites (Anon 1972).

A number of stromatolite forms are known from no more than two regions (*Boxonia gracilis* Korol., *B. knjasevi* Dol., *B. ingilica* Kom. et Semikh., *Gongylina nodulosa* Kom. et Semikh., *Jurusania sibirica* Jak., *J. tuructachica* Schenf., *Collumnaefacta minuta* Dol., *C. vulgaris* Sid.), in all cases except for *Collumnaefacta minuta* Dol., occurrences from the second regions have not been compared at the Colloquia on stromatolites with holotypes and often the latter also have not been discussed (see Fig. 22).

A large number of the Yudomian forms are endemics known only from one region. Among them are: *Dgerbia grumulosa* Dol., *Jurusania tumuldurica* Kryl., *J. aldanica sibirica* Kryl., *Collumnaefacta usatica* Schenf., *Patomia aldanica* Kryl., *Gongylina urbanica* Kom. et Semikh., *Colleniella hemisphaerica* Korol., *Stratifera irregularia* Korol., *Conophyton gaubitza* Kryl. and *C. circulum* Korol.

Of special interest are stromatolites of the groups *Patomia* Krylov and *Linella* Krylov.

Later, it became necessary to introduce a correction to the original statements about the assignment of *Patomia* to the Vendian or to the Upper Riphean boundary beds (Krylov 1967). Dolnik (Dolnik and Vorontsova 1975; Anon 1975) reported *Patomia ossica* Kryl. from the Tagul Formation of the Karagas Series, i.e. from the fairly lower horizons of the Upper Riphean (Dolnik and Vorontsova 1972; Anon 1975).

Representatives of the group *Linella* Krylov are very common in deposits underlying the Vendian. For example, *Linella ukka* Kryl. was first described from the pre-Vendian deposits of the Uksa Formation, the Urals (Krylov 1967). In Australia, it was reported from the Etina Limestone, the Amberetana Series, along with Upper Riphean *Inzeria tjomusi* Kryl. (Preiss 1973, 1976 and others), also from the pre-Vendian deposits much below the level yielding the *Ediacara* fauna. *Linella aff. ukka* Kryl. was described from the Johnny Formation, North America, along with *Boxonia cf. gracilis* Korol. (Cloud and Semikhatov 1969). Most scientists agree to the assignment of the Johnny Formation to the Vendian (Khomentovsky 1976 and others).

Linella simica Kryl. also was reported from the Uksa Formation (Krylov 1967). Its occurrence in the Yudoma Series of the Aldan Shield has not been confirmed because the discussion at the Colloquium on stromatolites showed, apart from a similar microstructure, some important morphological differences preventing its placement into the group *Linella. Linella simica* Kryl. was reported from the Vendian (Yudoma) deposits of southern Siberia: from the Ayankan Formation, Baikal area, the Zherbina Formation, the Patom Uplift and the Irkutsk Formation, the Sayany Mountains near Irkutsk (Korolyuk and Sidorov 1969; Dolnik and Vorontsova 1974; Vorontsova et al. 1975); however, only the microstructure of forms from the first locality has been discussed at this Colloquium (Anon 1978) (see Fig. 22).

Linella avis Kryl. was recorded from the Vendian deposits of the Chichkan Formation, the Karatau Ridge (Krylov 1967), and later it was described from fairly low beds of the Upper Riphean of Australia, the Bitter Springs Formation. It was dated at about 900 m.y. (Glaessner et al. 1969).

Some doubts have been expressed as to placement of *Linella zhuica* Schenf. into the group *Linella* Krylov (Anon 1972a, Plate I).

As mentioned, the Vendian stromatolite assemblage is rather peculiar and differs from those of the underlying and overlying deposits. Though some Vendian stromatolites are known from the Riphean (see above), actually all characteristic Upper Riphean forms from groups *Inzeria, Minjaria* and *Gymnosolen* disappear at its lower boundary. Quite new forms appear in groups *Jurusania* Krylov, *Boxonia* Korolyuk, *Paniscollenia* Korolyuk, *Colleniella* Korolyuk and *Columnaefacta.*

At the upper boundary of the Vendian, all forms of the Vendian stromatolite assemblage (except for *Collumnaefacta elongata* Korol.) disappear and new stromatolite groups, *Vetella, Charaulachia, Ilicta, Tunicata, Uricatella, Chaiella,* along with new forms in the groups *Boxonia, Parmites, Collumnaefacta, Colleniella* and *Paniscollenia,* show their appearance in the Lower Cambrian deposits. Representatives of two new groups were recorded from the lowermost Cambrian, whereby the remaining stromatolites start to appear from the Elgyan or Tolbachan Horizons.

Thus, the Vendian deposits contain a characteristic set of stromatolites; however, a thorough analysis shows that it should not be defined as a characteristic Vendian index assemblage, but rather as an association of forms found in the Vendian deposits. In fact, some forms are known from older deposits (groups *Patomia* and *Linella*) and they can be used for a reliable dating of Vendian rocks which contain them and, in contrast, they cannot be used for substantiation of the lower boundary of the Vendomian or Kudash because they occur together with Upper Riphean stromatolites. There is no evidence indicative of the possible range of most of the above listed endemic forms (see Fig. 22). Only some forms have a wide distribution (i.e. occurring in more than three regions) and they may form a basis for recognition of a Vendian stromatolite assemblage. Among them are *Boxonia grumulosa, B. allahjunica, Paniscollenia emergens, Colleniella singularis, Jurusania*(?) *judomica*. The complete assemblage is known only from the Uchur-Maya area and the presence here of *Boxonia glumulosa* Kom. is quite inconclusive. Furthermore, the assignment of the form to a certain group has not been approved (Fig. 22). The same holds true for an assemblage from the Staraya Rechka Formation, the Anabar Massif, in which *B. allahjunica* and *Jurusania*(?) *judomica* are absent as well. Three forms of the above five, two being tentatively accepted, are known from the Patom Highland. Two or three index forms were recorded from the Olenek Uplift, the Kharaulakh Ridge and Baikal area (Fig. 22).

Occurrences of one of the above forms are known from some regions, however, in most cases index forms are either absent or have not been found as yet (see Fig. 22). To a certain degree this gap is filled by the presence of forms occurring in other regions in the Vendian. For example, in the Porokhtakh Formation (Vendian), the Aldan Shield, *Jurusania tuructachica* is associated with *Boxonia allahjunica*, while on the Nepa Arch it was recorded from the Iktekh Formation. The geological position and paleontological data relevant to other fossil groups support the Vendian age of the Iktekh Formation. However, it would be incorrect to judge its Vendian age exclusively on the basis of *Jurusania tiructachica* without formation on the geological environment because occasional finds do not show the real range of its distribution. This statement can be illustrated by the history of the evolution of ideas on the position in a section of *Patomia ossica* and *Linella avis* described from the Chickhan Formation (Vendian) of the Karatau Ridge (Krylov 1967). Later, as mentioned above, both forms were found in the Riphean deposits (Glaessner et al. 1969; Dolnik and Vorontsova 1972).

It is difficult to give a more detailed subdivision of the Vendian by means of stromatolites. Above we showed that the occurrence of *Boxonia grumulosa* is related to the lower half of the Vendian but the available data at present preclude substantiation of this subdivision.

In summary, it should be noted that despite the presence of a peculiar stromatolite association related to the Vendian, it is difficult to use them for subdivision and correlation of Vendian deposits. However, the wide distribution of stromatolites in the Vendian, the peculiar features of the Vendian stromatolite association and the examples showing the usage of stromatolites for subdivision and correlation of the Precambrian (e.g. correlation of the Yudoma Series with Staraya Rechka, Kharayutekh and some other formations) are critical considerations, indicating their importance for Precambrian biostratigraphy. At present, many difficulties arise from the poor knowledge of Vendian stromatolites in our country and in western countries as well. This gap should be filled by the elucidation of the stratigraphic importance of stromatolites.

2 Part I

Collumnaefacta usatica Schenfil (Plate 76, *1–3*; Fig. 23)

Collumnaefacta usatica: Khomentovsky et al. 1972, p. 327, Plate XXXIV, Text Fig. 71.

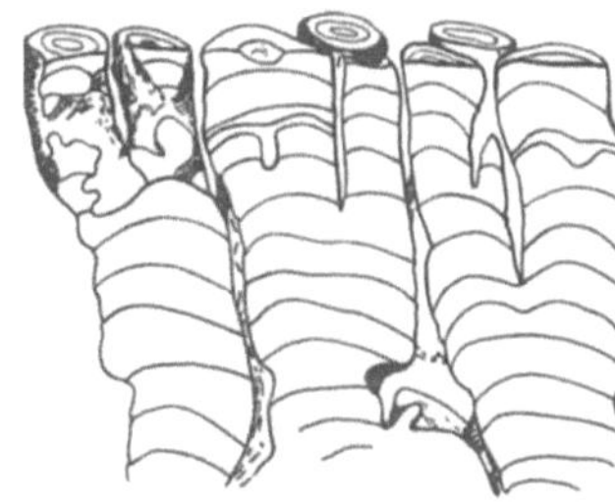

Fig. 23. Shape of columns of *Collumnaefacta usatica* Schenf., Usaty Formation, Chaya River, natural size

Holotype. IGiG, No. 310/13, North Baikal Highland, the Chaya River, 18 km above the mouth of the Limpeya River; the Usatov Formation.

Diagnosis. Collumnaefacta with an indistinct envelope of the column lateral surface.

Description. Straight, vertical, even columns locally slightly expanding upward. Cross-sections are circular or ovate. Columns are 5–15 cm high and 1-3 cm thick. Intercolumnar spaces are very narrow, 1-2 mm. Columns start growing from a thin (1-2 mm), stratiform stromatolite disintegrating upward into columns which, in the upper portion of the structure, divide in turn into two to three new columns but not increasing the entire width of the structure. The lateral surface of the column is even, smooth, with scarce, fine tubercles and individual lips. Microlaminae (lamellae) either break off sharply towards the intercolumnar space or bend slightly down enveloping the preceding microlaminae.

Lamination is distinct. Laminae are gently convex. The arch height to column width ratio is 0.2–0.4. Lamination is mainly inherited.

Light microlaminae (0.2–0.4 mm) are composed of fine (0.05–0.1 mm), pale, round dolomite grains.

Dark microlaminae are thinner (0.05–0.1 mm). They consist of even smaller (0.01 to 0.03 mm) dolomite grains and dark pelitomorphic carbonate filling interstices between grains.

Discussion. Differs from *Collumnaefacta elongata* Korol. in having an indistinct envelope of microlaminae and a peculiar microstructure.

Distribution and Age. North Baikal Highland, the Chaya River, the middle Usatov Formation.

Jurusania aldanica Schenfil (Plate 76, *4–6*; Fig. 24)

Jurusania aldanica: Khomentovsky et al. 1972, p. 324, Plate XXXVII, Text Fig. 72.

Holotype. IGiG, No. 310/9; the Aldan River, the Upper Yudoma Formation.

Diagnosis. Jurusania with a cloud-shaped microstrucutre.

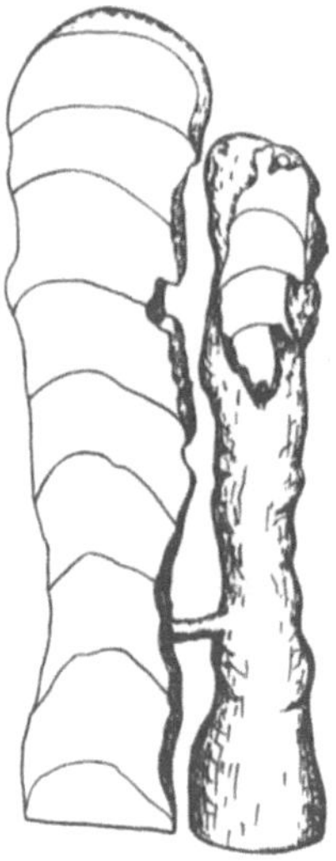

Fig. 24. Shape of columns of *Jurusania aldanica* Schenf., Yudoma Formation, Aldan River, × 2/3

Description. Even, subcylindrical, locally feebly torulose columns, vertically or slightly obliquely arranged in the bed. Columns are 10–20 cm high and 0.5–2.0 cm thick. The cross-section is circular or ovate. Disintegration into columns is multiple in the upper part of the structure. Commonly, an original column disintegrates into two to three smaller new columns. The lateral column termination is distinct. Microlaminae break off bluntly towards the column margins, sometimes forming small lips. Some columns are fringed by the light rim.

Lamination is irominent. Laminae are fairly consistent, gently convex, steeper at the column margins. The arch height to column width ratio is 0.2–0.4. Lamination is mainly inherited.

Pale grey microlaminae (0.5–3.0 mm) consist of fine (0.003–0.01 mm) carbonate grains. Locally, they become light due to weak, secondary recrystallization. In this case grains attain 0.05 mm in size.

Dark grey microlaminae (0.3–1.0 mm) consist of even much finer carbonate grains (0.001 mm and below). They either form consistent, dark laminae easily distinguishable from light ones or cloud-shaped clusters (0.05–0.2 mm); in this case the shape of laminae changes.

Discussion. Differs from *Jurusania sibirica* (Jak.) in microstructure pattern, all features being similar, and from *Jurusania cylindrica* Kryl in having less prominent lips and cornices.

Distribution and Age. Aldan Shield, the Aldan River, the upper Yudoma Formation.

Jurusania tuructachica Schenfil (Plate 77, *1,2*; Fig. 25)

Jurusania tuructachica: Khomentovsky et al. 1972,p. 325, Plate XXXVIII, Text Fig. 72.

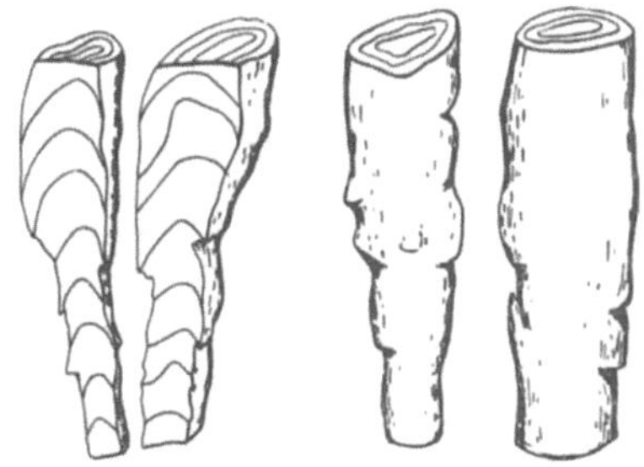

Fig. 25. Shape of columns of *Jurusania tuructachica* Schenf., Porokhtakh Formation, Olekma River, × 2/3

Holotype. IGiG, No. 310/10; the Olekma River, at the mouth of the Turuktakakh Creek; the Porokhtakh Formation.

Diagnosis. Jurusania with partially weak envelope of microlaminae and laminar-clotted microstructure.

Description. Even, subcylindrical or feebly flattened vertical columns 0.5–2.0 cm thick and 8–15 cm high. The cross-section is ovate with a complex, undulating outline. At the base, columns are thin (0.4–0.5 cm), widening slightly upward. Division into two columns of smaller diameter is rare. The lateral termination of columns is distinct due to the rim of pale calcite enveloping the column. In some portions, microlaminae envelop the lateral surface of the column, but more often they break off steeply towards the intercolumnar space or form overhanging lips. Lamination is distinct. Pale grey and brownish-grey carbonate laminae alternate and somewhat thicken in the axial portions of the columns. Laminae are convex in shape. The arch height to column thickness ratio is 0.5–1.0.

Microfabric is clotted.

Light lamellae (0.08–0.3 mm) are composed of fine (0.01–0.05 mm) grains of light carbonate. Grains sometimes slightly decrease in size from the bottom up to the lamella.

Dark lamellae (0.08–0.3 mm) consist of fine (0.005 mm and below) grains of pelitomorphic carbonate forming either consistent, platy lamellae thinning towards the column margins or continuous laminae and rounded clots (0.02–0.1 mm, rarely 0.2 mm).

Discussion. Differs from *Jurusania sibirica* in the character of microfabric and from structurally similar *Jurusania aldanica* in having a partial envelope and more convex lamellae.

Distribution and Age. Aldan Shield, the Olekma River, the Porokhtakh Formation; the Nepa Arch, the Iktekh Formation, Vendian.

Jurusania sibirica (Jak.) (Plate 77, *3*, 4; Fig. 26)

Gymnosolen sibiricus: Yakovlev 1934, p. 589–591.

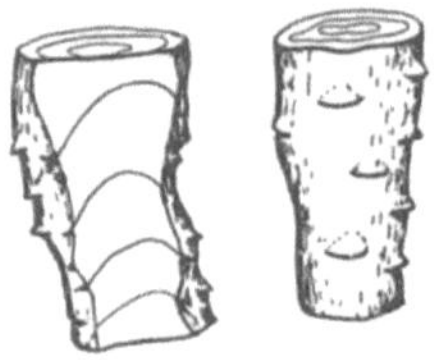

Fig. 26. Shape of columns *Jurusania sibirica* (Jak.), Yudoma Formation, Aldan River, × 2/3

Jurusania: Shenfil 1965b, p. 39.

Jurusania sibirica (Jak.): Semikhatov et al. 1967, p. 1119–1122.

Holotype. Lost, topotype-GIN, No. 3590/3; the Aldan River, Ugino Village, the upper member of the Yudoma Formation.

Diagnosis. Jurusania with rare lips and thin, consistent laminae composed of dark, pelitomorphic carbonate separated by wider, fine-grained, light lamellae.

Description. Even, subcylindrical columns arranged vertically in the bed, sometimes slightly inclined. The thickness of columns is 0.5–2.5 cm, height, 10–20 cm. The cross-section is ovate, often complicated with cornices. Division into columns is scarce, but simple with disintegration into two columns of smaller diameter. Lateral termination of columns is fairly distinct. Laminae break off abruptly towards the intercolumnar area or form lips. On some portions, dark lamellae do not reach the margins, ending 1–1.5 mm from it. As a result, columns have a thin envelope of light, fine-grained carbonate and, therefore, they are smooth on these portions.

Lamination is fairly distinct. Laminae are slightly convex, sometimes almost flat, turned down at the margin of a column. Laminae are inherited. The arch width to column width ratio is 0.2–0.3.

Light lamellae (0.2–1.0 mm) are composed of fine (0.02–0.05 mm), rounded grains of light carbonate.

Dark laminae (0.05–0.3 mm) consist of dark, pelitomorphic carbonate evenly scattered among fine (0.005–0.01 mm), rounded grains of light carbonate. Thicker lamellae are commonly lighter than thinner ones, sometimes they disintegrate along the strike into several thinner lamellae.

Discussion. Differs from *Jurusania cylindrica* Kryl. in having smoother walls of columns, less distinct rim and in the character of the microfabric.

Distribution and Age. Aldan Shield, the Aldan River, the upper Yudoma Formation; Baikal area, the Ayankan Formation.

3 Part II

Dgerbia grumulosa Dol. (Plate 78, *1, 2*; Plate 81, *1*)

Dgerbia grumulosa: Dolnik and Vorontsova 1974, p. 71–72, Plate VII.

Holotype. IGU, No. 572/10; Patom Highland, the Zhuya River; the Zherbina Formation.

Diagnosis. Dgerbia with fine, clotted fabric.

Description. Columns small, "shaggy", attain 1–7 cm in height and 0.2–4 cm in diameter. Branching is not always distinct, commonly there is disintegration into two, sometimes three short dactylic columns. Columns are often coalesced in the upper portion. The lateral surface of columns is very uneven with small processes, sometimes indented; wall-less. Arches are semiconvex, often indistinct. Laminae are not inherited.

Lamination is indistinct, not observed in each column, due to alternation of very uneven, corroded, dark laminae 0.1–0.2 mm wide, separated by lighter laminae approximately of the same width. Both laminae are often discontinuous lenticular or from irregular nests. Columns with indistinct lamination are made up of dark cellular tissue. Cells are composed of light carbonate. They are 0.05–0.1 mm in size. Small cells are usually ovate and oriented along the lamination. Large cells are irregular in shape and sometimes arranged across the lamination. Configuration and orientation of some cells may well be attributed to secondary transformations.

Structure of Laminae. Dark laminae are composed of pelitomorphic carbonate and exhibit a subclotted structure. Individual clots attain 0.01–0.03 mm. Light laminae are made up of fine-grained carbonate (0.01–0.01 mm).

Discussion. Dgerbia grumulosa is similar in microfabric to *Boxonia grumulosa* Kom., hence, the general specific name of the forms discussed. Moreover, the forms differ greatly in gross morphology, character of the lateral surface and in the absence of walls in *Dgerbia grumulosa.*

Age and Distribution. The lower Zherbina Formation, the Yudoma Complex, the Zhuya, Bolshoi Patom, Lena and Chencha Rivers.

Material. Twelve samples from six exposures.

Collumnaefacta minuta Dol. (Plate 79, *1*; Plate 81, *2*)

Collumnaefacta minuta: Dolnik and Vorontsova 1974, p. 72, Plate X.

Holotype. IGU, No. 48; Baikal area, Sokhatiny Creek (tributary to the Lena River); the Usolka Formation.

Diagnosis. Small *Collumnaefacta* with fine, simple lamination and multiple connecting laminae bridges.

Description. Columns are small, closely spaced subvertically; 0.7–1.5 in diameter, 2–3 cm high.

Branching is indistinct in the type specimen; in other specimens, columns disintegrate into two, subparallel ones.

Lateral surface is fairly even, distorted by transitional laminae often connecting adjacent columns. Wall is seen on many portions of the columns.

Arches are gently convex in shape. Laminae are generally inherited.

Lamination is distinct due to alternation of thin (0.02–0.05 mm), dark and wider (0.2–0.5 mm) light laminae.

Structure of Laminae. Dark and light laminae exhibit homogeneous, granular structure. Dark laminae are composed of pelitomorphic carbonate, light ones consist of fine-grained carbonate with scattered, small-sized grains.

Discussion. Differs from the typical form of the group (*Collumnaefacta elongata* Korol.) and also from other forms in having smaller dimensions and thinner lamination.

Age and Distribution. The lower Lower Cambrian, the Yudoma Complex. The Usolka Formation of the Baikal area (Lena River-Sokhatiny Creek, the Ayankan Formation of the Baikal area (Khidusa River), the lower member of the Usaty Formation of the North Baikal Highland (Chaya River).

Boxonia knjasevi Dol. (Plate 80, *1*; Plate 81, *3*)

Boxonia knjasevi: Dolnik and Vorontsova 1974, p. 74, Plate XII.

Holotype. IGU, No. 583/159. The lower member of the Nokhtuisk Formation, the Lena River at Nokhtuisk Village.

Diagnosis. Boxonia with distinct lamination and multilayered wall consisting mainly of light laminae.

Description. Columns are small, subcylindrical, sometimes feebly tuberculate. Some have a common base from which three four columns detach. Columns are 1.5–4 cm in diameter and over 7 cm high, branching into two equidiametric columns.

Lateral surface is even. Light laminae envelop the lateral surface and form a multilayered wall.

Arches are in general gently convex, rarely medium convex. As a whole, laminae are inherited.

Lamination is distinct: alternating light and dark laminae. The most consistent are light laminae (0.1–1.0 mm) commonly seen throughout the entire column and forming a wall. Dark laminae (0.1–1.5 mm) resemble discontinuous lenses and ribbons. Contacts between the dark and light laminae are uneven.

Structure of Laminae. Dark laminae are composed of pelitomorphic and fine-grained carbonate and exhibit an irregular, mottled fabric. There is a combination of fine-grained, dark carbonate and pockets of lighter, apparently, secondary carbonate. The pockets are mainly 0.2–0.5 mm in size and oriented mainly parallel to the lamination. Light laminae are composed of fine-grained carbonate and exhibit a more homogeneous fabric than the dark ones.

Discussion. Boxonia knjasevi differs from *Boxonia gracilis* Korol. and *B. lissa* Kom. in the lack of columnar microstructure and in having the wall mainly composed of light laminae. The form discussed differs from *Boxonia grumulosa* Kom. in having more distinct lamination and its columns lack clotted fabric.

Age and Distribution. The Yudoma Complex, the lower member of the Usaty Formation, North Baikal Highland (Chaya River).

Material. Fourteen samples from two exposures.

Boxonia zharkovi (Korol.) Dol., forma nov. (Plate 79, *2*; Plate 80, *2*; Plate 81, *4*)

Holotype. IGU, No. 339/31; Patom Highland, the Chuya River, below the Beryezovy Creek; the lower member of the Pokhtui Formation.

Diagnosis. Boxonia with fine-clotted fabric.

Description. Columns are 10–15 cm high and 3–5 cm in diameter, subcylindrical, often with a common base.

Branching into two to three equidiametric columns, sometimes there are short, dactylic daughter columns.

Lateral surface is even. The wall is seen on many portions, some portions are wall-less. The margin of a column is locally altered and recrystallized.

Arches are mainly medium, rarely steeply convex.

Lamination is fairly distinct, alternating thin, dark and wider, lighter laminae. The width of dark laminae is rather consistent and varies within the small range of 0.05–0.25 mm. The width of lighter laminae is inconsistent (0.5–2.0 mm) and varies both within the lamina and from lamina to lamina. The transition between dark and light lamina is gradual. Some laminae display quite indistinct lamination, dark laminae are inconsistent and form separate, irregular lenses and portions.

Structure of Laminae. Both dark and light laminae are composed of small (0.01–0.03 sometimes to 0.05 mm) clots of dark, fine-grained carbonate. The darker laminae differ from light ones in having clots feebly discernible, closely spaced, sometimes oriented along the lamination and aligned in chains. In light laminae, clots are more isolated and separated by light, fine-grained carbonate, commonly randomly arranged and not oriented. Some clots have a light nucleus.

Discussion. Fine-clotted microfabric of *B. zharkovi* strongly resembles *B. grumulosa* Kom. and *B. gracilis* Korol. Differs from *B. gracilis* in having no light, incrusted laminae and from *B. grumulosa* Kom. in smaller clots (almost one-half of the former).

Age and Distribution. The Yudoma Complex, the lower member of the Nokhtui Formation (Chuya River).

Material. Five samples from two exposures.

4 Part III

Group Aldania Krylov, in Krylov et al. 1969

Diagnosis. Transverse, costate columns arranged vertically or obliquely in the bed. The growth starts from a common stratiform crust pillow from which a great number of narrow, closely spaced columns with dense and multiple branching in the lower portion of the structure detach upward and sideways. In the marginal portion of a column, laminae are thin, bend down and terminate at different distances from the margins, forming small transverse ridges without large cornices or lips (Krylov et al. 1969 p. 200).

Aldania sibirica (Jakovlev) Krylov et al. 1969 (Plate 84, *1, 2*; Figs. 27, 28)

Gymnosolen sibiricus: Yakovlev 1934, p. 589–590.

Aldania sibirica: Krylov in Krylov et al. 1969, p. 202–203, Figs. 50, 51; Plate XXXVII, Figs. 3–5.

Holotype. GIN, specimen No. 3599/7, slide K-265; the left side of the Aldan River, 1.5 km below Vaskin Protoka (Vaskin Arm); upper horizons of the Yudoma Formation.

Description. Subcylindrical columns, straight or bent, arranged vertically or obliquely in the central part of nodular or subspherical bioherms up to 3–7 cm long and 2–3 cm high. The columns are 1–2, rarely 3 cm thick and 1 cm high. Cross-section is circular or ovate. Columns begin growth from a common stratiform "pillow" at the base of a bioherm, branching into three, four and more new columns which, in the upper part of the bioherm, are oriented subparallel and which rarely branch into new ones. The lateral surface with small transverse ridges is without prominent cornices or lips. There are connecting bridges. Columns are oriented fanlike; the lamination is distinct; on the whole, arches are domelike; wide columns are flattened and rectangular. Laminae are composed of alternating dense, fine-grained and coarse-grained limestones. Irregular distribution of iron oxides and clay gouge make lamination more prominent.

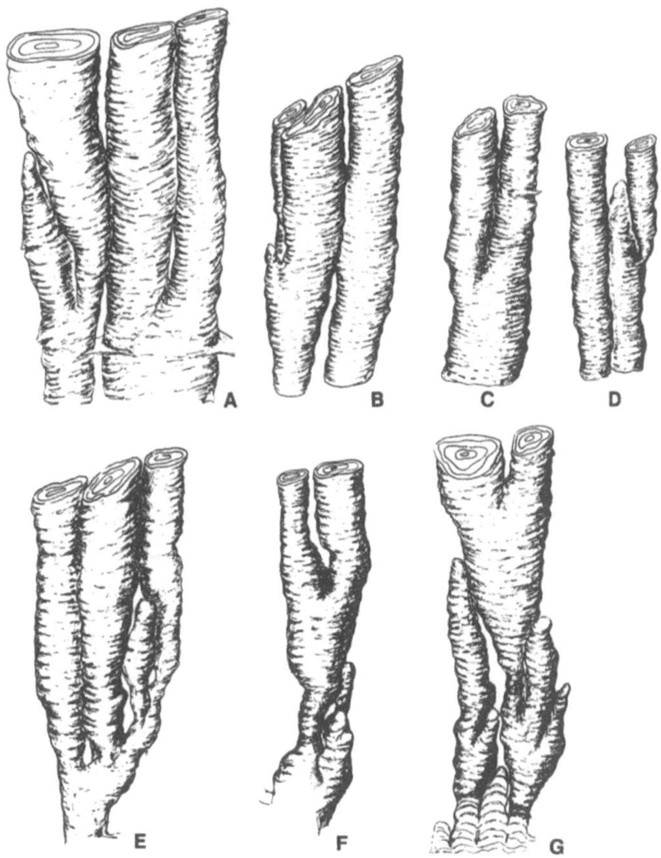

Fig. 27A-G. Shape of columns of *Aldania sibirica* (Jak.) Kryl. **A** Sample 3599/6; **E-F** sample 3599/7, Aldan River near Vaskin Arm; **B-D,G** sample 3599/8, Aldan River near Uugun Village

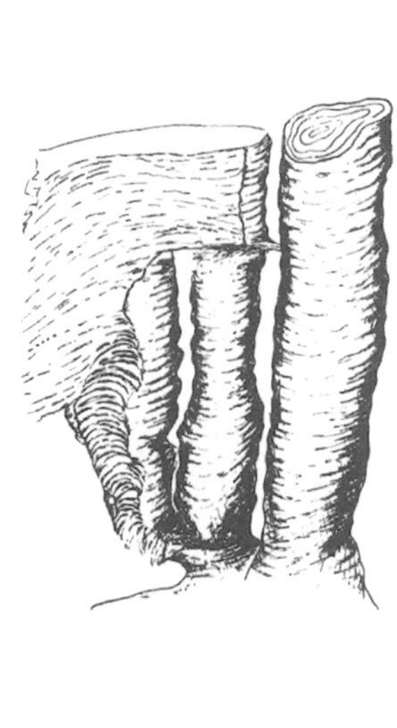

Fig. 28. Transition of columns of *Aldania sibirica* into stratiform stromatolites in the marginal part of the bioherm. Sample 3599/9, Aldan River near Uugun Village

In a bioherm, columns are closely spaced, stratiform structures similar in fabric to that of laminae of columns occur at the base, top and in marginal parts of a bioherm.

Remarks. The form was described by Yakovlev as *Gymnosolen sibiricus*; Raaben and Komar (in: Rozanov 1966, p. 15) had assigned it to *Jurusania*, while Krylov et al. (1969) recognized a separate group.

Age and Distribution. The upper Yudoma Formation, Vendian (?) of the Siberian Platform, the Aldan Basin.

Group Patomia Krylov 1967

Diagnosis. Subcylindrical columns with fine, tuberculate surface arranged vertically or slightly obliquely in the bed. Branching is complex. Many minor branches ramify from the major column sideways and upward, most of them resemble small, acute, dactylic processes. Laminae are commonly densely enveloped without lips and cornices, on the lateral surface of columns, however, laminae bridges appear in the marginal parts of the bioherm, at its base and top (Krylov 1967, p. 47).

Patomia ossica Krylov 1967 (Plate 83, *2–5*, Fig. 29)

Patomia ossica: Krylov 1967, p. 48–52, Text Figs. 26, 27, Plate VIII.

Holotype. GIN, specimen No. 3570/9; the left side of the Lena River opposite the mouth of the Bolshoi Patom; a transitional succession at the base of the Zherba Formation.

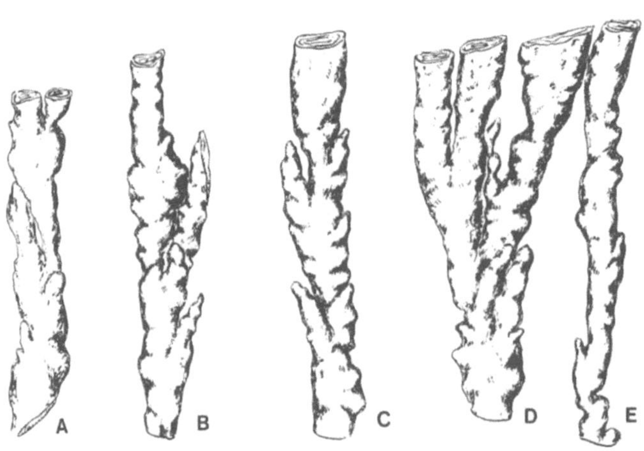

Fig. 29A-E. Shape of columns of *Patomia ossica* Kryl. **A** Sample 3599/17; **B-E** sample 3599/16, Lesser Karatau Range

Description. Thin, subcylindrical columns with fine-tubercular surface up to 2–3 cm wide (commonly 1–1.5 cm) and 20–30 cm high. Columns are usually vertically oriented, in the marginal parts of bioherms, they are tilted towards the margins. The cross-section is circular, ovate or irregular with rounded angles. A great number of small dactylic processes ramify from the major column sideways and upwards. The lateral surface of the columns is usually smooth with an envelope. There are separate lips. Laminae bridges occur at the top and in the marginal parts of the bioherms. Arches, as a whole, are dome-shaped with strongly variable convexity, sometimes finely undulated, corrugated. Microfabric is clotted; in samples from the Lesser Karatau and Tien Shan, laminae are calcified.

Remarks. Only stromatolites from Central Asia are considered here. Upper Riphean Patomia from the Lena River are not presented.

Age and Distribution. A transitional unit at the base of the Zherba Formation, the Patom Highland, and the Chichkan Horizon of the Tien Shan and Karatau.

Patomia aldanica Krylov in Krylov et al. 1969 (Plate 84, *3, 4*; Fig. 30)

Patomia aldanica: 1969, p. 198–200, Fig. 49, Plate XXXVII, Figs. 1, 2; Plate XXXVIII, Fig. 1.

Holotype. GIN, specimen No. 3599/5, slide K-266; the left side of the Aldan River, 1.5 km below the Vaskin Arm; upper horizons of the Yudoma Formation.

Description. Subcylindrical columns with tubercular surface 0.7–0.8 to 2–3 cm wide and up to 20 cm high. The cross-section is circular, ovate or complex "palmate" in outline. Columns exhibit many small palmate ramifications. The lateral surface is sometimes with a multiple envelope and individual lips and bridges. The lamination is indistinct due to alternating chains of clots and patches. The structure of bioherms is obscured. Perhaps they composed common bioherms with *Jurusania tumuldurica* found in the same outcrop.

Age and Distribution. Upper horizons of the Yudoma Formation, the Aldan Basin.

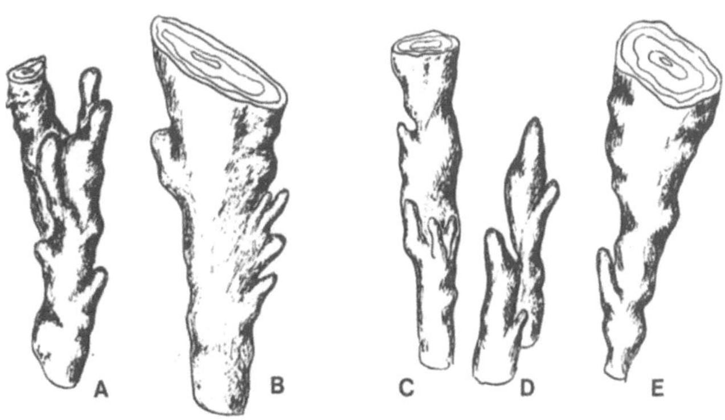

Fig. 30A-E. Shape of columns of *Patomia aldanica* near Vaskin Arm

Group Linella Krylov 1967

Diagnosis. Nodular, tuberculate, subcylindrical or tuberous columns of variable diameter with constrictions and swollen portions; branching is complex. Thin branches ramify from the major column upwards and sideways, most of them look like acute, fine tubercles, but further up some of them widen and develop into a new column. In the marginal portion of the columns, some laminae are butted up against the enclosing rock, forming locally small lips, or they repeatedly and very deeply envelop the lateral surface of columns (Krylov 1967, p. 37).

Linella avis Krylov 1967 (Plate 83, *1*; Fig. 31)

Linella avis: Krylov 1967, p. 43–47, Text Figs. 22–25, Pate VIII, Fig. 1.

Holotype. GIN, specimen No. 570/6, the Lesser Karatau, the Chichkan Formation.

Description. Nodular and tubercular columns of variable diameter (12 to 7–8 cm) with constrictions and swollen portions. The height attains 40 cm. Branching is fairly sparse, more common are acute processes of different size. Laminae densely envelop the lateral surface of columns. Columns make up bioherms or biostromes up to 1–1.5 m high. In biostromes, they are oriented vertically, in bioherms, fanlike. Columns are composed of dark chert with an obscured lamination in the form of alternating chains of lenticular pellets and patches. Arches vary in shape from flat to strongly convex, dome-shaped and conic.

Age and Distribution. The Chichkan Horizon of the Tien Shan and Karatau.

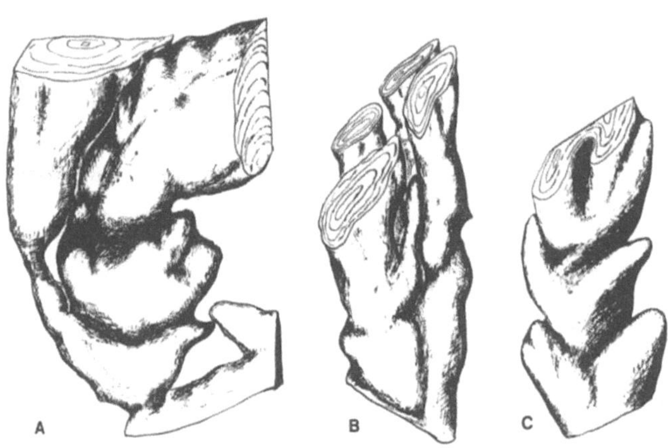

Fig. 31A-C. Shape of columns of *Linella avis* Kryl. **A** Sample 3599/11; **B,C** 3599/14, Tien Shan Mts., Inkel Tau Ridge

Group Jurusania Krylov 1963

Jurusania tumuldurica Krylov in Krylov et al. 1969 (Plate 85, *1–3*, Figs. 32–33)

Jurusania tumuldurica: Krylov in Krylov et al. 1969, pp. 195–198, Text Figs. 47, 48, Plate 38, Figs. 2–4.

Holotype. GIN, specimen No. 3599/1, slide K-268; the left side of the Aldan River, 1.5 km below the Vaskin Arm; upper horizons of the Yudoma Formation.

Description. Subcylindrical columns with circular to ovate cross-sections, arranged vertically within the bed. The thickness of columns is 3–7 cm; height, above 40 cm. Sparse branching into two new columns. Laminae composing a column terminate at a different distance from its margins sunken into a peculiar envelope around the column. The outer surface of the envelope is uneven, finely costate with individual lips. There are transitional bridges. Lamination is indistinct; arches are generally domelike to rectangular and flattened. Fabric is clotted.

Age and Distribution. Upper horizons of the Yudoma Formation, the Aldan Basin.

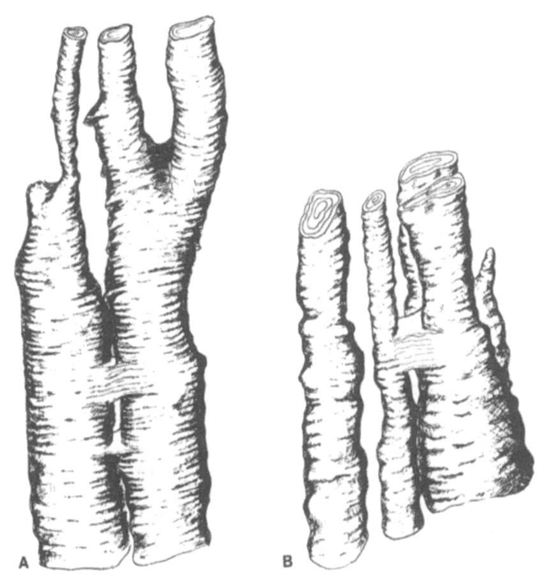

Fig. 32A,B. Shape of columns of *Jurusania tumuldurica* Kryl. **A** Sample 3599/2, Aldan River off Tumuldur Island; **B** sample 3599/1, Aldan River near Vaskin Arm

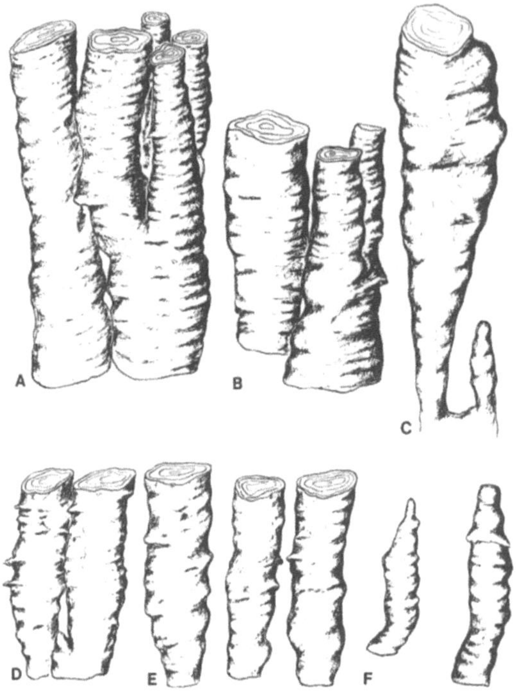

Fig. 33A-F. Shape of columns of *Jurusania tumuldurica* Kryl. **A,C** Aldan River near Vaskin Arm; **B** sample 3599/3; **D-F** sample 3599/4, Aldan River off Tumuldur Island

Group Conophyton Maslov 1937

Maslov placed (1937, p. 334) stromatolites composed of "high stratiform domes of circular cross-section, gradually tapering with acute ends up" into the "genus" *Conophyton*. Later, he stressed such an important feature as the presence of a characteristic nodular, axial zone in their central part. Other, more detailed diagnoses, in particular, by Komar et al. (1965), list about 20 characters, none being obligatory, and many inconsistent with those actually observed. Certainly, we cannot accept such a diagnosis. In the author's opinion, *Conophyton* are stromatolites with conic arches and an axial zone, representing columnar or stratiform buildups. As a rule, they are fragments of more complex structures, in particular, jakutophytons, but they occur as isolated columns as well.

Conophyton gaubitza Krylov 1967 (Plate 82)

Conophyton gaubitza: Krylov 1967, p. 57–58, Plate VII.

Holotype. GIN, specimen No. 3570/11; the Lesser Karatau, the Chabakty River, Ayu Sakan Village, the Chichkan Horizon.

Description. Subcylindrical columnar buildups lying nearly horizontally in the bed (5–10°). Columns are 5-6 to 20–25 cm in diameter and up to 1.5-2 m high (long). Column base is unknown. They are circular or ovate in plan outline. The enclosing bed bears resemblance to stacking logs. Lamination is indistinct, especially in the marginal portions of the columns, however, the conic shape of the laminae and the axial zone are readily seen. Columns are composed of black and grey chert.

Age and Distribution. The Chichkan Horizon of the Tien Shan and Karatau.

5 Part IV

Group Collumnaefacta Koroljuk 1960a (Plate 86, 1-3, Fig. 34)

Collumnaefacta vulgaris: Sidorov in: Tommotian Stage 1969 p. 207, Plate XXXXI, Figs. 1-3.

Holotype. IGiRGI, specimen No. 109/2, the Lena River, Rysiya Village; Lower Cambrian, the Pestrozvet Formation.

Diagnosis. Collumnaefacta with isolated, spongy laminae, some laminae with a whimsically dentate outline.

Description. Straight, terete columns with a smooth, lateral surface. Most columns disintegrate into two to three smaller ones. The total width of a stromatolite does not change significantly as the column disintegrates. Columns are often connected by transitional bridges. Columns are up to 10 cm high and 1-2 cm wide. Lamination is inherited, arches are predominantly simple, sometimes slightly bent. Arch convexity (or height to diameter ratio) ranges from 0.1 to 0.5, commonly 0.3. Lateral terminations are even. Thin, one-layered, discontinuous walls exhibit both dark and light laminae. Wall-forming laminae often envelop the underlying beds.

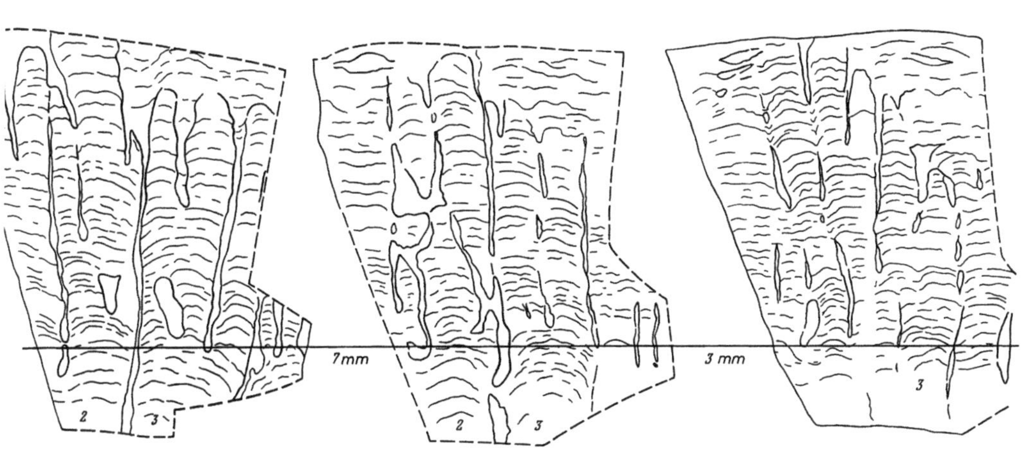

Fig. 34. *Collumnaefacta vulgaris* Sidorov, × 0.8. Holotype, Coll. IGiRGI No. 109/2, sample 7564. Lena River near Rysiya Village, Pestrotsvet Formation. A series of parallel slices through the structure. General view and character of column branching

Columns are clearly stratiform. Three types of laminae alternate: dark, homogeneous, very thin, light, thicker and spongy.

Dark laminae are 0.02–0.07 mm thick and composed of pelitomorphic or very fine-grained carbonate, more often they are structurally homogeneous. Most exhibit even boundaries and a consistent thickness. Laminae with a complex crenate outline are less common.

Light laminae (0.1–0.5 mm thick) are mainly composed of fine-grained carbonate (grain size, 0.02–0.06 mm) with clots and pellets of pelitomorphic carbonate with diffused outline.

Spongy laminae occur sporadically. The fine-grained matrix displays numerous, light vermicular and irregular, void canals filled by coarser crystalline carbonate. When poorly preserved, spongy laminae grade into inequigranular or indistinct, spotty laminae. Spongy laminae are 1–2 mm thick.

Columns exhibit an indistinct structure. There are zones (7–15 mm thick) of finely alternating, dark and light laminae and zones of more structurally complex laminae: spongy ones and dark laminae have a whimsical crenate outline.

Discussion. The form is similar to *Collumnaefacta elongata* Korol., but differs in having more distinct laminae, thinner, dark laminae, and spongy laminae, a more intricate pattern of individual, dark laminae and a larger number of connecting bridges.

Age and Distribution. The lower Pestrozvet Formation, Lower Cambrian, the Lena and Rysiya Rivers at Nokhtuisk Village where they make up the entire bioherms. They were reported also from the Usolka Formation and from a core of the Zima borehole, Kutulika, and from the Urik River.

Group Boxonia Koroljuk 1960a

Group *Boxonia*: Korolyuk 1960a, p. 139–140, Plates XI, XII; Komar 1964, p. 96; 1966, p. 79.

Diagnosis. Columnar stromatolites with distinct, multilayered "wall". Branching occurs without broadening. Laminae are structurally complex.

Holotype. Boxonia gracilis Koroljuk 1960, from the above bauxite unit of the Bokson Formation, East Sayan.

Group Composition. Boxonia gracilis Koroljuk, *B. lissa* Kom., *B. grumulosa* Kom., *B. krasivica* Gov., *B. allahjunica* Kom. et Semikh., *B. divertata* Sid., *B.(?) bianca* Raab., *B. ingilica* Kom. et Semikh., *B. togoica* Gol., *B. knjasevi* Dol., *B. tolbatchania* Dol., *B. zharkovi* (Korol.) Dol.

Discussion. The group is very similar to *Gymnosolen* Steinman and differs from it in having simpler branching, the absence of swollen portions at the point of branching and a characteristic set of microstructures.

Group Colleniella Koroljuk 1960a

Group *Colleniella*: Korolyuk 1960, p. 125, Plate III, IV; Komar 1966, p. 97; Krylov 1975, p. 204.

Diagnosis. Isolated, "wall-less" stromatolites. Lamellae extend through the entire structure. Lamination is wavy. No disintegration into individual columns or only in the marginal portion of a nodule.

Holotype. Colleniella idensis Koroljuk; the Angara Formation, Lower Cambrian; the Angara River.

Group Composition. Colleniella idensis Korol., *C. bella* Korol., *C. calix* Korol., *C. singularis* Kom., *C. evoluta* Shap., *C. uzorica* Kryl., *C. scidulata* Gol.

Discussion. Representatives of the group differ from other nodules in lacking a "wall" and in feeble, wavy lamination.

Age and Distribution. Most forms of the group are known from the Lower Cambrian of East Siberia. *Coloniella singularis* was recorded from the Vendian of East Siberia and *C. uzorica* was described from the Ordovician of East Siberia.

Group Collumnaefacta Koroljuk 1960a

Group *Collumnaefacta*: Korolyuk 1960a, p. 137, Plate X, Figs. 1–3; Krylov 1975, p. 201.

Diagnosis. Columnar stromatolites with thin, indistinct "walls", convex lamination, laminae "bridges", transitional from column to column. Laminae exhibit granular, indistinctly clotted fabric.

Holotype. Collumnaefacta elongata Korol.

Group Composition. Collumnaefacta elongata Korol., *C. (Schancharia) schancharia* Korol., *C. vulgaris* Sid., *C. minuta* Dol., *C. laudata* Sid., *C. shichanica* Sid., *C. devonica* Kryl., *C. ilica* Kryl., *C. erica* Kryl.

Discussion. Differs from other groups in having a thin, indistinct wall.

Age and Distribution. Group is widespread. It was reported from the Vendian, Cambrian, Ordovician, Devonian, Carboniferous and Permian of East Siberia, the Urals, Kazakhstan and Byelorusia.

Collumnaefacta elongata Korolyuk 1960 (Plate 87, *1–4*)

Collumnaefacta elongata: Korolyuk 1960, p. 137, Plate X, Figs. 1–3.

Holotype. IGiRGI, specimen 102/32; East Sayan, below the bauxite unit of the Bokson Formation.

Description. Small columns 5–23 mm in diameter, 40–150 mm high, somewhat irregular in shape, therefore, cross-sections vary from circular to roughly triangular and biscuitlike. Growing columns disintegrate into two to three smaller ones. Columns are formed by stacking of gently convex lamellae. Lamination is not strictly inherited. Configuration is more distinct due to individual laminae sinking down the column; "wall-forming" lamellae are both dark and light: (1) light, fine-grained; (2) grey, fine-grained; (3) dark grey, fine pelletal. Contacts between laminae are indistinct perhaps due to recrystallization.

Discussion. Differs from other forms of the group in outward appearance, shape of arches, but mainly in the granular fabric of the lamellae.

Age and Distribution. Widespread in the upper and lower bauxite units of the Bokson Formation (East Sayan), i.e. in the Vendian and Early Cambrian deposits: was reported from the Usolka Formation, the Sayany area, the Moty Formation, Baikal area.

Chapter 10

Microphytolites

The Vendian in its type area (Sokolov 1952) and its equivalents within the Russian Platform and at its eastern margin in fact yield no microphytolites. A microphytolite assemblage, interpreted as Vendian, was first described from the Patom Highland and from the western slope of the Aldan Shield (Reitlinger 1959). Zhuravleva (1964) recognized it as the fourth Precambrian assemblage. Microphytolites of the Yudoma Series (Formation), Uchur-Maya area, as the most representative, were used as the basis for recognition of the assemblage.

Units, recommended for the unified scheme at the All-Union meetings, held in Novosibirsk, 1979, of the Yudoma Series and its equivalents will be considered here as equivalents of the Vendian in Siberia.

The problems of the history and nature of microphytolites have been rather extensively elucidated in the literature and, therefore, they will not be discussed here in detail. It should only be noted that the study of microphytolites started as early as the end of the last century (see the review by Voronova and Radionova 1976), and prior to the late 1950s results obtained were widely published by Russian and foreign scientists. The early stage of the study of microphytolites culminated in numerous works by Maslov (see Voronova and Radionova 1976). The investigations of this lengthy period showed the complex nature of microphytolites (oncolites and catagraphs). The algal nature of microphytolites is obvious, however, carbonate material of different origin i.e., biochemical, chemogenic, mechanical (clastic products together with non-carbonate, terrigenous material), had participated to a variable degree in their formation.

The complex nature of microphytolites predetermines the variability, often rather significant, of the character of microstructures as dependent on environments (Khomentovsky and Yakshin 1973). In view of the facies control of microphytolites, it is important to recognize the features which reflect the biological aspect of the organisms studied and to take into consideration changes to which microphytolites were subject in the course of metamorphism of enclosing rocks.

All the above, despite the adequate study of microphytolites, both fossil and modern, causes substantial differences in the interpretation of biological aspects of their diagnostic characters (Korolyuk 1966; Zhuravleva 1968; Voronova and Radionova 1976; and others), which, in turn, lead to fairly unambiguous inferences regarding the stratigraphic importance of microphytolite forms and their assemblages.

The complex and ambiguous nature of microphytolites determined the approach to their classification, first proposed by Maslov (1937, 1950, 1952) and further developed by Reitlinger (1959); Zhuravleva (1964) and Korolyuk (1966).

Reitlinger (1959), who studied microphytolites from some Siberian sections, recognized some vertical distribution patterns of microphytolites and suggested that age variability in the composition of microphytolite assemblages (microproblematics) could be used as a basis for subdivision and correlation of Upper Precambrian deposits.

Zhuravleva (1964) gave a first account of data on the distribution of microphytolites in the old carbonate successions of Siberia and recognized four Precambrian microphytolite assemblages, namely, three Riphean and one Yudomian (Vendian) assemblage. The scheme, proposed by Zhuravleva (1964), has practically served as a basis for all recent studies of microphytolites.

Most collections of microphytolites from the Upper Precambrian key sections of Siberia were examined at some colloquia (Anon. 1972a,b, 1975, 1979); lists of microphytolites accepted at these meetings for the stratigraphic units of the Upper Precambrian of Siberia were recommended as characteristics of sections used for unified schemes. It was unfeasible to discuss material of all the collections of Upper Precambrian microphytolites at the meetings. Thus, to better understand the distribution pattern of microphytolites in the Vendian and underlying strata, a number of contributions should be taken into account (Zhuravleva 1964, 1968, 1974 and others; Rabotnov et al. 1971; Postnikova and Revenko 1969; Raaben and Zabrodin 1972, and others), especially as data on Precambrian microphytolites from the Russian Platform were not discussed at the colloquia, and material from the Ural sections was discussed only at Working Groups, whose resolutions have not been published. Moreover, not all the material available on Precambrian microphytolites of this region was taken as a basis for the decisions, recommended to be used for faunistic characterization of sections in a unified scheme for the Urals.

The Yudomian microphytolite assemblage, recognized by Zhuravleva (1964) as the fourth Precambrian assemblage, first consisted of six forms, namely, *Vesicularites bothrydioformis* (Krasnop.), *V. lobatus* Reitl., *V. concretus* Z. Zhur., *Vermiculites tortuosus* Reitl., *V. irregularis* (Reitl.), *Osagia minuta* Z. Zhur.

This microphytolite assemblage was proposed by Zhuravleva to faunistically characterize the fourth Upper Precambrian unit. It was used to substantiate the recognition and correlation of equivalents of the Yudoma Formation in many sections of the Siberian Platform. The Porokhtakh Formation in the Olekma River area, the Kerbi and Tinnovka Formations of the Patom Highland, the Kharayutekh Formation of Kharaulakh Mountains, the Turkukut Formation of the Olenek Uplift and the Starorechinskaya Formation of the Anabar Massif were correlated with the Yudoma Formation of the Uchur-Maya area (Fig. 35). On the basis of findings of *Osagia minuta* Z. Zhur., the Vorogovka Formation (Series) of the Yenisei Mountain Range was correlated with the Yudoma Formation.

During the ensuing 6 to 8 years, microphytolites were studied from practically all major sections of the USSR; the boundaries of the vertical distribution of many forms of this assemblage were drawn more precisely; new forms were added to the list of microphytolites: *Vesicularites reticulatus* Nar., *V. subinensis* Zabr., *V. scutulatus* Yaksch., *V. simplaris* Yaksch., *V. celluleus* Yaksch., *Vermiculites sucharicus* Milst., *Nubecularites abustus* Z. Zhur., *Osagia corticosa* Nar., *O. glolulosa* Korol., *O. monolamellosa* Z. Zhur., *O. svalbardica* Milst., *Radiosis punctatus* Nar., *Asteros-*

phaeroides radiatus Korol., *Ambigolamellatus horridus* Z. Zhur., *Volvatella zonalis* Nar., *V. obsoleta* Nar., *V. vadosa* Z. Zhur., *Medullarites lineolatus* Nar., *M. ovatus* Nar.

Moreover, among Yudomian forms, microphytolites regarded earlier as typical only of Riphean deposits, are at present found in a number of sections; they are *Vesicularites compositus* Z. Zhur., *V. flexuosus* Reitl., *Vermiculites angularis* Reitl., *Osagia grandis* Z. Zhur., *O. tenuilamellata* Reitl., *Glebosites glebosites* Reitl. and Cambrian *Nubecularites catagraphus* Reitl. (see Fig. 37).

The Yudomian microphytolite assemblage is one of the most consistent in composition and it is distinguished for its qualitative homogeneity of forms. Figure 35 shows the vertical distribution of the main forms of the Yudomian microphytolite assemblage (together with some Riphean survivers) for the Upper Precambrian key sections, characterized by representative microphytolites. However, in many areas, this has not enabled one to reliably draw the boundaries between the Yudoma Complex and the underlying and overlying beds. The upper boundary is rather distinct, but mainly due to the appearance of the Lower Cambrian skeletal fauna in the overlying deposits (Khomentovsky et al. 1972), however, the Yudomian microphytolites are replaced by Cambrian fauna within a rather wide zone. This is well illustrated by the examples when the upper Yudomian and lower Cambrian sections contain similar fauna: Yudomian *Vermiculites tortuosus* Reitl., *Medullarites liniolatus* Nar., *M. ovatus* Nar. and *Ambigolamellatus horridus* Z. Zhur. were found together with Cambrian forms in the Nokhtuisk section of the Patom Highland in the lower Macha Formation (Lower Cambrian).

The attempt to draw the lower boundary of the Yudoma Complex by means of microphytolites involves considerable difficulties. The problem is complicated by the fact that within the Siberian Platform and its margin, carbonate deposits yielding Yudomian microphytolites are separated by terrigenous deposits of variable thickness or by hiatus from those containing Riphean microphytolites. Besides, in some sections microphytolites of the Yudomian assemblage appear in the Riphean deposits together with or without Riphean forms (Krylov et al. 1971; Rabotnov et al. 1971; Anon 1975, 1979 and others).

Based on the unified, stratigraphic schemes for the Late Precambrian, we shall analyze the position of the Yudomian microphytolite assemblage, as a whole, in the section, along with the vertical distribution of separate forms.

In the Uchur-Maya area, the Yudoma Series (the Aim Formation and the lower Ust Yudoma Formation), containing typical Yudomian microphytolites, allowed the description of the most representative assemblage: *Vesicularites concretus* Z. Zhur., *V. bothrydioformis* (Krasnop.), *V. lobatus* Reitl., *V. reticulatus* Nar., *V. simplaris* Yaksch., *V. scutulatus* Yaksch., *V. celluleus* Yaksch., *Osagia corticosa* Nar., *Nubercularites abustus* Z. Zhur., *Vermiculites irregularis* (Reitl.), *V. tortuosus* Reitl., *Radiosus punctatus* Nar., *Volvatella zonalis* Nar., *V. vadosa* Z. Zhur., *Ambigolamellatus horridus* Z. Zhur., and two forms, *Vesicularites compositus* Z. Zhur., *V. flexuosus* Reitl., from the Riphean assemblage. One of the forms of the Yudomian assemblage, namely, *Vesicularites bothrydioformis* (Krasnop.) appears in the section starting from the lower Lakhandin Series.

The representativeness of the microphytolite assemblage in the Yudoma Series and its peculiar composition are beyond any doubt, when the series is recognized

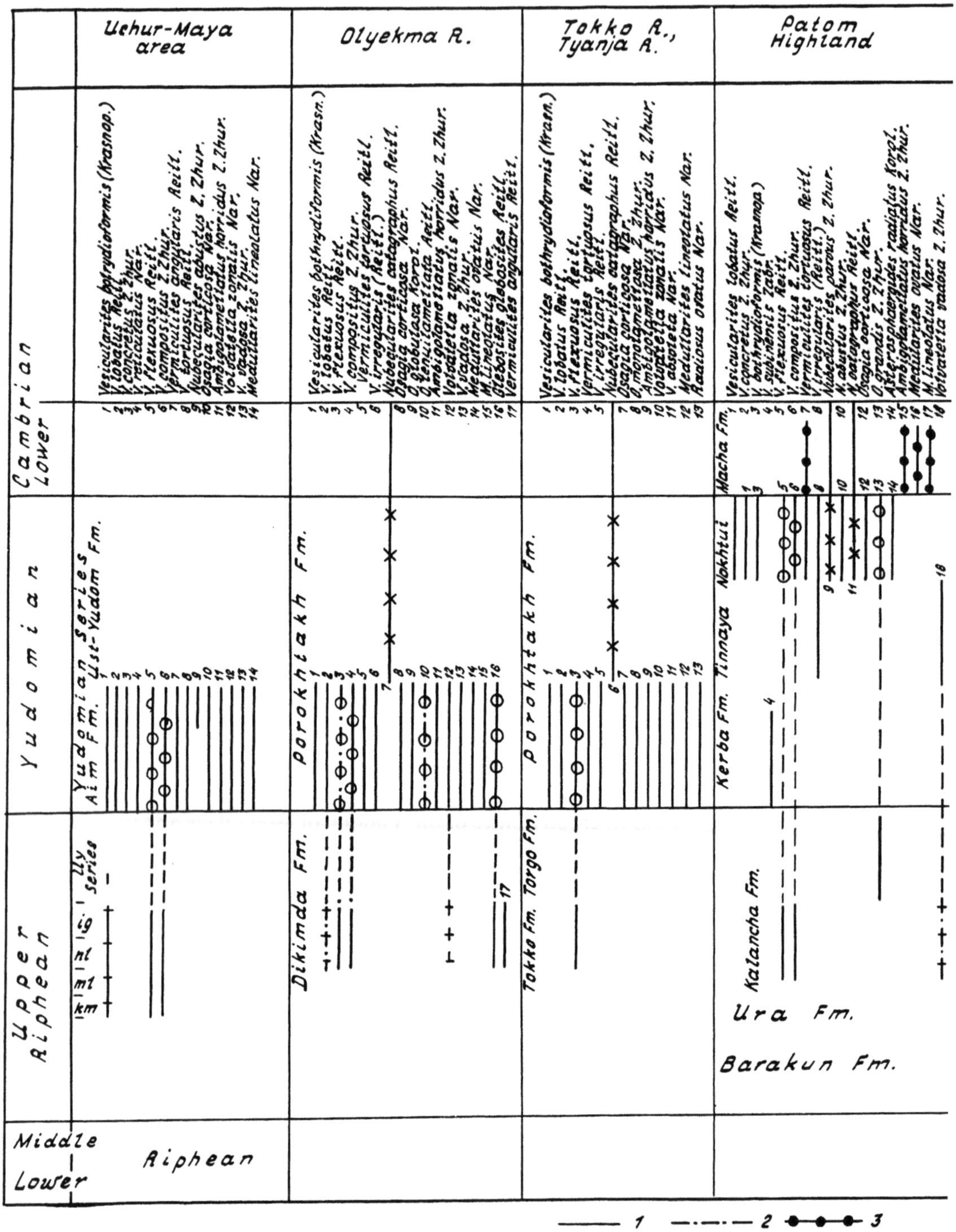

Fig. 35. Microphytolite assemblage of the Yudoma Complex from key sections of the Siberian and European Platforms and vertical distribution of forms of this assemblage. **1** Intervals recognized from material discussed at meetings on microphytolites; **2** published data; **3** Riphean and Vendian microphytolites continuing into the Lower Cambrian; **4** microphytolites from the Vendian and Cambrian; **5** microphytolites from the Vendian and Riphean deposits; **6** Yudomian microphytolites recognized from

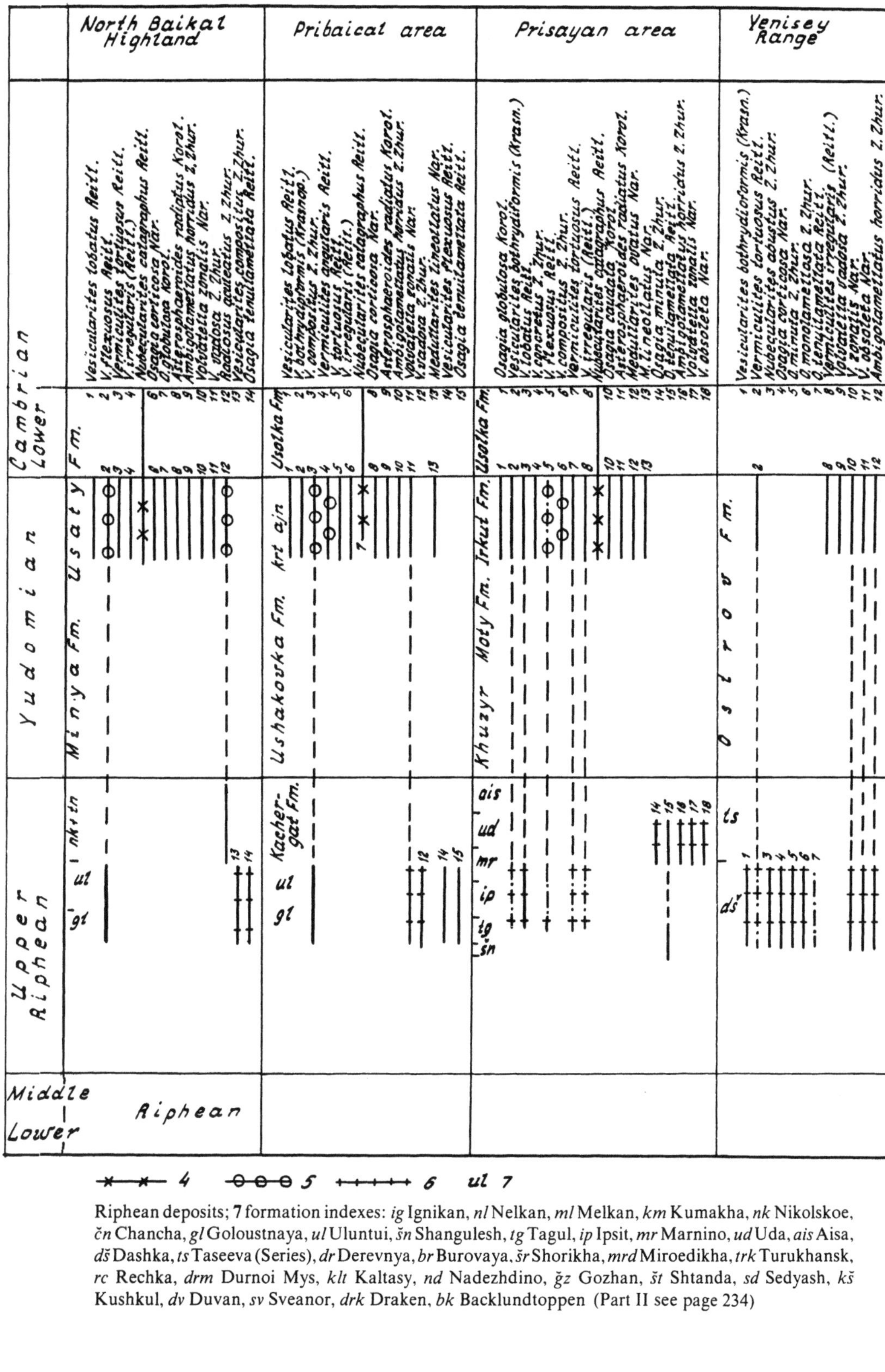

Riphean deposits; 7 formation indexes: *ig* Ignikan, *nl* Nelkan, *ml* Melkan, *km* Kumakha, *nk* Nikolskoe, *čn* Chancha, *gl* Goloustnaya, *ul* Uluntui, *šn* Shangulesh, *tg* Tagul, *ip* Ipsit, *mr* Marnino, *ud* Uda, *ais* Aisa, *dš* Dashka, *ts* Taseeva (Series), *dr* Derevnya, *br* Burovaya, *šr* Shorikha, *mrd* Miroedikha, *trk* Turukhansk, *rc* Rechka, *drm* Durnoi Mys, *klt* Kaltasy, *nd* Nadezhdino, *ğz* Gozhan, *št* Shtanda, *sd* Sedyash, *kš* Kushkul, *dv* Duvan, *sv* Sveanor, *drk* Draken, *bk* Backlundtoppen (Part II see page 234)

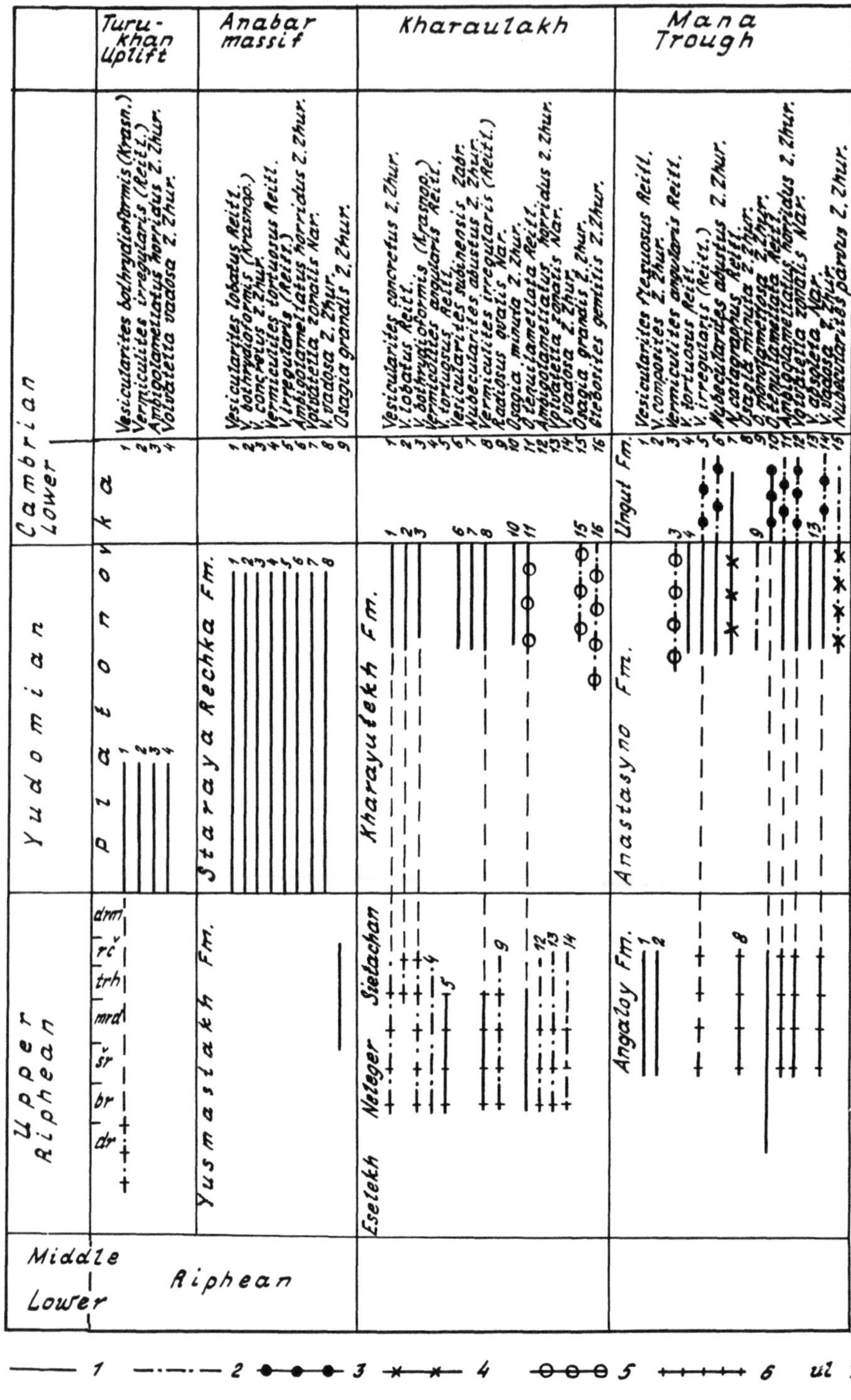

Fig. 35, Part II

as an age complex, though the Ui Series, underlying the Yudoma Series, is represented exclusively by terrigenous deposits devoid of microphytolites. Surviving forms do not disturb the general picture of contrasting microphytolite assemblages of the Riphean and of the Yudomian.

In the lower part of the series, the microphytolite assemblage is rather completely represented in the Porokhtakh Formation at the Olekma, Tyatya and Tokko Rivers. Similarly to the Yudoma Series, the whole assemblage occurs in the lower 50–60 m (Anon. 1972a). Microphytolites of the Porokhtakh Formation also contain Riphean *Vesicularites flexuosus* Reitl., *V. compositus* Z. Zhur., *Glebosites gentilis* Z. Zhur. (Anon. 1972a,b, 1979), *Osagia tenuilamellata* Reitl. (Khomentovsky et al. 1972), and the Riphean deposits contain Yudomian *Vesicularites lobatus* Reitl. and *Volvatella zonalis* Nar. (Khomentovsky et al. 1972). On the whole, the microphytolite assemblage of the Porokhtakh Formation contains all the main microphytolites of the Yudomian assemblage. At the Tokko River section, the Porokhtakh Formation is underlain by carbonate rocks of the Torgo Formation yielding microphytolites of the Upper Riphean Chencha assemblage (Anon. 1972a, 1979).

In the above mentioned sections the upper Yudoma Series is ubiquitously built up of carbonate rocks, but characteristic Yudomian forms have not been found at this level.

In the Patom Highland sections, the microphytolite assemblage discussed is represented rather completely, however, it occurs here in the uppermost part of the section, i.e. in the Nokhtuisk Formation. From the lower part of section (Zherba Formation) only *Vesicularites subinensis* Zabr. was reported (Khomentovsky et al. 1972), while *Volvatella vadosa* Z. Zhur. is known also from the Tinnovka Formation[7]. Here Riphean forms were also established, e.g. *Vesicularites flexuosus* Reitl., *V. compositus* Z. Zhur. and *Osagia grandis* Z. Zhur. and the Cambrian forms *Nubecularites catagraphus* Reitl. and *N. parvus* Z. Zhur. Some parts of the Yudomian forms continue into the Lower Cambrian (Macha Formation), they are *Vermiculites tortuosus* Reitl., *Ambigolamellatus horridus* Z. Zhur., *Medullarites lineolatus* Nar. and *M. ovatus* Nar.

Microphytolites of the Yudomian assemblage occur in the uppermost part of section, ubiquitously from the Patom Highland to the Sayany Mountains area (see Fig. 35). In the North Baikal Highland, they are associated with the lower Usaty Formation, in the Baikal area they occur in the Kurtun and Ayakan Formations, and in the Sayany Mountains area they are known from the Irkut and the upper Ust Tagul Formations. In most sections the Yudomian forms occur together with such Riphean forms as *Vesicularites flexuosus* Reitl., *V. compositus* Z. Zhur.; *Vermiculites angularis* Reitl. was reported from the Baikal area. In these areas, *Volvatella zonalis* Nar. and *V. vadosa* Z. Zhur. are also very common in the Riphean deposits.

---

[7]Resolutions of the colloquia Anon. (1972a, 1979) give a more complete list of microphytolites from the Tinnovka Formation, but a comparative study of the collections, presented at the colloquia, allowed the author to establish that these microphytolites occur in grey limestones, known in the region only from the Nokhtuisk Formation.

In the Turukhansk area, four Yudomian forms, such as *Vermiculites irregularis* Reitl., *Ambigolamellatus horridus* Z. Zhur., *Volvatella vadosa* Z. Zhur., *Vesicularites bothrydioformis* (Krasnop.), were recognized from the lower Platonovskaya Formation, the latter is known also from the Riphean (Derevnya Formation).

Yudomian microphytolites occur uniformly only in the Staraya Rechka Formation within the Anabar Massif and in the Moty and Irkut Formations within the Irkutsk Amphitheater.

Attention should be drawn to the Upper Precambrian sections of the Sayany area, Yenisei Mountain Range, Kharaulakh Mountains, and the Mana Trough, where rather rich assemblages of Yudomian microphytolites were described from two diachronic horizons (Anon. 1975).

Microphytolites from the Dashka Formation, Oslyanka Series, Yenisei Mountain Range, were first described by Zhuravleva et al. (1969). The authors consider that here microphytolites represent a "pure" Yudomian assemblage: *Vesicularites bothrydioformis* (Krasnop.), *Nubecularites abustus* Z. Zhur. and some new forms[8]. The list was supplemented by the following forms: *Volvatella zonalis* Nar., *Ambigolamellatus horridus* Z. Zhur., *Osagia cortiosa* Nar., *O. monolamellosa* Z. Zhur. and Riphean *O. udereica* Yaksch as the result of the examination of microphytolite collections from the Dashka Formation at the colloquia (Anon. 1972a, 1979). Almost all the scientists who studied this list of microphytolites agreed on the Yudomian age of the Dashka Formation (Anon. 1972a, 1979). However, although in the late 1960s the dating of the Dashka Formation was preferentially based on microphytolites without regard to finds of Riphean stromatolites (Khomentovsky et al. 1972), which were considered dubious because of their uncertain position in the Upper Precambrian section, in 1978 such a conclusion did not seem to be substantiated. By that time, a large-scale geological mapping of the Yenisei Mountain Range area had practically been completed, the Upper Precambrian sections had been studied in detail and correlated both with the Yenisei Range and the adjacent regions. It had been reliably established and reflected in the unified scheme accepted at the All-Union Conference, held in 1979, that the Oslyanka Series is overlain by the Taseeva Series, a direct equivalent of the Oselkovaya Series of the Sayany area. Hence, the Dashka Formation can by no means be younger than the Karagassky Formation, containing abundant Riphean microphytolites from Kalancha, Chencha and other assemblages, and, according to Narozhnykh (Rabotnov et al. 1971), subordinate Yudomian *Vesicularites flexuosus* Reitl. and *V. compositus* Z. Zhur. Besides, in the Sayany area, above the Oselkovaya Series, and in the Yenisei Range, above the Taseeva Series, occur respectively, the Moty Series and the Ostrovnaya Formation, yielding microphytolites of the Yudomian assemblage. Only in the Sayany area it contains the Riphean forms *Vesicularites flexuosus* Reitl. and *V. compositus* Z. Zhur., displaying a wide vertical range (Fig. 35, see also insert, Fig. 36). Correlation of the Moty Series with the Ostrovnaya Formation is reliably substantiated by direct geological data and is universally accepted. Thus, in the Yenisei Range there are diachronic levels, containing primarily Yudomian microphytolites.

---

[8]This and other lists of microphytolites presented in this chapter contain only forms which had been monographically described up to the time of writing.

The Kharaulakh Range section provides support to the above evidence. Unlike the Yenisei Range section, the Kharaulakh section is more completely characterized by microphytolites whose assemblages vary in composition in an ascending order. The complex composition allows extremely contradictory inferences.

The second (Kalancha) microphytolite assemblage was recognized by Zhuravleva (1964) in the Neleger Formation; the age of the assemblage was determined as Middle Riphean. The age was later confirmed (Zhuravleva 1968; Krylov et al. 1971), however, a further revision of the same collection made Zhuravleva (1974) establish a new list of microphytolites, represented exclusively by Yudomian forms: *Ambigolamellatus horridus* Z. Zhur., *Volvatella zonalis* Nar., *V. vadosa* Z. Zhur., *Vesicularites bothrydioformis* (Krasnop.), *V. concretus* Z. Zhur. and a number of new forms inferring a Yudomian age of the Neleger Formation. The study of new collections of microphytolites from the Kharaulakh section, carried out by the author, did not support the later inference of Zhuravleva (Yakshin 1979).

In 1978, microphytolites from the Upper Precambrian Kharaulakh section were discussed at the coloquium in Irkutsk, where the following list of microphytolites was accepted for the Neleger Formation: Riphean *Osagia columnata* var. *baicalica* Yaksch., *O. columnata* var. *ovsianisa* Yaksch., *O. ex gr. tenuilamellata* Reitl., *Vesicularites flexuosus* Reitl., *V. consuetus* Yaksch., *V. vapolensis* Zabr. and Yudomian *Vermiculites tortuosus* Reitl., *V. irregularis* (Reitl.), *Medullarites* Nar. (Anon. 1979). This is a mixed list, but shows the obvious predominance of Riphean forms, of which only two forms, namely, *Osagia tenuilamellata* and *Vesicularites flexuosus*, are known at the Yudomian level (Fig. 36 presents the ratio of Riphean to Yudomian forms). Higher in the section, Riphean forms also dominate the Sietachan Formation, but relatively representative microphytolites of the Yudomian assemblage were described from the uppermost Kharayutekh Formation. Like the Yenisei Mountain Range and the Sayany Mountain area, there are two levels containing microphytolites of the Yudomian assemblage, however, it is just the presence in this section of microphytolites, typical only of the Riphean deposists (but unknown in younger beds) that suggests that here Yudomian forms appear, starting from the Riphean.

A similar picture is observed at the folded margin of the Siberian platform. In the Mana Trough, typical Yudomian microphytolites are abundant in the carbonate beds of the Anastasieno Formation (Gudymovich 1970 and others), but in the underlying Riphean beds, the Ovsyankina (Angul) Formation, along with the Upper Riphean Kalancha microphytolite assemblage, the Yudomian forms *Ambigolamellatus horridus* Z. Zhur., *Volvatella zonalis* Nar., *V. vadosa* Z. Zhur., *Osagia minuta* Z. Zhur. were found (Anon. 1975).

If we now consider Fig. 37, which summarizes data on the occurrence of microphytolites of the Yudomian assemblage (including forms surviving from the Riphean) from all the discussed Siberian sections and at different levels, we see that most Yudomian forms appear starting from the Upper Riphean, and the number of finds of some forms [*Vesicularites bothrydioformis* (Krasnop.), *Nubecularites abustus* Z. Zhur., *Ambigolamellatus horridus* Z. Zhur., *Volvatella zonalis* Nar., *V. vadosa* Z. Zhur.] from the Riphean deposists is similar to that from the Yudomian level.

Analysis of the proportion of Riphean and Yudomian forms in microphytolite assemblages and reliably distinguish between the Riphean and the Vendian

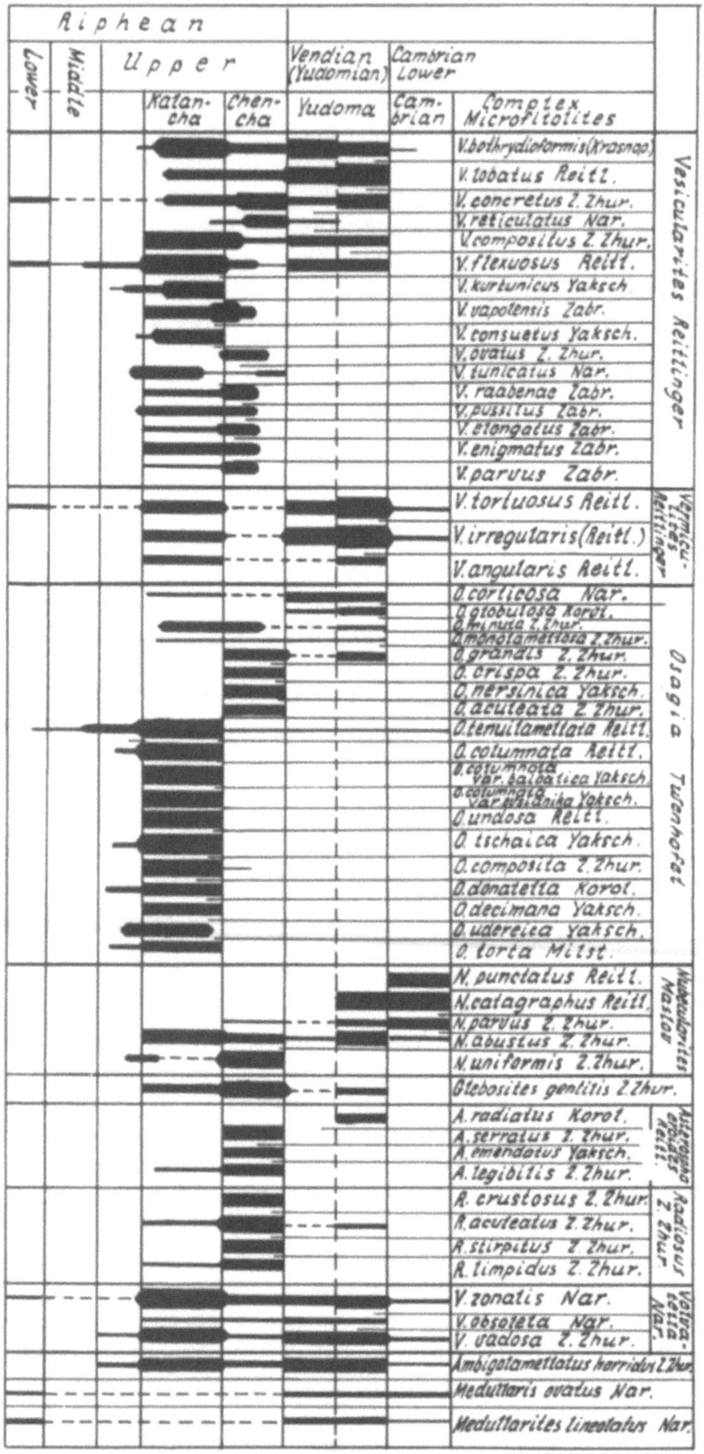

Fig. 36. Abundance ratios of microphytolite forms in different Riphean, Vendian and Lower Cambrian assemblages in key sections of the Siberian and European Platforms (width of line shows the number of forms recorded from a given stratigraphic level)

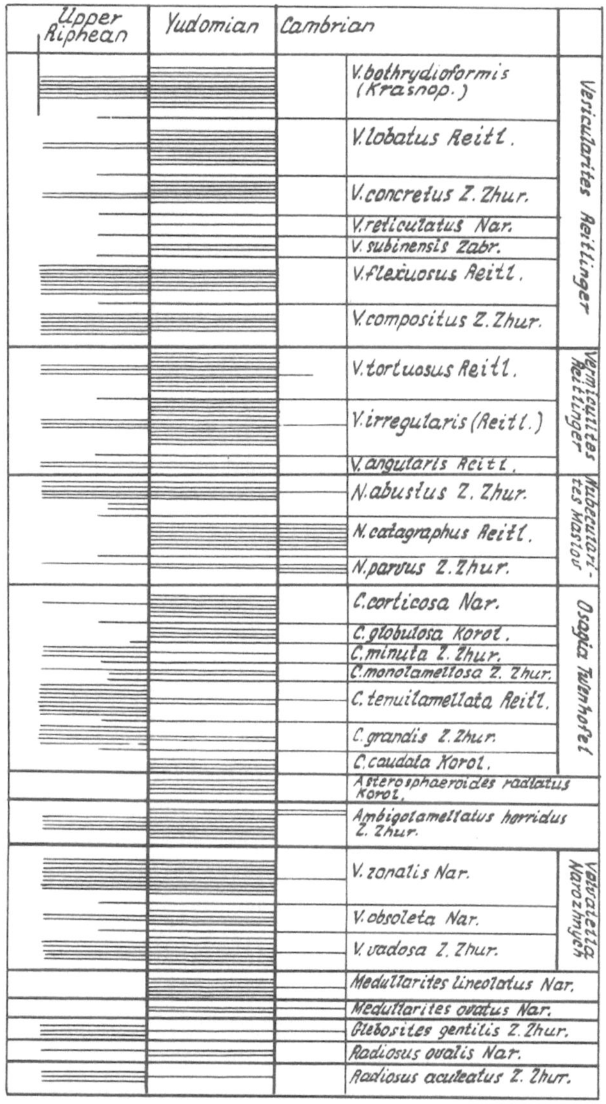

Fig. 37. Frequence of occurrence of microphytolites in Vendian beds and its equivalents in sections of the Siberian Platform at the Yudomian (Vendian) level, in underlying and overlying deposits (occurrence of forms in a key section is shown by a separate line)

deposits (Fig. 36). However, this is not always feasible. When occasional Yudomian forms occur in a fossiliferous Riphean assemblage, as in the Lakhanda Series of the Uchur-Maya area, the Kalancha Formation of the Patom Highland, or when the Yudomian microphytolites are subordinate in the Riphean assemblage, for example, in the Karagassky Series, Sayany area, and in the Neleger Formation, Kharaulakh Mountains (Fig. 35), they cannot affect our assumption of a true position of boundaries between the assemblages. But if the appearance of some Yudomian forms at the Riphean level is not accompanied by sufficiently representative Riphean microphytolites (Dashka Formation in the Yenisei Mountain

Range area), this can lead to a considerable increase in the volume of a unit yielding microphytolites.

Comparative analysis of the microphytolite distribution in various areas shows that it differs across a section, even in the Yudoma Formation and its equivalents. In the sections adjacent to the Aldan Shield all microphytolites occur in the lowermost Yudoma section (Uchur-Maya area, Aldan and Olekma Rivers). In the Staraya Rechka Formation of the Anabar Massif, microphytolites are evenly distributed throughout the section, but in the Patom Highland sections, at the southern and western margins of the Siberian Platform and in the Kharaulakh area, microphytolites occur in the uppermost part of the Yudoma section and locally, as stated above, they continue into the Cambrian (Figs. 35, 37).

The above examples indicate that the vertical distribution of practically all forms of microphytolites of the Yudomian assemblages surpasses the extent of the Yudoma Series (Fig. 37), and that the main microphytolite assemblage is the most complete in various parts of the section, which may be ascribed to environments favourable to their development. This does not provide evidence for the subdivision of the Yudomian assemblage into subassemblages used in a general stratigraphic scale.

A detailed subdivision of the Yudomian assemblage was proposed by Zhuravleva et al. (1969), who subdivided microphytolites of the Dashka Formation, Yenisei Mountain Range, into subassemblages. The fact that the Dashka Formation became a stratotype of the subassemblages of the Yudomian microphytolite assemblage makes this attempt abortive, since the Riphean age of the Dashka Formation is now proved unambiguously and it can be regarded as universally accepted. At the same time the analysis of microphytolite distribution in various sections (Fig. 35) indicates that microphytolites are mainly associated with lithology rather than with stratigraphic levels (Yakshin 1975).

From the association of the maximum abundance of microphytolites with this or that part of the section, the sections may be classified into certain types or provinces (in the sense of Yakshin 1979), and in a number of provinces the appearance of certain Yudomian microphytolites at the Riphean level could be regarded as characteristic. A more detailed stratigraphic subdivision within the provinces on the basis of microphytolite subassemblages appears reasonable.

A differentiated approach to the appraisal of microphytolite assemblages of some distant regions allows us to avoid many errors in stratigraphic subdivision. The Upper Precambrian sections of the Russian Platform and its margins, where some microphytolite provinces can be recognized, provide a good example of a region in terms of microphytolites.

In the European sections, the Vendian consists of terrigenous rocks void of microphytolites. Only in Spitsbergen, where rocks of the Polarisbreen Series are regarded as equivalents of the Vendian (Khomentovsky 1976), Yudomian microphytolites: *Vesicularites bothrydioformis* (Krasnop.), *V. lobatus* Reitl., *V. concretus* Z. Zhur., *Vermiculites irregularis* (Reitl.), *Nubecularites abustus* Z. Zhur., were described by Zabrodin from the carbonate layers and units (Raaben and Zabrodin 1972). However, the same forms of microphytolites (except for *Vermiculites irregularis*) were determined by Zabrodin in the Riphean starting almost from the lower Akademikerbreen Series (see Fig. 35).

In the southern Urals, the Yudomian forms start to appear from the upper Minyar Formation: *Vesicularites concretus* Z. Zhur., *V. bothrydioformis* (Krasnop.) (Raaben and Zabrodin 1972), *V. lobatus* Reitl., *Osagia nimia* Z. Zhur. (materials from the Sverdlovsk Conference on the unified stratigraphic schemes for the Urals), *Vermiculites tortuosus* Rеitl., *Nubecularites abustus* Z. Zhur., *Ambigolamellatus horridus* Z. Zhur., *Volvatella zonalis* Nar., *V. vadosa* Z. Zhur., *Vesicularites immensis* Z. Zhur. (Zhuravleva 1980). The Minyar Formation, as a whole, contains a rich assemblage of Riphean microphytolites.

Higher in the section, the Uk Formation contains a mixed microphytolite assemblage with some predominance of Yudomian forms over such Riphean microphytolites as *Vesicularites compositus* Z. Zhur., *V. flexuosus* Reitl., *V. tunicatus* Nar. (Zhuravleva 1968). The typical Uk form *Osagia monolamellosa* Z. Zhur. is known both from the Riphean (Dashka Formation, Yenisei Mountain Range) and from the Yudomian (Porokhtakh Formation, Tokko River Basin) deposits.

It should be noted that the earliest appearance of some microphytolite forms, considered as characteristic of the Yudomian assemblage of Siberia, is recorded from the European sections. In the Ural area, the Lower Riphean Kaltasy Formation contains *Vesicularites concretus* Z. Zhur., *Volvatella zonalis* Nar., *Medullarites ovatus* Nar., *M. lineollatus* Nar., while the lowermost Upper Riphean, i.e. the Gozhan and Shtanda Formations is, comprised of *Vesicularites bothrydioformis* (Krasnop.), *V. concretus* Z. Zhur., *Vermiculites tortuosus* Reitl., *Volvatella zonalis* Nar., *V. vadosa* Z. Zhur., *Nubecularites abustus* Z. Zhur., *Ambigolamellatus horridus* Z. Zhur. In the Kaltasy Formation, the Yudomian microphytolites form part of a complex association, also including Lower and Upper Riphean microphytolites, but the Gozhan and the Shtanda Formations contain no Riphean forms at all.

Quite identical Yudomian microphytolites, such as *Vesicularites lobatus* Reitl., *V. concretus* Z. Zhur., *V. reticulatus* Nar., *Nubecularites abustus* Z. Zhur., *Volvatella zonalis* Nar., were also described from the pre-Vendian strata, namely, from the upper part of the Orsha Series, Orsha Depression (Zhuravleva and Chumakov 1968) and from the upper part of the Polesie Series, Volyn Depression (Bessonova and Chumakov 1968, Bessonova and Narozhnykh 1970, and others (see Fig. 35).

Thus, on the Russian Platform and in adjacent areas, the Yudomian microphytolites are relatively abundant in the pre-Vendian deposits correlative with the Uk Formation, southern Urals. At the Conference in Ufa, 1976, the Uk Formation was accepted as a stratotype of the fourth Riphean unit, named the Kudash Horizon (Semikhatov et al. 1979). The recognition of this unit, first as the lower member of the "terminal Riphean", proposed by Keller (1973), is primarily substantiated by peculiar microphytolites dominated by the Yudomian forms. A great similarity of microphytolites of the Uk Formation, southern Urals, and those of the Yudoma Series, Uchur-Maya area, is the main argument for a direct correlation of the Yudomian of Siberia and the "terminal Riphean" represented by the Kudash and Vendian Horizons (Keller 1974; Bekker et al. 1979).

Such a correlation may be considered substantiated, if the notion that the Yudomian microphytolite assemblage had appeared simultaneously in all the known Upper Precambrian sections, is taken as an axiom. Is this the case, indeed? We see a contradiction in the stratotype of the Uk Formation itself – the upper Minyar Formation overlain by the erosion surface and the basal conglomerate of the Uk Formation (Becker et al. 1979; Semikhatov et al. 1979) yields predominantly Yudomian microphytolites. Zhuravleva (1980) assigns this part of the Minyar Formation to the Kudash Horizon as well. From microphytolites the Uk Formation may be correlated with the above discussed Gozhan and Shtanda Formations (Postnikov and Revenko 1969 and others). Using Yudomian microphytolites, some workers (Morozov and Revenko 1969) correlate the Kaltasy Formation (Lower Riphean, see Fig. 35) with the Uk Formation of the Urals area.

Attempts to find equivalents of the Kudash in the Siberian sections lead to numerous contradictions, and the substantiation of these possible equivalents often put aside microphytolites (although they were one of primary factors in the recognition of the Kudash). In the Yenisei Range the Taseeva Series is regarded as the Kudash (Semikhatov et al. 1979), but terrigenous rocks, composing the series, contain no microphytolites, while sufficiently representative pure Yudomian microphytolites are known from the Dashka Formation which underlies the Taseeva Series (Anon. 1979). As stated above, in the Sayany area, some Yudomian forms were described from the Tagul and Ipsit Formations of the Karagassky Series; in the Kharaulakh Mountains, Yudomian forms were also found just as low in the section as in the Sayany area (see Figs. 35, 36). In Siberia, the vertical distribution of Yudomian microphytolites also surpasses the extent of the units which could be regarded as equivalents of the Uk Formation. At the conference in Novosibirsk, in 1979, when drafts of the unified schemes were discussed, terrigenous rocks of the Ui Formation, which underlies the Yudoma Series, were considered as equivalents of the Kudash (i.e. Uk Formation) in the Uchur-Maya area. In Siberia, on a whole, the Kudash, as a unit of the standard in drafts discussed at the Novosibirsk meeting, has not been precisely paleontologically substantiated.

Considering the Yudomian microphytolite assemblage, as a whole, it should be noted that in Siberia the assemblage transcends the extent of the Yudomian, and the Yudomian forms may also appear here at various levels starting particularly from the Upper Riphean. However, within the entire area adjacent to the Siberian Platform, there are two sufficiently distinct levels containing Yudomian microphytolites, namely, the lower level which could be considered as a level of the first appearance in the Precambrian section of Yudomian forms, known in association with Riphean forms, and the upper level where abundant Yudomian forms sharply dominate the existing Riphean forms. But we cannot exclude possible anomalies at the lower stratigraphic level, e.g. in the Riphean, Yudomian forms may become predominant, as in the Dashka, Gozhan and Kaltasy Formations. The composition of microphytolite assemblages seems to be substantially affected by environmental control; this may explain the absence of microphytolites in the Vendian of Europe. As a whole, within the Russian Platform and in adjacent areas we deal with the lower level of development of Yudomian microphytolites, especially in the Uk section where they are most completely represented, they occur in association with Riphean forms.

Evaluating the proportion of the Upper Riphean and Yudomian microphytolite assemblages as a whole, it may be stated that in most of the Upper Precambrian sections of Siberia, the qualitative and quantitative analysis of particular assemblages allows us to establish a succession of assemblages (Fig. 36), previously recognized by Zhuravleva (1964) for most of the Upper Precambrian terrains. Of course, we cannot obtain sufficiently well-defined and synchronous boundaries for the units characterized by microphytolites due to their facies control.

In summary, it should be noted that the extent of the Vendian (Yudomian) is probably "insufficient" to draw its boundaries within the Upper Precambrian by means of microphytolites (Khomentovsky 1974, 1976, 1978; Yakshin 1975; Sokolov 1979); evidently, a microphytolite assemblage can help to locate it in the section.

Chapter 11

Vendian Polychaeta

The Polychaeta is the oldest class of annelids (type Annelida). However, the Polychaeta requires an extremely favourable fossilization environment in order to preserve the remains of body structure, or at least quite distinct impressions. Therefore, the paleontological record contains a fairly small number of reliable occurrences of the Polychaeta. Nevertheless, they confidently suggest a very high level of morphological diversity and taxonomic radiation of the Cambrian Polychaeta and articulate in general. Well-known annelid families (no less than six) reported long ago from the famous Middle Cambrian Burgess Shale of British Columbia strongly imply the beginning of Polychaeta differentiation as early as the Precambrian. Indirect evidence is provided by different burrows of "worms" richly represented in the Vendian paleoichnology and in products of gigantic biological processing of bottom mud and associated nutrient benthos. In some places, Vendian calcareous and ash-rich mudstones (Olenek Uplift, Podolia) represent a lithified sediment which passed through the digestive system of organisms. This is evidenced by peculiar structural features. The process seems to have been dominated by Polychaeta mud-eaters.

The presence of the Polychaeta in the Ediacaran biota of the Vendian Metazoa is directly evidenced by numerous records of the genus *Spriggina* (Glaessner 1958, 1962, 1976 and others). Many zoologists believe that its representatives (family Sprigginidae in general and the Cambrian family Canadiidae) are ancestors of older and Recent phyllodocemorphic polychaete worms (Ushakov 1972). In case we agree with those who think that a distinct cephalization of the *Spriggina* is a character indicative of trilobitomorphic articulates or just an ancestral or one of the ancestral trilobite forms (Fedonkin 1983a and others), *Spriggina* still retains affinity to pelagic annelids of the class Polychaeta.

The systematic position of a sedentary Sabellitida remains uncertain. These tubular organisms with an elastic chitinoid tube, which is brown or yellow in colour and sometimes parchmentlike or transparent, are preserved in a fossil state due to fine, filmy pyritization of the organic substance; it is a very important group for the Vendian, especially Late Vendian. Sabellitids consist of two families, viz. Sabelliditidae and Saarinidae (Sokolov 1965a, 1967, 1972 and others). The former incorporates typical Upper Vendian genera of *Sabellidites* (surviving in Lower Cambrian), *Parasabellidites, Paleolina, Sokoloviina;* the latter consists of two genera: *Saarina* (Redkino-Rovno Horizons) and *Calyptrina* (Redkino Horizon). Sokolov's hypothesis on the high taxonomic rank of sabelliditids and their affinity with the type Pogonophora has been supported by some zoologists (Zenkevich, Ivanov, Fedotov). However, a comparative submicroscopic study of fossil tubes of

Sabellidites and Recent tubes of Zenkevitchiana (Urbanek 1979) has neither confirmed nor contradicted the statement. Yanishevsky, in 1926, was the first to study the genus *Sabellidites* and assigned it to sedentary Polychaeta. Some workers do not reject even the affinity of the Pogonophora with annelids. In any case, the sabelliditids represent a major group of the oldest tubular fossils with an integument of organic origin, which is very important for Vendian paleontology and stratigraphy.

Paleontological data on mobile or vagile Polychaeta (subclass Errantia) are known to be based on well-preserved fossil elements of the jaw apparatus, scolecodonts, because the Polychaeta Errantia are preserved only in a unique fossilization environment. Scolecodonts have been recorded since the Ordovician and are fairly well kown in comparison to the jaw apparatus of Recent and vagile Polychaeta. It is noteworthy that those who studied the oldest Ordovician scolecodonts (Kozlowski 1956; Kielan-Jaworowska 1966) had reported two important features: a very high level of morphological diversity of the jaw apparatus of the Polychaeta as early as in Ordovician and their probable Precambrian-Cambrian evolution and at the same time a similarity of the jaw apparatus to pharyngial tubes of the modern Errantia.

The jaw apparatus performs two functions: capture of food and its grinding. It is situated just behind the oral cavity commonly within a strong pharynx, usually an everted one. The pharynx is covered by fine papilli and bears special chitinous lamellae of the jaw apparatus. The structure of the latter is very important for the systematics of Errantia. There is the lower jaw at the ventral side in the pharynx and the upper jaw at the dorsal side. The latter consists of a pair of mandibles (MI) with supports and mandibles of second maxillae (M II) and several pairs of fine lamellae (M III, M IV, etc.). All lamellae are denticulated at one side. The lower pair of maxillae (M II) is commonly made up of wider lamellae denticulated on the internal edge. Chitinous lamellae in the pharynx can be highly diverse. Polychaeta (Errantia) vary in size from several millimetres to tens of centimetres; the jaw elements are also quite variable in size, sometimes a fraction of a millimetre.

Rocks macerated by microfossils provide fairly fine denticulated or spiny, chitinous lamellae which apparently should be interpreted as jaw lamellae of the Polychaeta (Errantia). This fossil group had not been reported earlier from the Vendian, however, the Vendian scolecodonts undoubtedly develop quite diversely. The largest fragments of a probable Polychaeta jaw apparatus belong to the genus *Redkinia* (Sokolov 1977a) reported from the Redkino Series (Vendian) of the Moscow Syneclise. This enigmatic occurrence, found many years ago by Z.P. Ivanova, was examined by many scientists and was assigned either to conodonts, or to integuments of possible ancient Onichophora (resembling outwardly the outline of Aysheaia) or even to problematics. Below this very interesting and important form is briefly discussed.

Scolecodonta (Group Redkiniida gr. n.; Family Redkiniidae fam. nov.)

Diagnosis. The oldest Polychaeta (Errantia) which have preserved only dark, chitinous lamellae of different size bearing denticulated or smooth teeth on one edge; gapings between them are either smooth or have minute denticles-setules.

Genus *Redkinia* Sokolov 1977, Vendian, Redkino Series

Type Species. Redkinia spinosa, Sokolov (1977, p. 441, Fig. A).

Diagnosis. As for Type Species.

Redkinia spinosa Sokolov 1977 (Plate 88, *1–5*)

Redkinia spinosa Sokolov, gen, sp. n.: Sokolov 1977, p. 441, Fig. a; Sokolov 1984, p. 7, Fig. 6.

Holotype. PIN AN USSR No. 3993/3, Vendian, Redkino Series, Nelidovo Beds, Moscow Syneclise, Nepeitsino No. 1 borehole, depth, 1417–1428 m. Illustrated in Sokolov 1977, p. 441, Fig. a, left specimen.

Diagnosis. Based on lamellae which form clusters of fairly incomplete forms. Brownish-black chitinous lamellae, relatively large, up to 2–3.5 mm long in the lower maxillae (M II). Long teeth on one oval edge and minute denticles are separated by discernible interstices.

Description. Lower maxillae (M II) seem to be best preserved. Those are dense, brownish-black lamellae with a lighter, brownish edge. On the oval (inner) edge there are very prominent downward-directed and slightly turned down. hornlike, long denticles; apical margins of the lamellae are undenticulated. Teeth are slightly denticulated. Lamella edges between teeth (interstices) have minute denticles-setules up to six in number, usually low, but of different size. The number of large teeth on the largest lamellae attains 12–15; however, only 9–10 are well preserved. In all, there are 18 incomplete lamellae. Lamella are 2–3.5 mm long; teeth, 0.3–0.45 mm long; width at the base, 0.10–0.15 mm; teeth are spaced at 0.15–0.20 mm.

Remarks. Coeval forms (two fragments) undoubtedly belong to the same genus, but may belong to a different species; Aseeva (1983a, Plate XI, Figs. 9,10; 1983b, Plate XVIII, Figs. 8,9) determined them as scolecodonts of the *Hindeodella* sp., the Zinkov Beds, Vendian of the Ukraine. They are one-fifth of the above size, and large teeth are separated by numerous denticles-setules.

Distribution. Vendian, Redkino and Mogilev-, Podolian Series of the Russian Platform.

Explanation for plates see pages 336–351

Plate 1

Plate 2

Plate 3

Plate 4

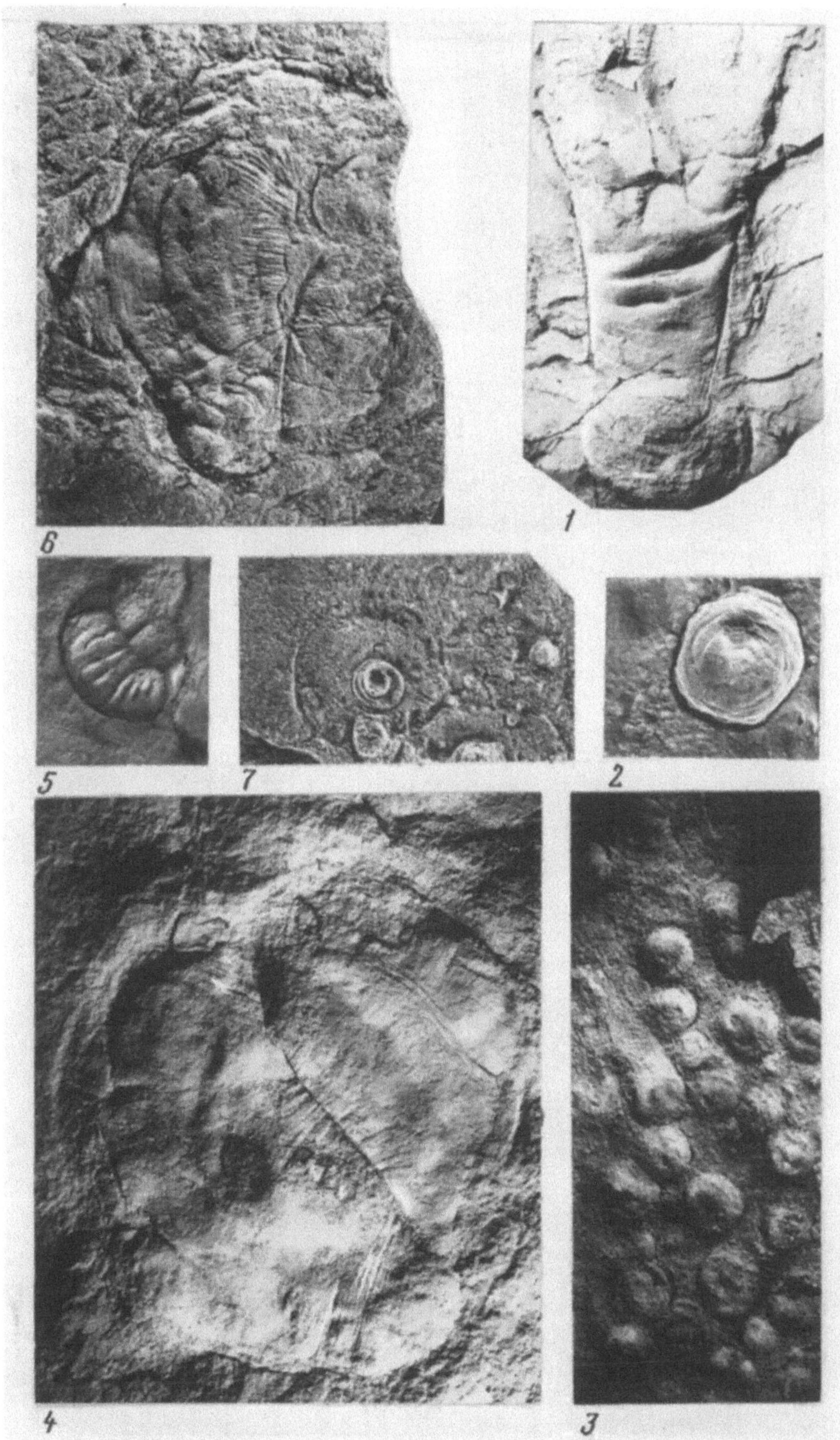

Plate 5

Plate 6

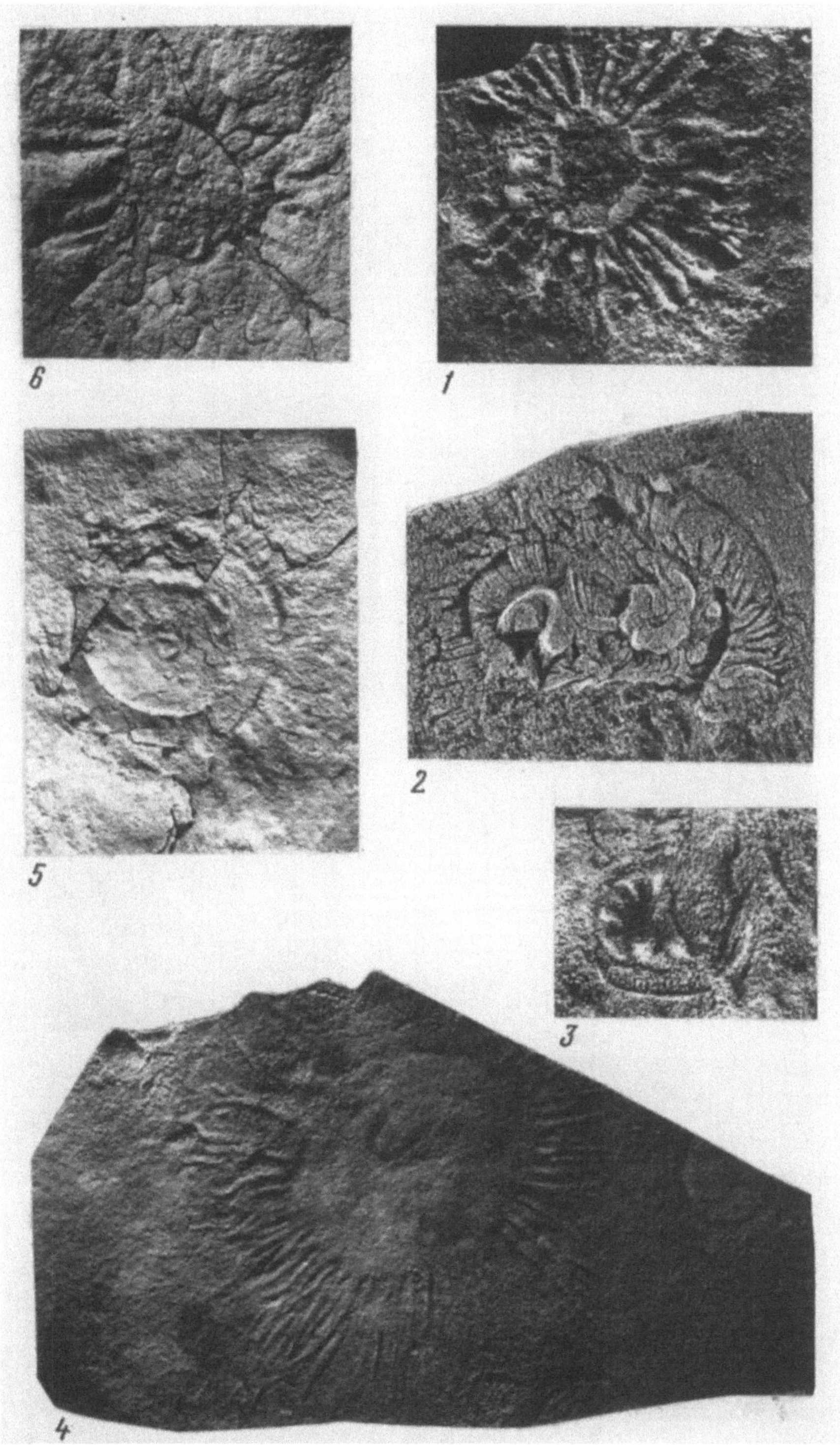

Plate 7

Plate 8

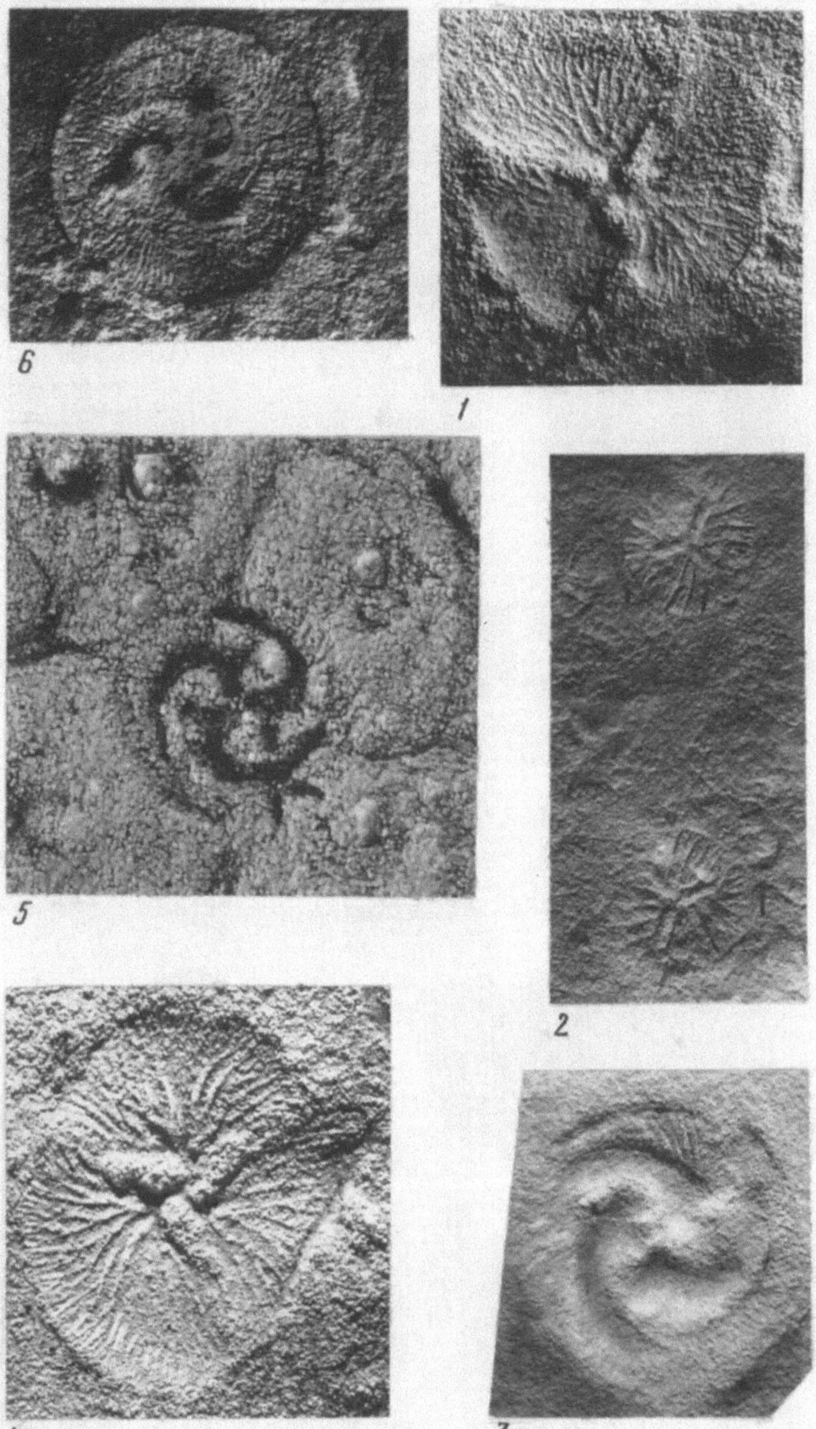

Plate 9

Plate 10

Plate 11

Plate 12

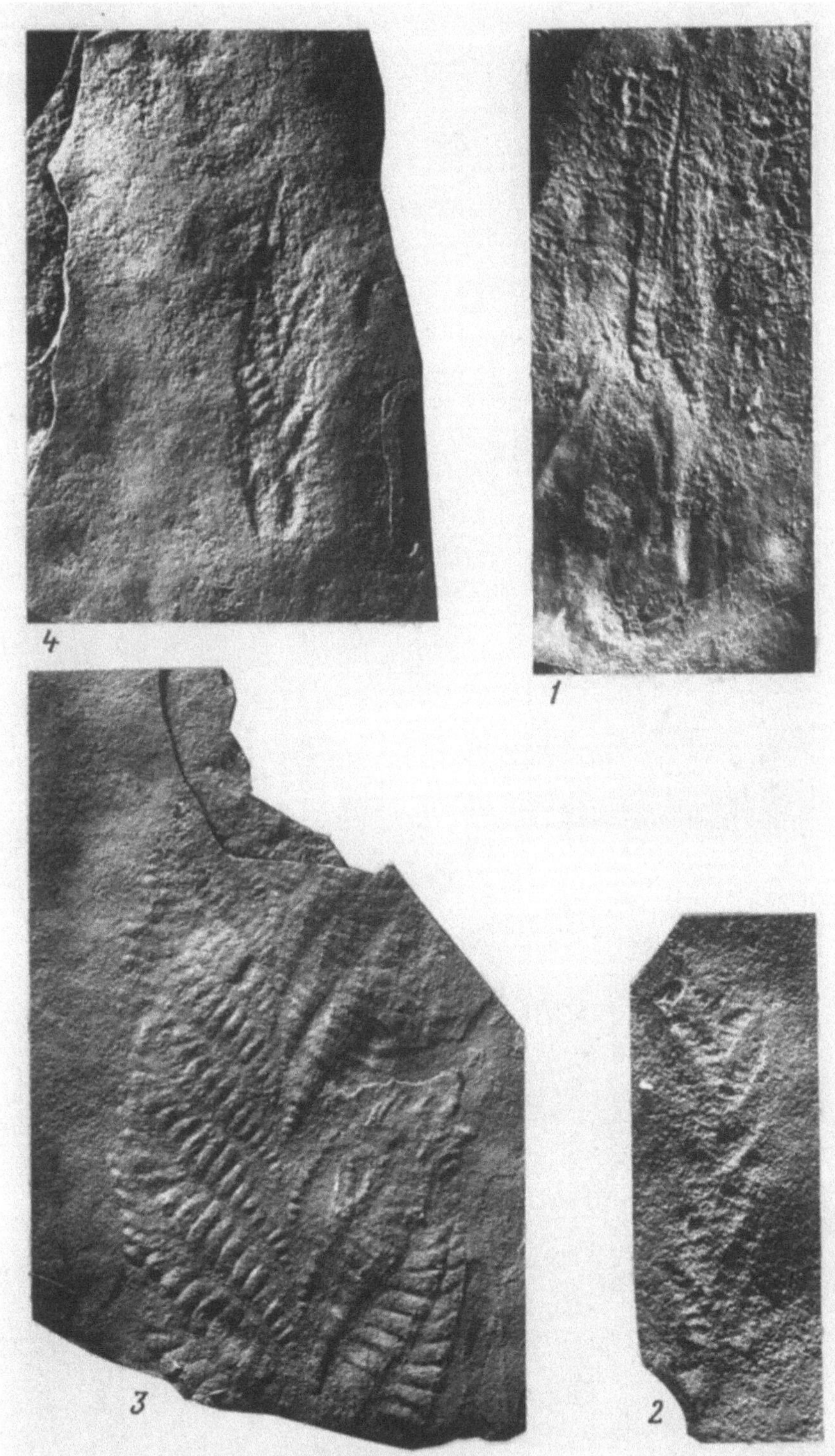

Plate 13

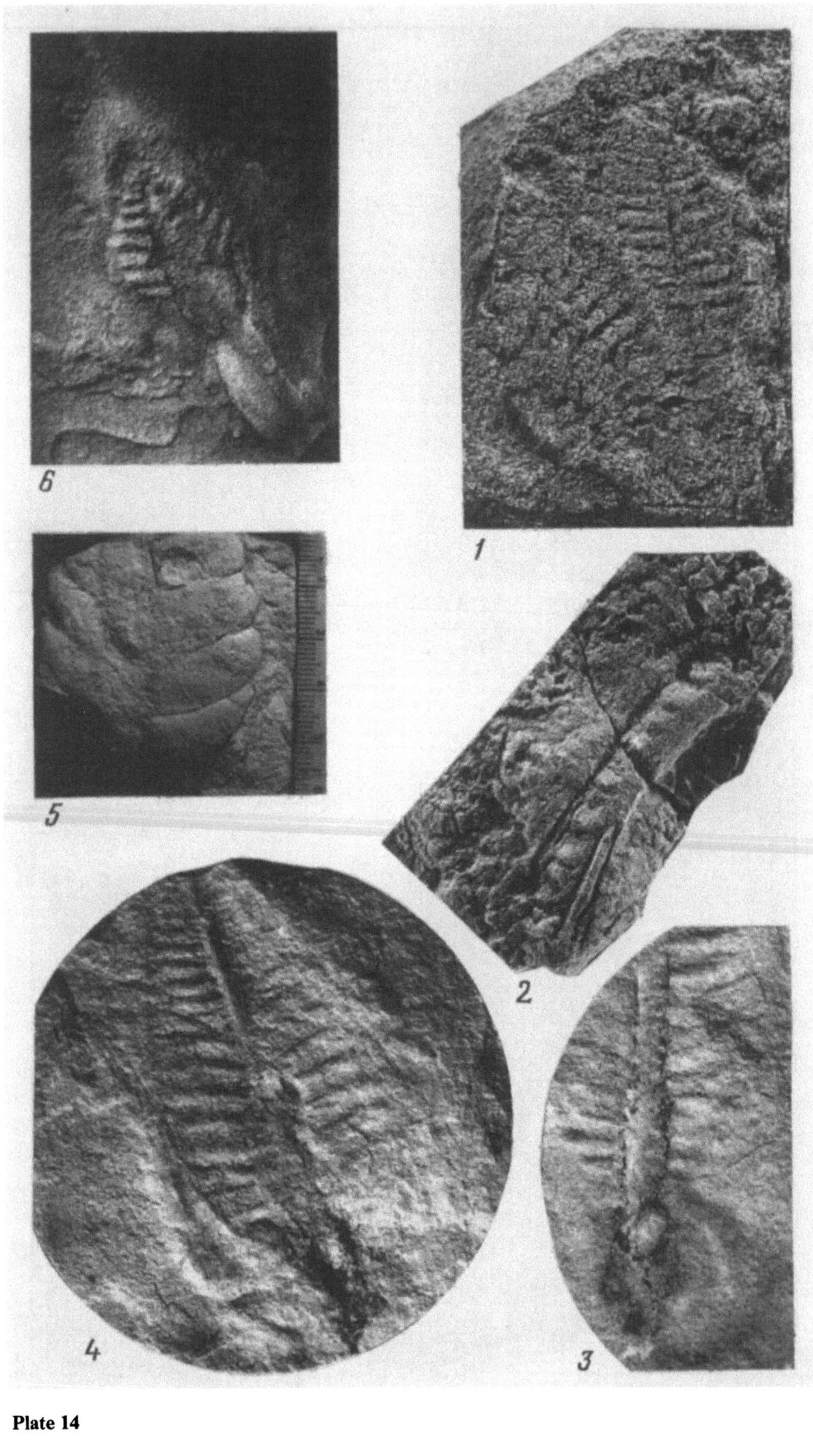

Plate 14

1

2

Plate 15

Plate 16

Plate 17

Plate 18

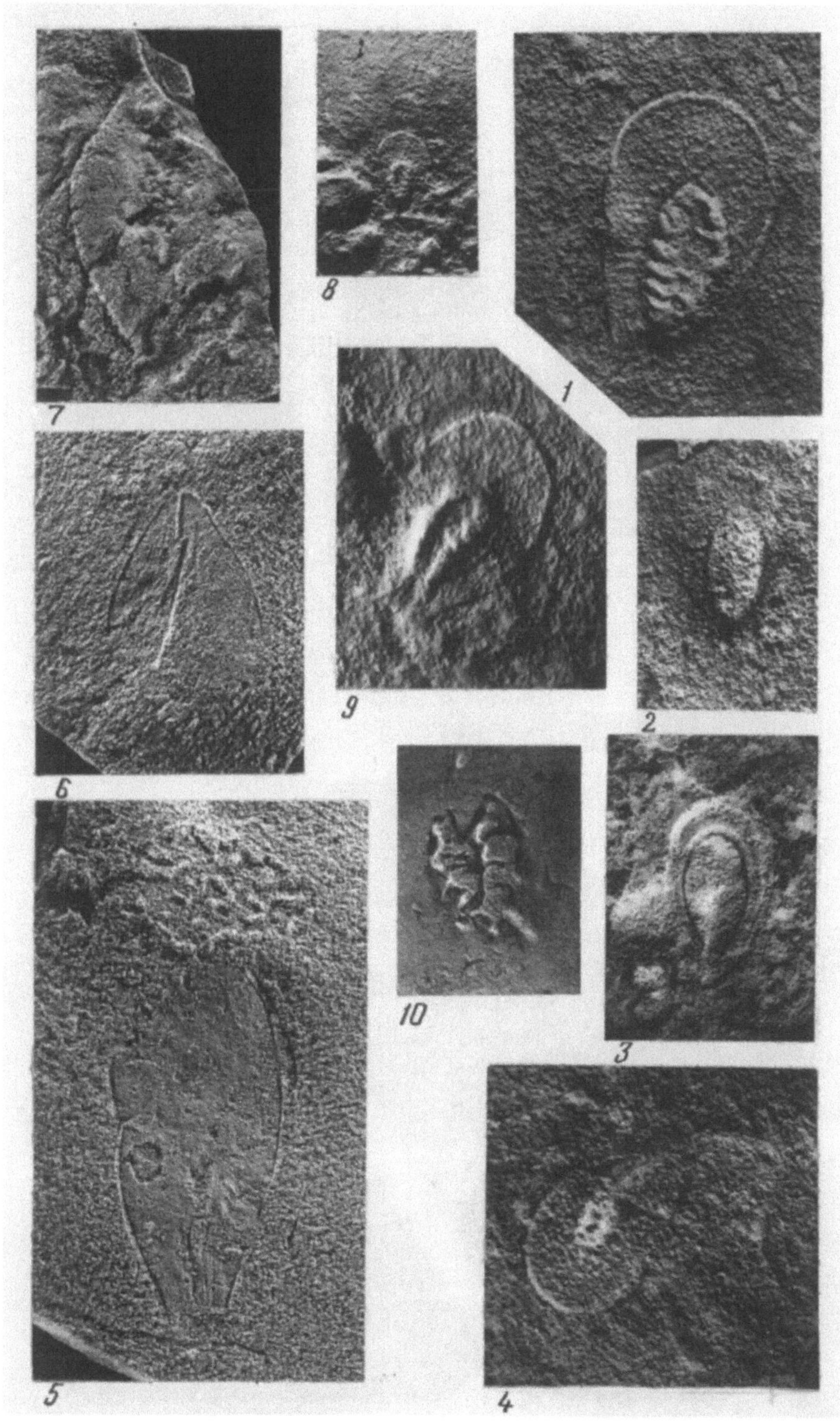

Plate 19

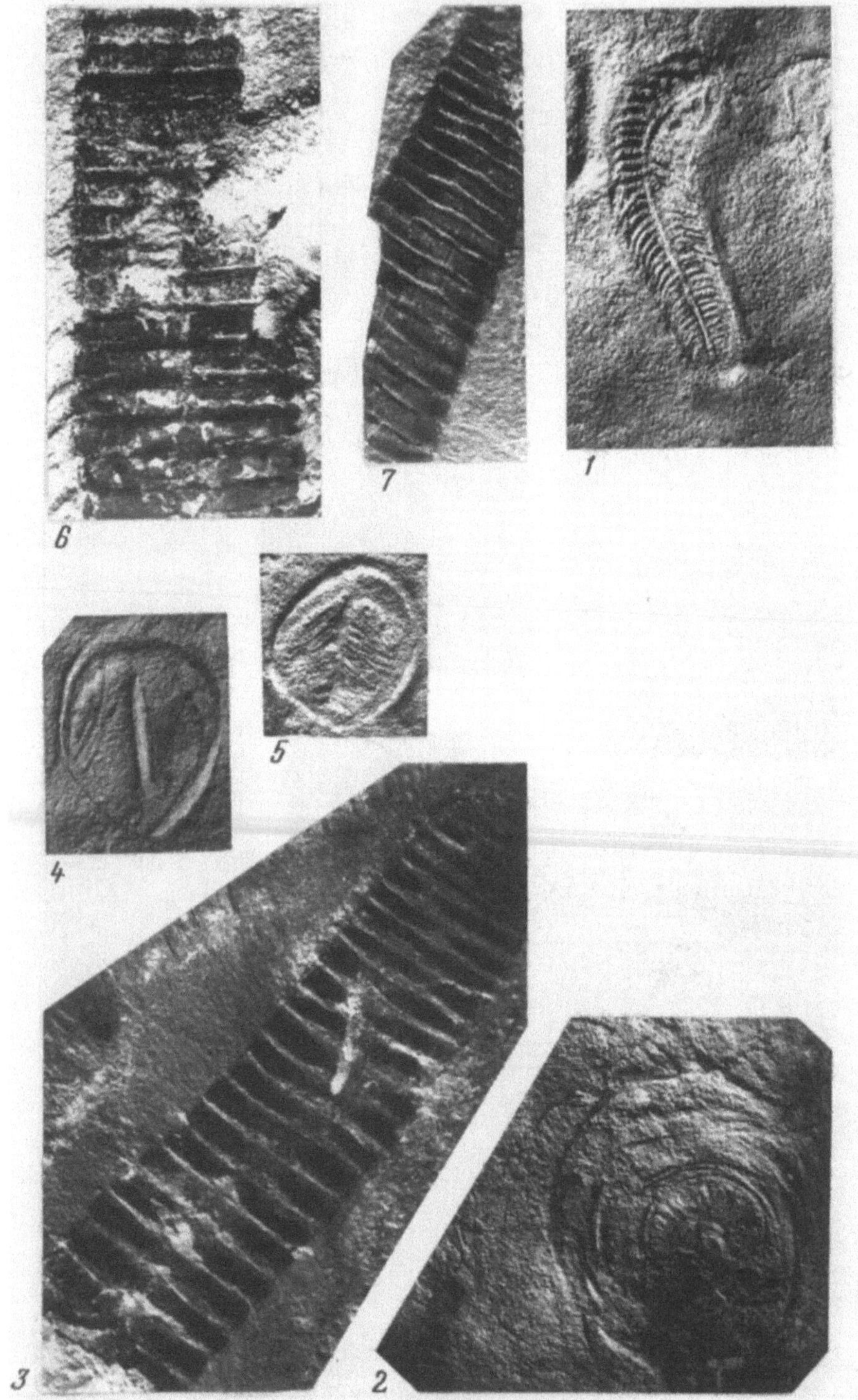

Plate 20

Plate 21

Plate 22

Plate 23

Plate 24

1

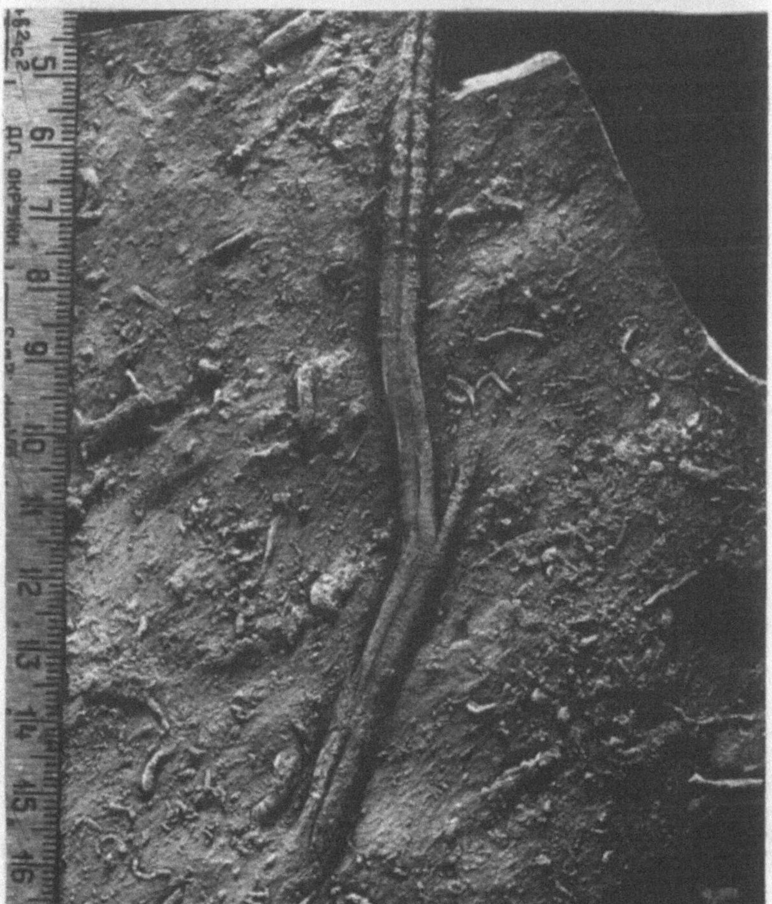

2

Plate 25

Plate 26

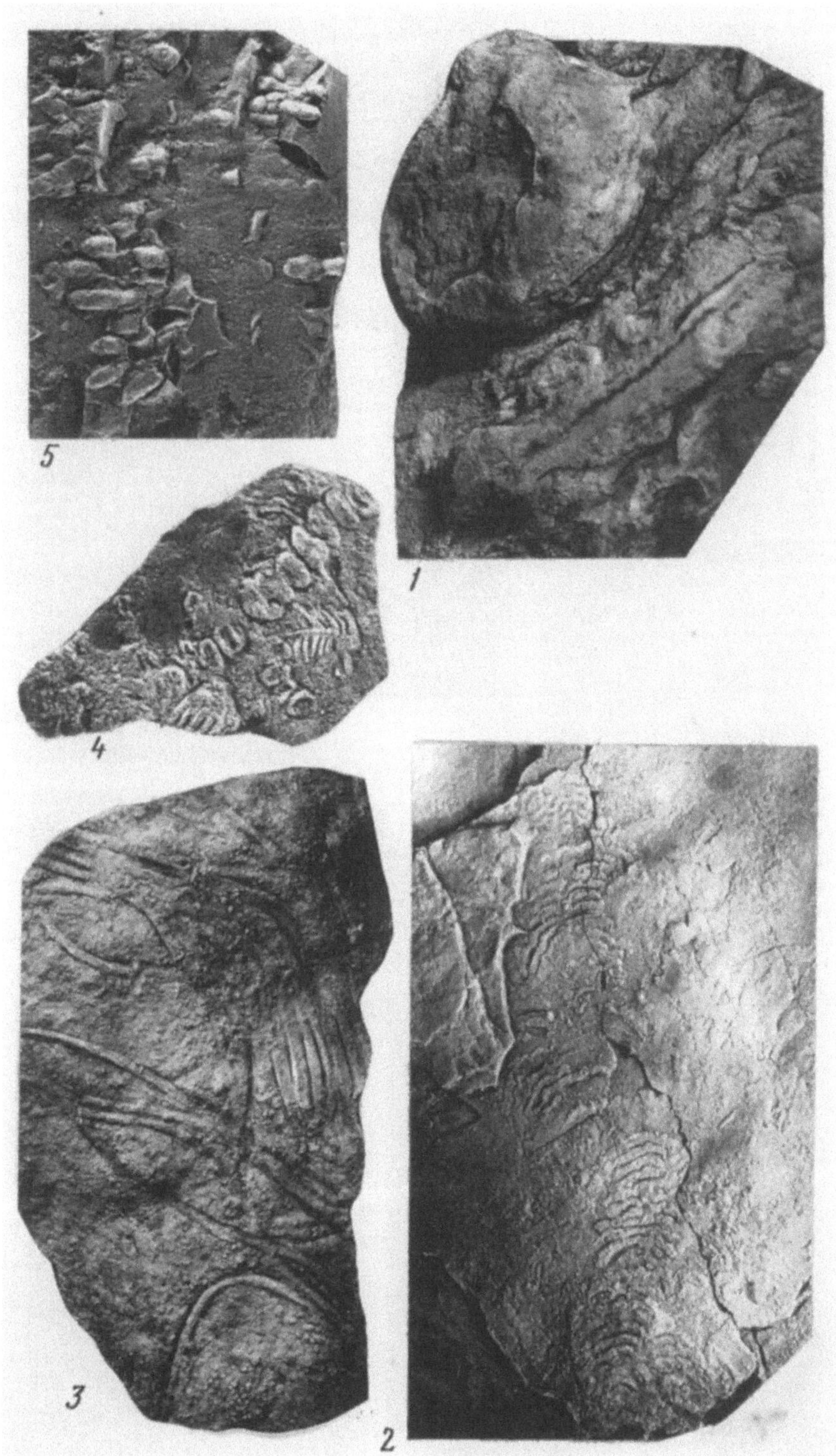

Plate 27

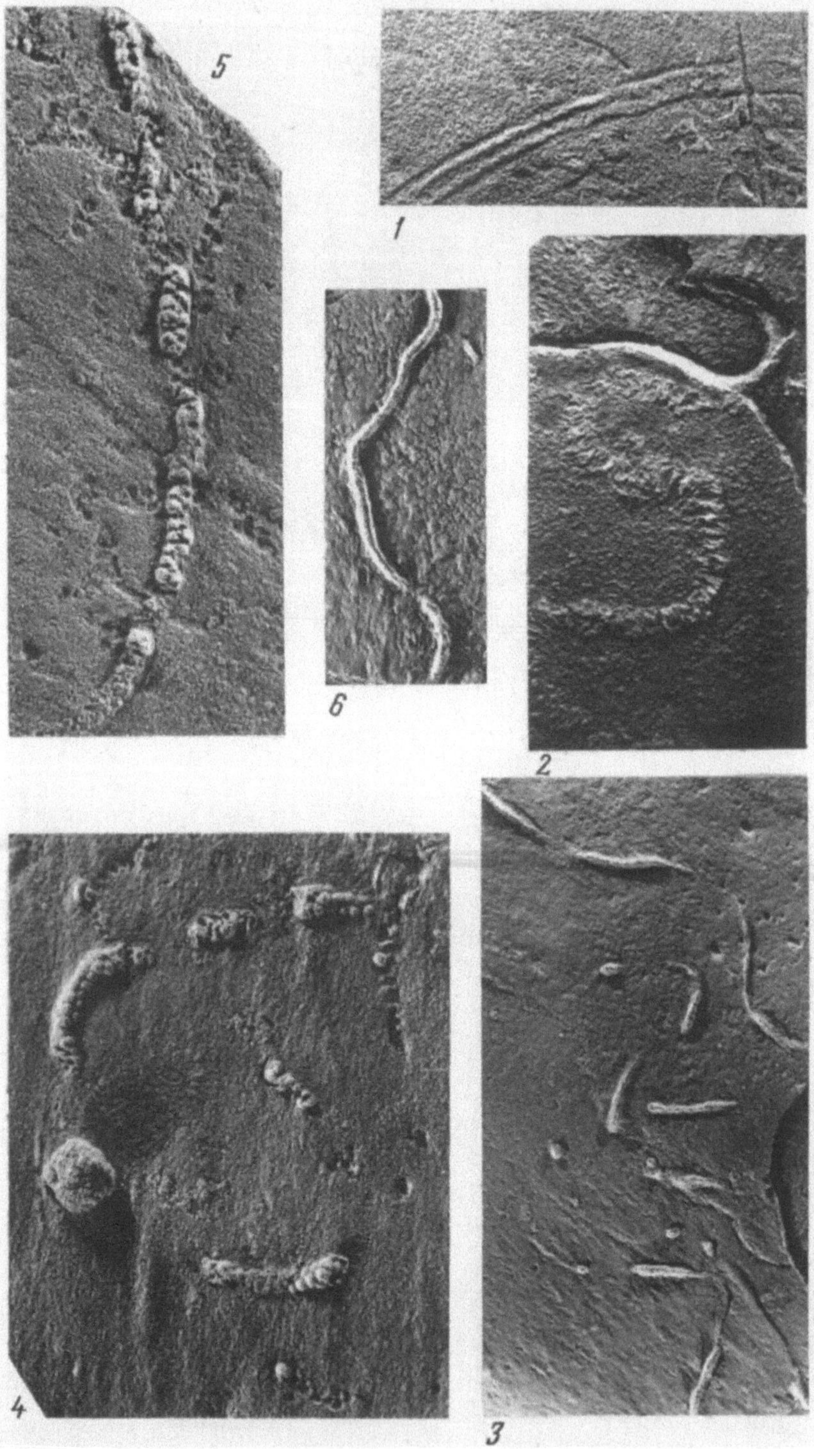

Plate 28

Plate 29

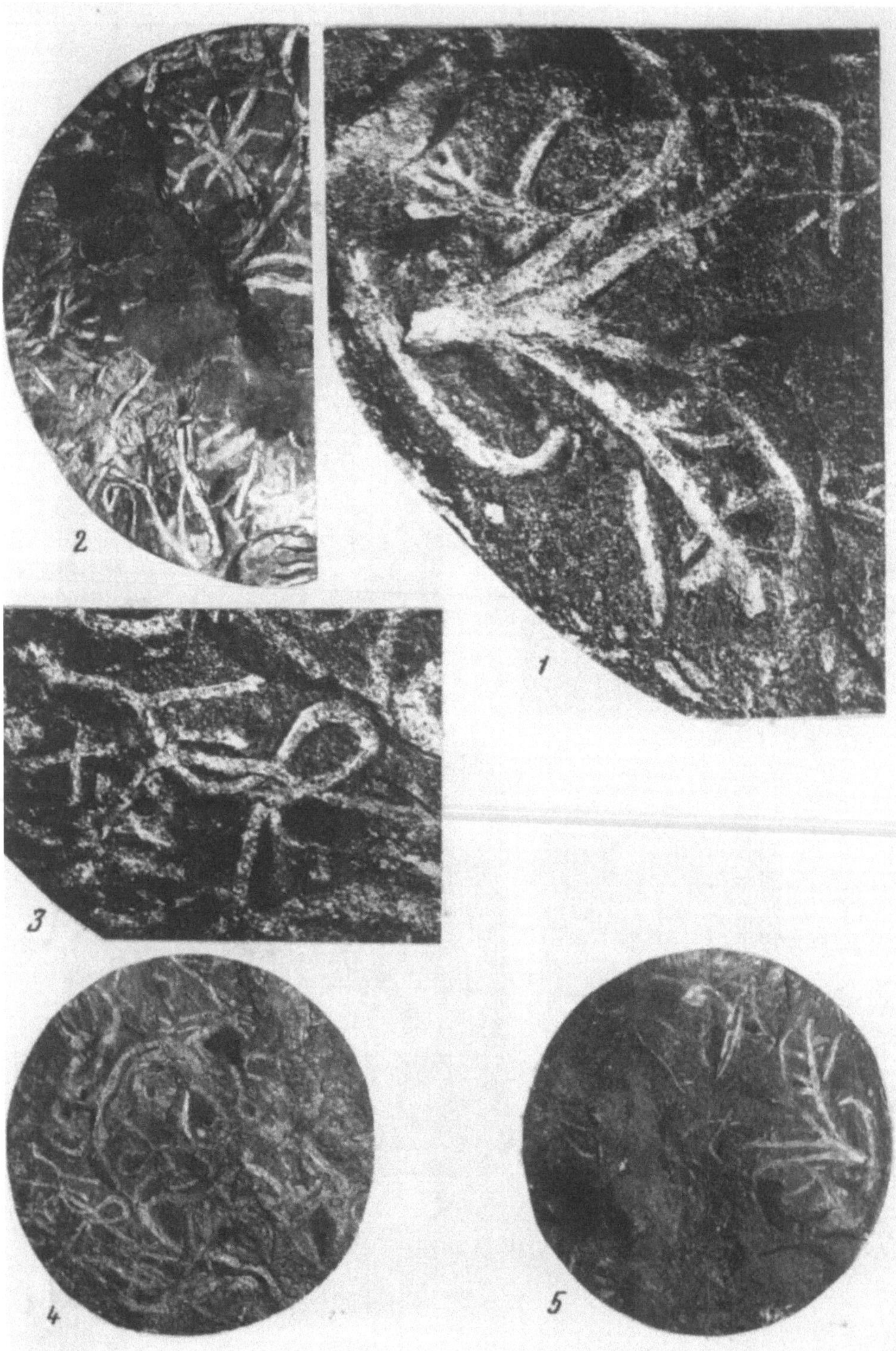

Plate 30

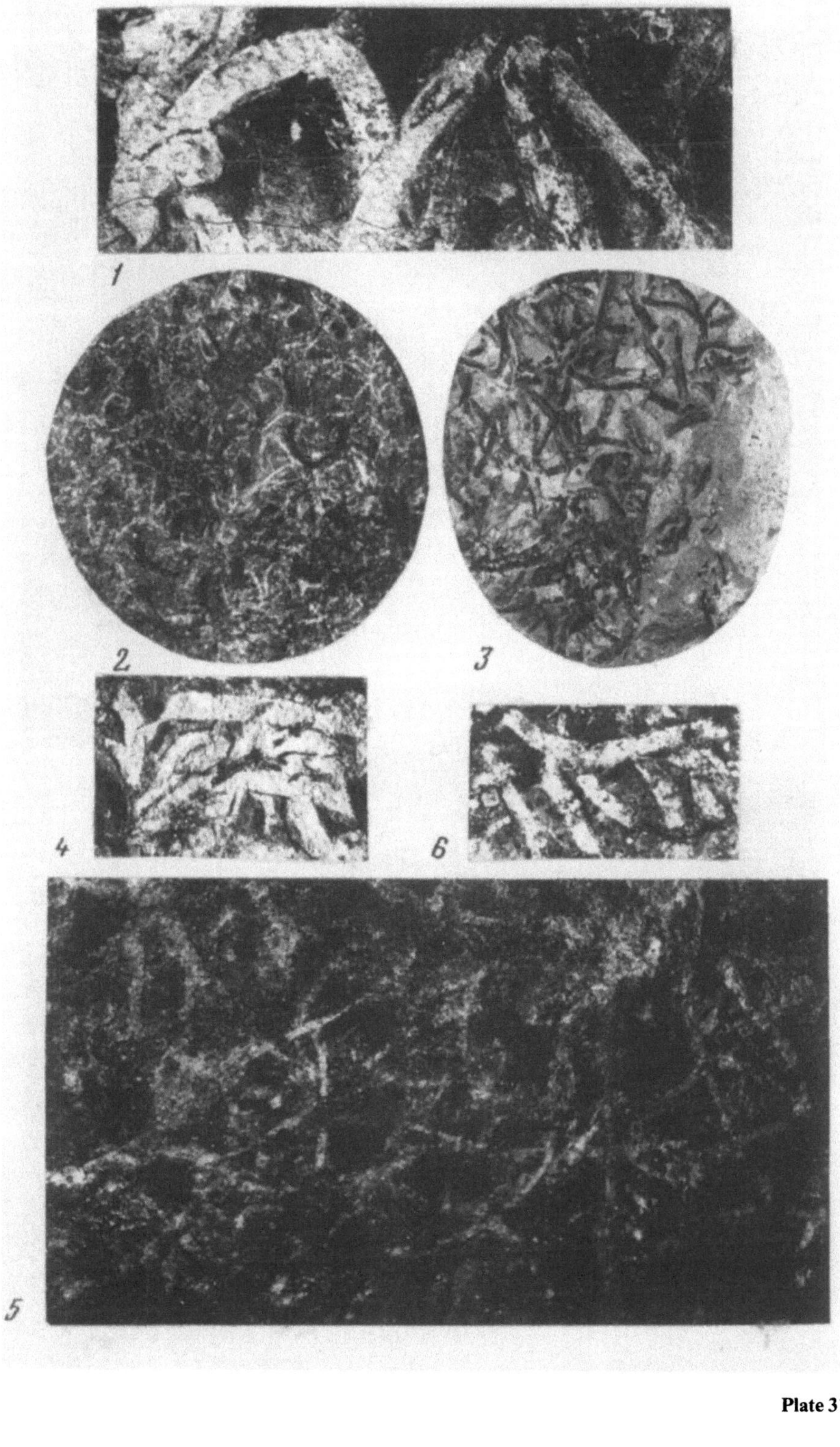

Plate 31

Plate 32

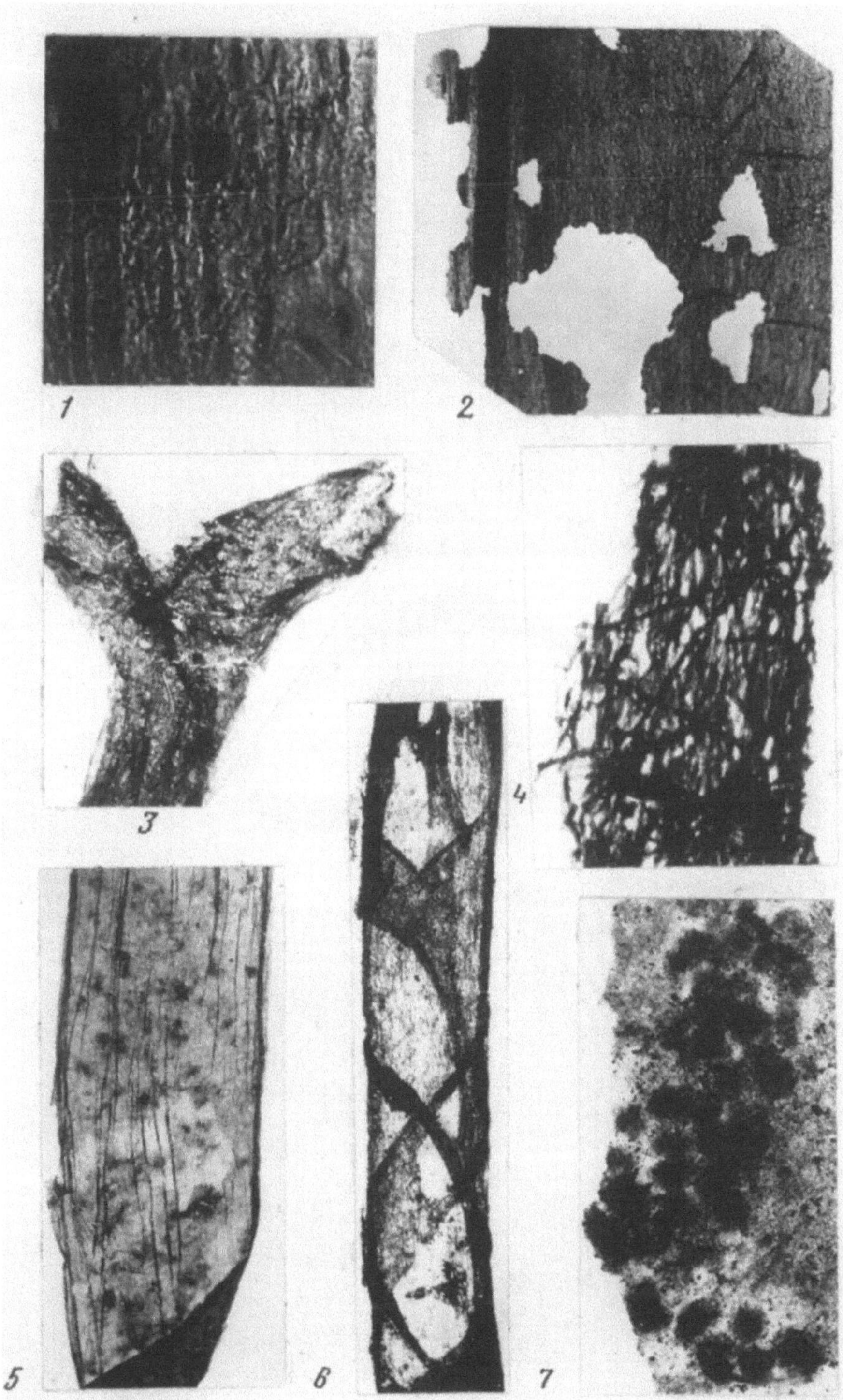

Plate 33

Plate 34

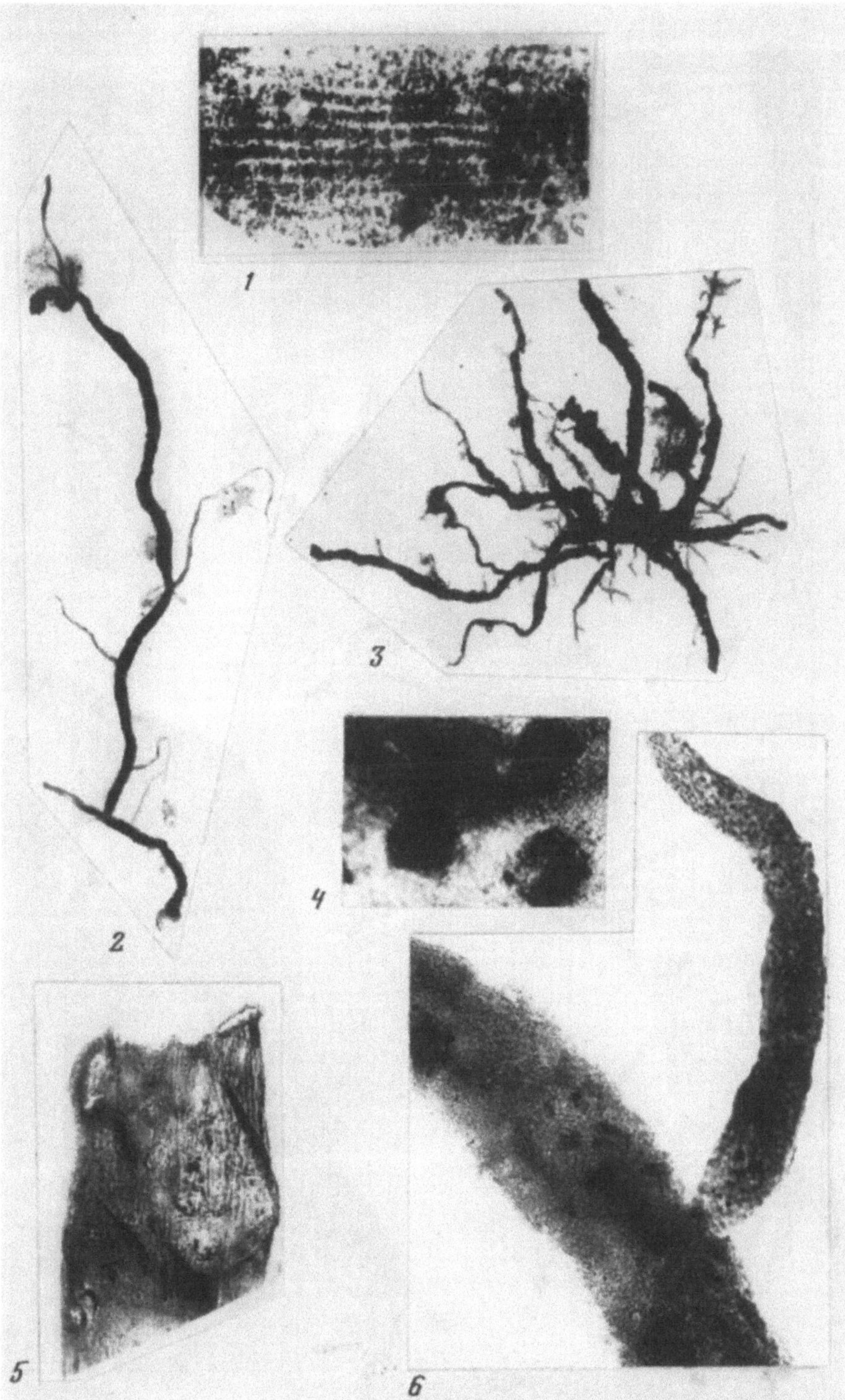

Plate 35

Plate 36

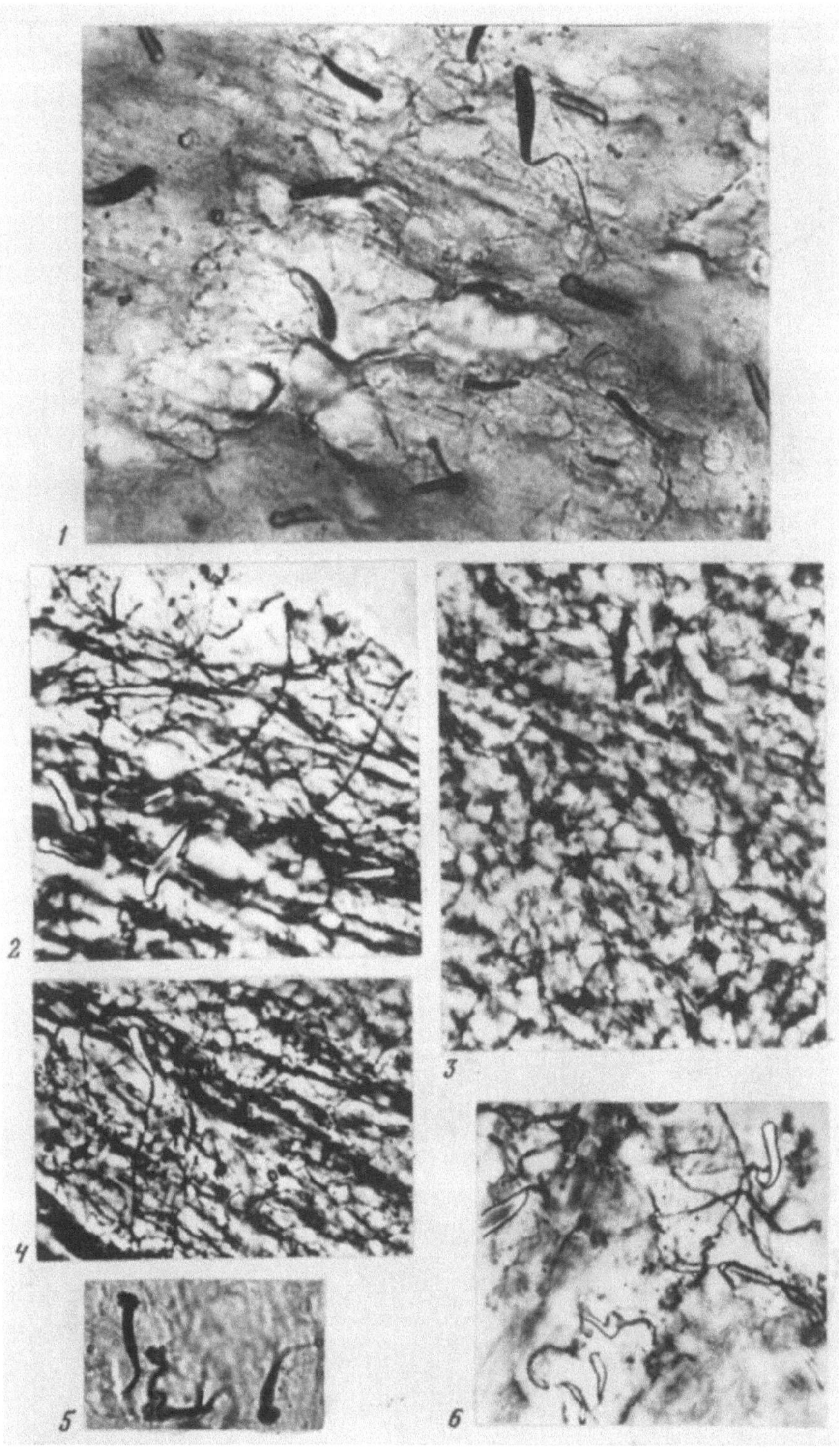

Plate 37

Plate 38

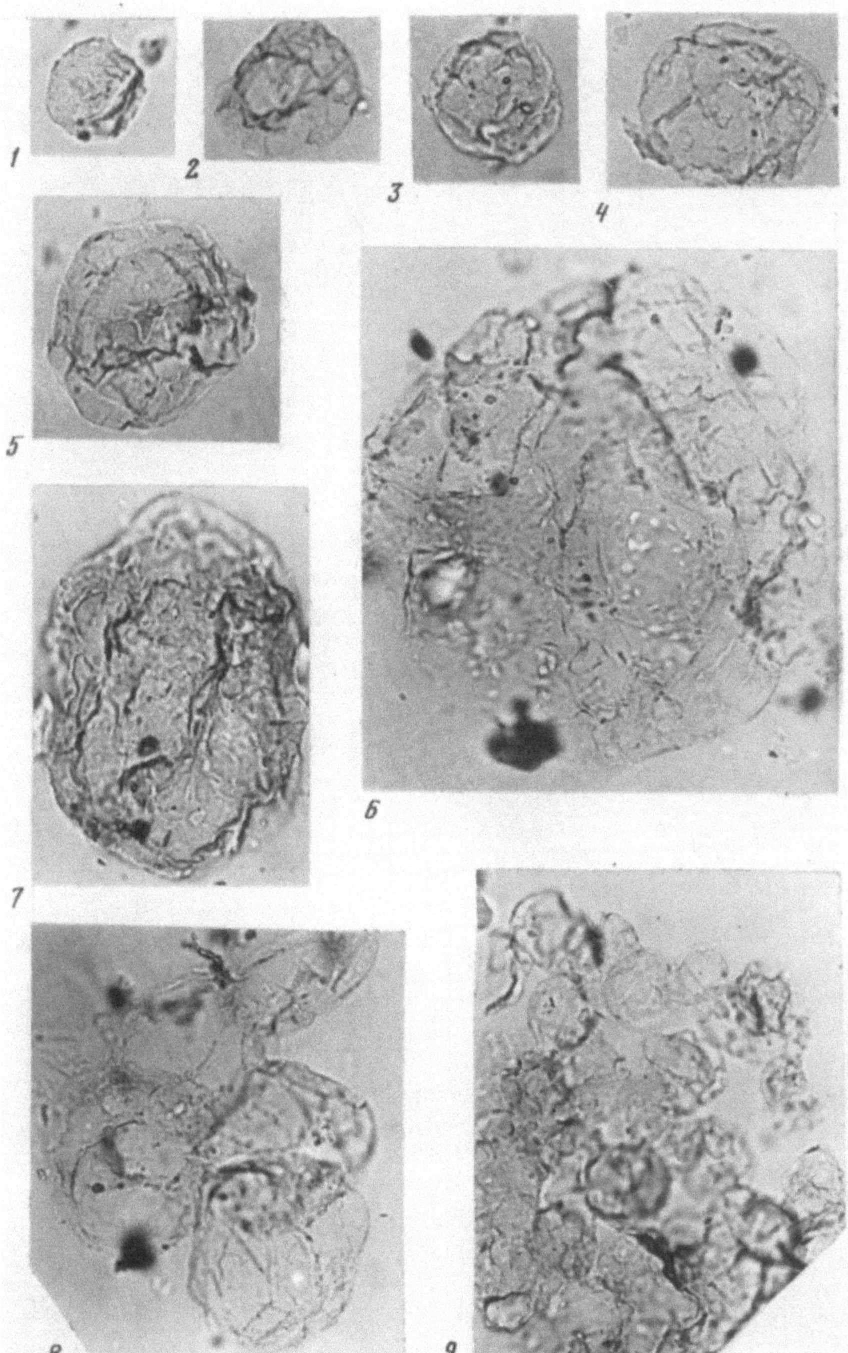

Plate 39

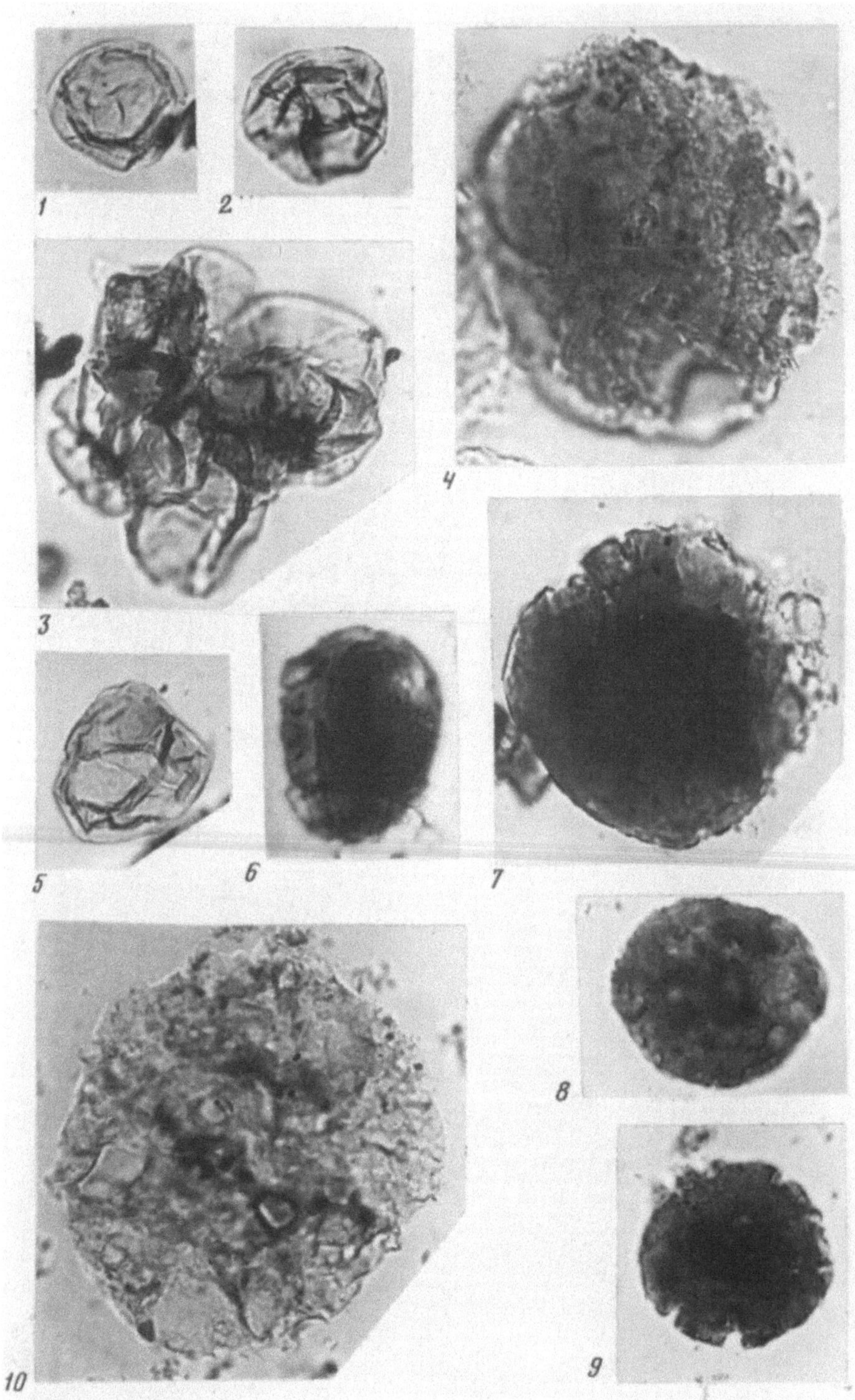

Plate 40

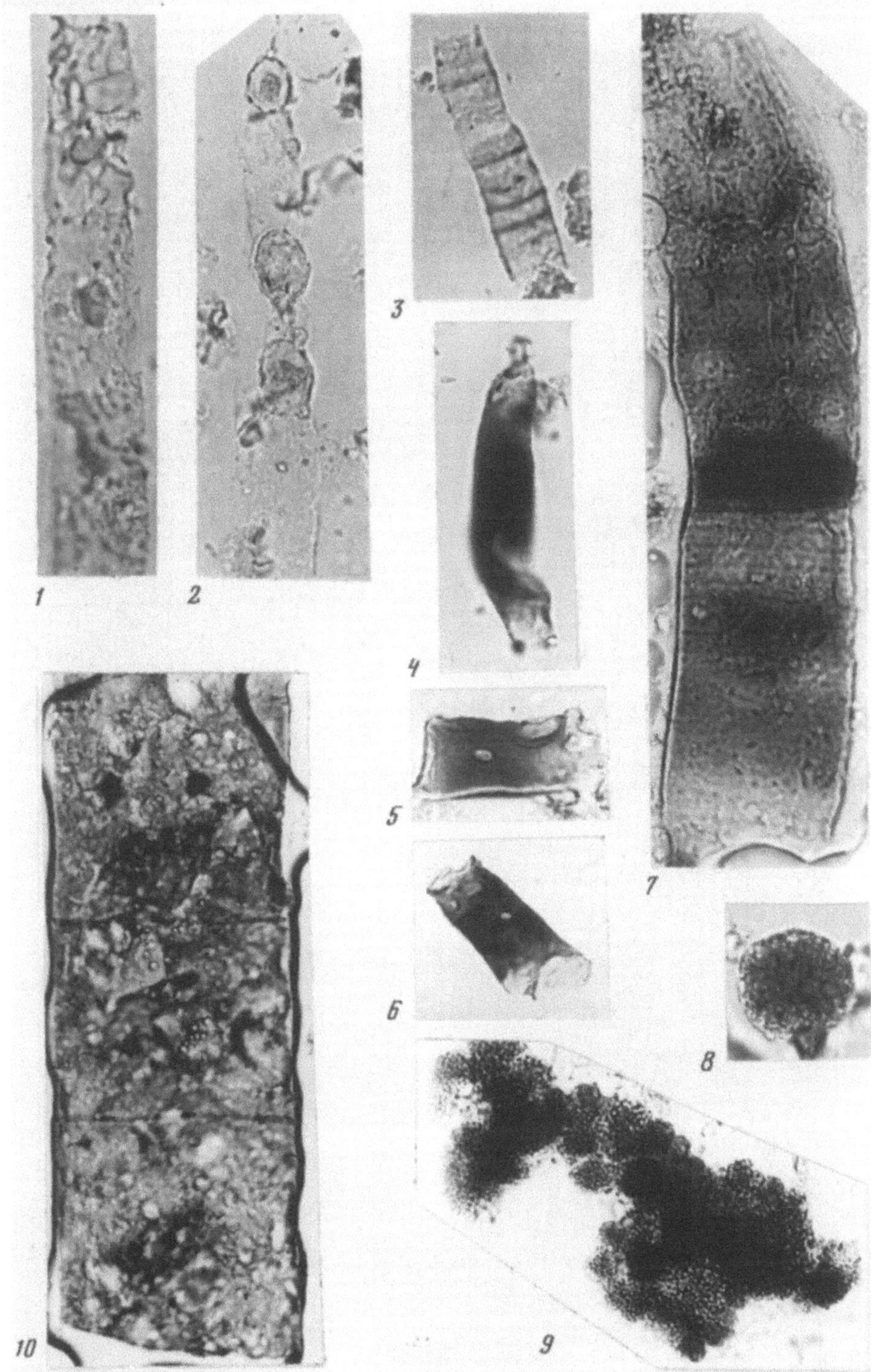

Plate 41

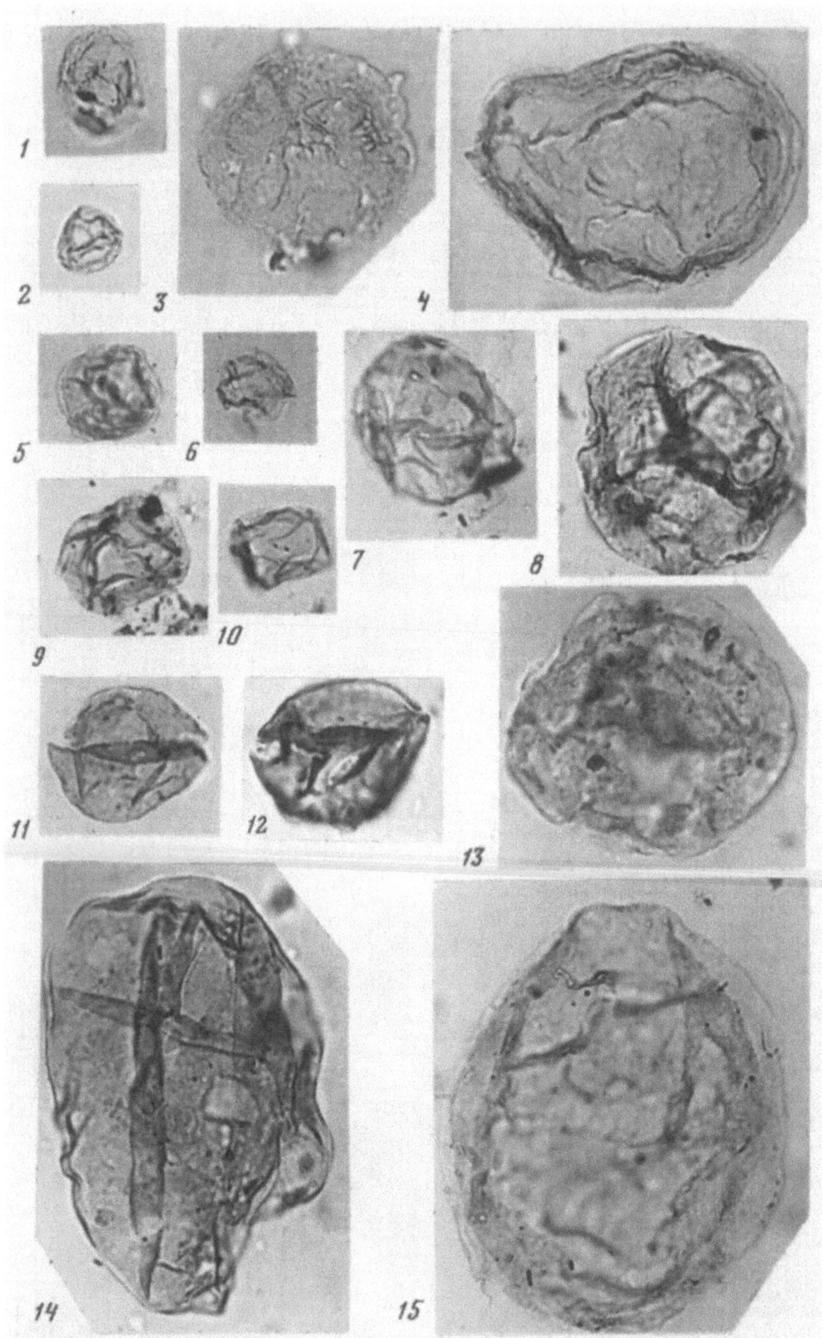

Plate 42

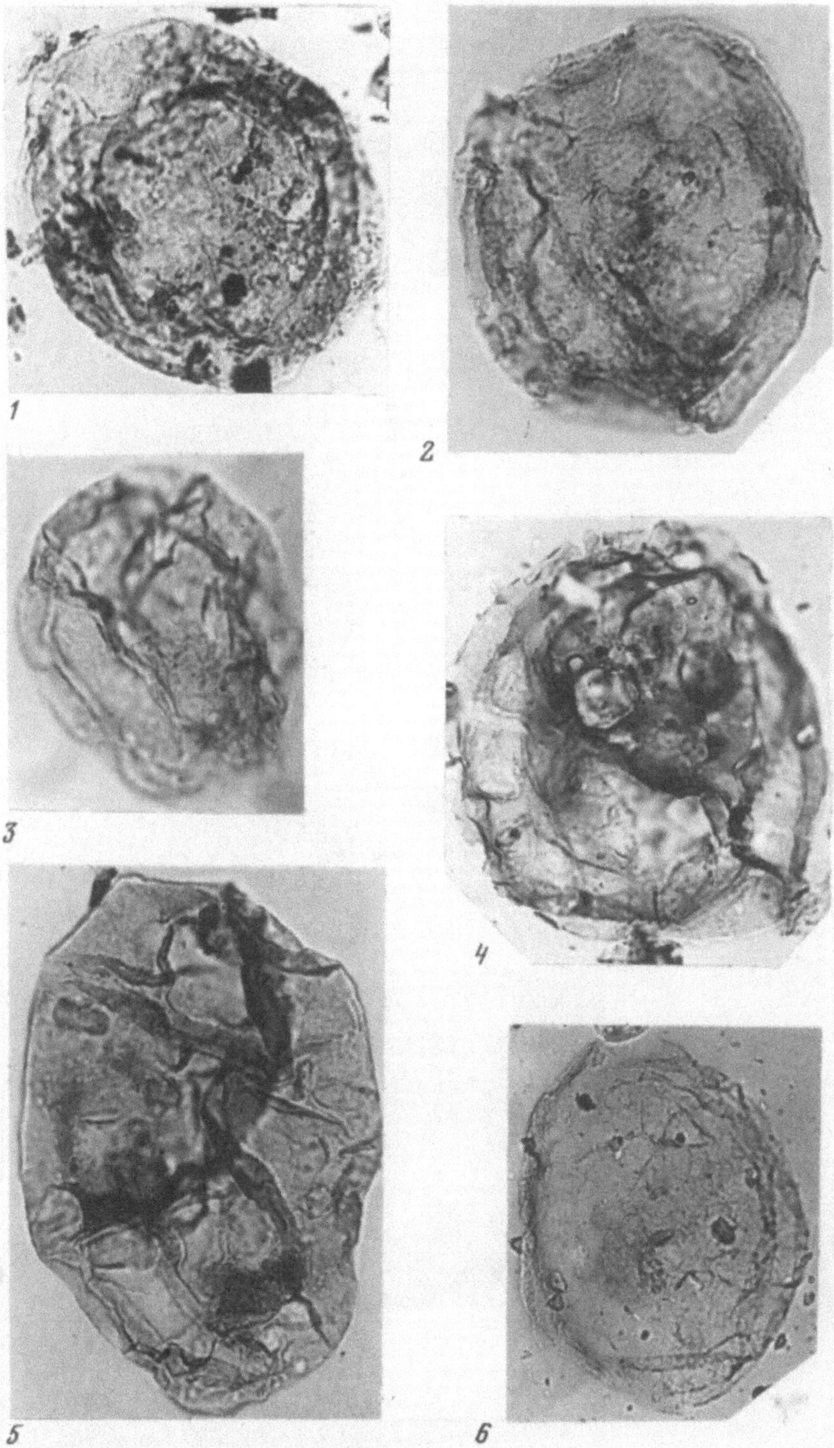

Plate 43

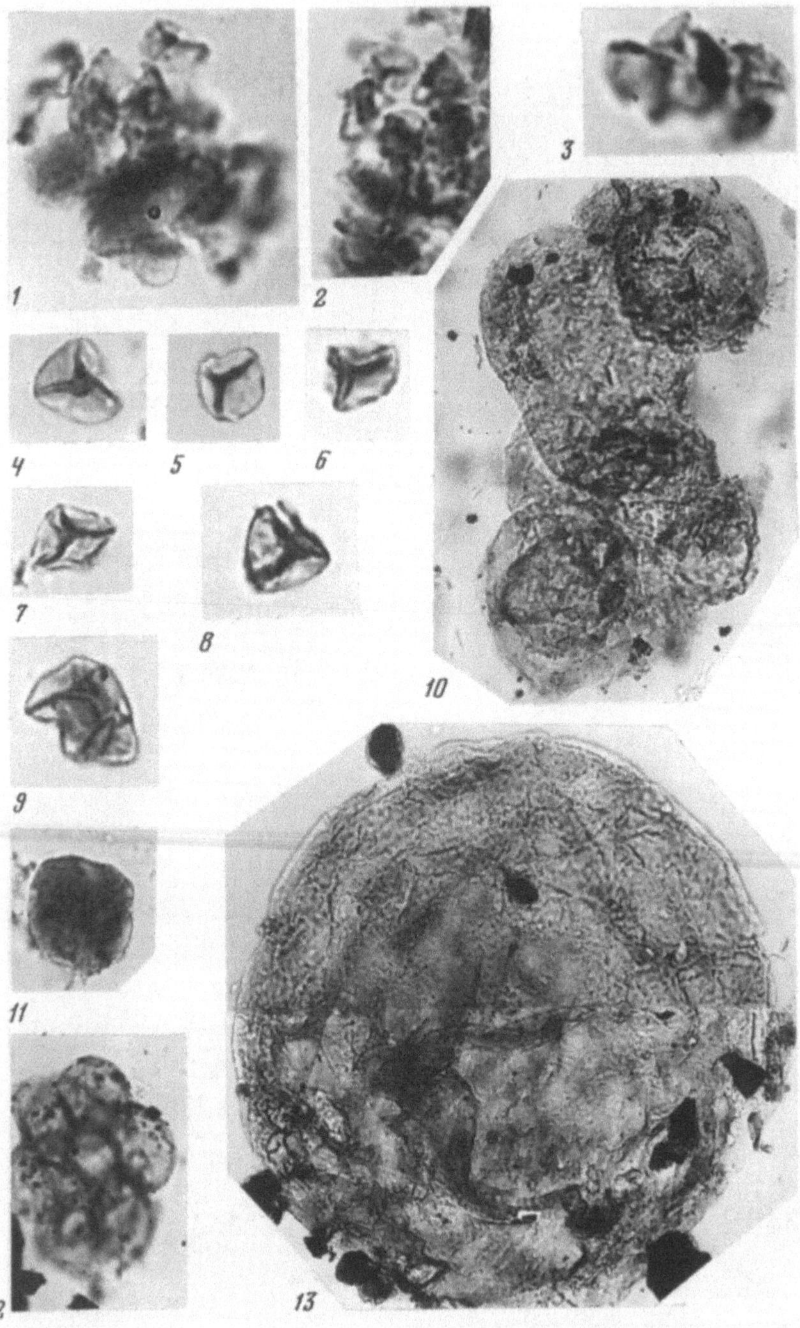

Plate 44

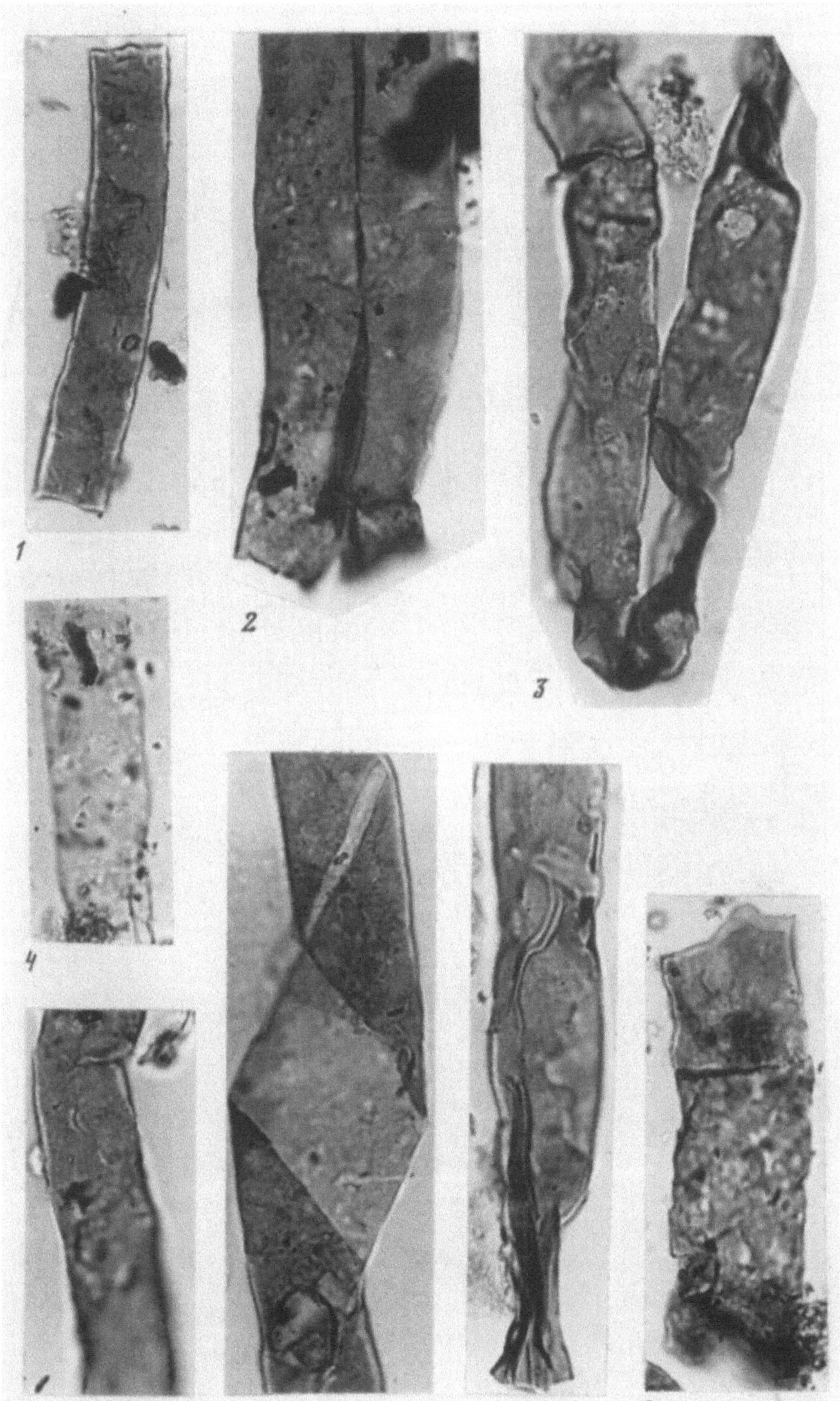

Plate 45

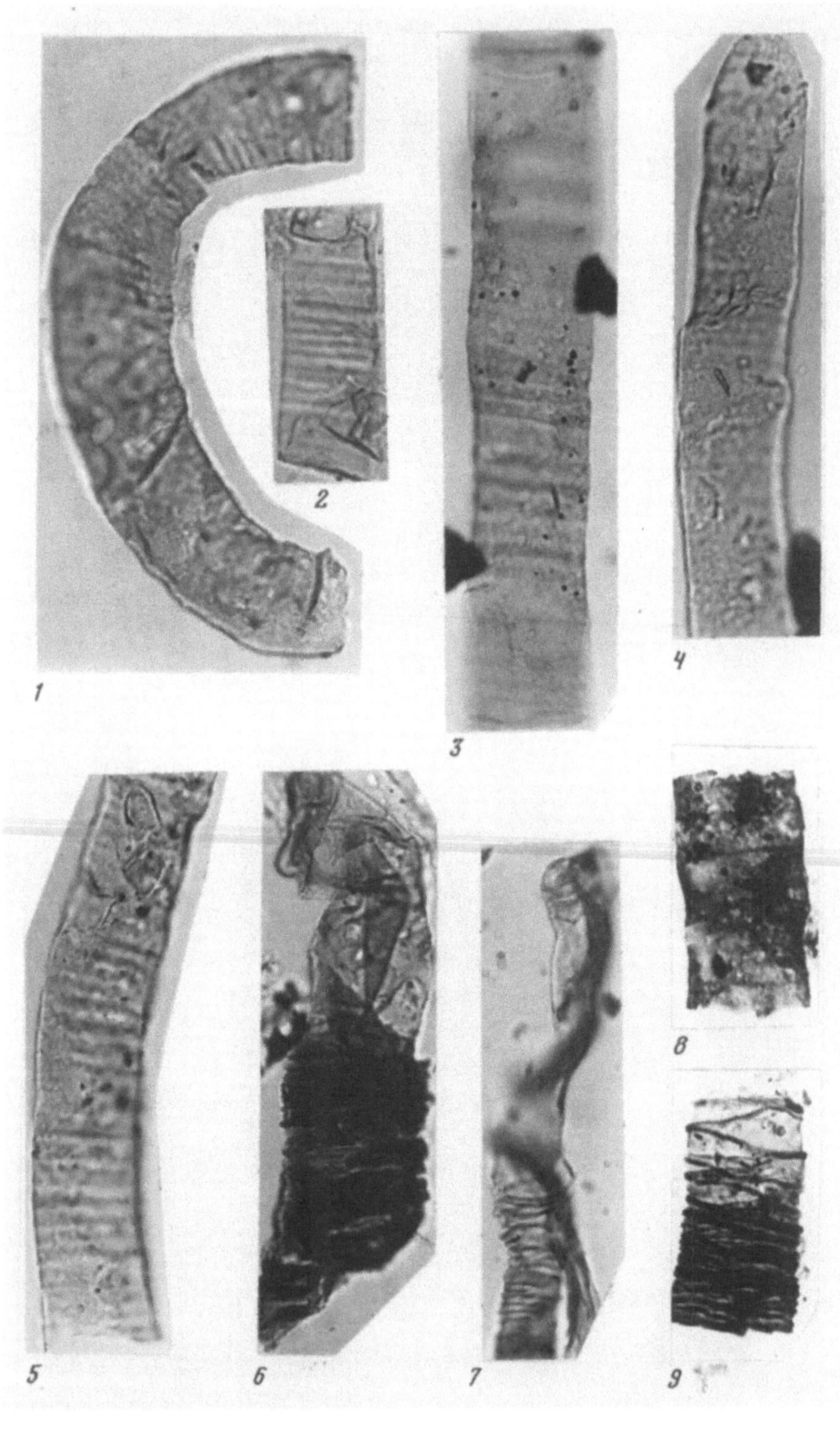

Plate 46

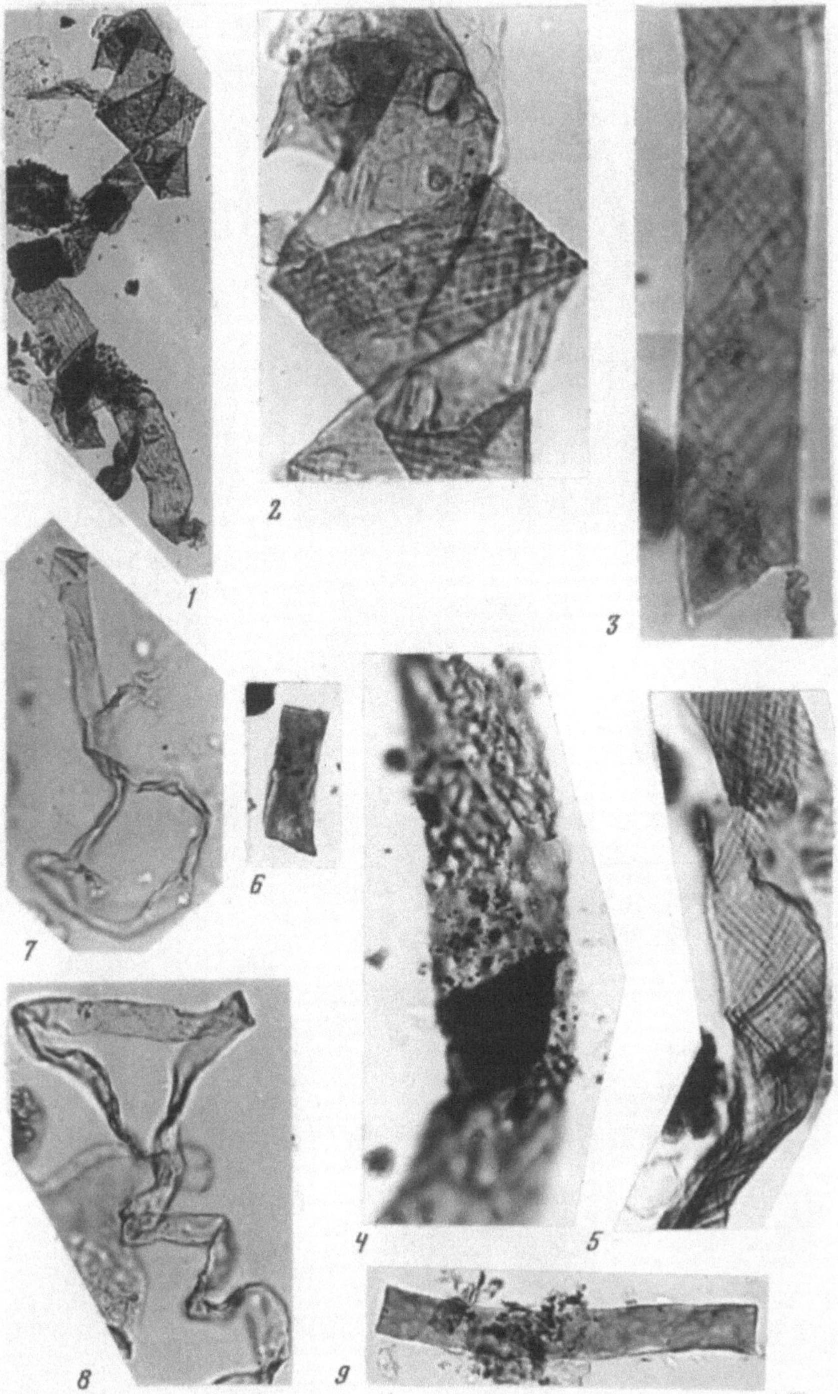

Plate 47

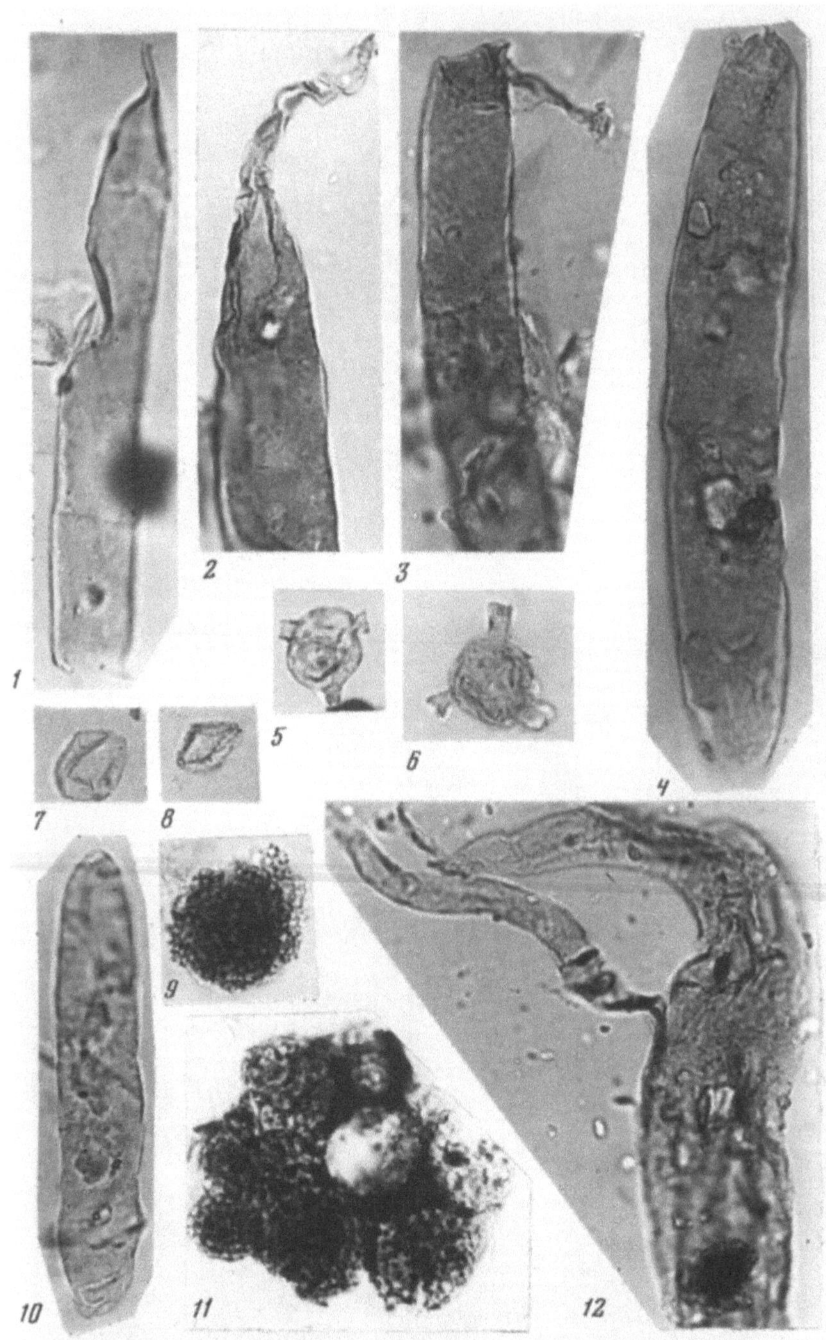

Plate 48

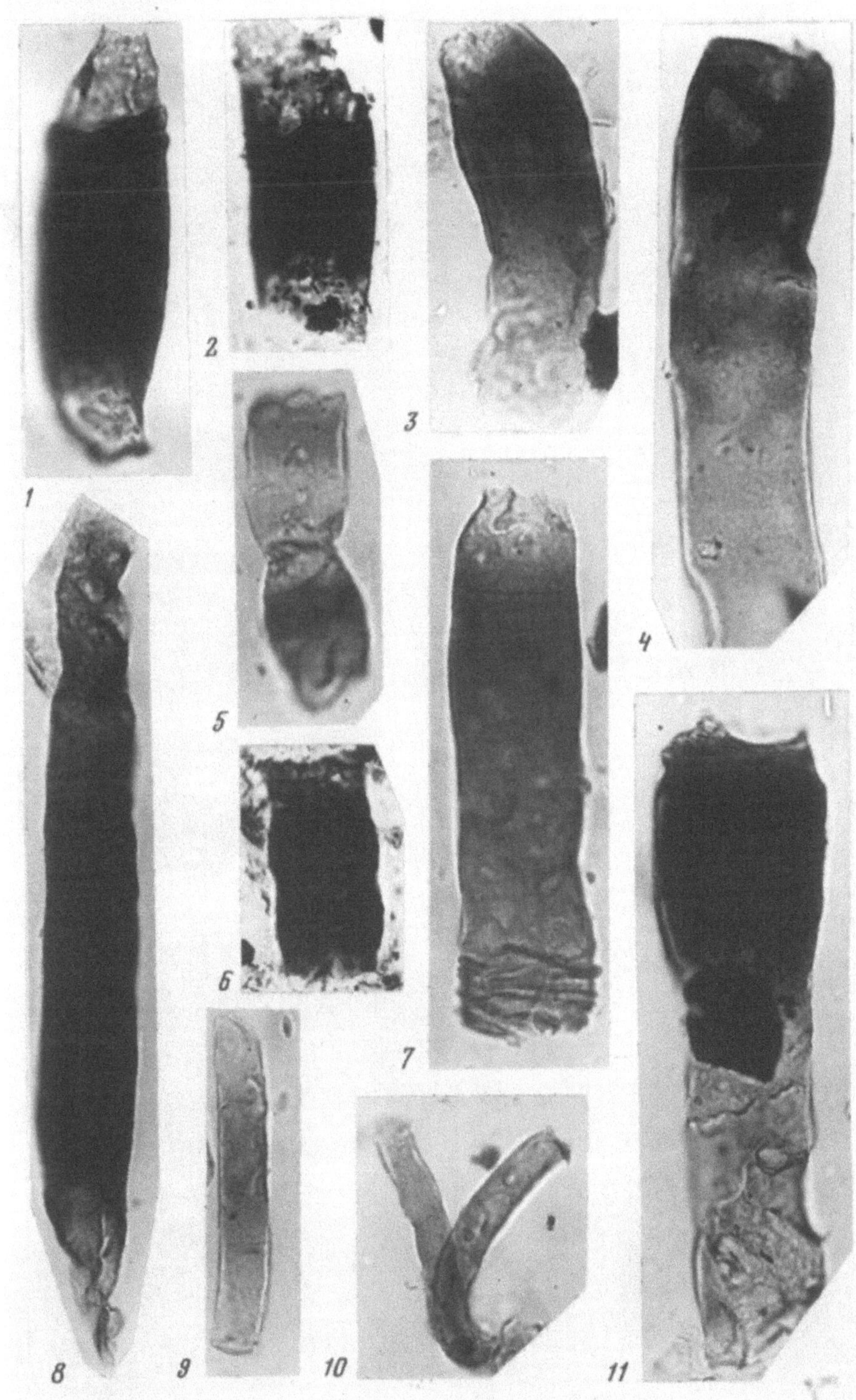

Plate 49

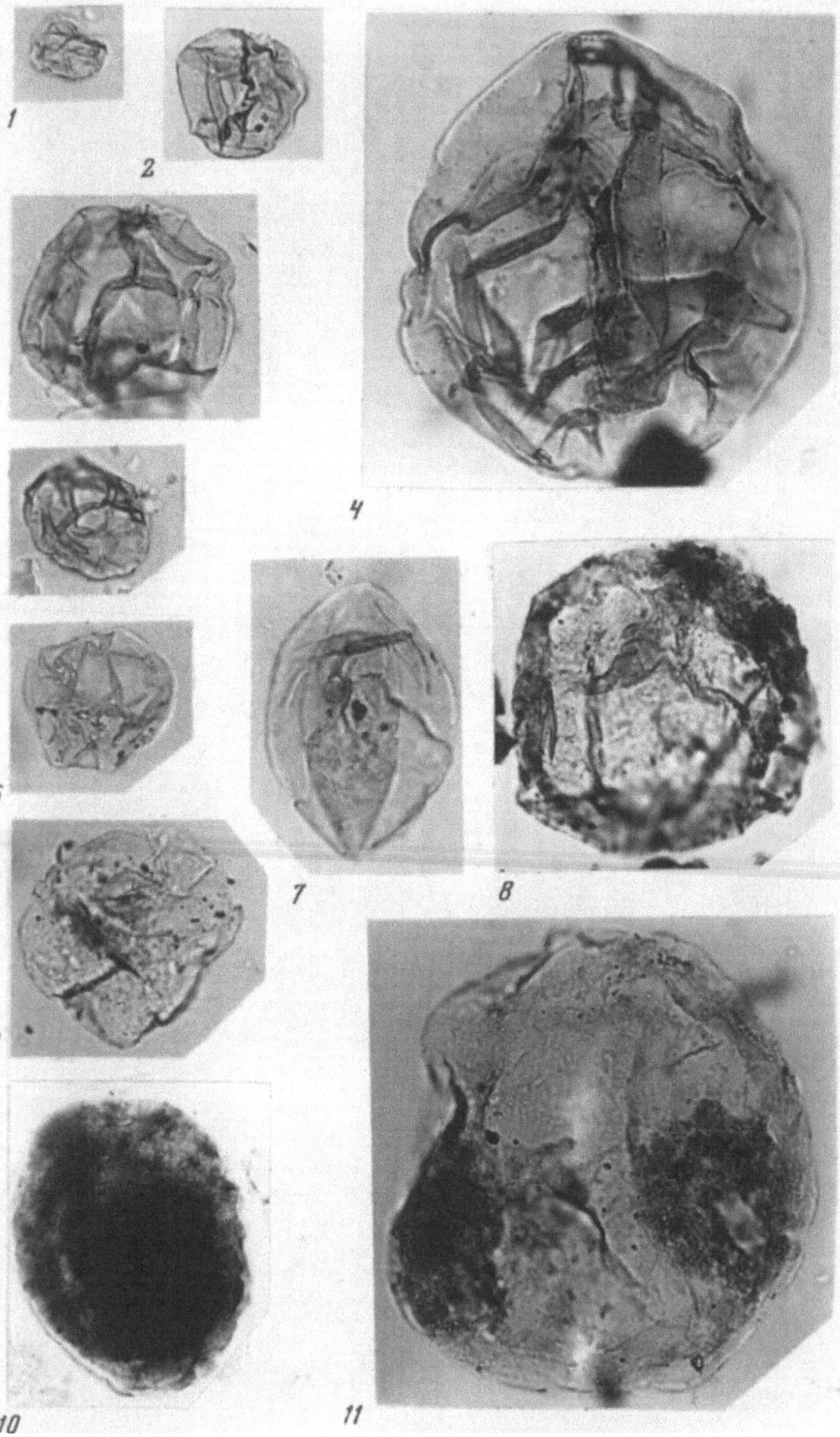

Plate 50

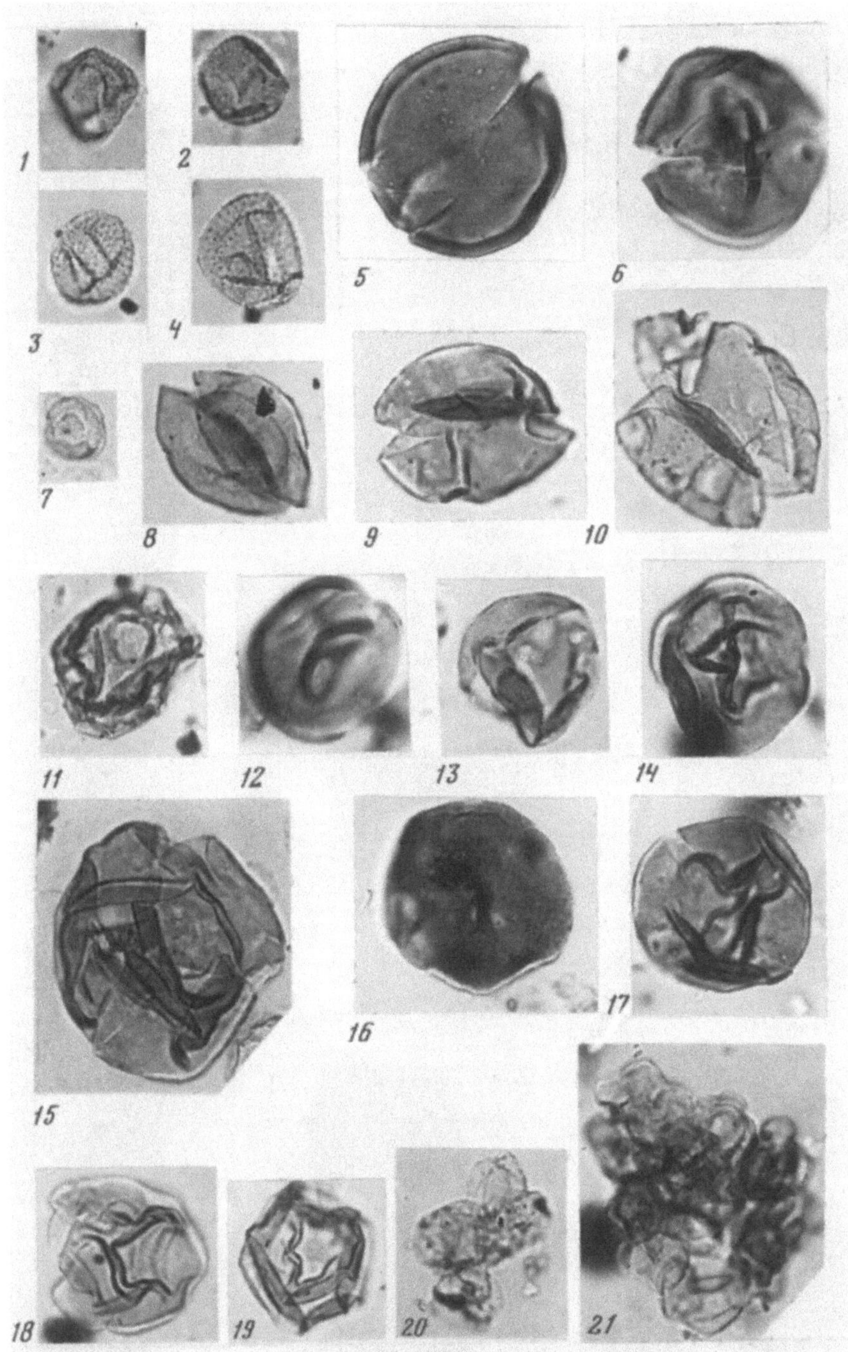

Plate 51

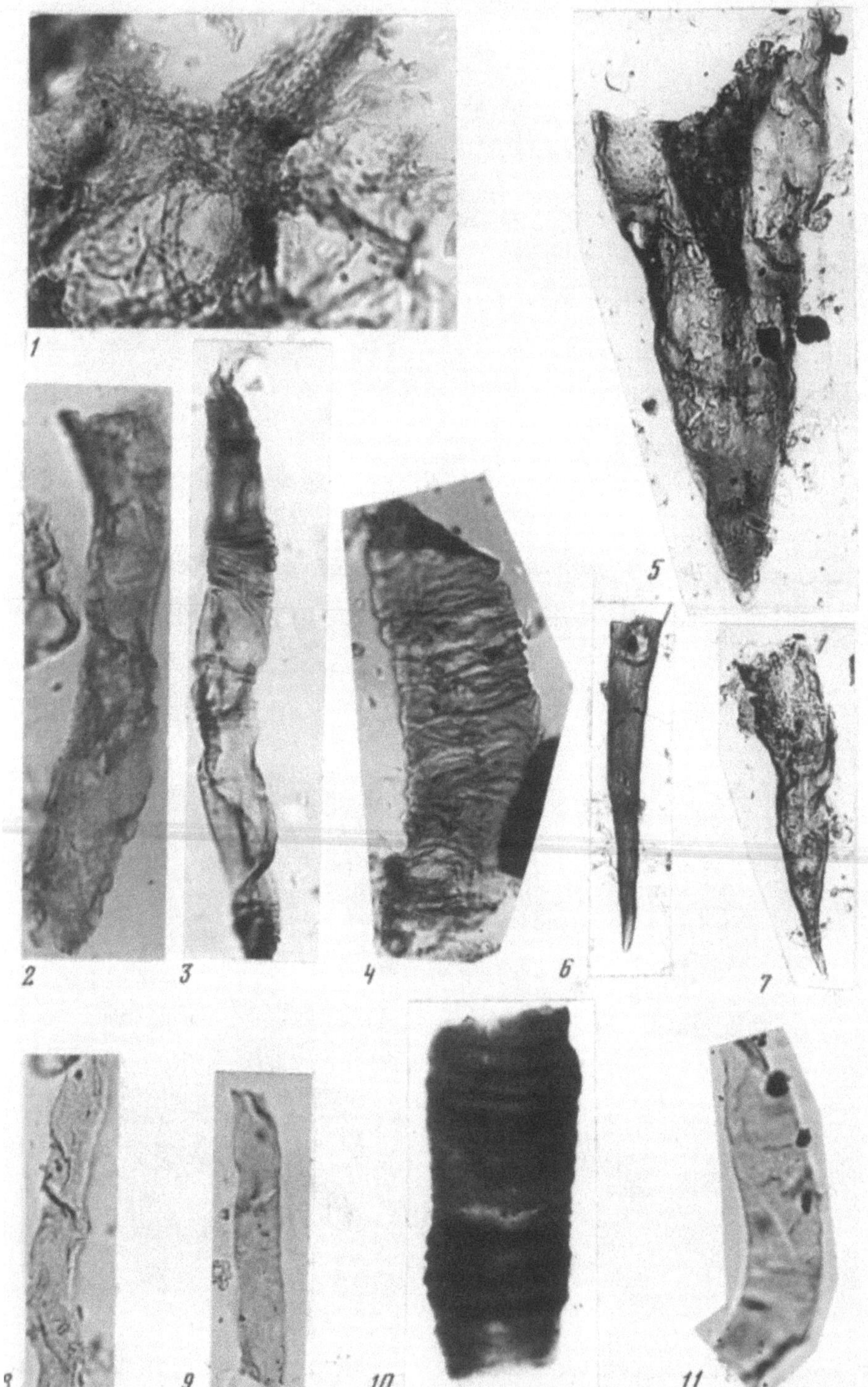

Plate 52

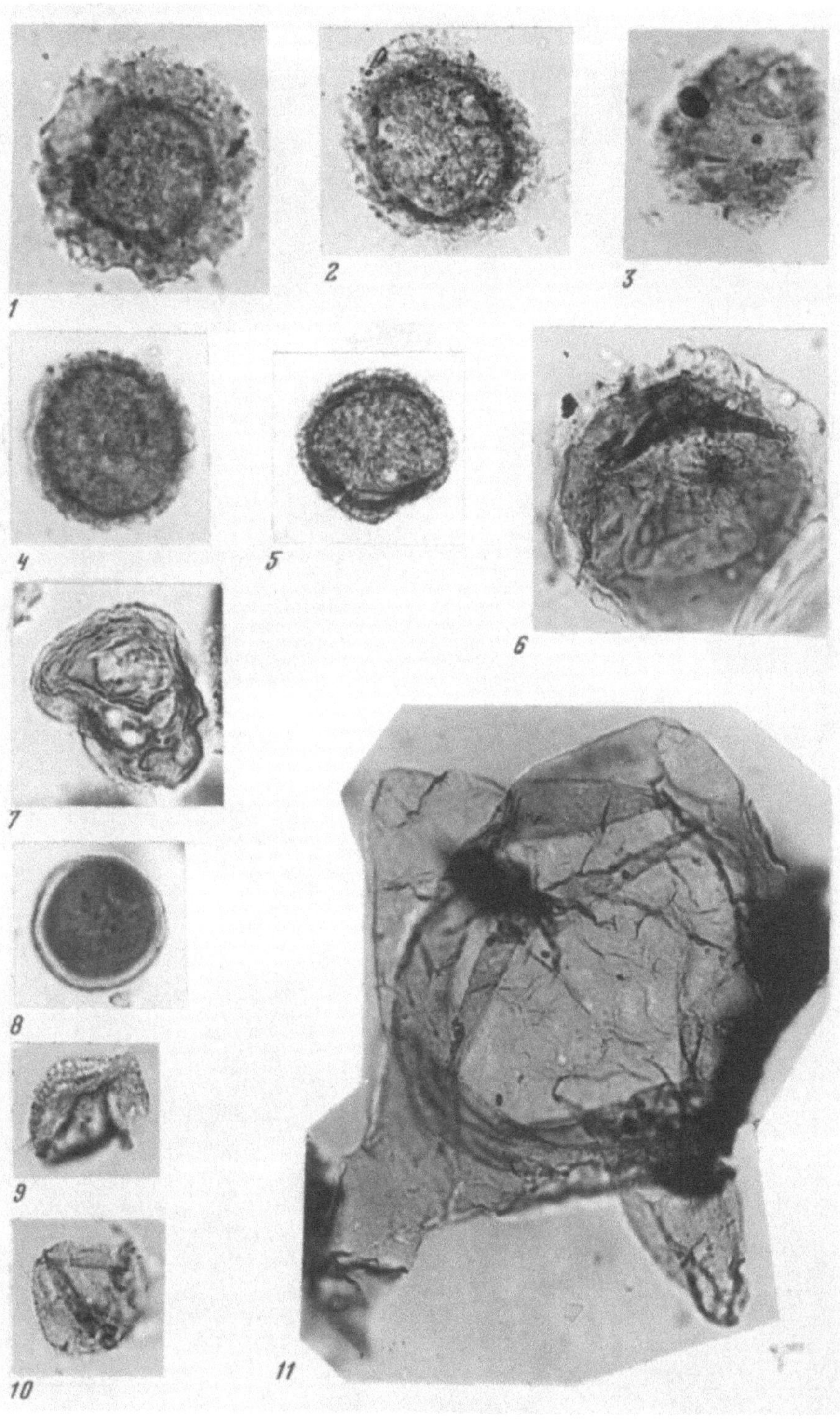

Plate 53

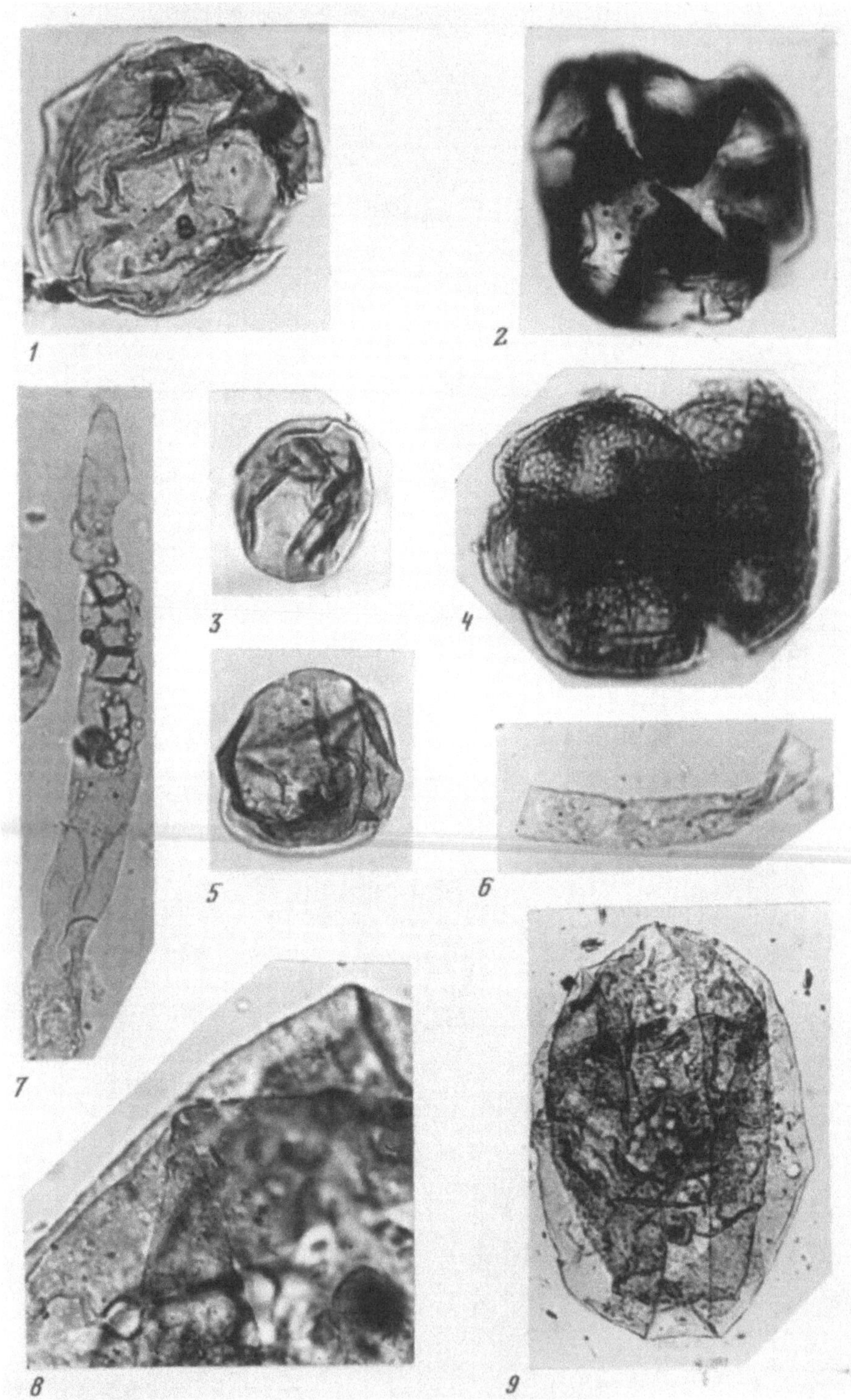

Plate 54

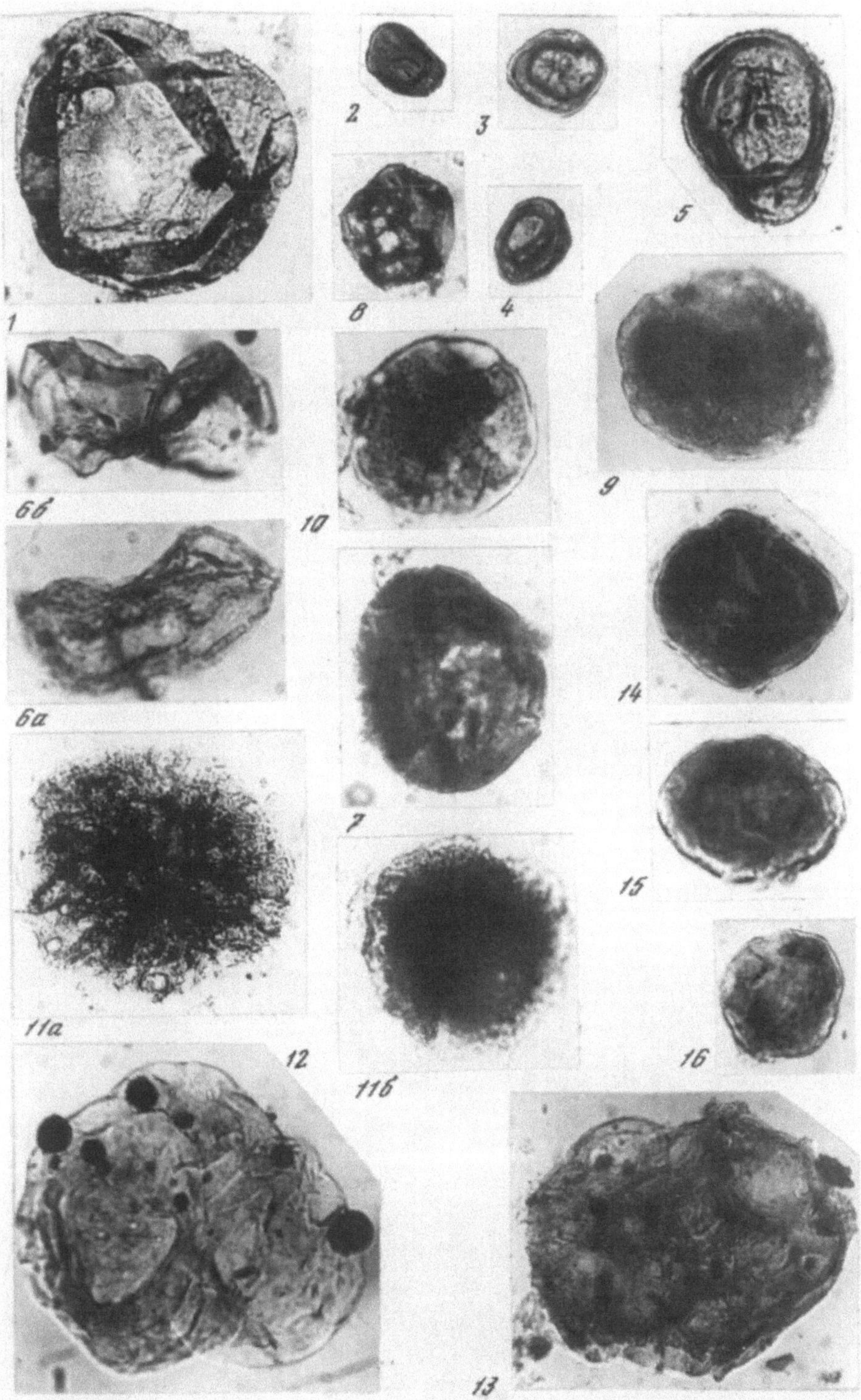

Plate 55

Plate 56

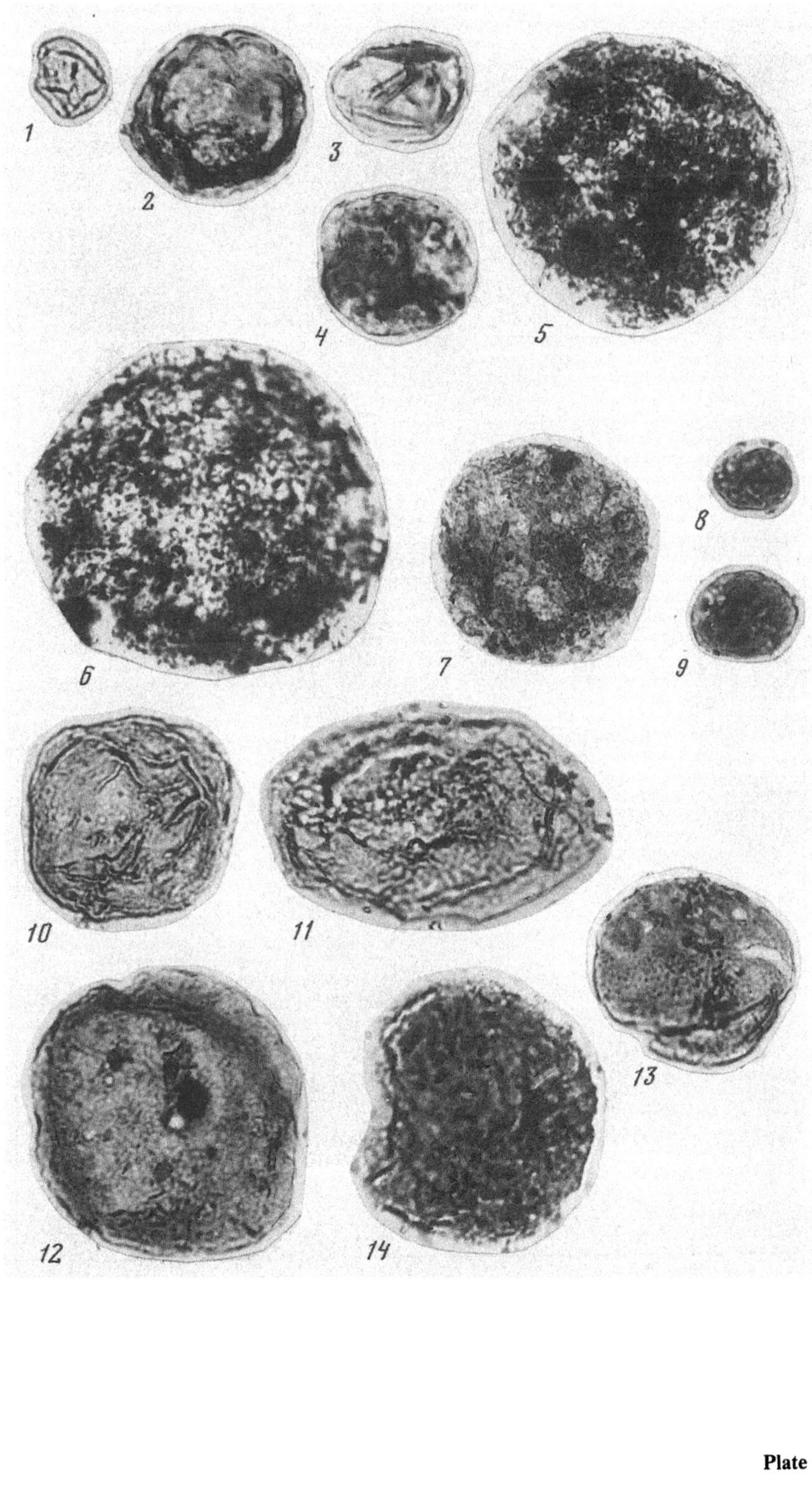

Plate 57

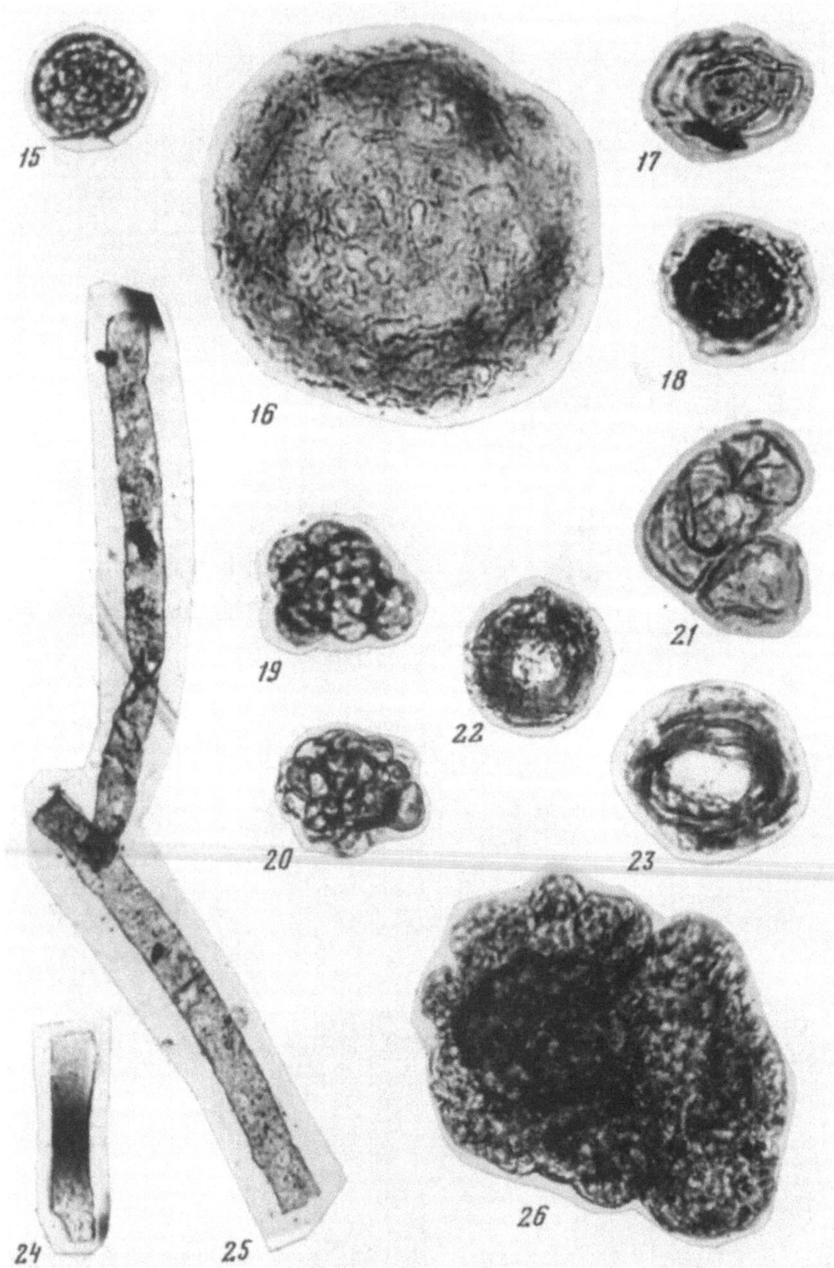

Plate 58

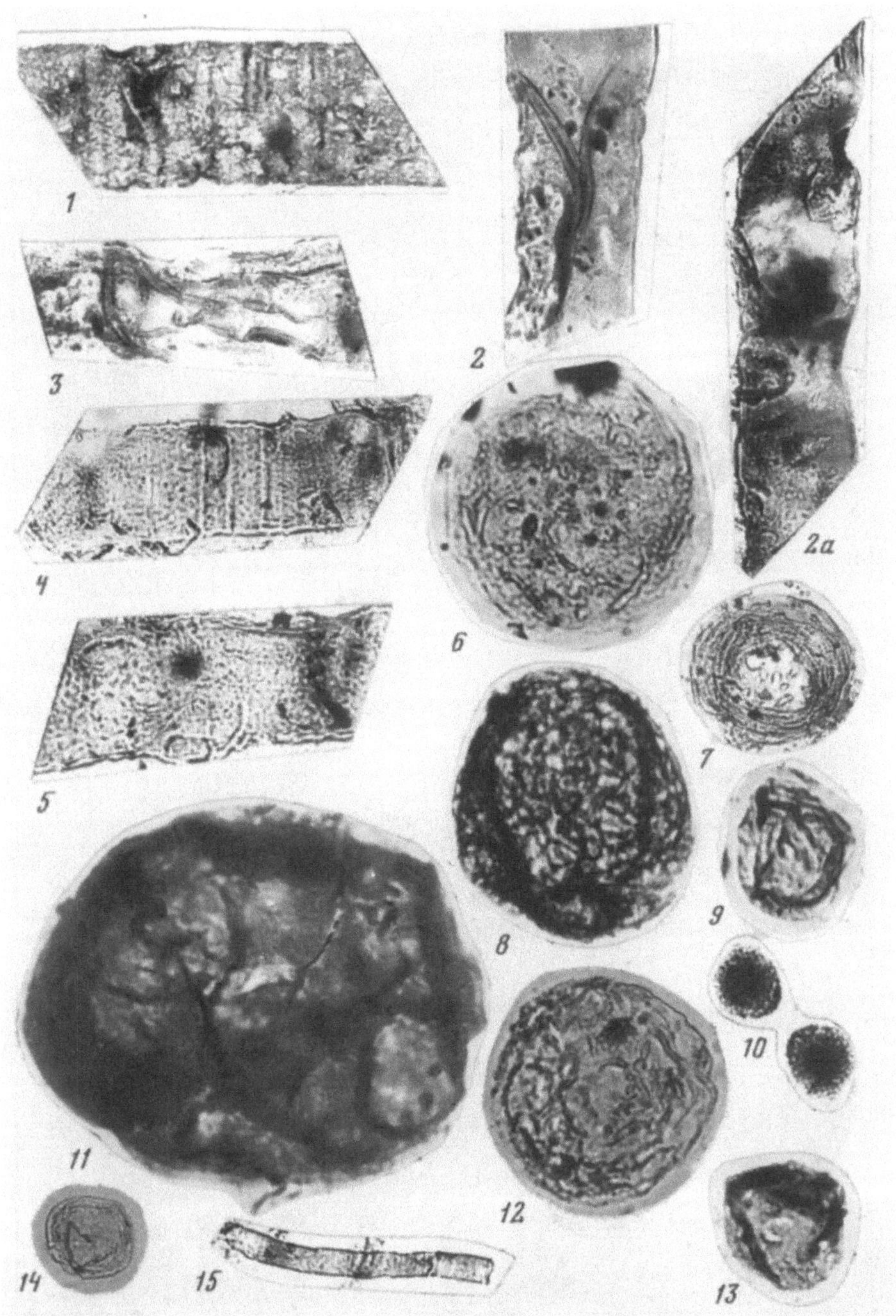

Plate 59

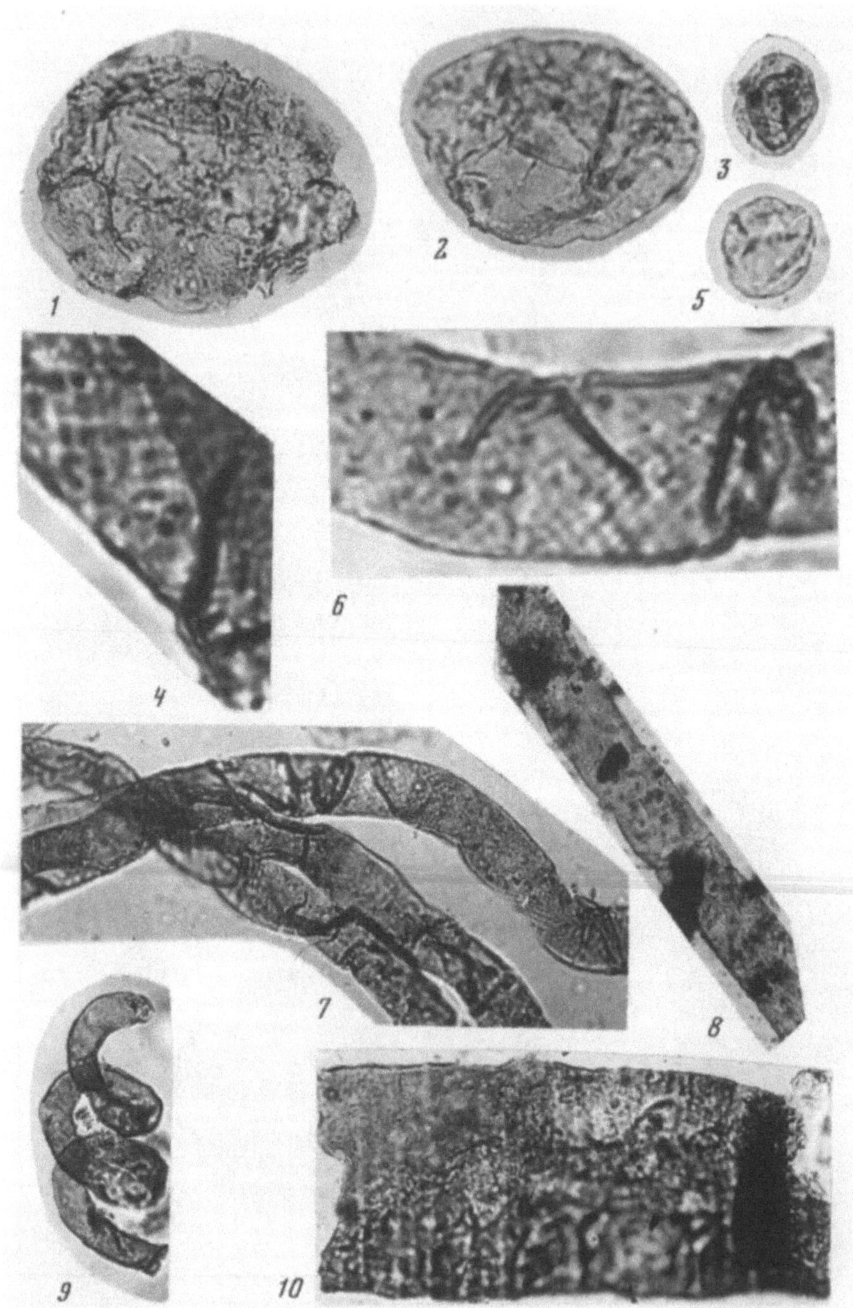

Plate 60

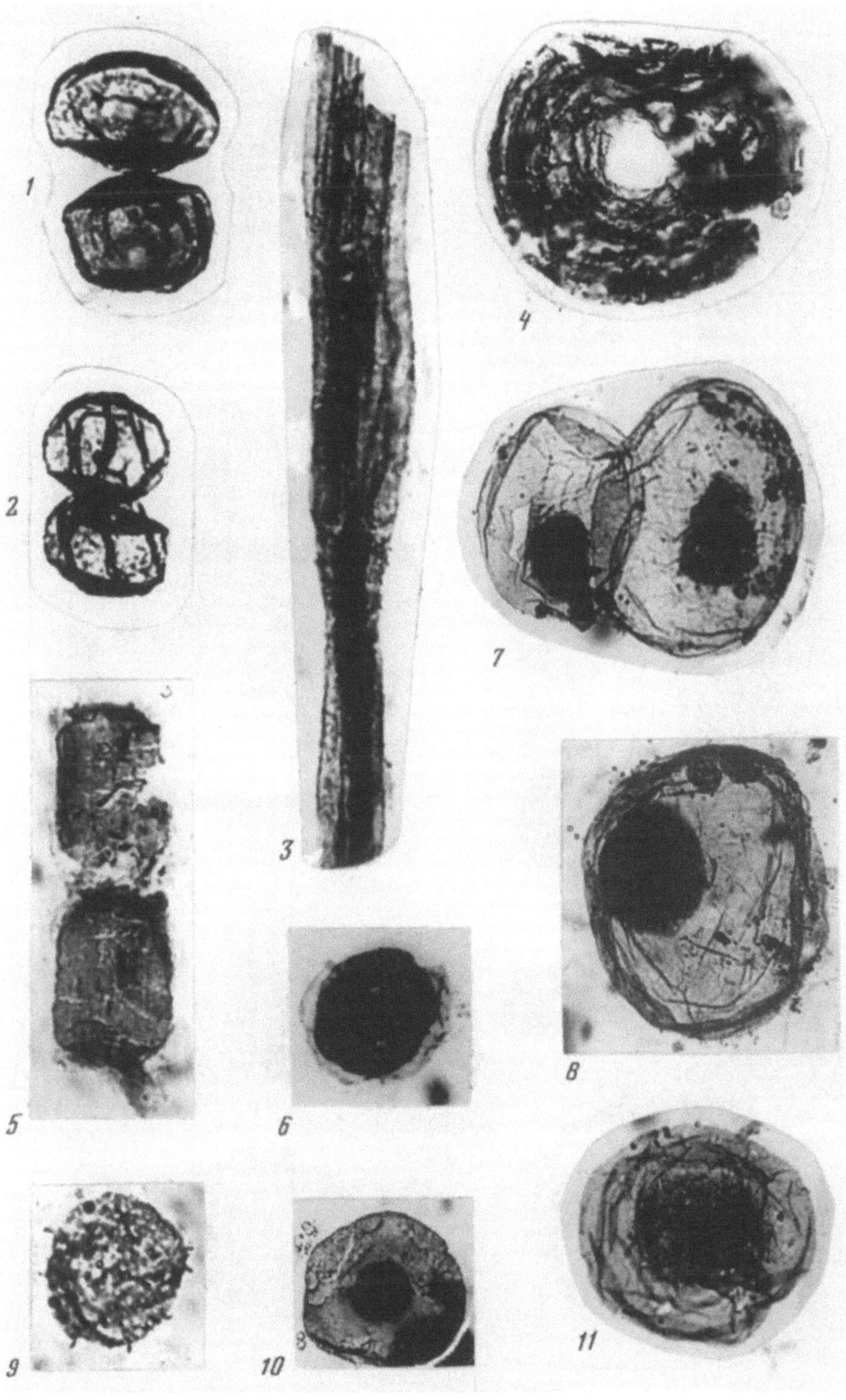

Plate 61

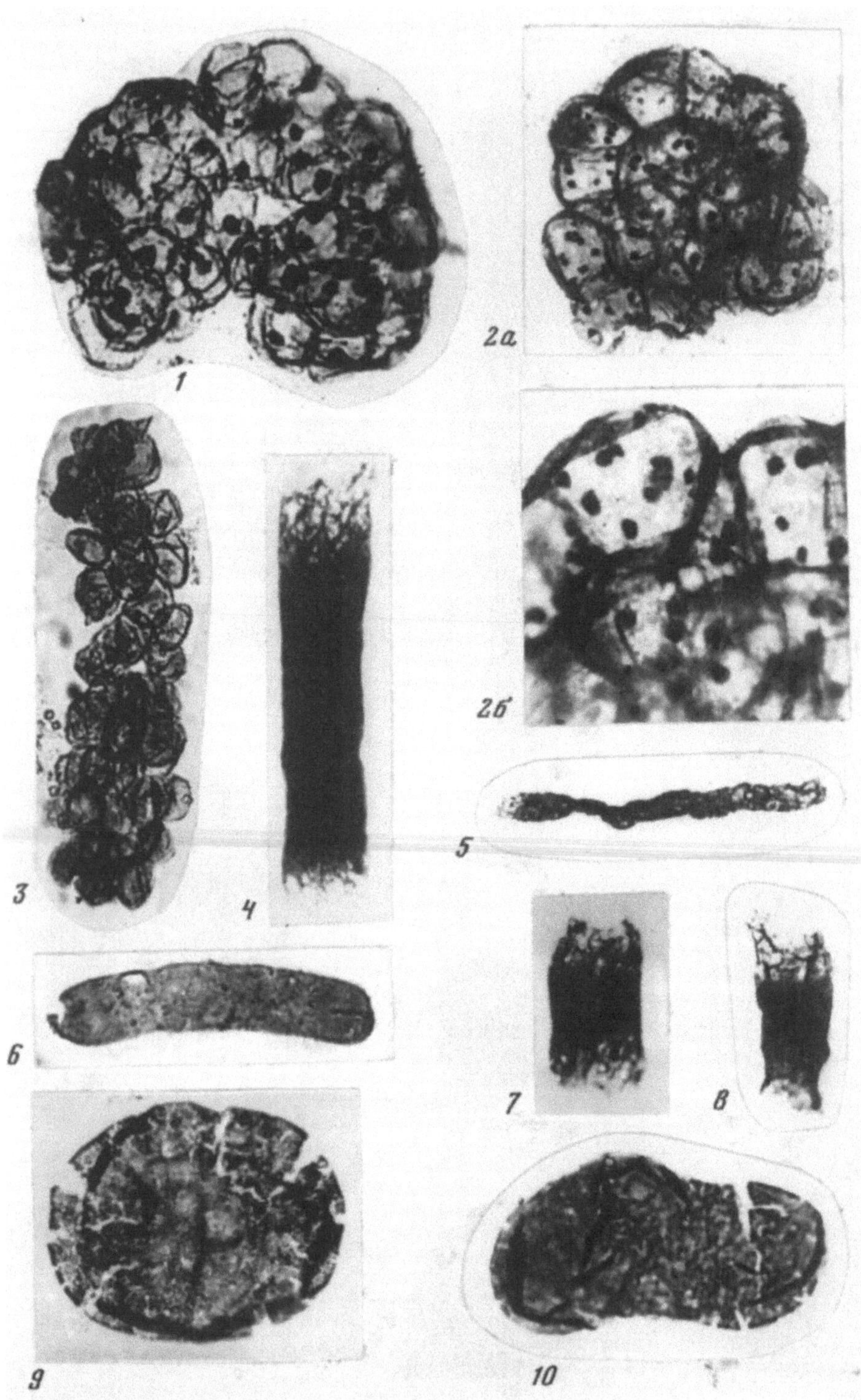

Plate 62

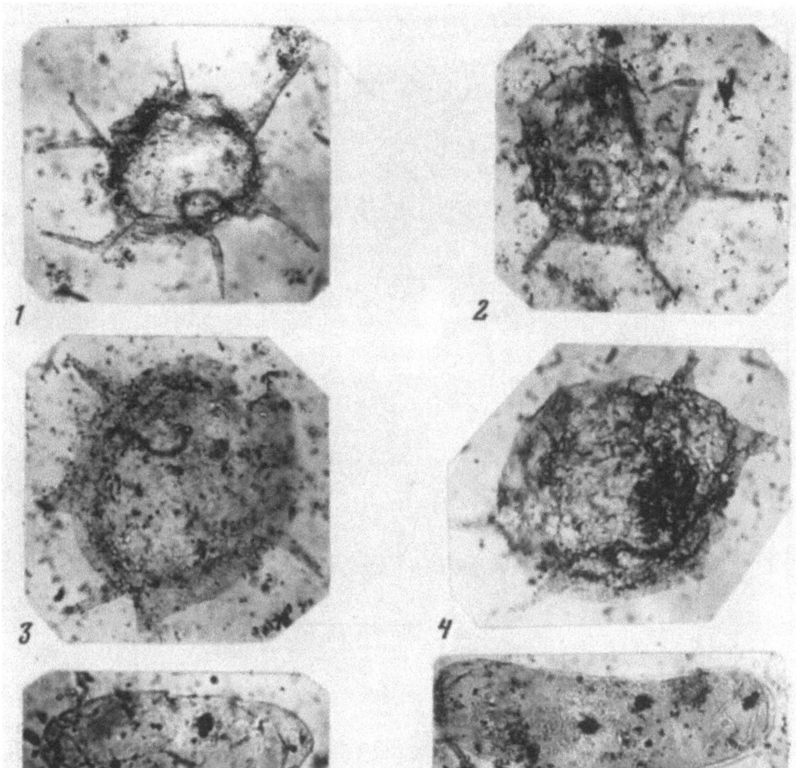

Plate 63

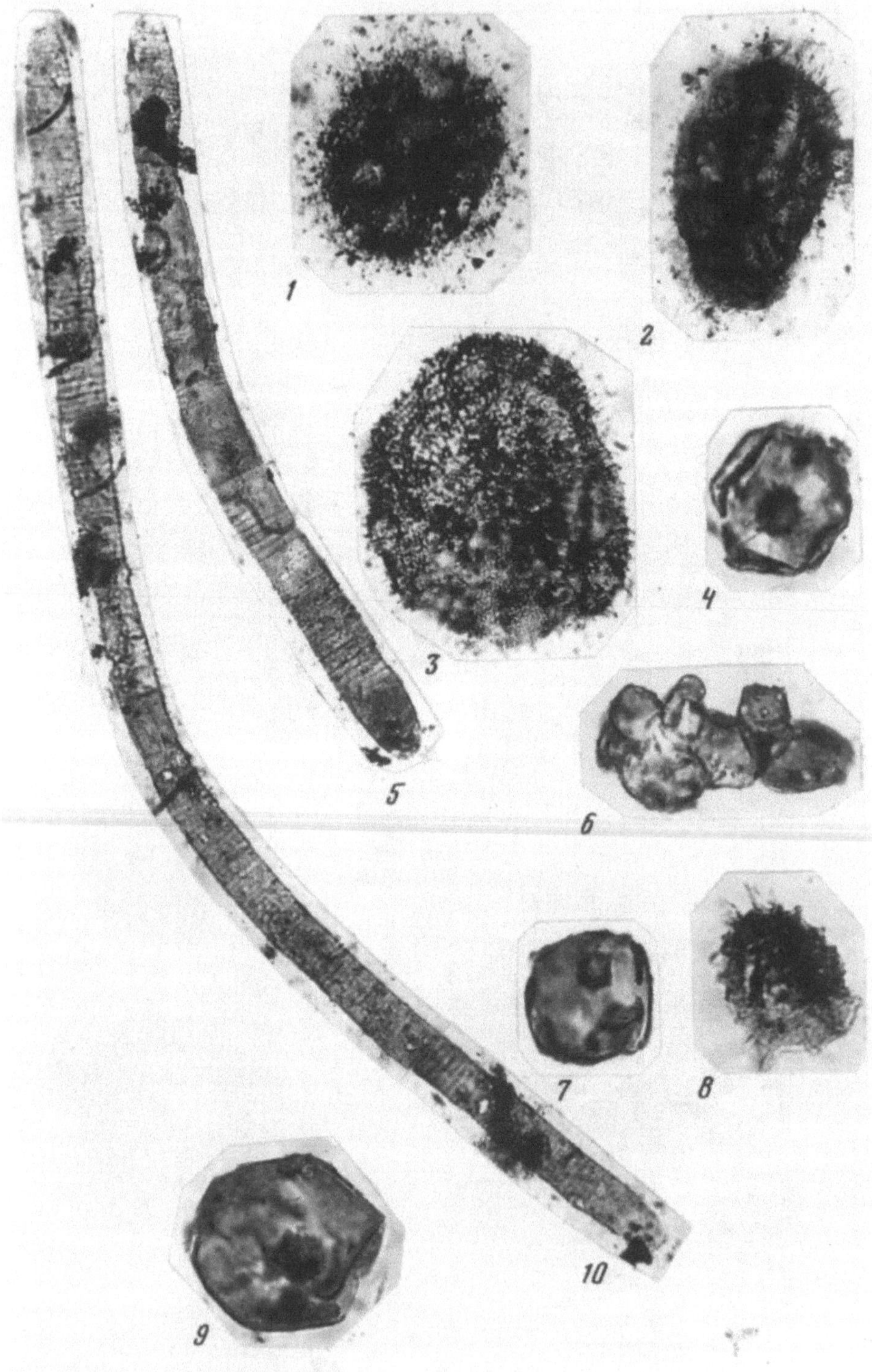

Plate 64

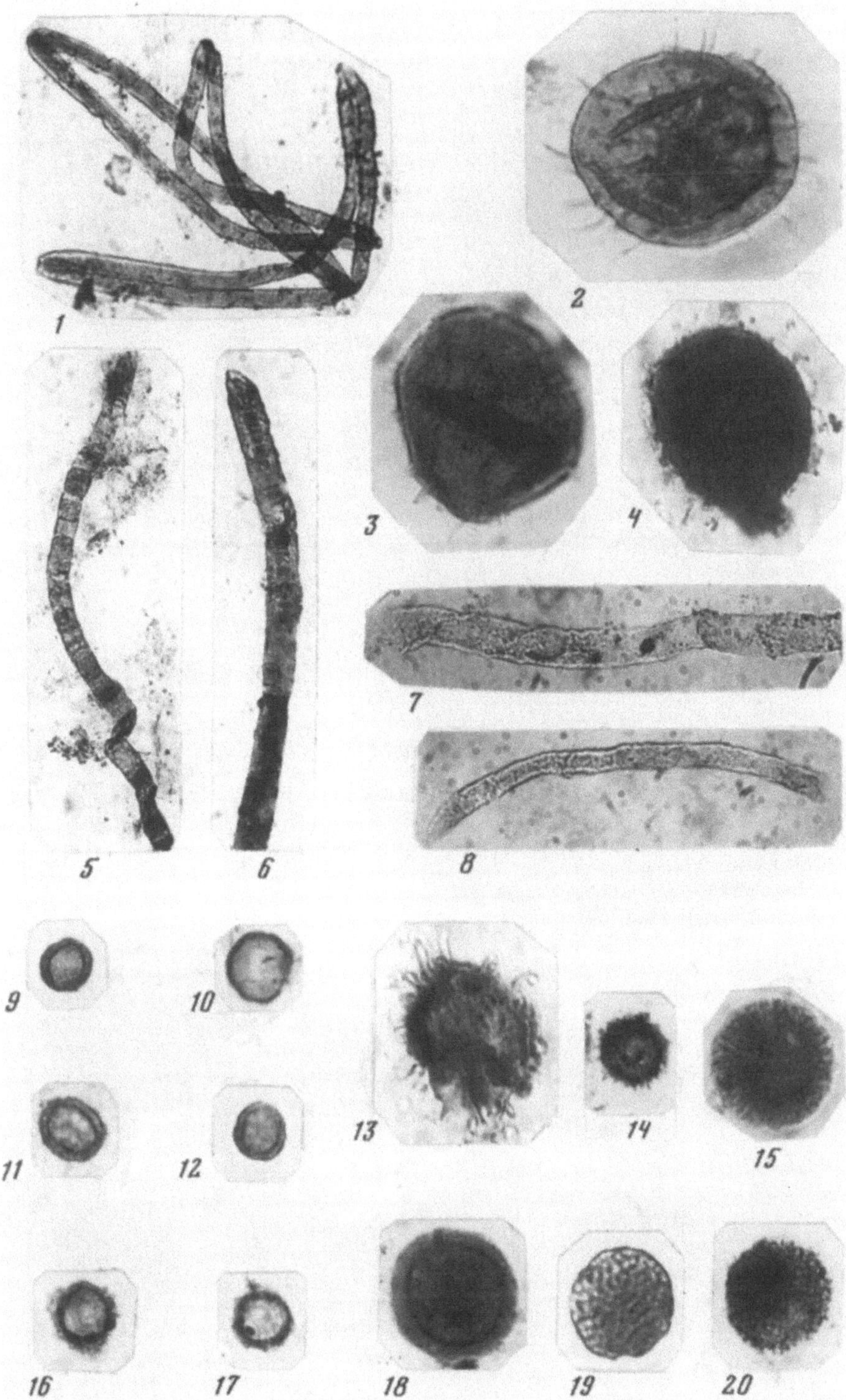

Plate 65

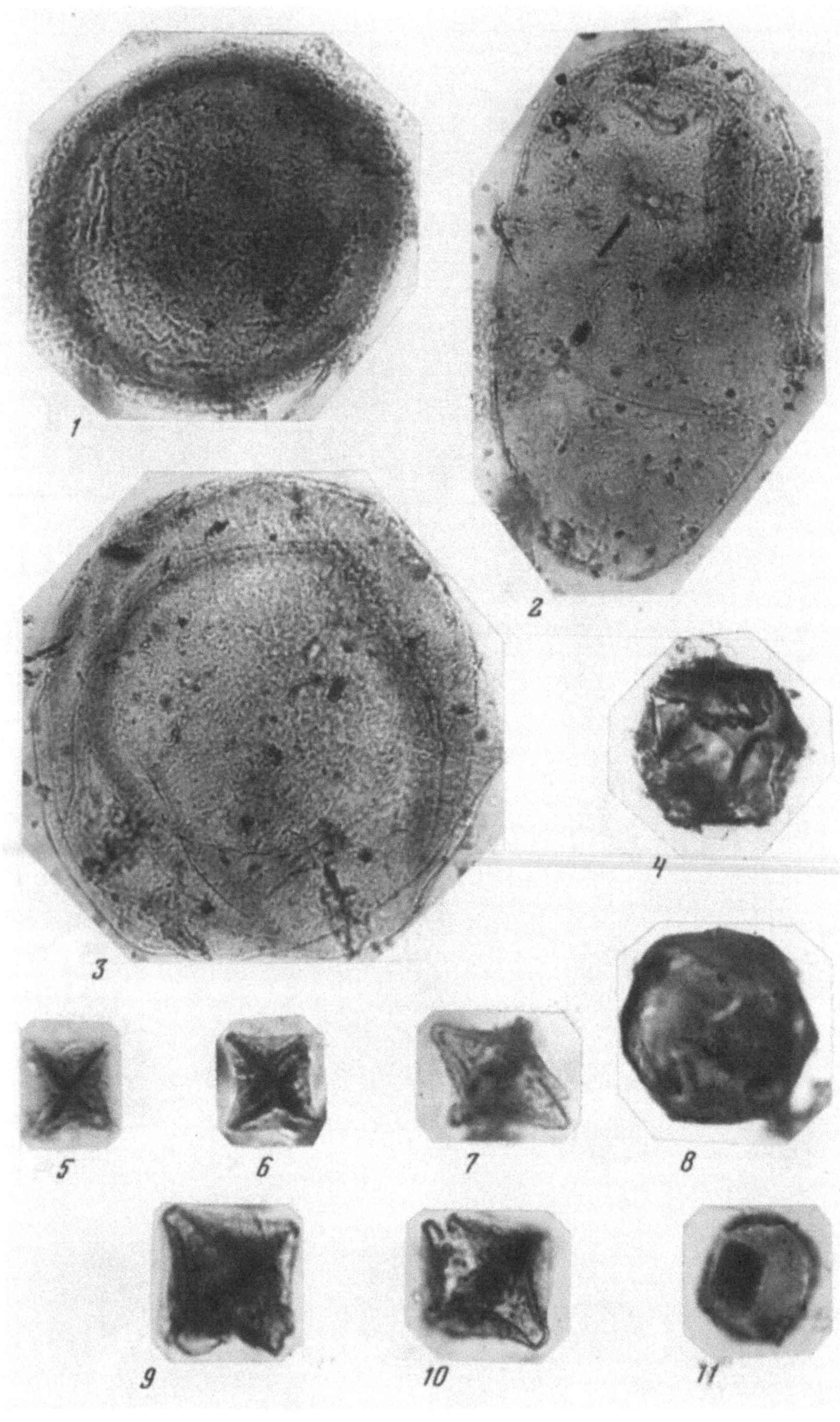

Plate 66

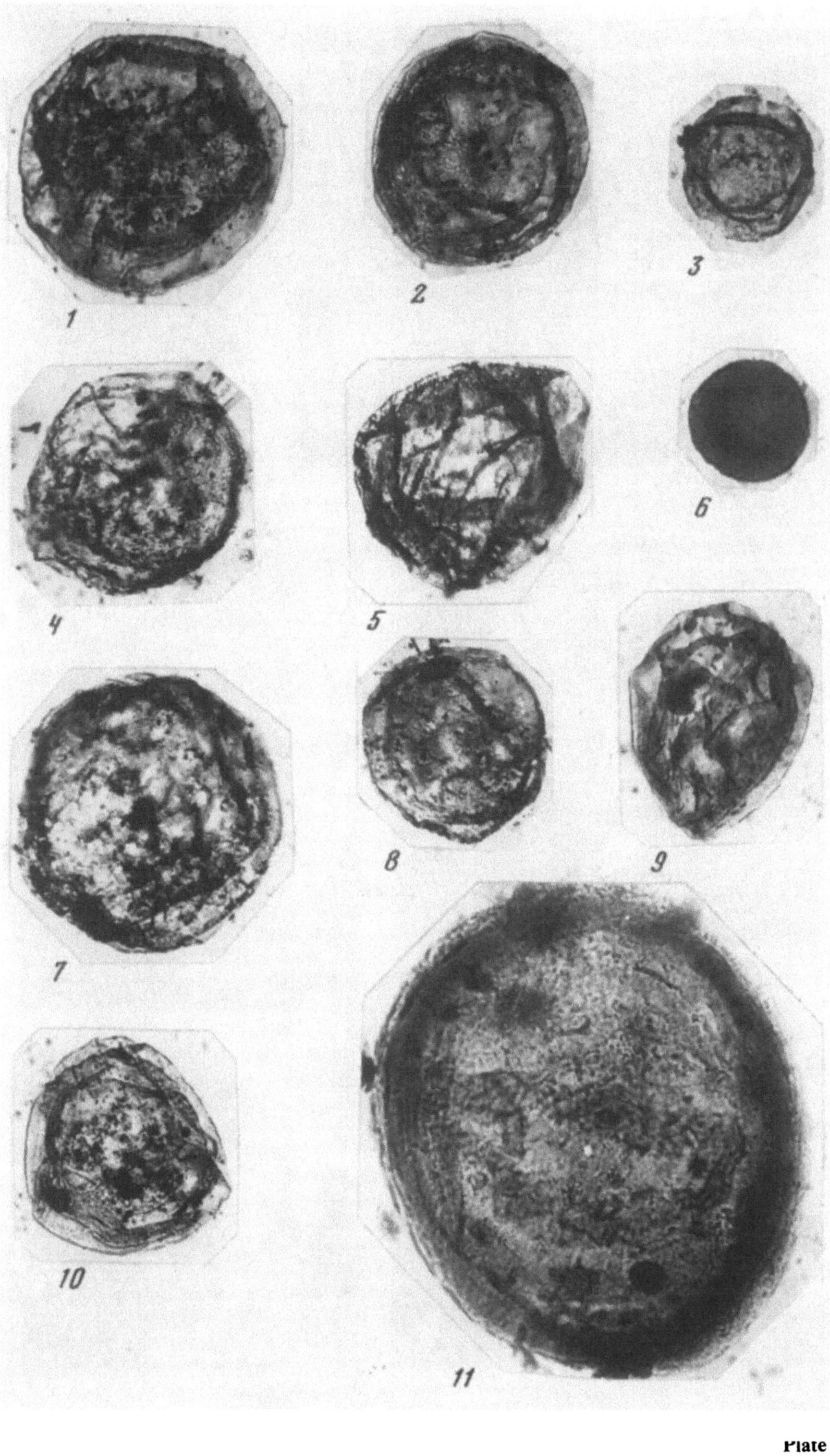

Plate 67

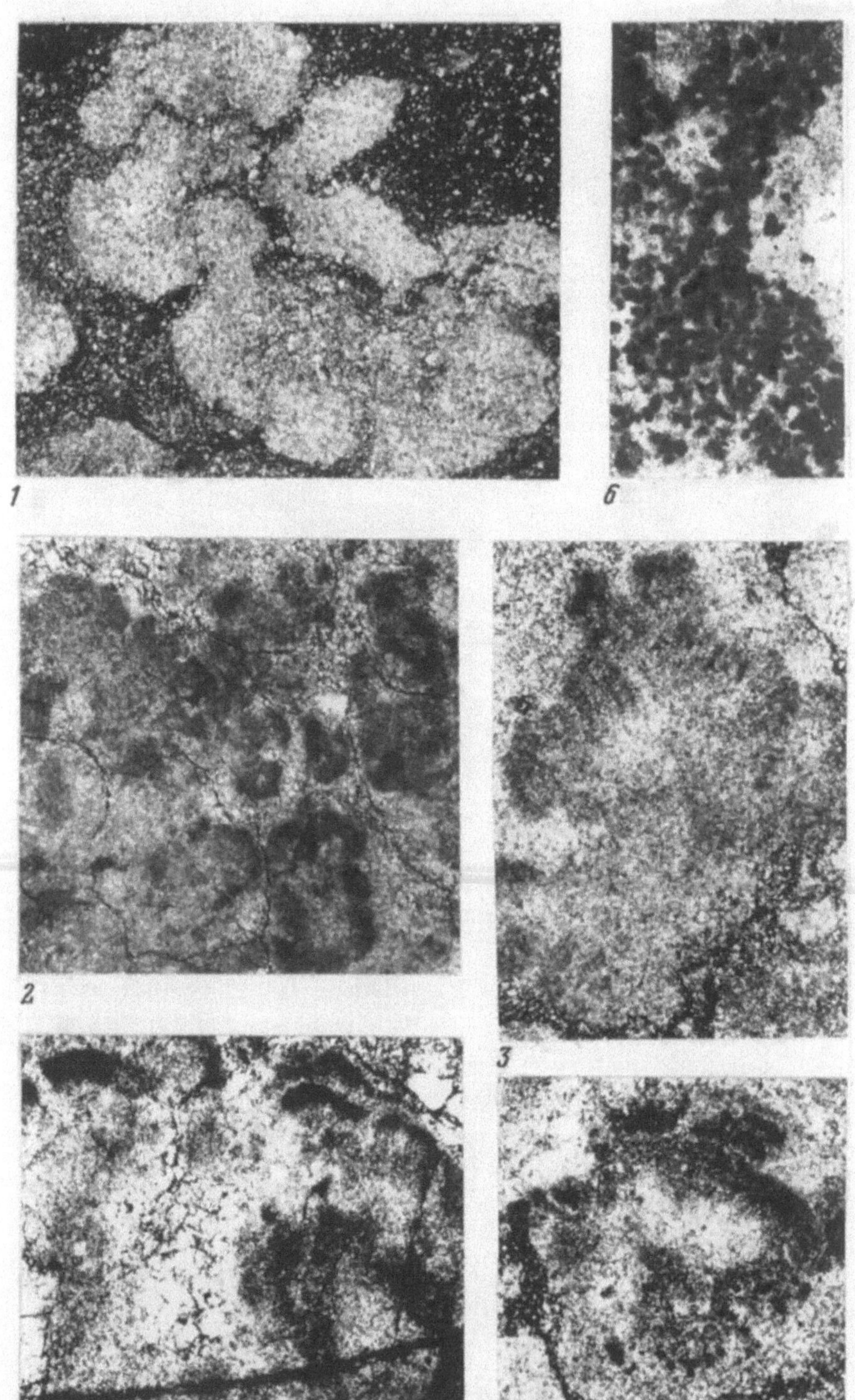

Plate 68

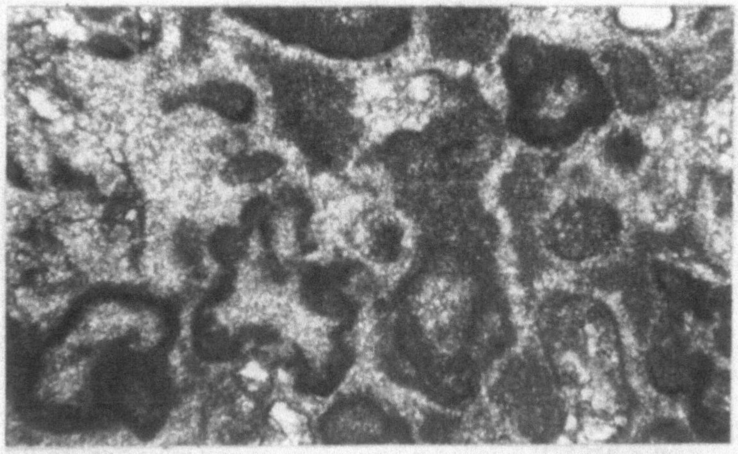

1

2

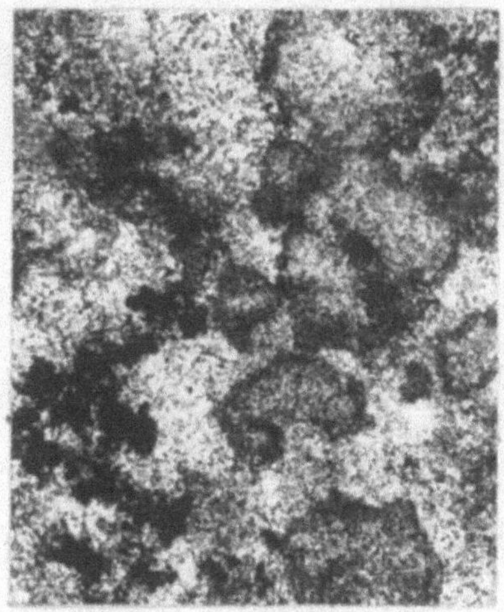

3

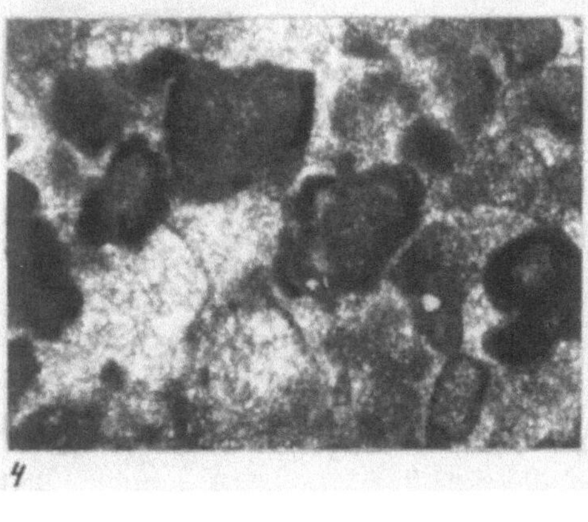

4

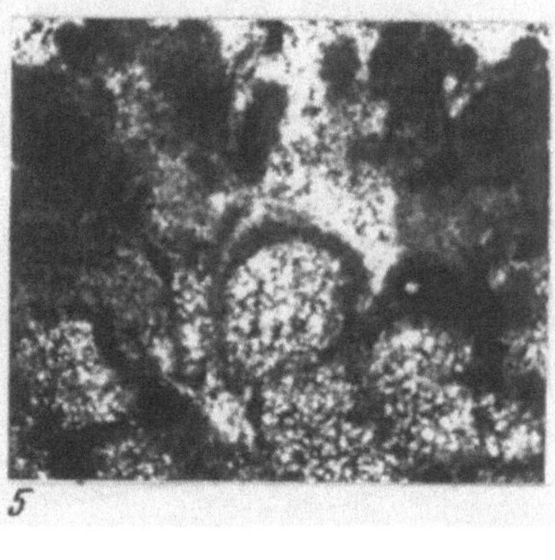

5

Plate 69

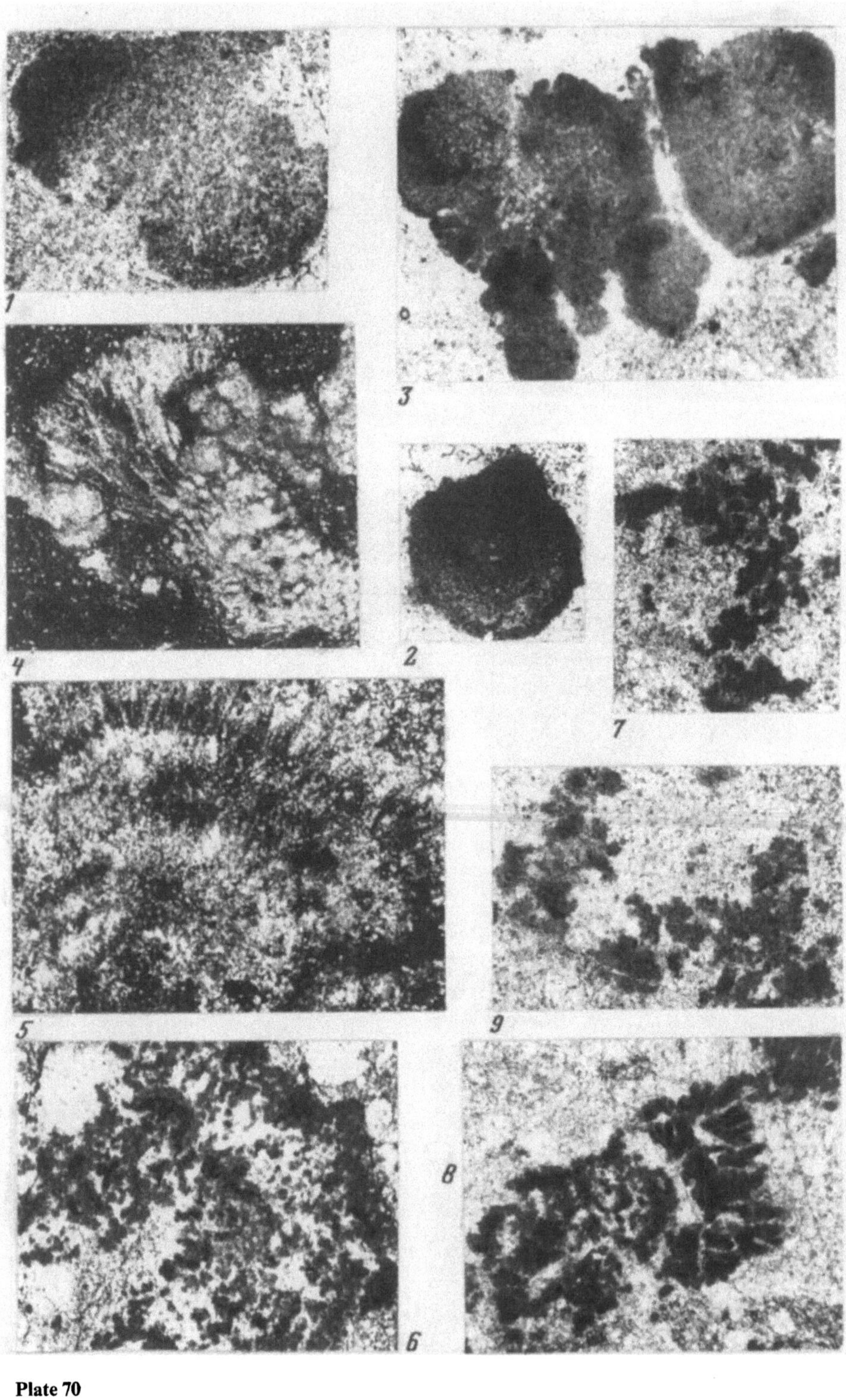

Plate 70

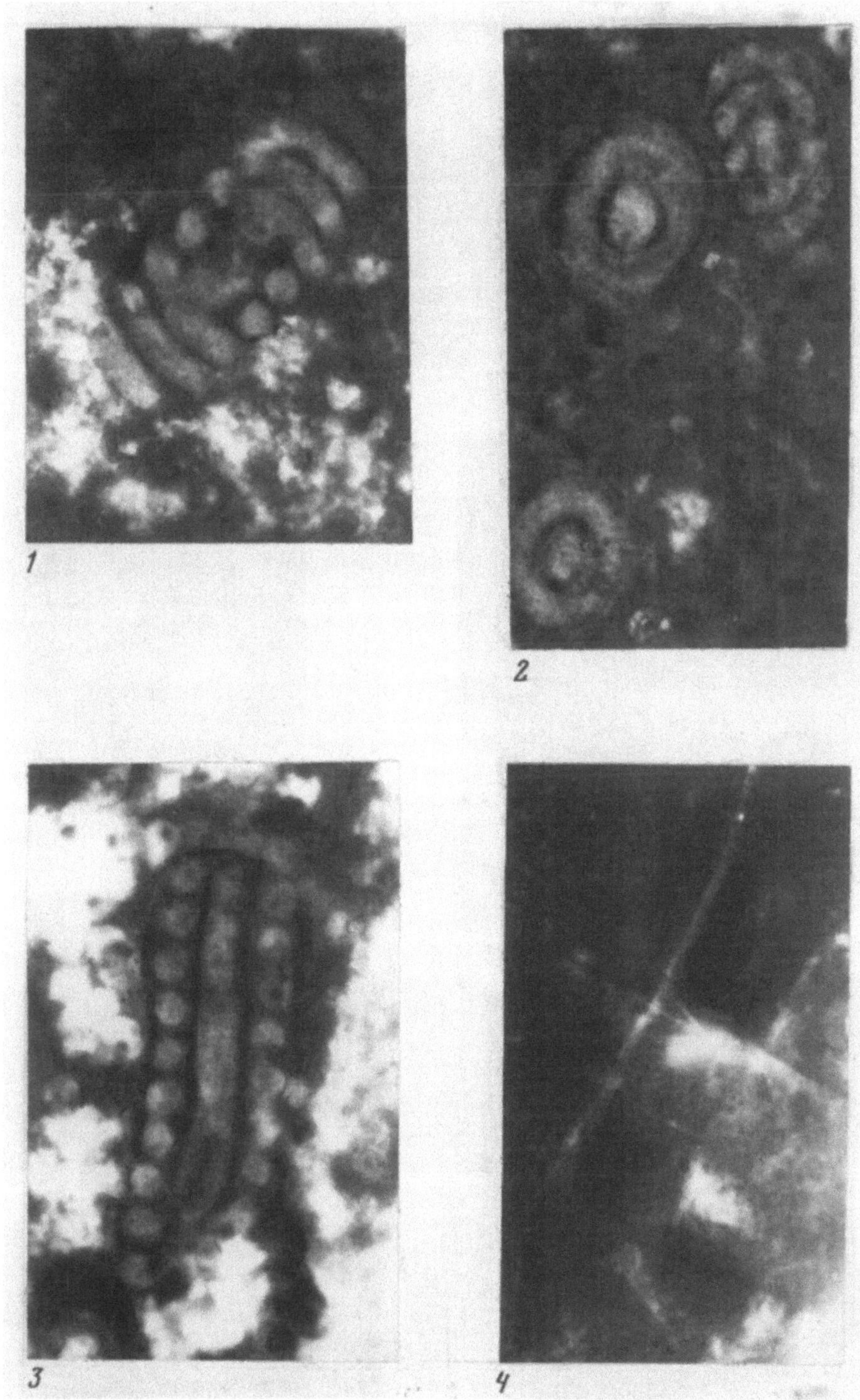

Plate 71

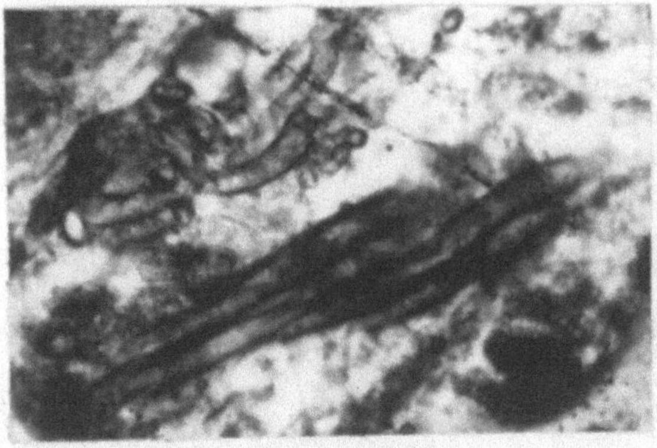
1

2

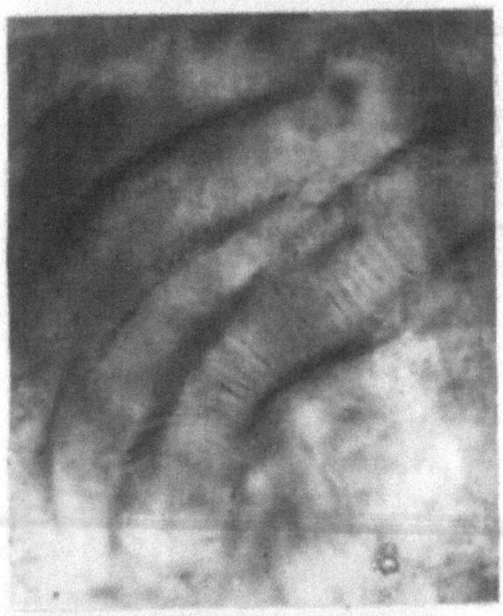
3а

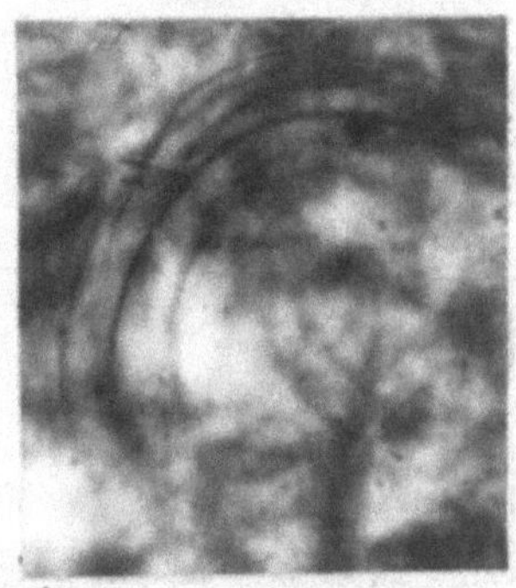
4

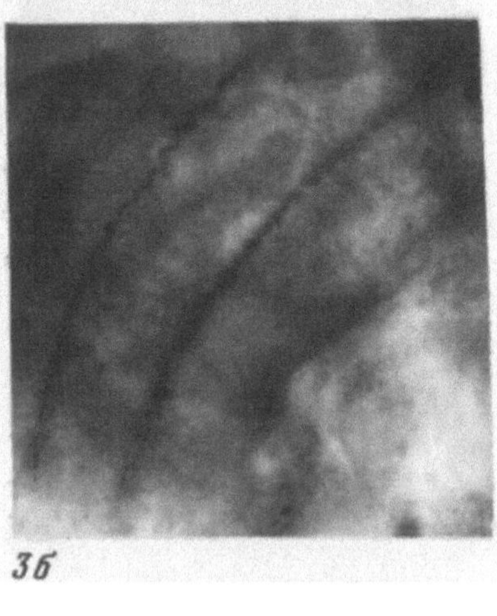
3б

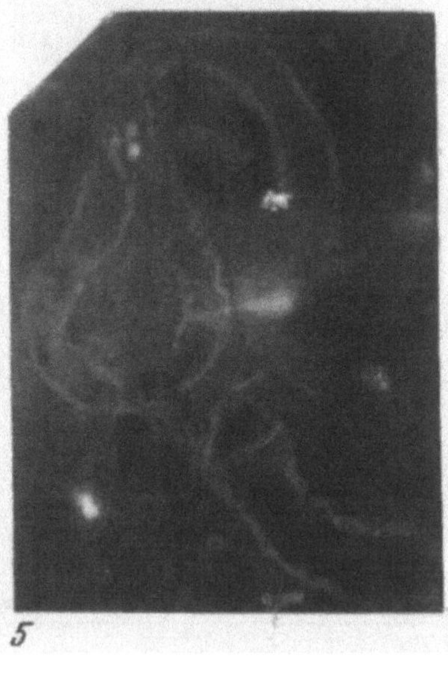
5

Plate 72

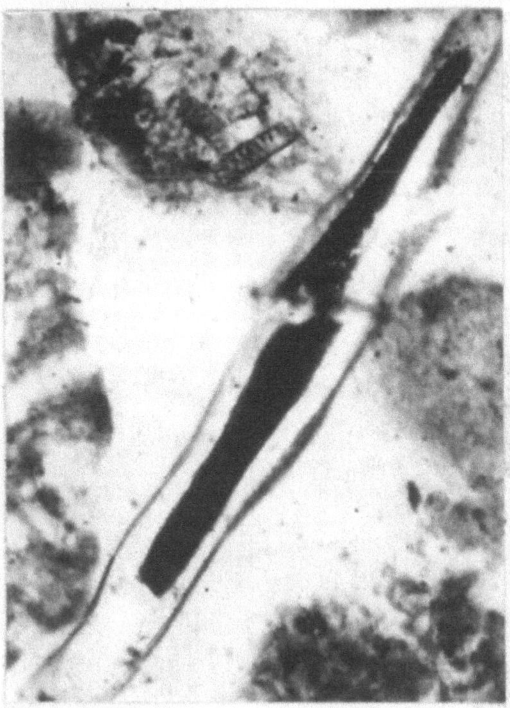
1

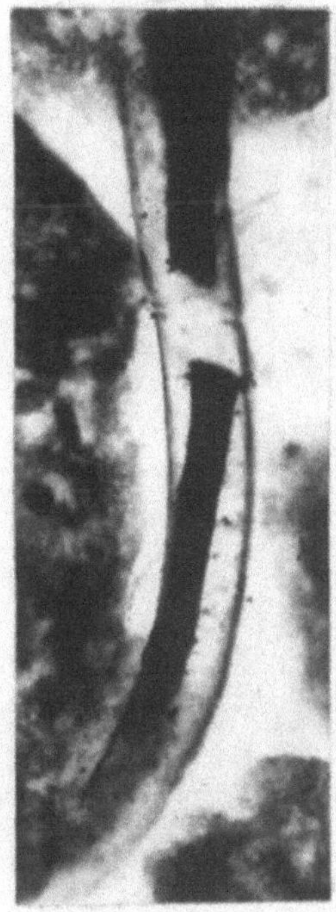
2

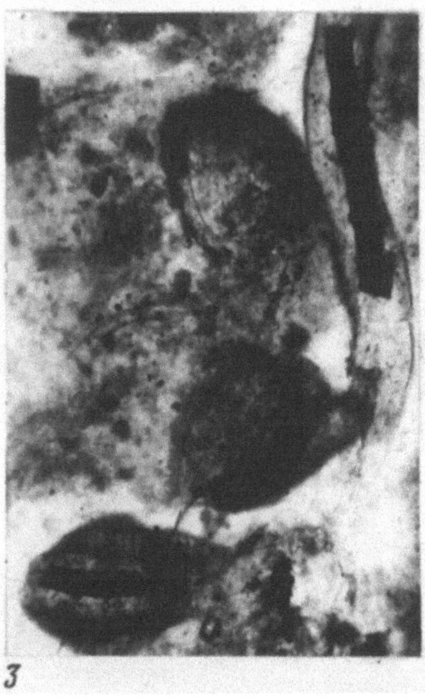
3

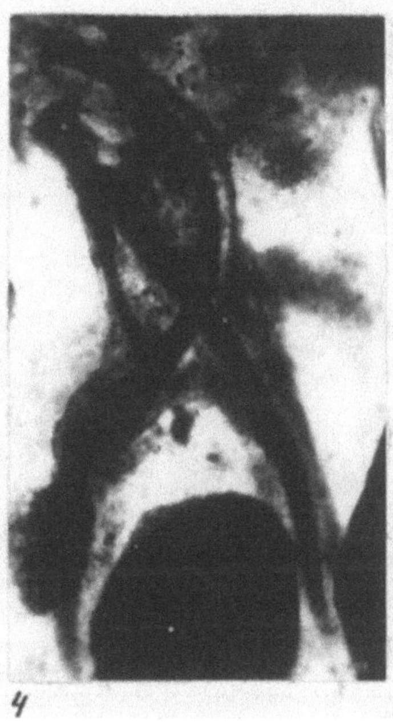
4

Plate 73

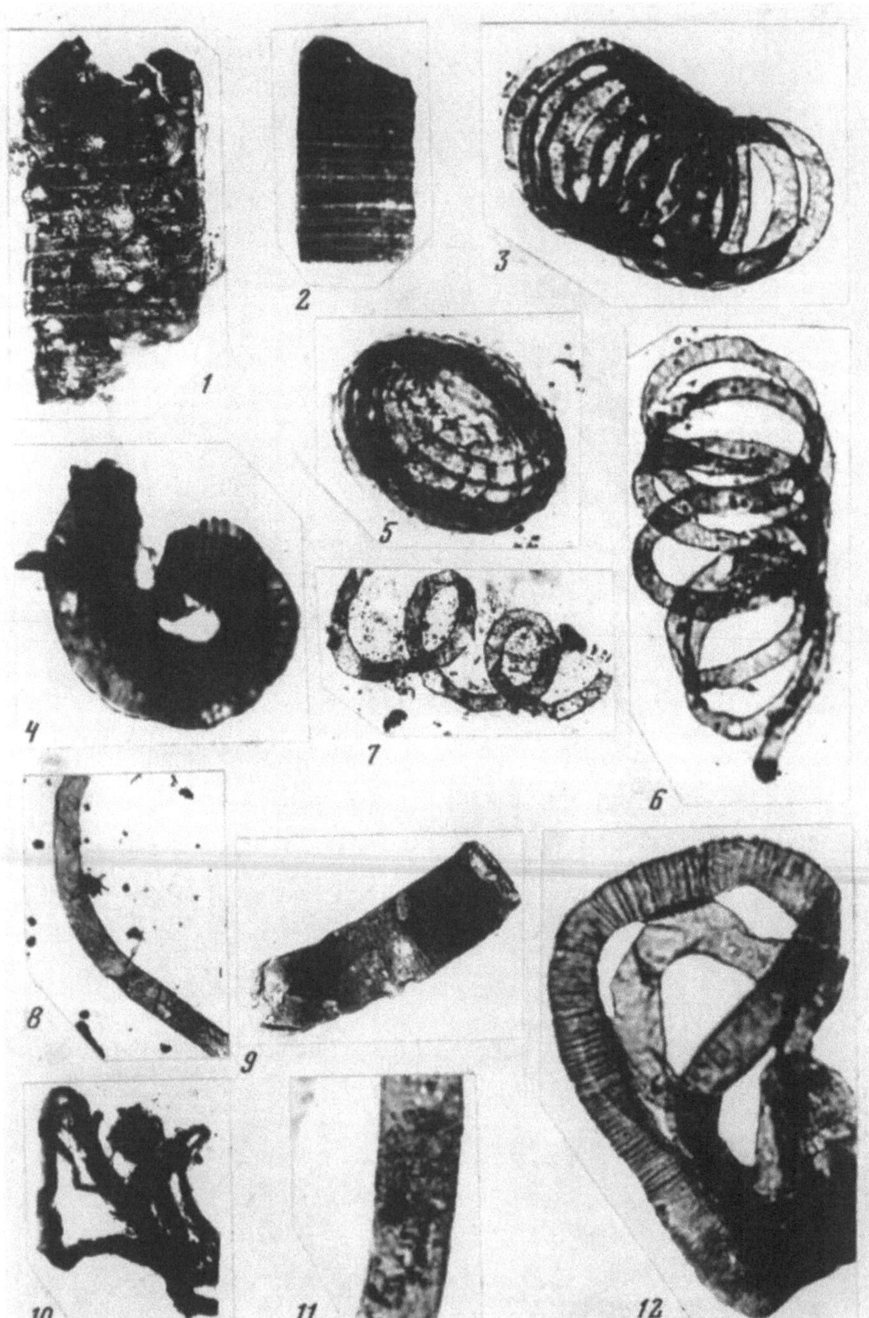

Plate 74

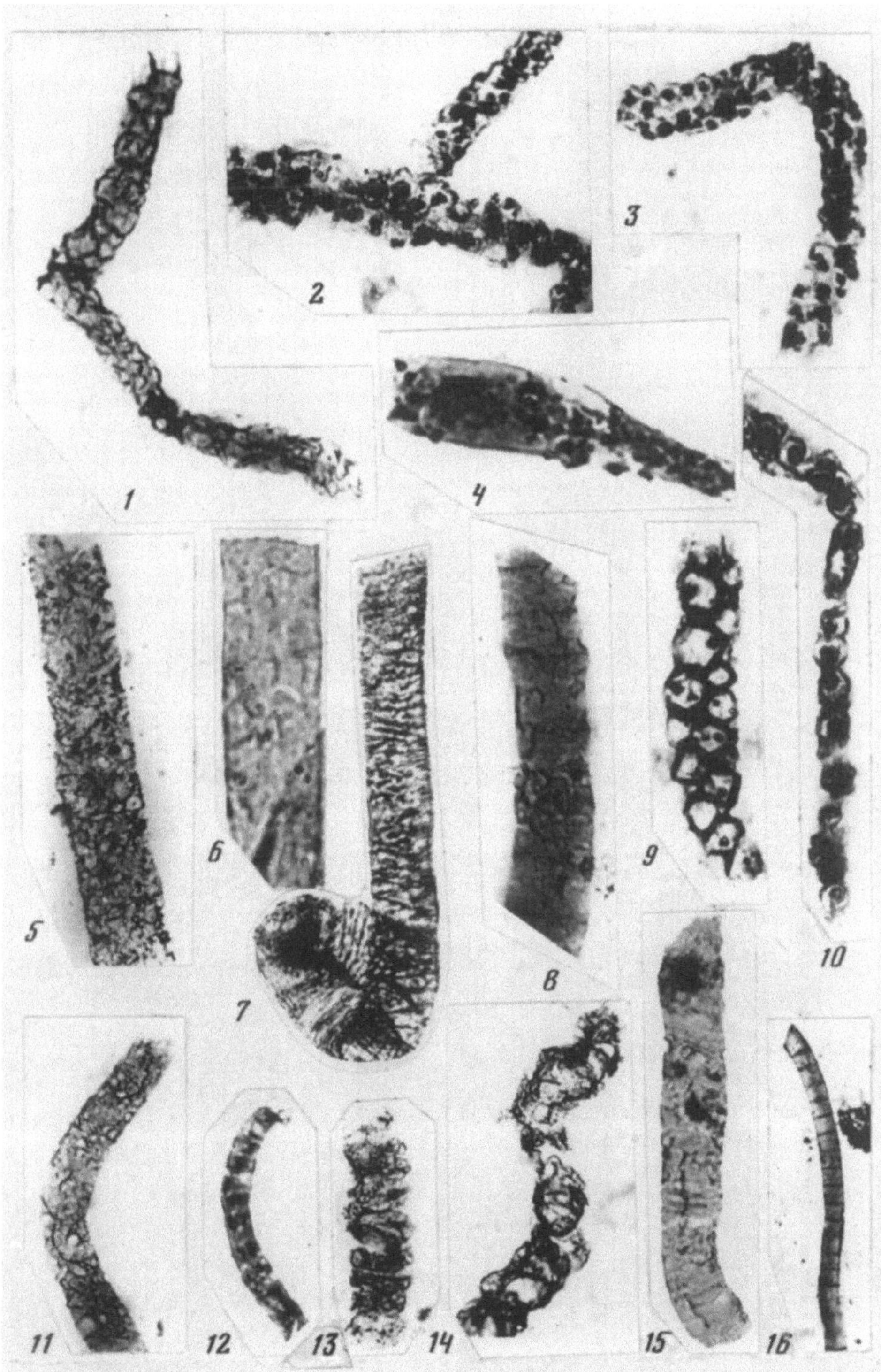

Plate 75

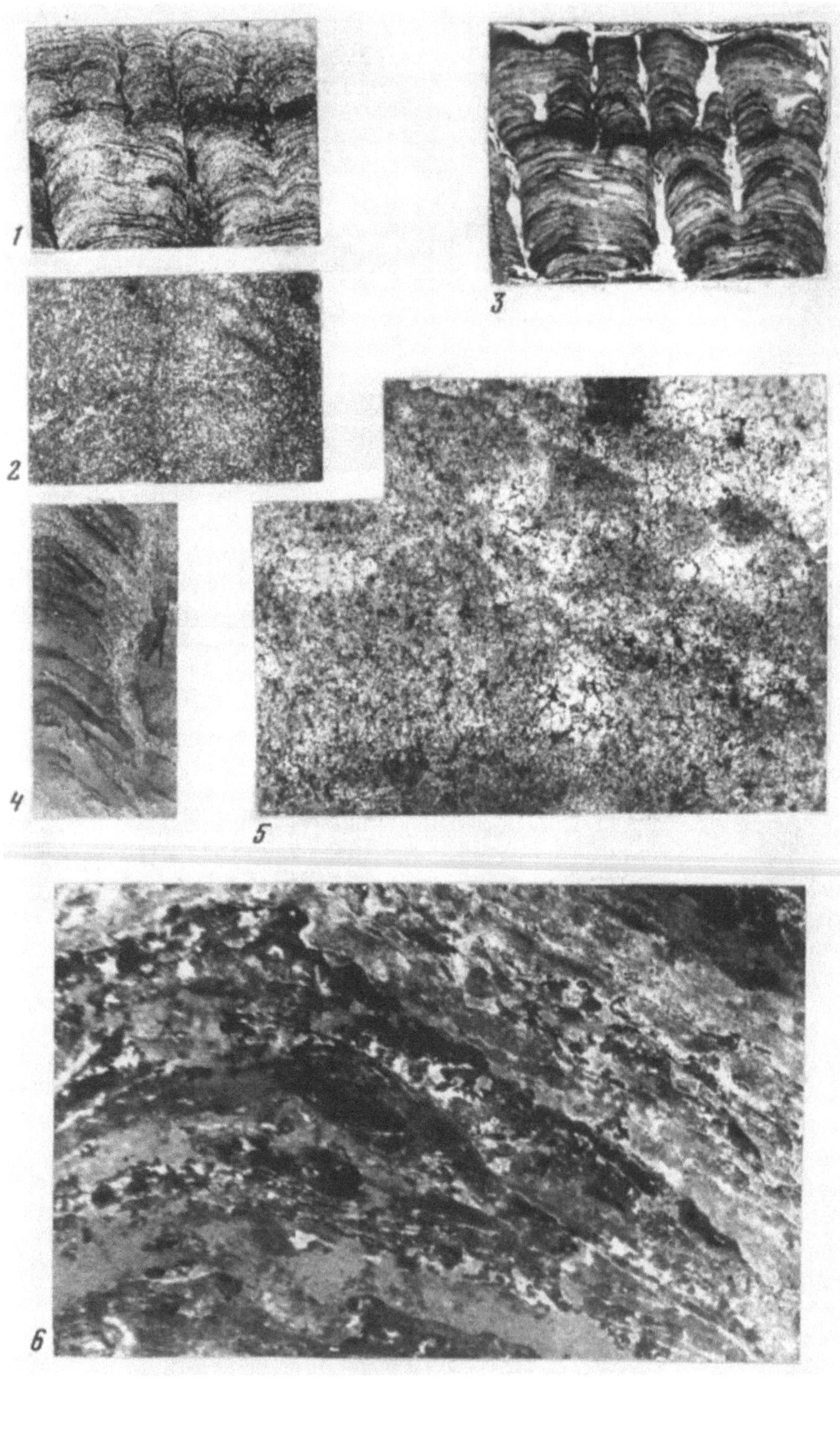

Plate 76

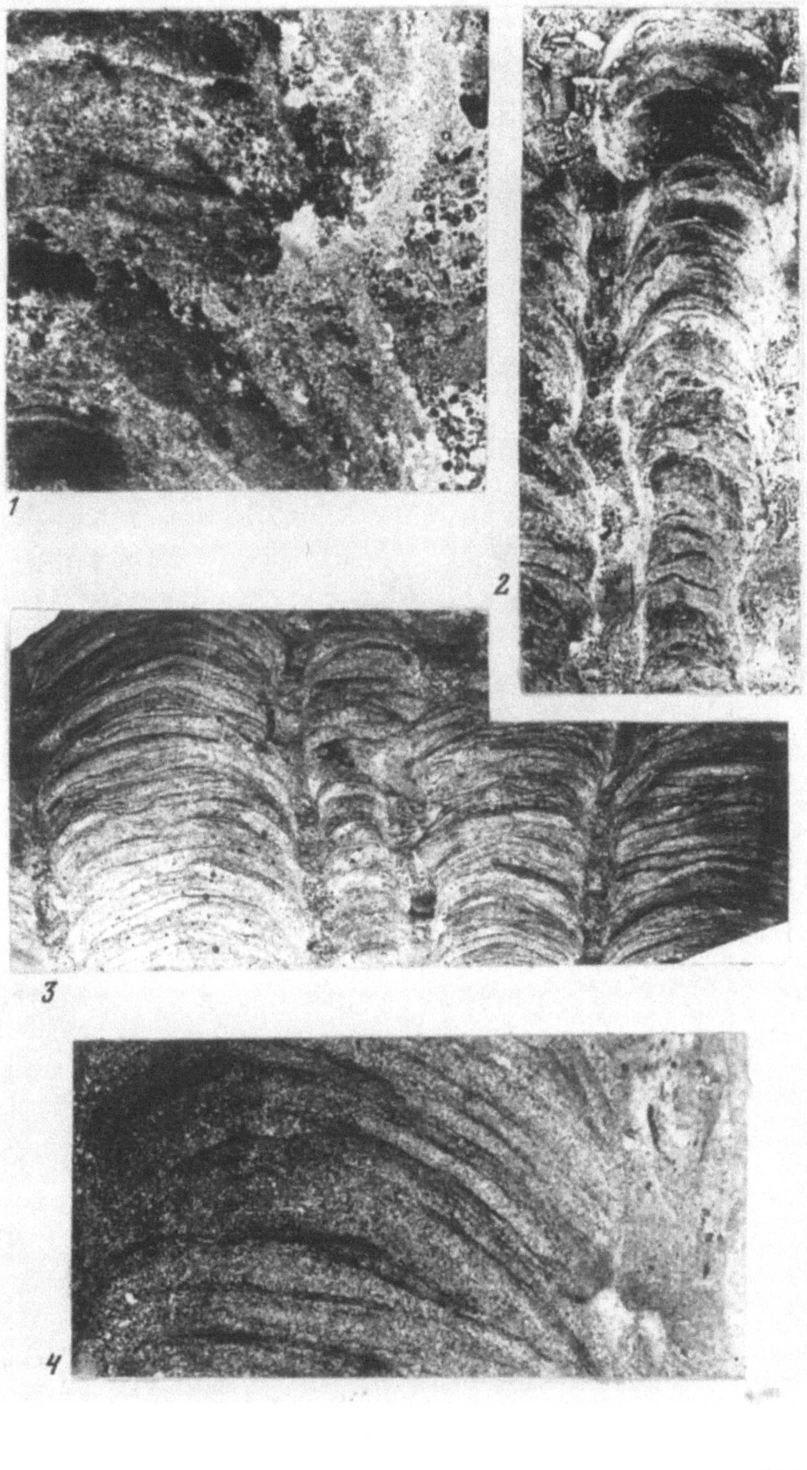

Plate 77

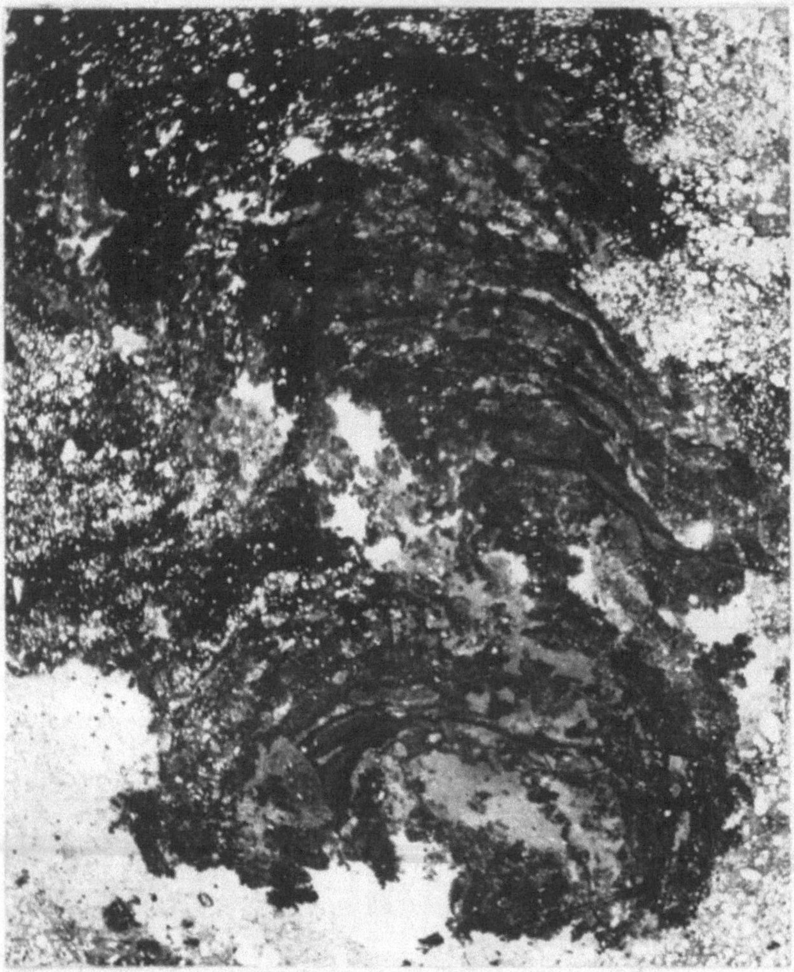

1

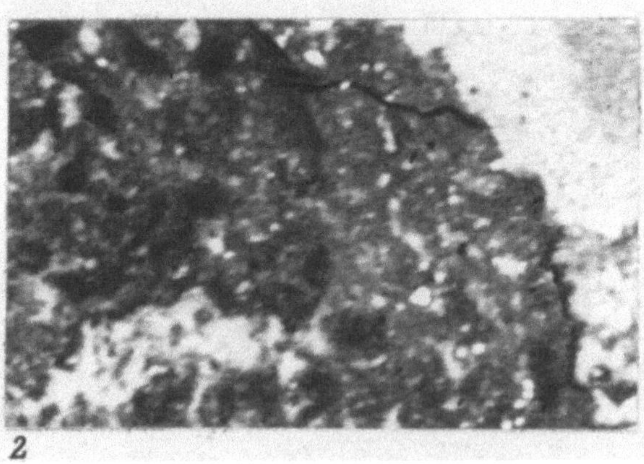

2

Plate 78

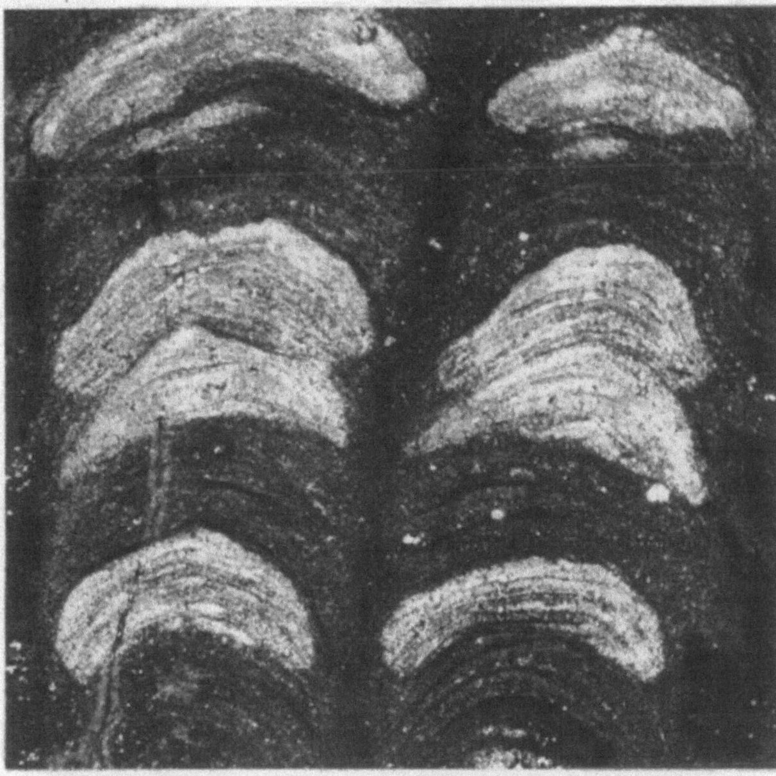

1

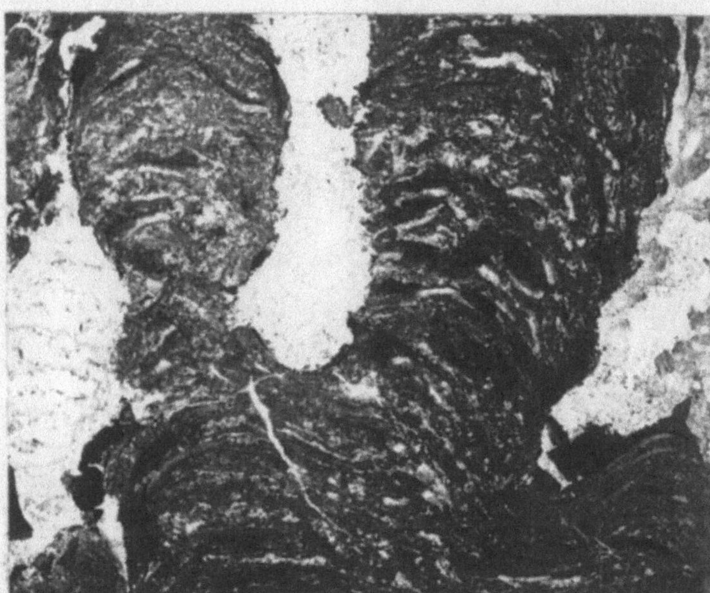

2

Plate 79

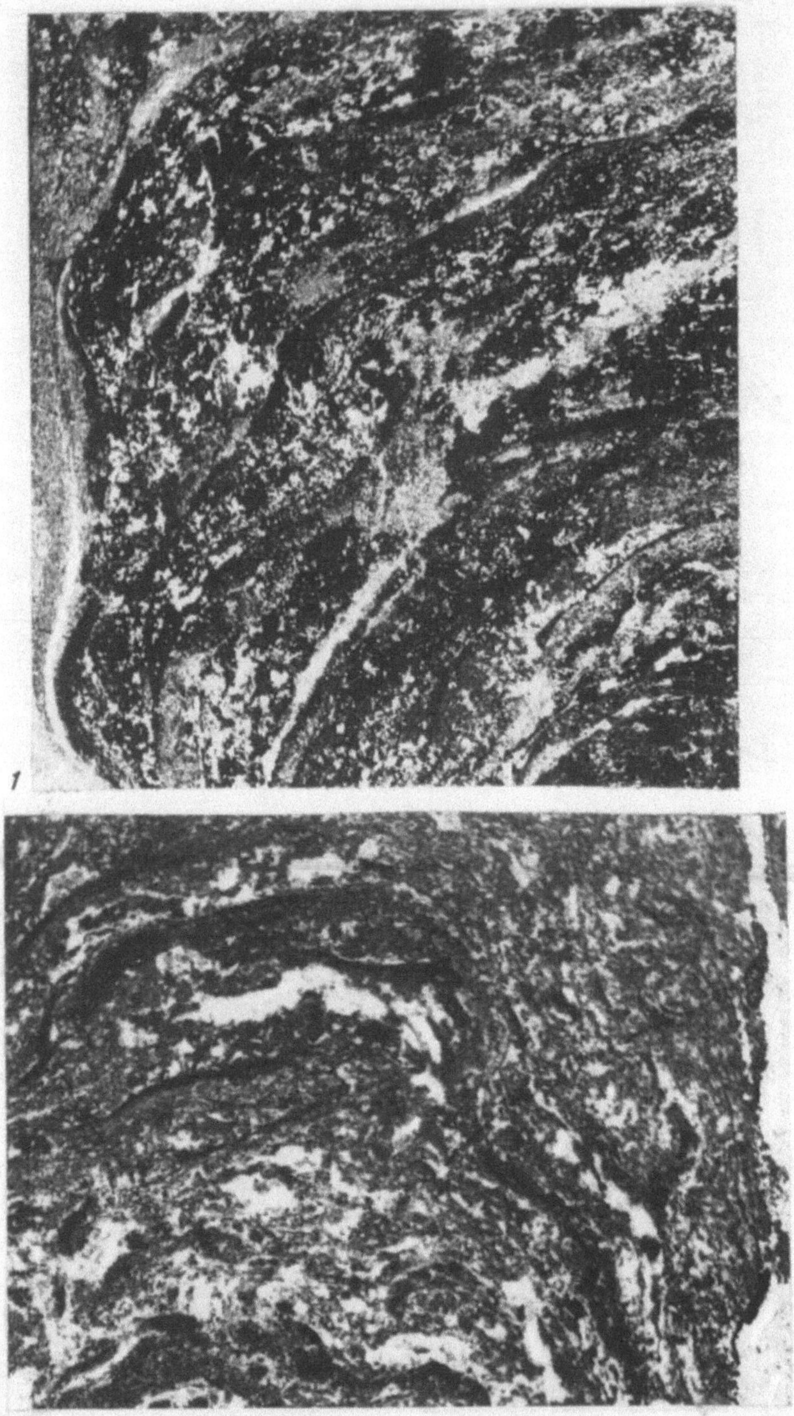

Plate 80

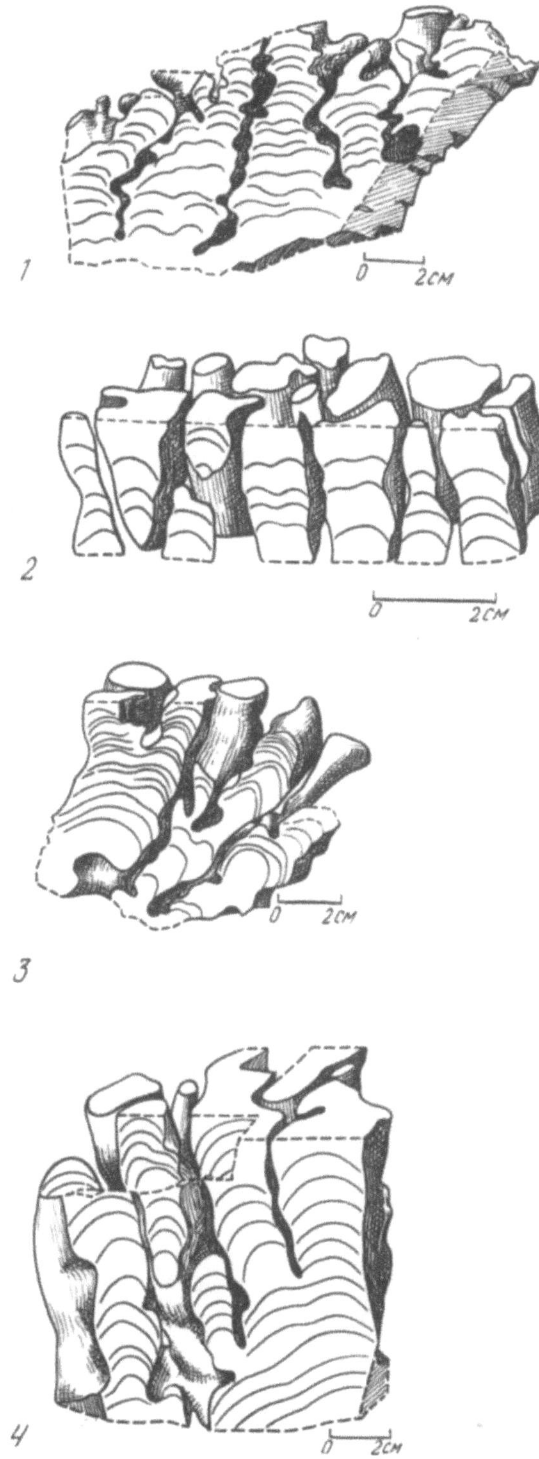

Plate 81

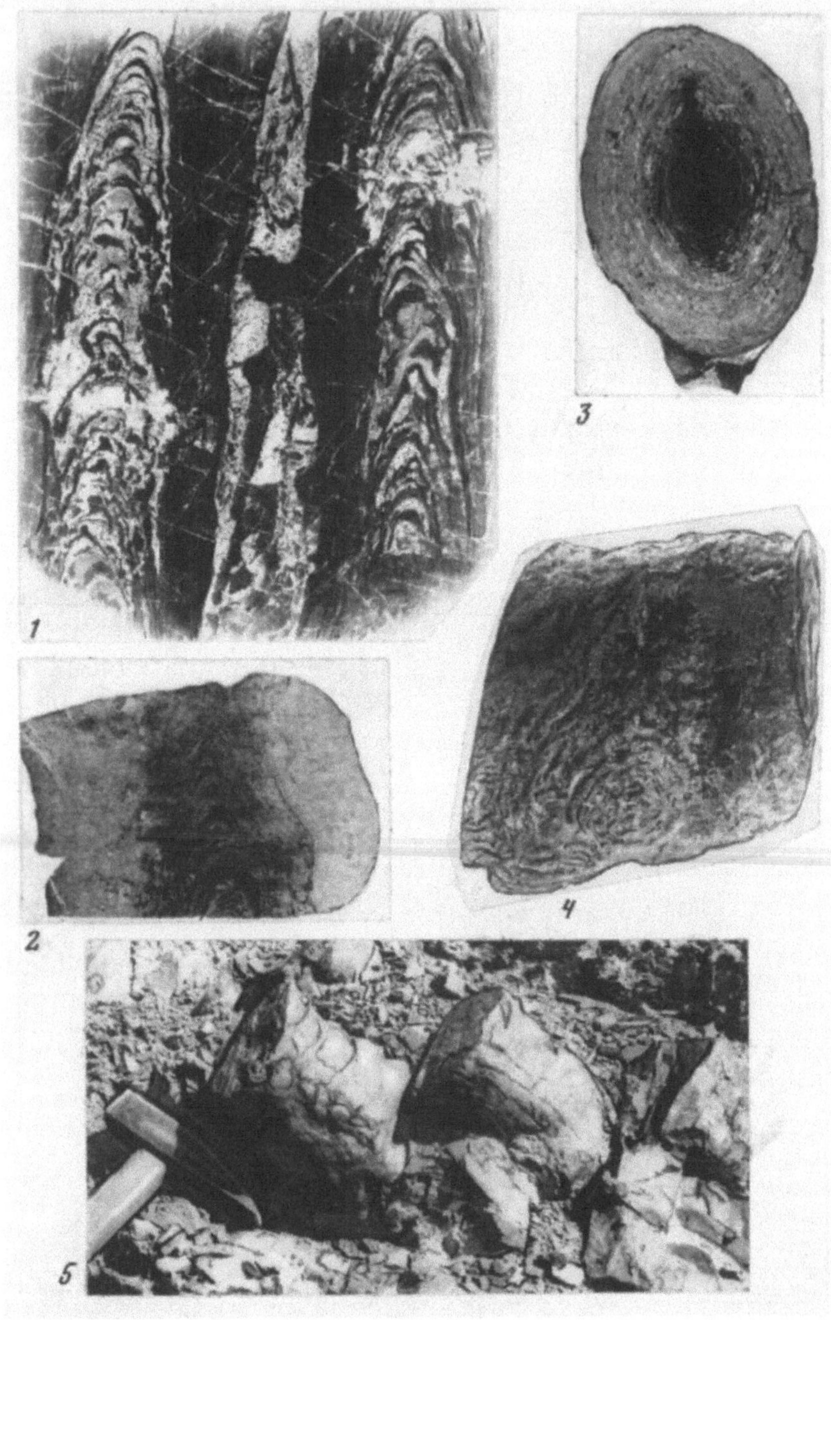

Plate 82

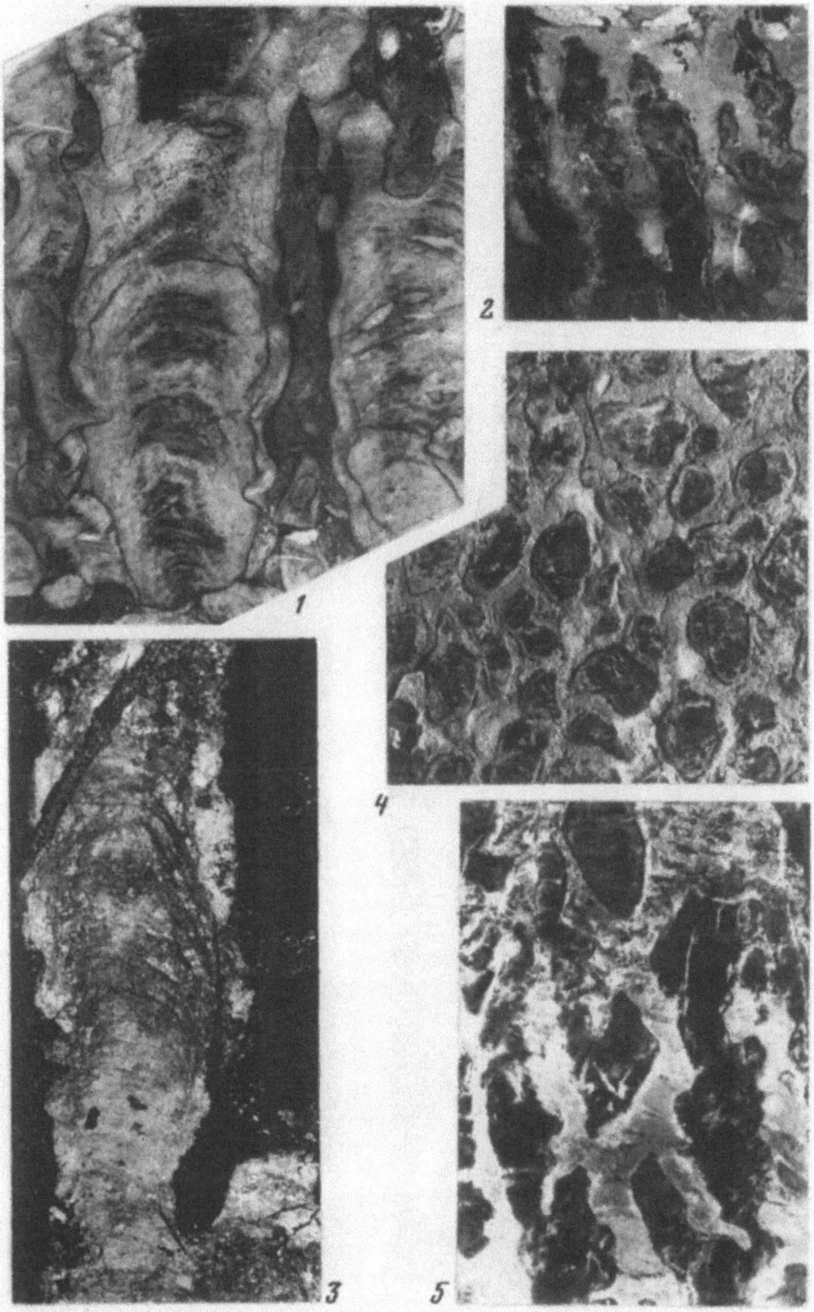

Plate 83

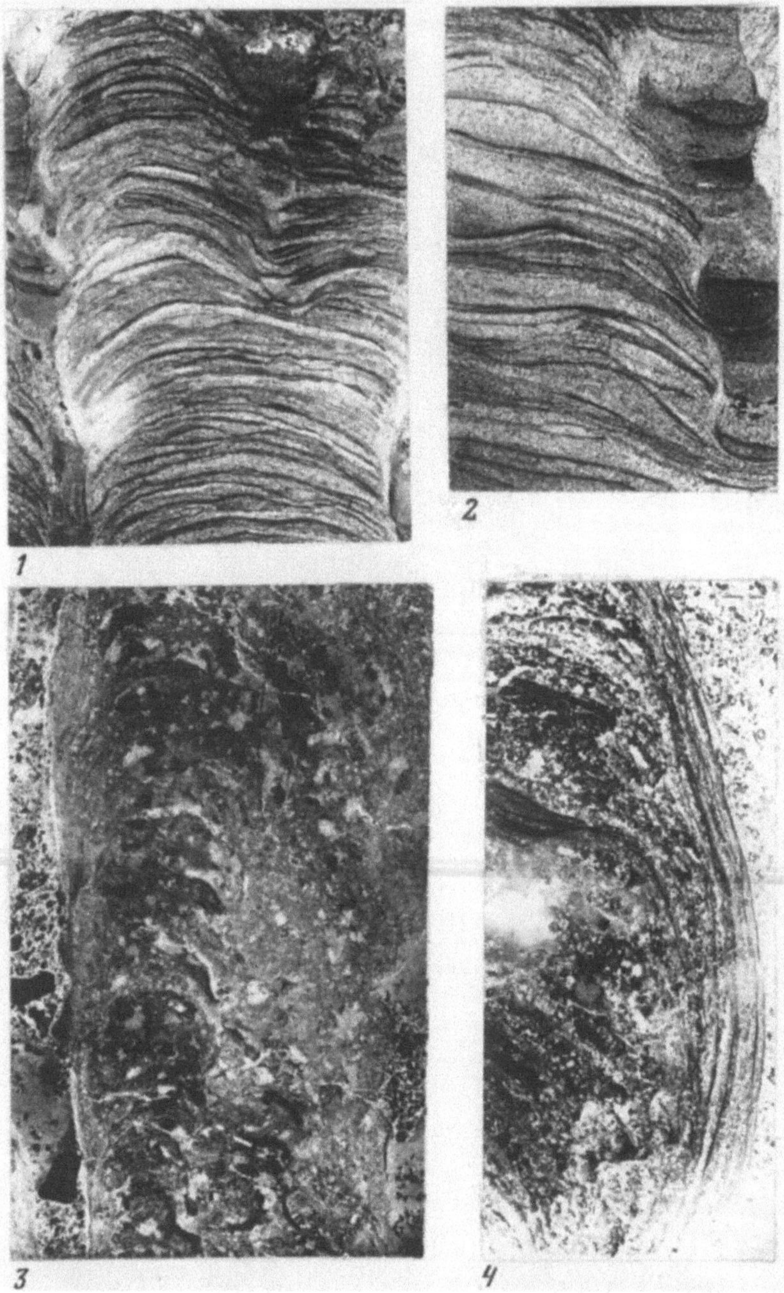

Plate 84

1

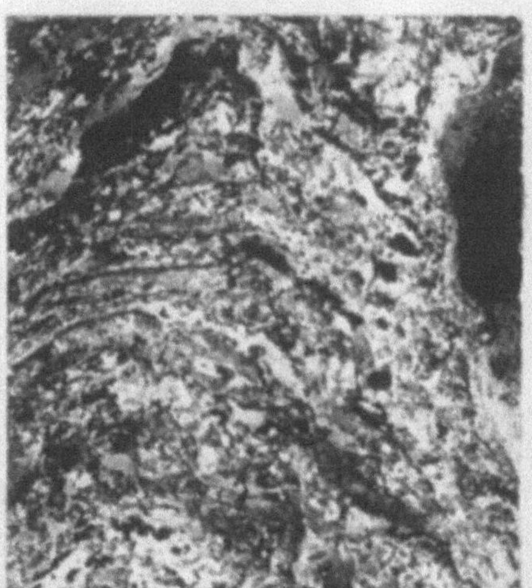

2

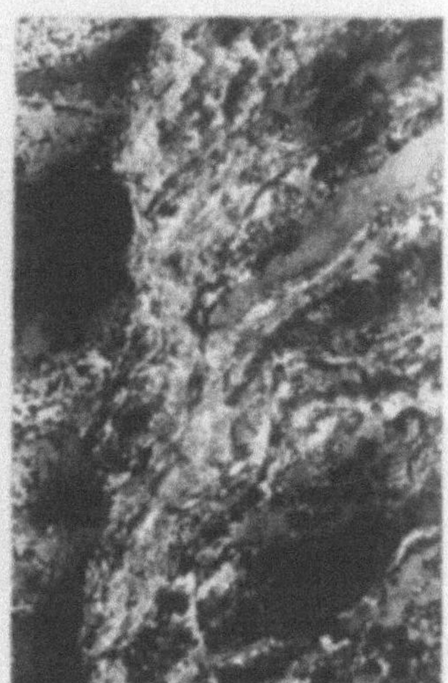

3

Plate 85

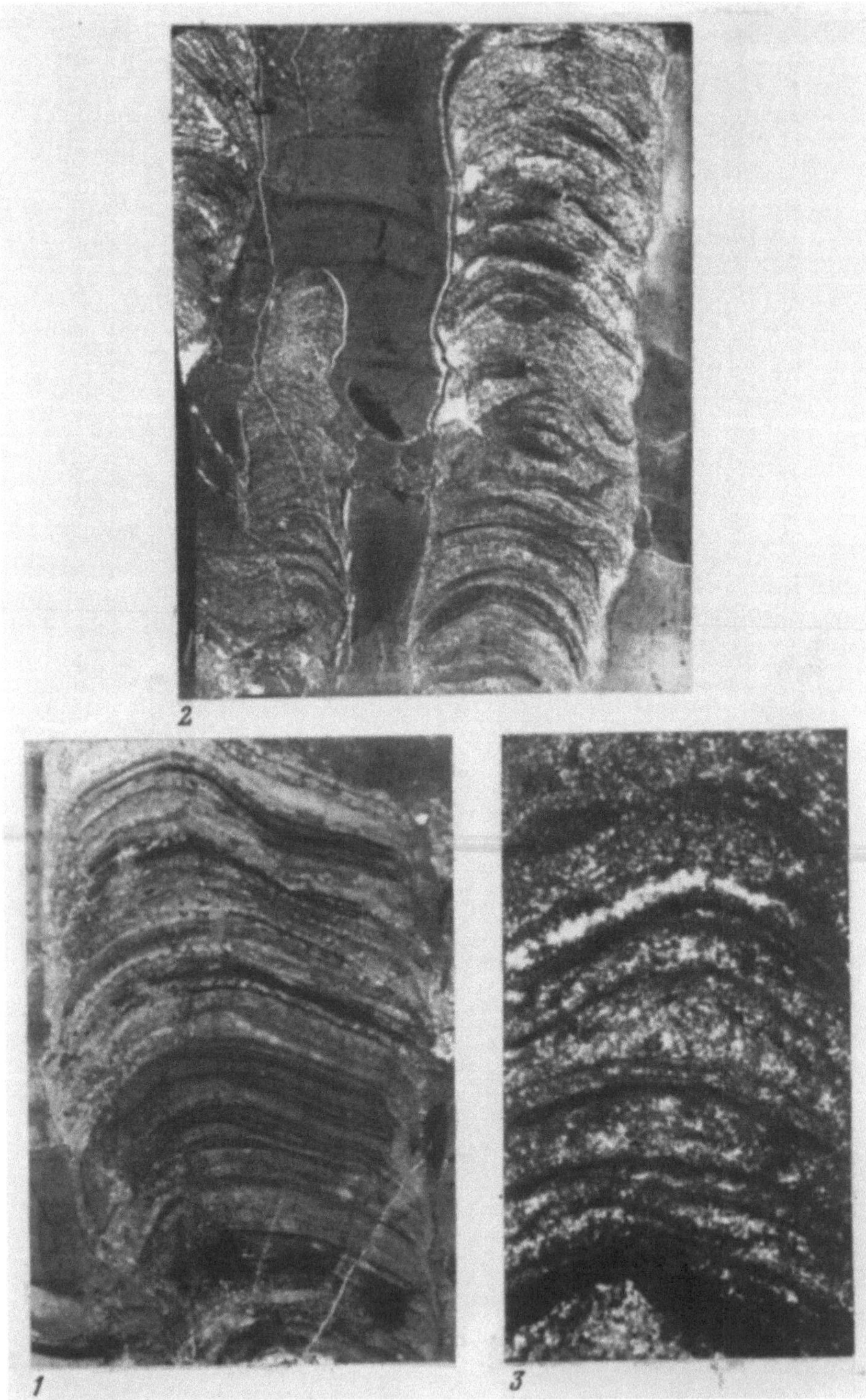

Plate 86

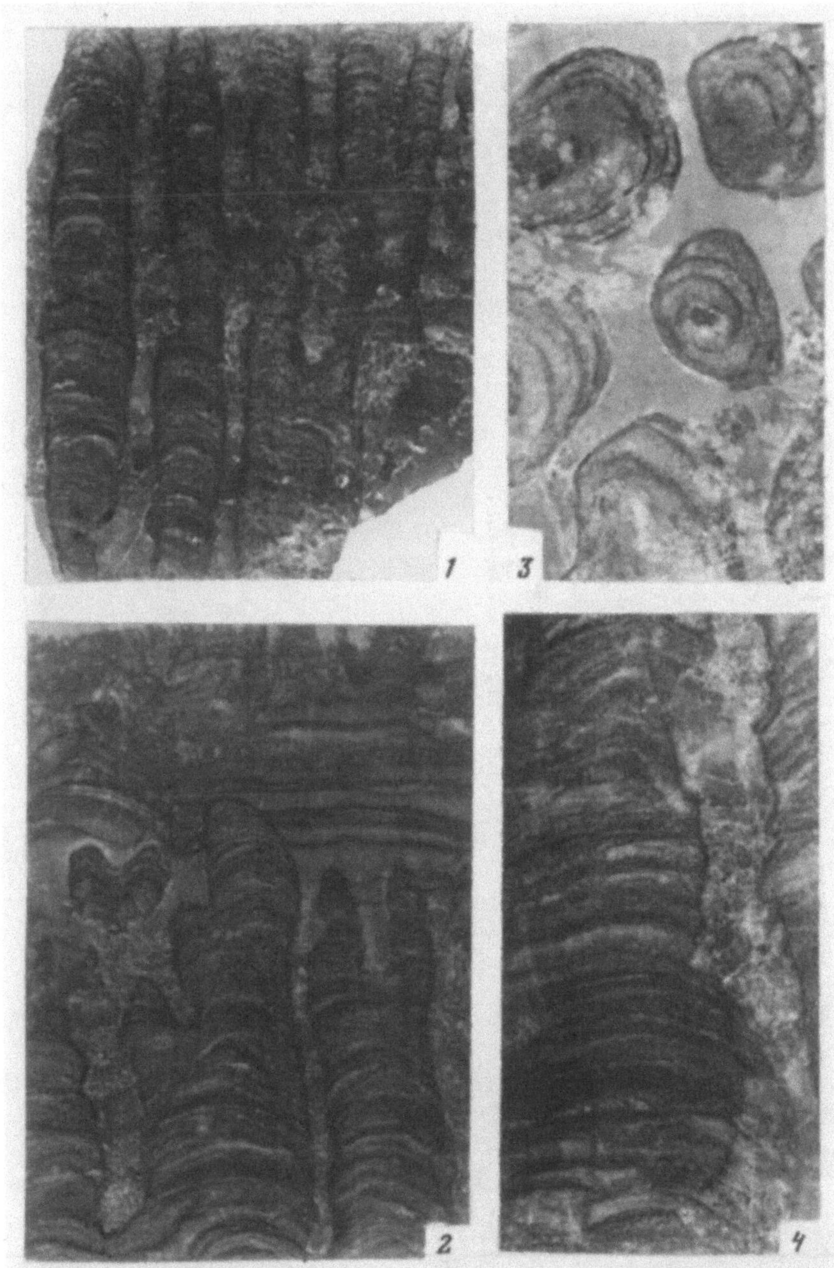

Plate 87

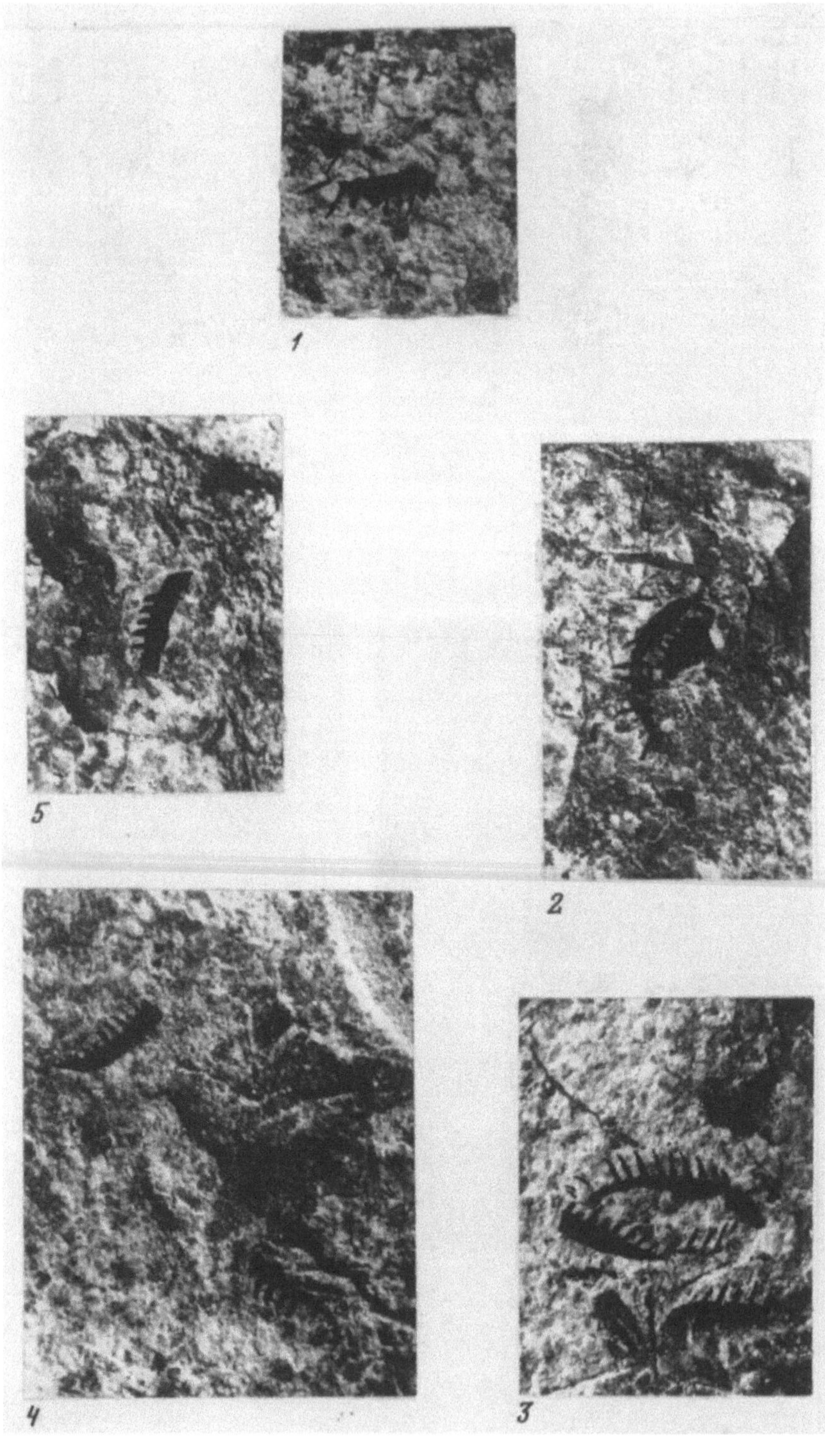

Plate 88

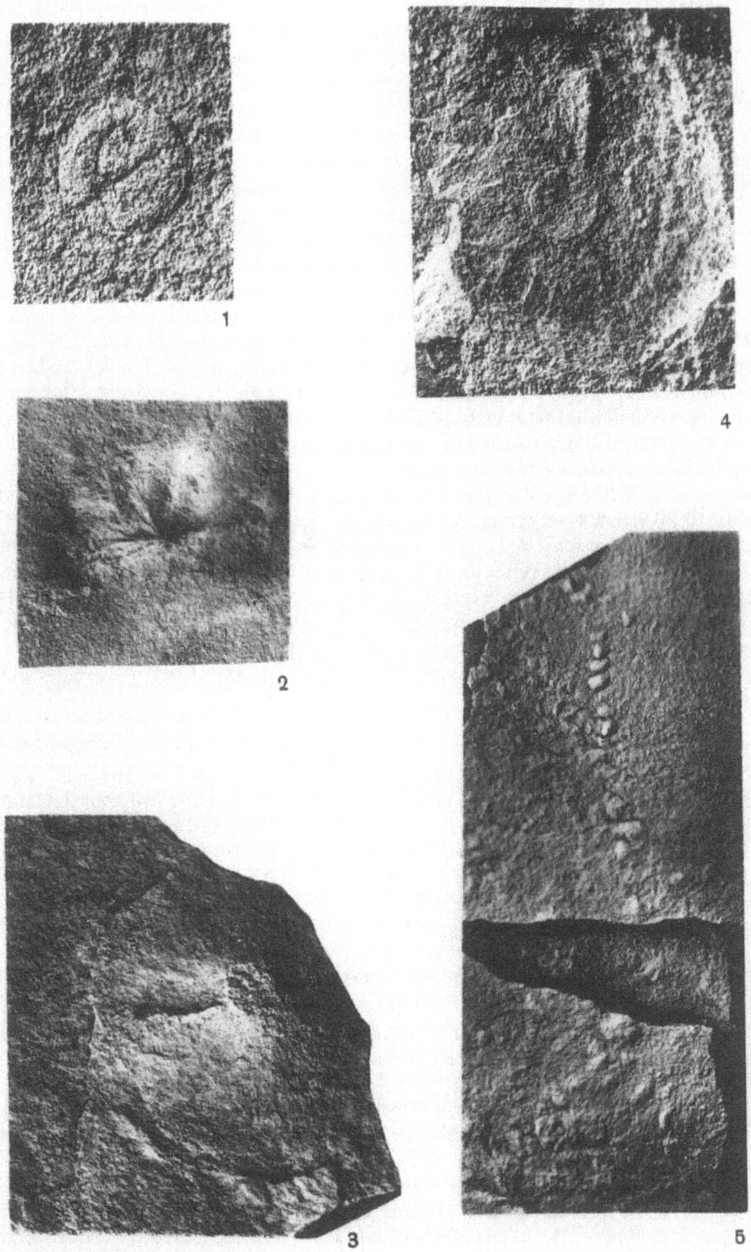

Plate 89

Explanation for Plates

Plate 1. *1 Ediacaria* sp., PIN No. 3994/668. To the *left* is a fragment of the fairly thick stolon without the narrow distal part, x 0.9; Dniester Basin of Podolia, Vendian, Mogilev Formation, Lomozov Beds. *2,5 Ediacaria flindersi* Sprigg. *2* PIN No. 3993/338, the central disc is elevated above the outer ring; the attaching organ is somewhat deformed and displaced towards the left, x 0.9; *5* PIN No. 4482/153, a larger specimen earlier erroneously assigned to *Tirasiana disciformis* Palij (Fedonkin 1981a; Plate 2, *4*). The rugae around the central disc and around the bell periphery result from post-mortem deformation, x 0.9; Zimny Coast of the White Sea, Vendian, Ust Pinega Formation. *3,8 Cyclomedusa cf. davidi* Sprigg. *3* PIN No. 3994/350, x 0.9; *8* PIN No. 3994/288, x 0.9; Dniester Basin of Podolia, Vendian, Mogilev Formation, Lomozov Beds. *4 Cyclomedusa delicata* Fedonkin. PIN No. 3992/355, a fragment of the aboral side: the central cone with a destroyed vertex and part of the wide outer zone with regular, concentric furrows, x 0.9; Letny Coast of the White Sea, Vendian, Ust Pinega Formation. *6 Kullingia* (?) sp., PIN No. 3995/1. To the *right*, *Nimbia occlusa* Fedonkin, No. 3995/2 *upper left*, x 0.9; Olenek Uplift, Khorbusuonka River, Vendian, Khatyspyt Formation. *7 Cyclomedusa radiata* Sprigg. PIN No. 4482/167, dense (closely spaced, radial furrows in the peripheral part of the body, x 0.9; Zimny Coast of the White Sea, Vendian, Ust Pinega Formation

Plate 2. *1 Ediacaria* sp. PIN No. 3993/417, a fragment of the aboral part, x 0.9; Zimny Coast of the White Sea, Vendian, Ust Pinega Formation. *2 Ediacaria* sp. PIN No. 3994/334, simultaneous longitudinal division into three parts with enclosed, median, triangular segment and two semicircular parts of the primary disc, x 0.9; Dniester Basin of Podolia, Vendian, Mogilev Formation, Lomozov Beds. *3 Cyclomedusa cf. plana* Glaessner et Wade. PIN No. 3993/462, a fragment of the medusoid bell with preserved central part and peripheral, feebly dissected marginal zone, the presumed oral side; radial furrows are possible ruptures; central disc may be equated to mouth cavity, x 0.9; Zimny Coast of the White Sea, Vendian, Ust Pinega Formation. *4 Ediacaria flindersi* Sprigg. PIN No. 3994/287, a fragment of the aboral side, x 0.9; Dniester Basin of Podolia, Vendian, Mogilev Formation, Lomozov Beds. *5 Cyclomedusa plana* Glaessner et Wade. PIN No. 3993/602, a fragment of the aboral side with preserved, concentrically rugate central part and smooth zone of a wide outer ring, x 0.9; Zimny Coast of the White Sea, Vendian, Ust Pinega Formation. *6 Paliella* (?) sp. PIN No. 3993/1301, aboral side, x 0.9; Zimny Coast of the White Sea, Vendian, Ust Pinega Formation

Plate 3. *1,6,7 Nimbia occlusa* Fedonkin. *1* PIN No. 3994/533, an individual preserved at the budding stage: *lower right*, an unseparated individual; *upper right*, a separated individual with still straight peripheral part of the body along which budding proceeded from a large parental individual, x 0.9; *6* PIN No. 3994/479-B, a specimen with well-preserved, central boss and sculptural outer ring, x 0.9; *7* PIN No. 3994/644, small specimens, x 0.9; Dniester Basin of Podolia, Vendian, Mogilev Formation. *2,8 Nimbia dniestri* Fedonkin. *2* Holotype PIN No. 3994/384-A, in division stage, two small, central circular dents on either side of the constriction line, may be equated to two newly formed mouth cavities, x 0.9; *8* PIN No. 3994/374, x 0.9; distribution and age as above. *3,4 Paliella* sp. *3* PIN No. 3994/840. To the *left*, a large specimen; to the *right*, two small individuals, probably in the division stage, x 0.9; *4* PIN No. 3994/710, flexible stolon detaches from central part of the disc, deformed during burial, x 0.9; distribution and age as above. *5 Cyclomedusa* sp. PIN No. 3994/308, a stolonal reproduction is evidenced by visible, thin, long stolons connecting individuals of different size, x 0.9; Zimny Coast of the White Sea, Vendian, Ust Pinega Formation. *9 Paliella patelliformis* Fedonkin. PIN No. 3994/16, an early division stage; constriction begins to show, central bosses are separated, individuals not yet separated, x 0.9;

Dniester Basin of Podolia, Vendian, Mogilev Formation. *10 Cyclomedusa cf. davidi* Sprigg. PIN No. 3994/370, early stage of longitudinal division; two very prominent, drifted-apart central bosses, nearly separated, still undivided, inner concentric rings envelope, x 0.9; distribution and age as for 9

Plate 4. *1,5 Stauridinia crucicula* Fedonkin, gen. et sp. nov. The bell is preserved as a circular depression (vegetative hyporelief) with sculptural, cruciform, radial grooves leading to ovate bosses (gonads). *1* PIN No. 3993/392-A; *5* PIN No. 3993/392-B; Zimny Coast of the White Sea, Vendian, Ust Pinega Formation. *2,3 Beltanelloides sorichevae* Sokolov. PIN No. 3992/501, 502; impressions in argillitelike clay; x 0.9; Ust Pinega Formation, Yarnema borehole, depth 118.4 m. *4 Paliella* sp. PIN No. 3993/897, a large specimen, aboral part; x 0.9; Zimny Coast of the White Sea, Vendian, Ust Pinega Formation. *6 Veprina undosa* Fedonkin. Holotype, GIN No. 4482/29, x 0.9; Zimny Coast of the White Sea, Ust Pinega Formation

Plate 5. *1* An organism of uncertain systematic position. PIN No. 3993/1301, basal part with attached disc. At the *top*, three branches diverge from the main stem, x 0.9; Zimny Coast of the White Sea, Vendian, Ust Pinega Formation. *2 Baltanelliformis brunsae* Menner. PIN No. 4464/115, aboral side, x 0.9; Letny Coast of the White Sea, Vendian, Ust Pinega Formation. *3 Nemiana simplex* Palij. PIN No. 3995/104, peculiar clusters, positive hyporelief, x 0.9; Olenek Uplift, Khorbosuonka River, Vendian, Khatyspyt Formation. *4,5,6 Protodipleurosoma regulosum* Fedonkin. Aboral side of the bell. *4* PIN No. 3993/304, x 0.9; *6* holotype, GIN No. 4482/21, x 0.9; large specimens with prominent mid-ridge from which thinner, sometimes branching rugae diverge sideways. Around the bell periphery there are ovate bodies separated by grooves from the bell (reproductive organs or budded daughter individuals); *5* PIN No. 3993/1302, a juvenile, with deeper and less numerous rugae, x 0.9; Zimny Coast of the White Sea, Vendian, Ust Pinega Formation. *7 Cyclomedusa cf. plana* Glaessner et Wade. GIN No. 4482/43, positive hyporelief, oral side of the bell, funnellike depression in the centre of the bell, whose concentric structure may be equated to the mouth cavity, x 0.9; distribution and age as *4*

Plate 6. All specimens except for *1* and from the Zimny Coast of the White Sea, Vendian, Ust Pinega Formation. *1–3 Ovatoscum concentricum* Glaessner et Wade. *1* Holotype cast, x 0.9; Pound Quartzite, Flinders Ranges, South Autralia; *2,3* smaller specimens, deformed during burial; *2* PIN No.3993/381. *3* PIN No. 3993/382, x 0.9. *4 Chondroplon* sp. PIN No. 3993/78, a fragment, x 0.9. *5 Eoporite medusa* Wade. PIN No. 3993/1302, positive hyporelief, aboral side showing numerous tentacles (polypides) forming concentric zones embracing the central oral cone, x 0.9. *6 Kaisalia mensae* Fedonkin gen. et sp. nov. Holotype, PIN No. 3993/284, positive hyporelief, aboral side; among concentric rings embracing the entire bell there are central, median and outer zones separated by low projections, x 0.9

Plate 7. *1,6 Hiemalora stellaris* Fedonkin. *1* PIN No. 3993/1308, positive hyporelief, structural details of the bell are not seen, the thickness of tentacles suggests that they were hollow, x 0.9. *6* PIN No. 3993/309-A, negative hyporelief, bell cast showing a large, central boss fairly often preserved in these forms and fine bosses closely spaced over the bell area (for explanation se text), x 0.9; Zimny Coast of the White Sea, Vendian, Ust Pinega Formation. *2 Pomoria corolliformis* Fedonkin. GIN No. 4482/203, positive hyporelief, oral side, holotype to the *right*, a compressed, short, proboscis to the *left*, x 0.9, Letny Coast of the White Sea, Vendian, Ust Pinega Formation. *3 Armillifera parva* Fedonkin. GIN No. 4482/109, negative hyporelief, x 0.9; age and distribution as above. *4 Hiemalora* sp. PIN No. 3995/00, an extensive, flat bell of a fairly large medusa, x 0.9; Olenek Uplift, Khorbusuonka River, Vendian, Khatyspyt Formation. *5 Ediacaria cf. flindersi* Sprigg. PIN No. 3994/626, negative hyporelief, oral side embraced by distinct marginal tentacles, x 0.9; Dniester Basin of Podolia, Vendian, Mogilev Formation

Plate 8. All specimens except for *4* are from the Dniester Basin of Podolia; Vendian, Mogilev Formation. *1 Conomedusites lobatus* Glaessner and Wade. PIN No. 3994/609, positive hyporelief , aboral side without marginal tentacles, x 0.9. *2 Medusinites asteroides* (Sprigg). PIN No. 3994/263, positive hyporelief, aboral side, x 0.9. *3 Medusinites* sp. PIN No. 3994/363-B, positive hyporelief, outer whorl exhibits relatively regular, radial structures, x 0.9. *4 Bonata septata* Fedonkin. GIN No. 4482/52-7, positive hyporelief, holotype, x 0.9; Zimny Coast of the White Sea, Vendian, Ust Pinega Formation. *5,6 Irridinitus multiradiatus* Fedonkin. *5* Holotype, PIN No. 3994/524, positive hyporelief, central part feebly pressed in with numerous prominent, radial grooves, x 0.9; *6* PIN No. 3994/127, negative hyporelief, probably showing internal structure of the animal: radial structures seem more ordered, x 0.9. *7 Elasenia* sp. PIN No. 3994/583, positive hyporelief, aboral side to the *left*, most prominent are three, often rounded bosses of the outer whorl, they may be equated to gonads. *8 Elasenia aseevae* Fedonkin. PIN No. 3994/338, positive hyporelief, aboral side, very prominent, central disc (pouch) embraced by fairly large, round bosses (gonads), x 0.9.

Plate 9. *1,4 Albumares brunsae* Fedonkin. *1* Holotype, GIN No. 4464/14, x 0.9; *4* PIN No. 3992/510, x 0.9, negative hyporelief, concave impression showing sculptural, large, elongate ridges (gonads) and numerous, fine branching grooves (canals); Letny Coast of the White Sea, Vendian, Ust Pinega Formation. *2 Anfesta stankovskii* Fedonkin. PIN No. 3993/260-A, negative hyporelief, three large furrows (ridges) diverging from the central depression, may be equated to gonads; holotype at the *bottom*, x 0.9, Zimny Coast of the White Sea, Vendian, Ust Pinega Formation. *3,5,6 Tribrachidium heraldicum* Glaessner. *3* PIN No. 3994/580-A, negative hyporelief, x 0.9, Dniester Basin of Podolia, Vendian, Mogilev Formation. *5* PIN No. 4482/150, negative hyporelief, better preserved are "bullae" departing from the proximal parts of the "arms", x 0.9, Zimny Coast of the White Sea, Vendian, Ust Pinega Formation. *6* PIN No. 3993/654, positive epirelief, oral side, abundant, fine radial grooves entering three arcuate dents distally ending with circular depressions (oral cavities), x 0.9, Zimny Coast of the White Sea, Vendian, Ust Pinega Formation

Plate 10. *1 Valdainia plumosa* Fedonkin. Holotype, PIN No. 3994/276, x 0.9, Dniester Basin of Podolia, Vendian, Mogilev Formation. *2,3 Pseudorhizostomites howchini* Sprigg. *2* GIN No. 4482/27; *3* GIN No. 4482/50, negative hyporelief, x 0.9, Zimny Coast of the White Sea, Vendian, Ust Pinega Formation. *4 Podolimirus mirus* Fedonkin. Holotype, PIN No. 3994/5, x 0.9, Dniester Basin of Podolia, Vendian, Mogilev Formation. *5 Paliella patelliformis* Fedonkin. PIN No. 3994/5; negative epirelief, bushy type of asexual reproduction, x 0.9. Distribution and age as above

Plate 11. *1-4 Pteridinium nenoxa* Keller. *1* PIN No. 3992/309, a small specimen lying in aleuropelite slab vertically and across bedding (probably, life position), x 0.9. *2* PIN No. 3992/400, latex acts, near mid-line, proximal part of each segment exhibits structures which may be interpreted as pocketlike organs with a wider, open side facing the outer margin of the petaloid, x 0.9. *3* PIN No. 3993/168, a fragment of a large specimen with smooth segments with worn and displaced distal ends (at the *bottom* of photograph), x 0.9. *4* PIN No. 3992/167, a fragment of the petaloid with sculptural segments, x 0.9; Letny Coast of the White Sea, Vendian, Ust Pinega Formation. *5 Archangelia valdaica* Fedonkin. Holotype, PIN No. 3992/59, x 0.9; age and distribution as above

Plate 12. *1 Vavelikia velikanovi* Fedonkin. Holotype, PIN No. 3994/581, x 0.9; Dniester Basin of Podolia, Vendian, Mogilev Formation. *2 Charnodiscus* sp. PIN No. 3993/251, a well-preserved, relatively large attaching disc, above it a less sculptural, frondlike body with prominent mid-range and transverse furrows (Lateral outgrowths) diverging from it on both sides, x 0.9; Zimny Coast of the White Sea, Vendian, Ust Pinega Formation. *3,5 Inkrylovia lata* Fedonkin. *3* PIN No.3992/507, x 0.9. *5* PIN No. 3992/156, x 0.9; Letny Coast of the White Sea, Vendian, Ust Pinega Formation. *4 Charnia masoni* Ford. PIN No. 3993/1-15, x 0.9; Zimny Coast of the White Sea, Vendian, Ust Pinega Formation *6 Zolotytsia biserialis* Fedonkin. Holotype, GIN No. 4482/158, x 0.9. Zimny Coast of the White Sea, Vendian, Ust Pinega Formation

Plate 13. *1 Khatyspytia grandis* Fedonkin gen. et. sp. nov. Holotype, PIN No. 3995/132, a colony with a strong attaching organ, thick base and relatively narrow, frondlike (pinnate) part, which gradually becomes thin; monopodial type of growth; lateral processes (outgrowths), thick at the base and thin towards the distal ends, which bend towards the apex, x 0.9; Olenek Uplift, Khorbusuonka River, Vendian, Khatyspyt Formation. *2-4 Charnia masoni* Ford. *2* PIN No. 3995/126, a juvenile colony with prominent, sympodial branching of lateral outgrowths, x 0.9. *3* PIN No. 3995/125, a fragment of a large form, x 0.9. *4* PIN No. 3995/173, a juvenile with preserved attaching disc, x 0.9; Olenek Uplift, Khobusuonka River, Vendian, Khatyspyt Formation

Plate 14. *1 Ramellina pannata* Fedonkin. Holotype, GIN No. 4482/47, x 0.9; Zimny Coast of the White Sea, Vendian, Ust Pinega Formation. *2 Zolotytsia* sp. PIN No. 3993/343, x 0.9, distribution and age as above. *3,4 Vaizitsinia sophia* Sokolov et Fedonkin. *3* PIN No. 3993/1306, a fragment of the lower part of a colony. A fairly broad, axial part of the colony grades at the bottom into ellipsoidal bullae (attaching organ), x 0.9. *4* Holotype, PIN No. 3993/1305, x 0.9; borehole No. 202 (Vaizitsa), depth 328 m, Vendian,

Ust Pinega Formation. *5* A fragment of a large pinnate form of uncertain systematic position, lateral outgrowths thick and smooth, x 0.9; left side of Onega River, 1 km below the mouth of Somba River, Vendian, Ust Pinega Formation. *6 Vaizitsinia* sp. PIN No. 3993/65, a juvenile form, x 0.9, Zimny Coast of the White Sea, Vendian, Ust Pinega Formation

Plate 15. *1,2 Lomozovis malus* Fedonkin. *1* Holotype, PIN No. 3994/418-A, photograph rotated 180°, therefore, the base of the basal part is at the *upper edge* of the sample (without attaching organ), right process twisted near the base, its distal, obtuse end bears a brushlike part; at the *bottom*, feebly compressed *Dickinsonia cf. tenuis* Glaessner et Wade, x 0.9. *2*PIN No. 3994/418-B, two long processes with long, brushlike ends, x 0.9; Dniester Basin of Podolia, Vendian, Mogilev Formation, Lomozov Beds

Plate 16. *1,3,6 Palaeoplatoda segmentata* Fedonkin. *1* GIN No. 4464/103, a fragment with worn end part, x 0.9; Letny Coast of the White Sea, Vendian, Ust Pinega Formation. *2 Dickinsonia costata* Sprigg. GIN No. 4464/17, x 0.9; age and distribution as above. *3 Dickinsonia cf. tenuis* Glaessner et Wade. PIN No. 3994/418, x 0.9; Dniester Basin of Podolia, Vendian, Mogilev Formation. *5,7 Dickinsonia lissa* Wade. *5* PIN No. 3993/77, most of the right side is absent; at the *top left*, the body is compressed, x 0.9. *7* PIN No. 3993/256, no segments, x 0.9; Zimny Coast of the White Sea, Vendian, Ust Pinega Formation

Plate 17. All specimens are from the Zimny Coast of the White Sea, Vendian, Ust Pinega Formation. *1,2,5,6,7 Dickinsonia costata* Sprigg. *1* PIN No. 3993/1305, x 0.9; *2* PIN No. 3993/1304, x 0.9, adults, positive epirelief. *5,7* PIN No. 3993/536: *5*, x 3,7, x 0.9; *6* PIN No. 3993/537, x 0.9; juveniles. *3 Dickinsonia lissa* Wade. PIN No. 3993/451, a fragment, negative hyporelief, x 0.9. *4 Dickinsonia elongata* Glaessner and Wade. PIN No. 3993/119, a fragment of impression of the middle part of the body, x 0.9

Plate 18. *1–3 Dickinsonia costata* Sprigg. Latex cast from original (Runnegar 1982). *1* General view, x 0.9. *2,3* Fragments magnified; distinct changing of position of the proximal ends of the segment at junction point with the median zone; South Australia, Vendian, Rawnsley Quartzite. *4 Dickinsonia elongata* Glaessner and Wade. PIN No. 3993/119-A, a fragment showing to the *right*, a part of the mid-ridge with staggering, narrow segments diverging from it; marginal line of impression is preserved to the *left*, x 0.9; Zimny Coast of the White Sea, Vendian, Ust Pinega Formation

Plate 19. *1,2,4,8,9 Onega stepanovi* Fedonkin. Latex casts from impressions preserved in negative hyporelief. *1* PIN No. 3992/401, x 5. *2* PIN No. 3992/504, x 5. *4* PIN No. 3992/505, x 5. *8,9* PIN No. 4464/57B. *8*, x 2; *9* The same specimen, a photograph at higher magnification with different lighting; Letny Coast of the White Sea, Vendian, Ust Pinega Formation. Three small bilateral shield-like organisms with numerous fine poorly preserved segments. PIN No. 3993/841, x 5; Zimny Coast of the White Sea, Vendian, Ust Pinega Formation. *5,6 Platypholinia pholiata* Fedonkin. *5* PIN No. 3993/191, x 0.9. *6* Holotype, PIN No. 3993/195, x 0.9; Zimny Coast of the White Sea, Vendian, Ust Pinega Formation. *7 Vladimissa missarzhevskii* Fedonkin. Holotype, PIN No. 3993/204, distribution and age as above x 1

Plate 20. *1 Spriggina floundersi* Glaessner. Holotype cast, x 0.9; South Australia, Vendian, Rawnsley Quartzite. *2 Evmiaksia aksionovi* Fedonkin. Holotype, PIN No. 3993/657, x 0.9; Zimny Coast of the White Sea, Vendian, Ust Pinega Formation. *3,6,7 Paleolina* sp. PIN No. 3993C/1,2,3, fragments of long, wide tubes, x 0.9; Malinovka borehole (Arkhangelsk District), depth 360 m, Vendian, Ust Pinega Formation. *4,5 Parvancorina minchami* Glaessner. Cast of originals, x 0.9; South Australia, Vendian, Rawnsley Quartzite

Plate 21. *1 Vendia sokolovi* Keller. Holotype, Yarensk borehole, depth 1,552 m; Vendian, Ust Pinega Formation, Redkino Horizon. *2,3 Mialsemia semichatovi* Fedonkin gen. et sp. nov. PIN No. 3993/401, x 0.9. *2* Imprint. *3* Cast (holotype); Zimny Coast of the White Sea, Vendian, Ust Pinega Formation. *4 Pseudovendia* sp. PIN No. 3993/501, latex cast of imprint preserved as negative hyporelief showing to the *right*, large, elongate head part followed by segmented body, x 0.9; Letny Coast of the White Sea, Vendian, Ust Pinega Formation. *5 Spriggina borealis* Fedonkin. Holotype, GIN No. 4464/110, x 0.9, distribution and age as above. *6 Bomakellia kelleri* Fedonkin, Holotype, PIN No.3992/508, x 0.9, most of the pleuralike appendages to the *left* are absent (not preserved); distribution and age as above. *7 Vendomia menneri* Keller. Latex cast of holotype (GIN No. 4464/57A); distribution and age as above

Plate 22. *1 Palaeopascichnus delicatus* Palij. PIN No. 3994/295, positive hyporelief, x 0.9; Dniester Basin of Podolia, Vendian, Mogilev Formation. *2 Bergaueria* sp. PIN No. 3994/306, positive hyporelief, small rounded bosses; casts of basal part of polyps of uncertain status (systematic position); the *upper part* shows trails of polyp lateral movement, forming a band of arcuate furrows with the concave side facing the terminal boss; the large, circular groove may be a medusa imprint (cast), x 0.9; distribution and age as above. *3* Accumulations of fine round caprolites. PIN No. 3994/342, positive hyporelief, x 0.9; distribution and age as above. *4 Phycodes pedum* Seilacher. PIN No. 3994/26, positive hyporelief, x 0.9; Dniester Basin of Podolia, mouth of Ternava River, Baltic Series (Rovno Horizon), Khmelnitsk Formation. *5 Treptichnus bifurcus* Miller (positive hyporelief) with elements of *Phycodes pedum*, PIN No. 3994/26-A, x 0.9; distribution and age as above, *4*. *6 Palaeopascichnus sinosus* Fedonkin, GIN No. 4310/202-6A, x 0.9; Letny Coast of the White Sea, Vendian, Ust Pinega Formation. *7 Treptichnus triplex* Palij. PIN No. 3994/38, x 0.9; Dniester Basin of Podolia, mouth of the Ternava River, Baltic Series (Rovno Horizon), Khmelnitsk Formation

Plate 23. Characteristic trace fossils from the lower Kessyuse Formation (Nemakit-Daldyn Horizon), Olenek River, left side. *1 Gordia* sp. PIN No. 3994/9, x 0.9. *2 Olenichnus irregularis* Fedonkin ichnogen et ichnosp. nov. PIN No. 3995/12A, negative hyporelief, x 0.9. Subhorizontal, sinuous burrow trails forming a sort of irregularly meshed network. Presence of narrow and wider burrows within a dwelling-burrow system suggests that they were repeatedly used. "Meshes" are circular to elongate, irregularly shaped "meshes" range widely in size. *3 Palnispiralichnus grandis* Fedonkin ichnogen. et ichnosp. nov. Holotype, PIN No. 3995A/101, x 0.9. Large horizontal system of feeding burrows in the form of flat spiral with coils, which close together, often intersecting. Burrows are from a circular section. *4 Protospiralichnus circularis* Fedonkin ichnogen. et ichnosp. nov. PIN No. 3995A/103, x 0.9. Horizontal feeding burrows on bedding plane between underlying mudstones and overlying silts. Burrows resemble a flat spiral originating from the centre: attaining a certain radius, the spiral widens and burrows develop into a circle. There are often several coils of the same radius; in this case, burrows can intersect and circles can be displaced sideways with each turn. *5 Phycodes* sp. PIN No. 3995A/226, x 0.9

Plate 24. *1 Planolites* sp. PIN No. 399A/19, x 0.9. Ichnocoenosus typical of Vendian-Cambrian transitional zones. Sediments were subject to severe biological processing due to mud-eaters. Olenek River, Kessyuse Formation, Nemakit-Daldyn Horizon. *2 Olenichnus irregularis* Fedonkin ichnogen. et ichnosp. nov. PIN No. 3995A/12B, x 0.9. Distribution and age as above. *3,4 Didymaulichnus meanderiformis* Fedonkin ichnosp. nov. *3* Holotype, PIN No. 3995A/232, x 0.9, positive hyporelief. *4* PIN No. 3995A/230, a form with less sprawling meanders, x 0.9; large meandering burrows with a characteristic, longitudinal furrow and smooth surface; burrows look like more or less broad meanders, sometimes looplike; Olenek River, Lower Cambrian, Tommotian, upper Kessyuse Formation

Plate 25. *1 Planolites* sp. and burrows with meniscoid fabric; PIN No. 3995A/256, x 0.9; Olenek River, Lower Cambrian, Tommotian, Kessyuse Formation. *2 Didymaulichnus cf. tirasensis* Palij. PIN No. 4482/185, x 0.9; Olenek River, Kessyuse Formation, Nemakit-Daldyn Horizon

Plate 26. All specimens are from the Zimny Coast of the White Sea, Vendian, Ust Pinega Formation. *1,8 Aulichnites* sp. Positive epirelief. *1* GIN No. 4482/185; *upper right*, an ovate depression like a coffee bean, possibly a resting track of a small, bilaterally symmetrical animal. *2 Neonereites renarius* Fedonkin. GIN No. 4482/118, positive hyporelief, x 0.9. *3,4 Vimenites bacillaris* Fedonkin. *3* GIN No. 4482/61, x 0.9, circular fecal pellets are seen in the *middle to the right*; positive hyporelief. *5 Skolithos declinatus* Fedonkin ichnosp. nov. Holotype, PIN No. 3994/640, x 0.9. Fine, cylindrical, straight burrows with smooth walls inclined to bedding plane at 45°, burrows are 50 mm long and 2–3 mm in diameter. The number of burrows attains 2–3 cm^{-2}. *6 Bilinichnus simplex* Fedonkin et Palij. GIN No. 4482/66, positive hyporelief, x 0.9. *7 Intrites punctatus* Fedonkin. Holotype, GIN No. 4482/81, positive hyporelief, x 0.9

Plate 27. All specimens are from the Zimny Coast of the White Sea, Vendian, Ust Pinega Formation. *1 Medvezhichnus pudicum* Fedonkin, ichnogen. et ichnosp. nov. Holotype, PIN No. 3993/1310, x 0.9. Burrow of circular section preserved in semirelief at the base of sandy silt slab; a peculiar feature is the narrow mid-ridge extending along the basal part of the burrow and limited on both sides by regular, wavy (zigzag) grooves; *upper left*, a large nodular cast of an animal of uncertain systematic position. *2*

Yelovichnus gracilis Fedonkin ichnogen. et ichnosp. nov. Holotype, PIN No. 3993/1309, x 0.9. A narrow trail (positive hyporelief) showing arcuate meanders closely spaced, hence a prominent, wide band of the sediment; each transverse burrow within the band is irregularly undulating, though there are no intact areas in the feed track due to closely packed, transverse burrows. *3 Helminithoida* sp. PIN No. 3993/874, cylindrical burrow preserved in positive hyporelief, locally covering the entire bedding plane (*top right*) due to closely spaced burrows, x 0.9. *4 Neonereites renarius* Fedonkin (holotype) and *Palaeopascichnus delicatus* Palij. Positive hyporelief, GIN No. 4482/163, x 0.9. *5* Feeding tracks close in the mode of formation to *Phycodes palmatum*. GIN No. 4482/49, positive hyporelief, x 0.9

Plate 28. All specimens were recorded from the Letny Coast of the White Sea, Vendian, Ust Pinega Formation. *1* Double crawling trail left by an animal in the course of passing by ventral peristaltics (pedal wave), GIN No. 4464/184, x 0.9. *2 Nenoxites curvus* Fedonkin. Holotype, GIN No. 4310/202-4, x 0.9. *3,6 Planolites cf. serpens* (Webby). Positive hyporelief. *3* GIN No. 4464/173. *6* GIN No. 4464/174, x 0.9. *4,5 Neonereites biserialis* Seilacher. Positive hyporelief. *4* GIN No. 4464/179. *5* GIN No. 4464/180, x 0.9

Plate 29. *1 Tirasiana disciformis* Palij. CGM, No. 1/74, x 0.9; Shirokovskoe Reservoir near Nyar Village, Chusovaya Basin, Sylvitsa Series, Cherny Kamen Formation. *2 Tirasiana cf. coniformis* Palij. CGM, No. 2/74, x 0.9; distribution and age as above. *3 Tirasiana concentralis* Becker. CGM, No. 232/72, x 0.9; distribution and age as above. *4 Tirasiana cocarda* Becker, sp. nov. Holotype, CGM, No. 2/72, x 0.9, distribution and age as above. *5a-c Arumberia banksi* Glaessner and Wade. CGM No. 264b/1969, x 0.9, Chusovaya River Basin, mouth of the Koiva River, Sylvitsa Series, Ust Sylvitsa Formation. *6 Nemiana simplex* Palij, CGM No. 7/76, x 0.9, Shirokovskoe Reservoir near Nyar Village, Chusovaya River Basin, Sylvitsa Series, Cherny Kamen Formation

Plate 30. *1-5 Vendotaenia antiqua* forma tertia Gnilovskaya. Thalli on the rock, core from boreholes: *1* No. L-80-1, x 0.9, East Poland, Leponnic borehole, depth 5448.4–5449.3 m, Upper Vendian, Lyubel Series. *2* No. B-754-4, x 0.9, Moldavia, right side of the Dniester River, Bolotino-1 borehole, depth 754.1-755.2 m, Valdai Series, Kaushany Formation, Kalyus Beds. *3* No. L-80-2, x 0.9, a core surface fragment shown in *4*, East Poland, Lopennic borehole, depth 5448–5449.3 m, Upper Vendian, Lyubel Series. *4* No. L-80-2, core surface, distribution and age as above. *5* No. L-80-1, core surface of the fragment shown in *1*, East Poland, Lopennic borehole, depth 5448.4–5449.3 m, Upper Vendian, Lyubek Series

Plate 31. *1-6 Vendotaenia antiqua* Gnilovskaya. Specimens on the rock, drill core: *1 Vendotaenia antiqua* forma quarta Gnilovskaya, No. B-768-1, x 0.9, Moldavia, right side of the Dniester River, Bolotino-1 borehole, depth 766.2–768.5 m, Mogilev-Podolian Series, Kaushany Formation, Kalyus Beds. *2 Vendotaenia antiqua* forma prima Gnilovskaya, No. L-80-3, x 0.9, East Poland, Lopennic borehole, depth 5448.4–5449.3 m, Upper Vendian, Lyubel Series. *3 Vendotaenia antiqua* forma tertia Gn., No. Dr.-537, x 0.9, Byelorussia, Drissa borehole (Verkhnedvinsk), depth 537–541.9 m, Valdai Series, Kotlin Horizon. *4 Vendotaenia antiqua* forma secunda Gn., No. B-768-2, x 0.9, Moldavia, right side of the Dniester River, Bolotino-1 borehole, depth 766.2–768.5 m, Mogilev-Podolian Series, Kaushany Formation, Kalyus Beds. *5 Vendotaenia antiqua* forma secunda Gn., No. B-676, x 0.9, distribution and age as above; depth 676–679.5 m, Avdarmin, Sokolets Formation. *6 Vendotaenia antiqua* forma prima Gnilovskaya, No. B-679, x 0.9, distribution and age as above; depth 679.0–681.8 m, Avdarmin Series, Sokolets Formation

Plate 32. *1-4 Vendotaenia antiqua* Gnilovskaya. Macerated fragments of thallus with reproductive organs. *1* Slide Rov. 29-4, a fragment of thallus with three sporangia, x 0.9; Rovno-2g borehole, depth 29–32 m, Vendian, Valdai Series, Kotlin Horizon. *2* Slide Rov. 29–1, a fragment of thallus with six sporangia aligned in series, x 0.9; distribution and age as above. *3* Slide Rov. 29–5, sporangia in organic film surface, x 0.9, distribution and age as above

Plate 33. *1-3,5,6 Vendotaenia antiqua* Gn. Macerated thallus fragments. *1* Slide K-19-9, x 0.9, East Poland, Kaplonosy borehole, depth 1731 m; Vendian, Lyublin Series. *2* Slide K-16-8, x 0.9, distribution and age as above. *3* Slide Bol.-145-9, thallus at point of branching, x 0.9, Moldavia, Bolotino-1 borehole, depth 524.7-527.3 m, Vendian, Avdarmin Series. *5* Slide Tor.-842-8, a fragment of thallus covered by scattered, tiny organic accumulations, x 0.9, Toropets borehole, depth 842.3–846.3 m, Vendian, Kotlin

Formation. *6* Slide Ol.-145-1, a fragment of thallus with crossed-strand structure, x 0.9, Moldavia, Bolotino-1 borehole, depth 524.7–527.3 m, Vendian, Avdarmin Series. *4* Ribbonlike fragment of knit filaments. Slide Bol.-130-3, x 0.9, Moldavia, Bolotino-1 borehole, depth 462.3–462.6 m, Avdarmin Series. *7 Orbisiana* Sokolov. Slide Dor. 873-2, macerated fragment of a colony on organic film, x 0.9, Dorogobuzh borehole, depth 873–881 m, Redkino Series, Nelidovo Formation

Plate 34-56. Microfossils are from the Upper Vendian of the northwestern and central East European Platform

Plate 34. *1–4 Tyrasotaenia* Gnilovskaya. *1 T. podolica* Gn., No. 6931-1, thalli on the rock, sample from outcrop, x 0.9, Dniester Basin of Podolia, Kitaigorod Village, left side of the Ternava River, Vendian, Kanilovka Formation. *2 T. tungusica* Gn., slide PV-16-4a-2, a macerated fragment of thallus, x 0.9, Siberian Platform, Turukhan Uplift, Sukhaya Tunguska River, Platonovka Formation, Nemakit-Daldyn Horizon. *3 T. podolica* Gn., slide Bol.-132-2, macerated fragment of thallus longitudinally folded, x 0.9, Moldavia, Bolotino-1 borehole, depth 479.6–482.7 m, Avdarmin Series, upper Ferapontievo Formation. *4 T. podolica* Gn., slide Bol.-135-4, a macerated fragment of thallus, transversely folded, x 0.9, distribution and age as above

Plate 35. *1 Aataenia reticularis* Gn., slide A-138-7, multiseriate, multicellular thallus of sapropel film, x 0.9, Estonia, Kohtla-Järve, Aa F-165 borehole, depth 138–140 m, Vendian, Kotlin Horizon. *2–4,6 Eoholynia mosquensis* Gn. Macerated fragments of thalli. *2* Slide Sol. 1983-2, monopodial branching of a major axis of thallus, x 0.9, Soligalich borehole, depth 1983–1990 m, Vendian, Redkino Horizon. *3* Slide Ros. 1604-2, a point of thallus branching with several major branches, x 0.9, Roslyatino borehole, depth 1605–1608 m, the same age. *4* Slide Sol. 1983-1, a portion of a major branch of thallus with sporangia on sapropel film, x 0.9, Soligalich borehole, depth 1983–1990 m, the same age. *6* Slide Gal. 3641-13, a portion of a major branch with side branch, x 0.9, Galich borehole, depth 3641–3642 m, Redkino Series. *5 Dvinia fibrosa* Gn., slide Dr. 454-1, a fragment of macerated thallus, x 0.9, Byelorussia, Drissa borehole (Verkhnedvinsk), Lower Cambrian, Baltic Series, Lontova Horizon

Plate 36. *1a,b,3,5 Caudinia cauda* Gn. Macerated thalli. *1* Holotype, slide Dor. 813-1: a general view of the organism, x 0.9; b upper part, x 50, Dorogobuzh borehole, depth 813.3–816 m, Redkino Series, Borodino Formation. *3* Slide Dor. 813-2, general view of organism, upper part bent down and twisted at a right angle, x 0.9, distribution and age as above. *5* Slide 813-3, x 0.9, lower end of organism, distribution and age as above. *2 Vendotaenia* sp. Slide Tor. 842-1, a fragment of thallus with sporangia aligned in longitudinal rows, x 0.9, Toropets borehole, depth 842.3–846.3 m, Povarov Series, Seliger Formation. *4 Sarmenta cepitula* Gnilovskaya. Slide Vor. 1029-1, tubular epibionts on the substrate, x 0.9, Vorobyevo borehole, depth 1029.5–1033.5 m, Povarov Series, Makariev Formation

Plate 37. *1–6 Primoflagella speciosa* Gnilovskaya. Macerated organisms on thalli of *Vendotaenia* and on organic films. *1* Slide Rov. 27-5, sporangia, sometimes sporangiophores are seen, x 0.9, amplitude microscope; Rovno-2 borehole, depth 27.2–28 m, Valdai Series, Kotlin Horizon. *2* Slide Kupa-330-2, site 1, mycelium, sporangiferous hyphas, sporangia, mycelium branching is seen, x 0.9, positive phase contrast, Byelorussia, Kupa borehole, depth 330–338 m, Valdai Series, Kotlin Horizon. *3* Slide Kupa-330-5, intensely branching mycelium, x 0.9, positive phase contrast; distribution and age as above. *4* Slide Kupa-330-2, site 2, mycelium, sporangiferous hyphas, sporangia, x 0.9, positive phase contrast; distribution and age as above. *5* Slide Rov.-2-27-5, mycelium, sporangiferous hyphas and sporangia, hypha branching easily discernible on thallus margin of *Vendotaenia*, x 0.9, positive phase contrast; Rovno borehole, depth 27.2–28 m, age as above. *6* Slide Dr. 537-3, sporangia poorly visible, sporangiferous hyphas, x 0.9, amplitude microscope, Drissa borehole, depth 537.0–541.9 m, age as above

Plate 38. *1,3,6 Primoflagella speciosa* Gnilovskaya. Sporangia of *Primoflagella speciosa* Gn. *1* Slide Kupa-330-3, mature sporangium with numerous spores(?) aligned in chains, indistinct contacts between spores in chains in the middle portion of sporangium, x 0.9, interference phase contrast, negative image; Byelorussia, Kupa borehole, depth 330–338 m, Valdai Series, Kotlin Horizon. *3* Slide Kupa 330-13, sporangium and sporangiferous hypha, spores(?) within sporangium aligned in series, one above the other, tubular sporangiferous hypha, x 0.9, electron microscope, glutaraldehyde treatment, distribution and age as above. *4* Slide Kupa 330-14, spores(?) in sporangium aligned in chains, one above the other, x 0.9, electron microscope, glutaraldehyde treatment, distribution and age as above. *5* Slide Kupa 330-15, racquet mycelium, x 0.9, electron microscope, glutaraldehyde treatment; distribution and age as above. *6* Slide Tor. 428-26, sporangia on sapropel film knit by racquet mycelium, x 0.9, scanning microscope, Toropets borehole, depth 842.3-846.3 m, Povarov Series, Seliger Formation. *2* Spore(?) spilled out of mature sporangia. Slide Kupa 330-5, small elongate bodies occurring in pairs or singly, x 0.9, interference phase contrast, Kupa borehole, depth 330-338 m, Valdai Series, Kotlin Horizon

Plate 39. Acritarchs from the Redkino Horizon, all figures x 0.9. All specimens are from the Redkino borehole, Kalinin District. *1-4, 8-9 Leiosphaeridia minor* (Schep.). *1,3* Slide 174/1, depth 1636.0-1671.0 m. *4* Slide 178/1, depth 1636.0-1671.0 m. *8,9* Several forms, slide 196/1, depth 1678.0-1687.0 m. *5-7 Leiosphaeridia pelucida* (Schep.). Slide 196/1, depth 1678.0-1687.0 m

Plate 40. Acritarchs from the Redkino Horizon, all figures x 0.9. All specimens are from the Redkino borehole, Kalinin District. *1-3,5 Leiosphaeridia aperta* (Schep.) *1* Slide 182/1, depth 1678.0-1687.0 m. *5* Slide 192/1, depth 1636.0-1671.0 m. *4,6,7-9 Orygmatosphaeridium rubiginosum* Andr. *4* Slide 169/1, depth 1636.0-1671.0 m. *9* Slide 175/1, depth 1637.0-1671.0 m. *10 Leiosphaeridia* sp. Slide 178/1, depth 1636.0-1671.0 m

Plate 41. Filamentous forms and acritarchs of the Redkino Horizon, *7* x 330, all others x 1000. All specimens are from the Redkino borehole, Kalinin District. *1,2* Fragments of smooth, filamentous forms (*2* shows round hollows resulting from a solution of framboidal pyrite). *1* Slide 180/1, depth 1636.0-1671.0 m. *2* Slide 173/1, depth 1636.0/1671.0 m. *3 Oscillatorites* sp. Slide 174/1, depth 1636.0/1671.0 m. *4-6* Microfossils of group *Rudnjana* Golub. *4* Slide 188/1, depth 1636.0-1671.0 m. *5-6* Slide 204/1, depth 1716.0-1720.0 m. *7* A fragment of a filamentous form (cellular structure is seen in the *centre*). Slide 170/1, depth 1636.0-1671.0 m. *8,9 Bavlinella faveolata* (Schep.). *8* Slide 169/1, depth 1636.0-1671.0 m. *9* (Group of forms), slide 204/1, depth 1716.0-1720.0 m. *10* A fragment of a filamentous form (easily discernible cellular structure). Slide 173/1, depth 1636.0-1671.0 m

Plate 42. Acritarchs of the Kotlin Horizon, all figures x 1000. *1 Leiosphaeridia minor* (Schep.). Slide 448/1, Estonia, Kaagvere borehole, depth 442.6 m, Kotlin Formation. *2,5-7,9,10 Leiosphaeridia aperta* (Schep.). *2,7* Slide 163/1, Kalinin District, Redkino borehole, depth 1328-1338 m. *5* Slide 1227/2, near Leningrad. *6* Slide 448/1, Estonia, Kaagvere borehole, depth 442.6 m. *9* Slide 157/1, Kalinin District, Redkino borehole, depth 1229.5-1236.0 m. *10* Slide 165/1, Redkino borehole, depth 1338.0-1356.0 m, Kotlin Formation (Laminarite Beds). *3,4 Leiosphaeridia pelucida* (Schep.). *3* Slide 1228/1, near Leningrad. *4* Slide 163/1, Kalinin District, Redkino borehole, depth 1328.0-1338.0 m, Kotlin Formation (Laminarite Beds). *8,13-15 Leiosphaeridia effusa* (Schep.). *8* Slide 491/1, Estonia, Ranna-Pungerja, depth 267.0 m. *13* Slide 1227/1. *14* Slide 1230/2, near Leningrad. *15* Slide 163/2, Kalinin District, Redkino borehole, depth 1328.0-1338.0 m, Kotlin Formation (Laminarite Beds). *11,12 Leiosphaeridia bicrura* Jankauskas. *11* Slide 1227/2, near Leningrad. *12* Slide 2736/1, Latvia, Baltanava borehole, depth 921.5 m, Kotlin Formation (Laminarite Beds)

Plate 43. Acritarchs of the Kotlin Horizon, *6* x 330, all others x 1000. *1-5 Leiosphaeridia effusa* (Schep.). *1* Slide 166/3, Kalinin District, Redkino borehole, depth 1338.0-1356.5 m. *2* Slide 492/1, Estonia, Ranna-Pungerja, depth 274.8 m. *3* Slide 2728/1, Latvia, Baltanava borehole, depth 883.5 m. *4* Slide 491/1, Estonia, Ranna-Pungerja, depth 267.0 m. *5* Slide 448/1, Estonia, Kaagvere borehole, depth 442.6 m, Kotlin Formation (Laminarite Beds). *6 Leiosphaeridia gigantea* (Schep.). Slide 1230/2, near Leningrad, Kotlin Formation (Laminarite Beds)

Plate 44. Acritarchs and spores of the Kotlin Horizon; *1-9* x 2000, *10* x 330, *11-13* x 1000. *1-9 Ambiguaspora parvula* Volk. L-3. Group of spores, slide 2731/2, Latvia, Baltanava borehole, depth 899.6 m, Kotlin Formation. *4* Slide 1244/1. *5* (Holotype), slide 1244/2. *6,7* Slide 1244/4. *8* Slide 1244/5. *9* (Group of spores), slide 1244/4, near Leningrad, Kotlin Formation (Laminarite Beds). *10,13 Leiosphaeridia pelucida* (Schep.). *10* (Group of forms), slide 165/1, Kalinin District, Redkino borehole, depth 1338.0-1356.5 m. *13* Slide 2729/1, Latvia, Baltanava, depth 889.0 m, Kotlin Formation (Laminarite Beds). *11 Orygmatosphaeridium rubiginosum* Andr. Slide 279/1, Estonia, Narva borehole, depth 149.3 m, Kotlin Formation. *12 Leiosphaeridia aperta* (Schep.). A group of forms. Slide 157/1, Kalinin District, Redkino borehole, depth 1229.5-1236.0 m, Kotlin Formation (Laminarite Beds)

Plate 45. Filamentous forms of the Kotlin Horizon; all figures x 0.9. *1–8* Fragments of smooth, filamentous, non-septate forms. *1,4,8* Slide 166/3, Kalinin District, Redkino borehole, depth 1338.0–1356.5 m. *2* (Double filaments), slide 157/1, Kalinin District, Redkino borehole, depth 1229.5–1236.0 m. *3* Slide 1244/4, near Leningrad. *5* Slide 2728/1, Latvia, Baltanava borehole, depth 883.5 m. *6* (A filament unrolled on a spiral), slide 1227/2, near Leningrad. *7* Slide 278/1, Estonia, Narva borehole, depth 144.15 m, Kotlin Formation (Laminarite Beds)

Plate 46. Filamentous forms of the Kotlin Horizon; all figures x 0.9. *1–5,8 Oscillatorites* div. sp. *1* Slide 1227/2, near Leningrad. *2* Slide 279/1, Estonia, Narva borehole, depth 149.3 m. *3* Slide 157/1, Kalinin District, Redkino borehole, depth 1229.5–1236.0 m. *4* Slide 2729/1, Latvia, Baltanava borehole, depth 889.0 m. *5* Slide 2728/1, Latvia, Baltanava borehole, depth 883.5 m. *8* Slide 166/3, Kalinin District, Redkino borehole, depth 1338.0–1356.5 m, Kotlin Formation (Laminarite Beds). *6,7,9* Fragments of filamentous forms with annular bands. *6* Slide 1235/1. *7* Slide 1227/1, near Leningrad. *9* Slide 166/3, Kalinin District, Redkino borehole, depth 1338.5–1356.5 m, Kotlin Formation (Laminarite Beds)

Plate 47. Filamentous forms of the Kotlin Horizon; *1* x 330, all others x 1000. *1–5* Fragments of filamentous forms with longitudinal thickenings, forming an intricate pattern. *1,2* (Upper portion of filament shown in *1*), slide 1227/2, near Leningrad. *3* Slide 278/1, Estonia, Narva borehole, depth 144.5 m. *4* Slide 157/1, Kalinin District, Redkino borehole, depth 1229.5–1236.0 m. *5* Slide 2729/1, Latvia, Baltanava borehole, depth 889.0 m, Kotlin Formation (Laminarite Beds). *6–9* Fragments of smooth, filamentous, non-septate forms. *6,9* Slide 166/3, Kalinin District, Redkino borehole, depth 1338.0–1356.5 m. *7* Slide 279/1, Estonia, Narva borehole, depth 149.3 m. *8* Slide 1244/4, near Leningrad, Kotlin Formation (Laminarite Beds)

Plate 48. Filamentous forms of the Kotlin Horizon; all figures x 1000. *1,2,3* Fragments of filamentous, non-septate forms with strongly tapering and elongated terminations. *1* Slide 1228/1, near Leningrad. *2* Slide 2729/1, Latvia, Baltanava borehole, 889.0 m. *3* Slide 1232/1, near Leningrad, Kotlin Formation (Laminarite Beds). *4,10 Tortunema* sp. *4* Slide 1232/1. *10* Slide 1228/1, near Leningrad, Kotlin Formation (Laminarite Beds). *5,6 Micrhystridium* sp. *1* Slide 1227/1, near Leningrad, Kotlin Formation (Laminarite Beds). *7,8 Micrhystridium tornatum* Volk, Slide 163/1, Kalinin District, Redkino borehole, depth 1328.0–1338.0 m, Kotlin Formation (Laminarite Beds). *9,11 Bavlinella faveolata* Schep. *9* Slide 448/1, Estonia, Kaagvare, depth 442.6 m. *11* (Group of forms), slide 165/1, Kalinin District, Redkino borehole, depth 1338.0–1356.5 m, Kotlin Formation (Laminarite Beds). *12* Bundle of filaments (*left*, a separated filament). Slide 1235/1, near Leningrad, Kotlin Formation (Laminarite Beds)

Plate 49. *Rudnjana* of the Kotlin Horizon, all figures x 0.9. *1–11* Microfossils of group *Rudnjana* Golub. *1,8* Slide 1227/2, near Leningrad. *2,6* Slide 166/3, Kalinin District, Redkino borehole, depth 1338.0–1356.5 m. *3,5,7,10* Slide 1227/1, near Leningrad. *4,11* Slide 2729/1, Latvia, Baltanava borehole, depth 889.0 m. *9* Slide 2728/1, Latvia, Baltanava borehole, depth 883.5 m, Kotlin Formation (Laminarite Beds)

Plate 50. Acritarchs of the Rovno Horizon; all figures x 0.9. *1,2,5,6 Leiosphaeridia aperta* (Schep.). *1* Slide 2723/1, Latvia, Baltanava, depth 850.0 m. *2* Slide 2724/1, Latvia, Baltanava borehole, depth 855.0 m. *5* Slide 1223/1, near Leningrad, Rovno Horizon. *6* Slide 159/1, Kalinin District, Redkino borehole, depth 1190.0–1192.0 m, blue clay. *3,4,7,8 Leiosphaeridia effusa* (Schep.). *3,7* Slide 162/1, Kalinin District, Redkino borehole, depth 1179.0–1190.0 m. *4* Slide 159/1, Kalinin District, Redkino borehole, depth 1190.0–1192.0 m, blue clay. *8* Slide 2721/1, Latvia, Baltanava borehole, depth 844.9 m, Rovno Horizon. *9,11 Leiosphaeridia pelucida* (Schep.). *9* Slide 2721/1, Latvia, Baltanava borehole, depth 844.9 m, Rovno Horizon. *11* Slide 161/2, Kalinin District, Redkino borehole, depth 1179.0–1190.0 m, blue clay. *10 Orygmatosphaeridium rubiginosum* Andr. Slide 162/2, Kalinin District, Redkino borehole, depth 1179.0–1190.0 m, blue clay

Plate 51. Acritarchs of the Rovno Horizon; all figures x 1000. *1–4,7 Micrhystridium tornatum* Volk. *1,2* Slide 1223/1, near Leningrad, Rovno Horizon. *3,4* Slide 2721, Latvia, Baltanava borehole, depth 844.9 m, Rovno Horizon. *7* Slide 160/1, Kalinin District, Redkino borehole, depth 1190.0–1192.0 m, blue clay. *5,6,8–10,15 Leiosphaeridia bicrura* Jankauskas. *5,8* Slide 161/2, depth 1179.0–1190.0 m. *6* Slide 160/1, depth 1190.0–1192.0 m, Kalinin District, Redkino borehole, blue clay. *9–15* Slide 1226/2, near

Leningrad, Rovno Horizon. *10* Slide 2723/1, Latvia, Baltanava borehole, depth 850.0 m, Rovno Horizon. *11,12 Leiosphaeridia* sp. with circular pylome. *11* Slide 1223/1, near Leningrad, Rovno Horizon. *12* Slide 161/2, Kalinin District, Redkino borehole, depth 1179.0–1190.0 m, blue clay. *13,14,16–19 Leiosphaeridia dehisca* Paškevičiene. *13,16,18* Slide 161/2, Kalinin District, Redkino borehole, depth 1179.0–1190.0 m, blue clay. *14* Slide 2724/1, Latvia, Baltanava borehole, depth 855.0 m, Rovno Horizon. *17* Slide 1223/1. *19* Slide 1226/2, near Leningrad, Rovno Horizon. *20,21 Leiosphaeridia minor* (Schep.). Group of forms. *20* Slide 2721/1, Latvia, Baltanava borehole, depth 844.9 m. *21* Slide 1225/1, near Leningrad, Rovno Horizon

Plate 52. Acritarchs and filamentous forms of the Rovno Horizon; *6,7* x 330, all others x 1000. *1 Leiosphaeridia typicus* Hermann. Slide *225s1,* near Leningrad, Rovno Horizon. *2,8,9* Fragments of smooth, non-septate, filamentous forms. *2* Slide 1226/2, near Leningrad. *8* Slide 2721/1, Latvia, Baltanava borehole, depth 844.9 m, Rovno Horizon. *9* Slide 160/1, Kalinin District, Redkino borehole, depth 1190.0–1192.0 m, blue clay. *3,4* Fragments of filamentous forms with annular bands. *3* Slide 1225/1. *4* Slide 1223/1, near Leningrad, Rovno Horizon. *5 Ceratophyton duplicum* Paškevičiene. Slide 1226/2, near Leningrad, Rovno Horizon. *6 Ceratophyton vernicosum* Kirjanov. Slide 1226/2, near Leningrad, Rovno Horizon. *7 Ceratophyton* sp. Slide 1226/2, near Leningrad, Rovno Horizon. *10* A tube with annular bands. Slide 1225/1, near Leningrad, Rovno Horizon. *11 Oscillatorites* sp. Slide 161/2, Kalinin District, Redkino borehole, depth 1179.0–1190.0 m, blue clay

Plate 53. Acritarchs of the Lontova Horizon; all figures x 0.9. *1,2 Granomarginata squamacea* Volk. *1* Holotype, slide 270/1, Estonia, Narva borehole, depth 89.35 m. *2* Slide 473/1, Estonia, Ranna-Pungerja borehole, depth 179.0 m, Lontova Formation. *3 Granomarginata* sp. Slide 2720/1, Latvia, Baltanava borehole, depth 843.5 m, Lontova Horizon. *4,5 Granomarginata prima* Naum. *4* Slide 411/1, Estonia, Uljäste, depth 112.9 m. *5* Slide 473/1, Ranna-Pungerja, depth 179.0 m, Lontova Formation. *6 Pterospermopsimorpha* sp. Slide 442/1, Estonia, Kaagvere borehole, depth 399.8 m, Lontova Formation. *7* Type B. Slide 473/1, Estonia, Ranna-Pungerja borehole, depth 179.0 m, Lontova Formation. *8 Leiomarginata simplex* Naum. Slide 411/1, Estonia, Uljäste borehole, depth 112.9 m, Lontova Formation. *9,10 Leiosphaeridia* sp. 1. *9* Slide 272/1, Estonia, Narva borehole, depth 100.35 m, Lontova Formation. *10* Slide 2720/1, Latvia, Baltanava borehole, depth 843.5 m, Lontova Horizon. *11 Pulvinosphaeridium antiquum* Paškevičiene. Slide 473/1, Estonia, Ranna-Pungerja, depth 179.0 m, Lontova Formation

Plate 54. Acritarchs and filamentous forms of the Lontova Horizon; *4,9* x 330, all others x 1000. *1,4,5 Leiosphaeridia effusa* (Schep.). *1* Slide 312/1, Estonia, Palamuuse, depth 350 m. *4* (Group of forms), slide 444/1, Estonia, Kaagvere, depth 408.8 m. *5* Slide 310/1, Estonia, Palamuuse borehole, depth 337.65 m, Lontova Formation. *2* Type A. Slide 254/2, Estonia, Narva borehole, depth 25.8 m, Lontova Formation. *3 Leiosphaeridia aperta* (Schep.). Slide 411/1, Estonia, Uljäste borehole, 112.9 m, Lontova Formation. *6,7* Fragments of smooth, filamentous, non-septate forms (*7* shows conical termination of the filament). Slide 270/1, Estonia, Narva borehole, depth 89.35, Lontova Formation. *8,9 Tasmanites tenellus* Volk. *8 Upper left* portion of specimen shown in *9*. Slide 413/1, Estonia, Uljäste borehole, depth 125.9 m, Lontova Formation

Plate 55. Acritarchs of Drevlyane, Redkino and Kotlin Horizons of Volyn-Podolia (Aseeva's collection). *1 Leiosphaeridia effusa* (Schep.). Slide 242, x 0.9 Podolia, Gusyatin-11673 borehole, depth 574 m, Mogilev-Podolian Series, Yaryshev Formation, Zinkov Beds. Smooth, flat-spherical vesicle with thickened walls with lancet-shaped plicae, 40–70 μm in size. *2 Leiosphaeridia aperta* (Schep.). Slide 52/1, x 0.9. Podolia, Kurazhin-14671 borehole, depth 171.2–175.0 m, Mogilev-Podolian Series, Yaryshev Formation, Zinkov Beds. Smooth, flat-spherical, thick-walled vesicle with very fine, prominent plicae, 20–40 μm in size. *3,4 Leiosphaeridia parva* Ass. Slide 265, x 0.9, Podolia, Gusyatin-11673 borehole, depth 646 m, Volyn Series, Grushka Formation. Flat-elongate, thick-walled vesicle with fine wrinkles forming fine-tuberculate, worm-shaped surface. Plicae are narrow, sinuous, crested; open by splitting, 18–20 μm in size. *5 Leiosphaeridia pruniformis* Ass. Slide 242, x 0.9, Podolia, Gusyatin-11673 borehole, depth 574 m, Mogilev-Podolian Series, Yaryshev Formation, Zinkov Beds. Flattened, spheromorphic vesicle of circular-ovate outline with thickened walls and granular surface. Plicae are narrow, sculptural, locally crested, arranged around the circumference; 30–50 μm in size. *6a,b Stictosphaeridium sinapticuliferum* Tim. Slide 1066/1, x 0.9, Lvov District, Peremyshlyany borehole, depth 4080.8–4084.4 m, Mogilev-

Podolian Series, Mogilev Formation. Fine vesicles having shagreen surface and ornamented by thin, wavy, irregularly arranged crests, 30–80 μm in size. *7 Leiosphaeridia laccata* (Tim.). Slide 102/2, x 0.9, Volyn, Kolki-130 borehole, depth 325–328 m Chartoryisk Formation. Vesicles are flat, circular, feebly angular in outline, smoothly thickened, with diffused plicae; 35–70 μm in size. *8,9 Orygmatosphaeridium induratum* Ass. *8* Slide 200, x 0.9, Byelorussia, Boguchevsk-2 borehole, depth 744.3 m, Liozno (Smolensk) Formation. *9* Slide 129, x 0.9, Volyn, borehole 300, depth 279–301 m, Chartoryisk Formation. Thick-walled vesicle 40–59 μm in diameter; flattened, circular or ovate in outline with finely pitted surface. Pits are circular, 2–3 μm across. No plicae or wide ribbonlike plicae. *10 Orygmatosphaeridium rubiginosum* Andr. Slide 205, x 0.9, Byelorussia, Boguchevsk-2 borehole, depth 727 m, Gdov Formation. Multilayered, flattened bodies with cavernous, pitted surface. Outline is circular; margin uneven. 40–110 μm in size. The species can be placed into the genus *Stratimorphis* Ass. *11a,b Polycavita concentrica* Ass. Slide 1/3, x 0.9, Dniester Basin of Podolia, exposure in Lomozov Village, Mogilev-Podolian Series, Mogilev Formation, Lomosov Beds. Colonial organism consisting of fine, feebly differentiated, round, concentrically whorled cells 2–10 μm across, coalesced in a common vesicle. Mature cells split, forming annular scars. Binary forms result from metabolism. Colonies are 50–150 μm in size. *12,13 Polycavita bullata* (Andr.) Ass. *12* Slide 610, x 0.9, Podolia, borehole 4269, depth 32.5 m, Mogilev Formation, Lomozov Beds. *13* Slide 249/1, x 0.9, Volyn, Grabov-350 borehole, depth 273.9–278.3 m, Chartoryisk Formation. Colonial organism consisting of semicircular cell 10–30 μm across, coalesced by a common vesicle. Mature cells split, forming annular scars. Colonies are 60–100 μm and above in size. *14–16 Stratimorphis plana* Ass. *14* Slide 257/1, x 0.9, Podolia, Gusyatin-11673 borehole, depth 630 m, Mogilev Formation, Lomozov Beds. *15* Slide 200, x 0.9, Byelorussia, Bogushevsk-2 borehole, depth 744.3 m, Liozno (Smolensk) Formation. *16* Slide 1066/1, x 0.9, Lvov District, Peremyshlyany borehole, depth 4080.8–4084.4 m, Mogilev Formation. Multilayered, discoid bodies, opaque, smooth or feebly undulating with distinct margin; 35–100 μm in size

Plate 56. Filamentous forms from the Redkino Horizon of Volyn-Podolia (Aseeva's collection). *1 Arctacellularia* sp. Slide 232, x 0.9, Volyn, Grabov-350 borehole, depth 186.0–191.4 m, Roznich Formation. Multicellular, individual trichomes consisting of closely packed cells with a thin sheath; cells are 7–10 μm in size. *2 Taenithrichoides jaryschevicus* Ass. Slide 1796/1, x 0.9, Odessa District, Perekrestovo-9 borehole, depth 385.5 m, Yaryshev Formation. Thin walled, smooth or shagree, non-septate translucent tubes consisting of organic substance, 35–100 μm in diameter. Tube thins slightly towards the end. Walls of the tubes are randomly folded. *3 Rogosoopsis* sp. Slide 756/1, x 0.9, North Podolia, borehole 4291, depth 31.5 m, Yaryshev Formation, Zinkov Beds. Ribbonlike, non-septate trichomes folded in transverse plicae. *4,5 Volyniella valdaica* (Schep.) Ass. *4* Slide 242, x 0.9, Podolia, Gusyatin-11663 borehole, depth 574 m, Yaryshev Formation, Zinkov Beds. *5* Slide 15/2, x 0.9, Dniester Basin of Podolia, borehole 14274, depth 124 m, Mogilev Formation, Yampol Beds. Cylindrical spiral of a filament 10–15 μm wide. Coil diameter is 50–100 μm. Folded along both the spiral axis (*5*) and normal to it (*4*). *6 Trachythrichoides ovalis* Hermann. Slide 597/1, x 0.9, North Podolia, borehole 4275, depth 36.5 m, Mogilev Formation, Lomozov Beds. Multicellular trichomes consisting of ellipsoidal, interconnected cells. Growing trichomes disintegrate into separate cells. Cells are 12–17 μm wide and up to 50 μm long. *7 Paleolyngbia catenata* Hermann. Slide 55/1, x 0.9, Podolia, borehole 14671 (Kurazhin), depth 174 m, Yaryshev Formation, Zinkov Beds. Tubular vesicle with thick, smooth wall containing trichome consisting of closely coalesced, opaque cells circular in outline. Apparently, they attach to the substratum by out growths at the end of the tube, 15–20 μm in diameter. *8 Striatella coriacea* Ass. Slide 1656/1, x 0.9, Podolia, Krushanovka-2 borehole, depth 320.0 m, Yaryshev Formation, Zinkov Beds. Straight or slightly curved tubes 50–90 μm wide, consisting of organic substance and having a finely pitted surface; some portions bear transverse bands about 2 μm wide, spaced at 5–7 μm. Tube contains trichome 30–50 μm wide, locally coalesced with the wall by a tubular vesicle. *9,10 Rudnjana* Golub. Slide 241/1, x 0.9, Volyn, borehole 350 (Grabov), depth 247.7–252.3 m, Chartoryisk Formation. Organic fossils of uniaxial, tubular structure with inner cavity, walls are locally fusiformlike, broadened and intensely coloured, tips are always thin. May be considered as stages of vegetative reproduction of filamentous algae. *11 Tubulosa corrugata* Ass. Slide 232, x 0.9, Volyn, borehole 350 (Grabov), depth 186.0–191.4 m, Roznich Formation. Tubular microfossils consisting of organic substance, wall of non-septate tube smoothly thickened, corrugated for the entire length or locally. Tip of tube tapers slightly, orifice is circular. There are specimens with trichomes inside of the tubular vesicle. *12 Polythrichoides* sp. Slide 232, x 0.9, Volyn, borehole 350 (Grabov), depth 186.0–191.4 m, Roznich Formation. Bundles of parallel, closely spaced filaments 10–15 μm wide. Circular holes at the end of tubular filaments are visible in some specimens.

13 Circumiella mogilevica Ass. Slide 263/1, x 0.9, North Podolia, borehole 16941, depth 164 m, Mogilev Formation, Toroidal bodies; vesicle is thin, toroidal, and has an internal toroid with a plicated wall

Plate 57-58. Redkino Horizon, assemblages I-II, magnification x 500. *1 Leiosphaeridia minor* (Schep.). Borehole 41, depth 100-101 m, Arkhangelsk Beds. *2 Leiosphaeridia effusa* (Schep.). Ust Pinega borehole, depth 602-607 m, Syuzma Beds. *3 Leiosphaeridia aperta* (Schep.). Ust Pinega borehole, depth 602-607 m, Syuzma Beds. *4 Leiosphaeridia laccata* (Schep.) As. Borehole 45, depth 133 m, Arkhangelsk Beds. *5 Trachysphaeridium laminaritum* Tim. Borehole 45, depth 75 m, Arkhangelsk Beds. *6 Trachysphaeidium laminaritum* Tim. Borehole 45, depth 215.5-217 m, Tamitsa Beds. *7 Trachysphaeridium bavlense* (Schep.). Borehole 219, depth 287 m, Vaizitsa Beds. *8 Protosphaeridium densum* Tim. Borehole 41, depth 155 m, Lyamtsa Beds. *9 Protosphaeridium densum* Tim. Borehole 41, depth 94-95 m, Arkhangelsk Beds. *10 Stictosphaeridium pectinale* Tim. Borehole 41, depth 109-110 m, Arkhangelsk Beds. *11 Stictosphaeridium sinapticuliferum* Tim. Borehole 41, depth 110-111 m, Arkhangelsk Beds. *12 Orygmatosphaeridium rubiginosum* Andr. Ust Pinega borehole, depth 635-640 m, Syuzma Beds. *13 Orygmatosphaeridium rubiginosum* Andr. Ust Pinega borehole, depth 664.8-669.5 m, Verkhovsk Beds. *14 Orygmatosphaeridium rubiginosum* Andr., borehole 202, depth 241.6 m, Syuzma Beds. *15 Favososphaeridium favosum* Tim. Borehole 41, depth 107-108 m, Arkhangelsk Beds. *16 Favososphaeridium favosum* Tim. Borehole 41, depth 107-108 m, Arkhangelsk Beds. *17 Pterospermopsimorpha* sp. Borehole 41, depth 150 m, Arkhangelsk Beds. *18 Pterospermopsimorpha binata* Tim. Borehole 41, depth 100-101 m, Arkhangelsk Beds. *19 Synsphaeridium bullatum* (Andr.). Ust Pinega borehole, depth 656-664.64 m. *20 Synsphaeridium bullatum* (Andr.). Borehole 41, depth 106-107 m, Arkhangelsk Beds. *21 Synsphaeridium sorediforme* Tim. Borehole 41, depth 136-137 m, Arkhangelsk Beds. *22-23 Volyniella valdaica* As. Borehole 5, depth 213-214 m, Verkhovsk Beds. *24 Rudnjana* Golub. Borehole 45, depth 73 m, Arkhangelsk Beds. *25 Leiothricoides* sp. Ust Pinega borehole, depth 592-597 m, Syuzma Beds. *26 Gloeocapsomorpha* sp. Ust Pinega borehole, depth 635-640 m, Verkhovsk Beds

Plate 59. Redkino Horizon, assemblage III. *1,4,5 Oscillatoriopsis constricta* Tynni et Bonner. Zimmy Coast of the White Sea, Zimnie Gory Beds, x 420. *2,2a,3 Oscillatoriopsis funiformis* sp. nov. Zimny Coast of the White Sea, x 420. *6 Trachysphaeridium bavlense* (Schep.), x 500. *7 Volyniella valdaica* (Schep.) As. Zimny Coast of the White Sea, x 500. *8 Favososphaeridium* sp. Ust Pinega borehole, depth 468 m, x 500. *9 Leiosphaeridia aperta* (Schep.). Borehole 219, depth 281 m, Zimnie Gory Beds, x 500. *10 Bavlinella faveolata* (Schep.). Zimny Coast of the White Sea, x 500. *11 Leiosphaeridia giganthea* (Schep.). Borehole 202, depth 409 m, Zimnie Gory Beds, x 500. *12 Stictosphaeridium sinapticuliferum* Tim. Zimny Coast of the White Sea, 63 m x 500. *13 Kildinella hyperboreica* Tim. Borehole 235, depth 155 m, Zimnie Gory Beds, x 500. *14 Leiosphaeridia minor* (Schep.). Borehole 235, depth 229 m, Zimnie Gory Beds, x 500. *15 Oscillatorites wernadskii* Schep. Zimny Coast of the White Sea, x 500

Plate 60. Kotlin Horizon, assemblage IV. *1 Leiosphaeridia pelucida* (Schep.). Borehole 235, depth 94 m, x 500. *2 Stictosphaeridium sinapticuliferum* Tim. Borehole 235, depth 94 m, x 500. *3 Nucellosphaeridium* sp. Borehole 235, depth 94 m, x 500. *4,6,7 Oscillatoriopsis rhomboidae* sp. nov. Borehole 247, depth 57 m: *4,6* x 1350; *7 x 500*. *5 Leiosphaeridia minor* (Schep.). Borehole 235, depth 94 m, x 500. *8 Oscillatoriopsis* sp. Borehole 235, depth 162 m, x 500. *9 Leiothrichoides* sp. Borehole 235, depth 162 m, x 500. *10 Oscillatoriopsis magna* Tynni et Bonner. Borehole 235, depth 115 m, x 1350

Plates 61-62. All microfossils shown are kept in LitNIGRI, Coll. No. 16

Plate 61. All specimens are from the Cis-Urals in the Bashkiria area, Sergeevsk-800 borehole, depth 2942.5-2946.4 m, Baikibashev Formation, Vendian, Drevlyane (?) Horizon. All x 0.9. *7* x 270, *8* x 320, all others x 1000. *1,2 Arctacellularia ellipsoidea* Herm. Slide 16-800-2942/7, specimens and 1 and 2. *3 Polythrichoides lineatus* Herm. Slide 16-800-2942/2, specimen 2. *4* Ringlike alga Toromorpha. Slide 16-800-2942/101, specimen 1. *5 Arctacellularia aff. doliiformis* Herm. Slide 16-800-2942/3, specimen 1. *6 Pterospermopsimorpha* sp. Slide 16-800-2942/2, specimen 3. *7,8,10,11 Leiosphaeridia incrassatula* Jank. Slide 16-800-2942/8. *8* Holotype, specimen 1. *7,10,11* Other specimens. *9 Baltisphaeridium perrarum* Jank. Holotype, slide 16-800-2942/1, specimen 1

Plate 62. *1,2a,b* Glenobo*trydion*(?) sp. I. Slide 16-800-2942/10, specimen 1, x 0.9; *2* Slide 16-800-2942/10; specimen *2a*, x 0.9; *2b* fragment x 0.9; Cis-Urals in the Bashkiria are a, Sergeevsk-800

borehole, depth 2942.4–2946.4 m, Baikibashev Formation, Vendian, Drevlayne(?) Horizon. *3 Chologloaeopsis* (?) sp. Slide 16–800–2942/2, specimen 4, x 0.9, distribution and age as above. *4,5,7,8* Fragments of tubular algae from group *Rudnjana,* similar to *Omalophyma gracilis* Golub. Slide 16–100/2, specimens 1–4, x 0.9, South Urals, Zigan River, Asha Series, Zigan Formation. *6 Navifusa* (?) sp. Slide 16–800–2942/3, specimen 2, x 0.9, Cis-Urals in the Bashkiria area, Sergeevsk-800 borehole, depth 2942.4–2946.4 m, Vendian, Baikibashev Formation. *9.10 Satka granulosa* Jank. Slide 16–800–2942/2. *9* Holotype, specimen 1. *10* Different specimen, x 0.9, distribution and age as above

Plate 63. *1–4 Baltisphaeridium primarium* Jankauskas. *1* Slide 821/6, x 0.9. *2* Slide 821/15, x 0.9. *3* Slide 821/22, x 0.9. *4* Slide 821/64, x 0.9, Siberian Platform, South Yakutia, Byuk-715 borehole, depth 1968.8 m, Yudomian, upper Kursov Formation. *5,6 Leiovalia tenera* Kirjanov. *5* Slide 813, x 0.9, Siberian Platform, South Yakutia, Byuk-715 borehole, 1987 m. *6* Slide 821, x 0.9, Byuk-715 borehole, depth 1968.8 m, Yudomian, upper Kursov Formation. *7–9 Baltisphaeridium*(?) *strigosum* Jankauskas. *7* x 0.9, Siberian Platform, South Yakutia, Peledui-750 borehole, depth 1835–1837 m. *8,9 Slide 819s23. 7* x 0.9, Byuk-715 borehole, depth 1964.8 m, Yudomian, upper Kursov Formation

Plate 64. *1–3 Baltisphaeridium*(?) *strigosum* Jankauskas. *1* Slide 818/27, x 0.9, Siberian Platform, South Yakutia, Byuk-715 borehole, 1963.7 m. *2* Slide 819/10, x 0.9. *3* Slide 819/21, x 0.9, Byuk-715 borehole, depth 1964.8 m, Yudomian, upper Kursov Formation. *4,9 Polyedryxium neftelenicum* Rudavskaja, x 0.9, East Siberia, middle Lena River, Yudomian, lower Nepa Formation, Bezymyanny Horizon. *5,10 Botuobia vermiculata* Pjatiletov. Slide 805/11, x 0.9, Siberian Platform, South Yakutia, Peledui-750 borehole, depth 1794.5 m, Yudomian, Kursov Formation. *6,7 Polyedryxium pritulae* Rudavskaja, x 0.9, East Siberia, middle Lena River, Yudomian, lower Nepa Formation. *8 Baltisphaeridium* sp. x 0.9, East Siberia, middle Lena River, Lower Cambrian, Moty Formation

Plate 65. *1 Leiothrichoides typicus* Hermann. Slide 805/3, x 0.9, Siberian Platform, South Yakutia, Peledui-750 borehole, depth 1794.5 m, Yudomian, Kursov Formation. *2,3 Baltisphaeridium ciliosum* Volkova, x 0.9, East Siberia, middle Lena River, Nepa-1 borehole, depth 2386–2421 m, Lower Cambrian, Moty Formation. *4 Baltisphaeridium cerinum*(?) Volkova, x 0.9, East Siberia, middle Lena River, Kazarka-4 borehole, depth 2997–3000 m, Lower Cambrian, Moty Formation. *5 Peleduica doliiformica* Pjatiletov. Slide 805/2, x 0.9, Siberian Plaftrom, South Yakutia, Peledui-750 borehole, depth 1794.5 m Yudomian, Kursov Formation. *6 Palaeolyngbya catenata* Hermann. Slide 805/10, x 0.9, Siberian Platfrom, South Yakutia, Peledui-750 borehole, depth 1794.5 m, Yudomian, Kursov Formation. *7,8 Leiothrichoides gracilis* Pjatiletov. *7* Slide 803, x 0.9, Siberian Platform, South Yakutia, Peledui-750 borehole, depth 1786.4 m. *8* Slide 804, x 0.9, Peledui-750 borehole, depth 1794.1 m, Yudomian, Kursov Formation. *9–12 Leiomarginata sinplex* Naumova. *9,10,12* x 0.9, Siberian Platform, Yudoma-Maya Trough, Yudoma Series. *11* Slide 622/16, x 0.9, western East Sayan, Mana Trough, Yudomian, Anastasjina Formation. *13,14 Micrhystridium* sp. 13 x 0.9, Siberian Platform, middle Lena River, Kazarka borehole, depth 2997–3000 m, Yudomian, Moty Formation. *14* Slide 522/403, x 0.9, western East Sayan, Mana Trough, Yudomian, upper Zhistyk Formation. *15 Paracrassosphaera dedalea* Rudavskaja in Trestshetenkova, x 0.9. East Siberia, middle Lena River, Yudomian, Moty Formation. *16–18 Granomarginata prima* Naumova. *16, 17* x 0.9, Siberian Platform, Yudoma-Maya Trough, Yudoma Series. *18* Slide 622/79, x 0.9, western East Sayan, Mana Trough, Yudomian, Anastasjina Formation. *19–20 Bavlinella faveolata* Schepeleva. *19* x 0.9, Siberian Platform, middle Lena River, Chastin No. 1 borehole, depth 2318–2337 m, Yudomian, Moty Formation. *20* x 0.9, Yudoma-Maya Trough, Yudoma Series

Plate 66. *1–3 Leiosphaeridia* div. sip. *1–3* x 0.9, Siberian Platform, South Yakutia, Byuk-715 borehole, depth 1923–1929 m, Yudomian, Kursov Formation. *4,8,11 Polyedryxium neftelenicum* Rudavskaja, x 0.9, East Siberia, middle Lena River, Lower Nepa Formation, Bezymyanny Horizon. *5–7,9,10 Octoedryxium truncatum* Rudavskaja. x 0.9, East Siberia, middle Lena River, Lower Cambrian, Moty Formation (*9* holotype, slide 93/1, Sosnino-1 borehole, depth 2070–2083 m)

Plate 67. *1–11 Leiosphaeridia* div. sp. *1* Slide 821/47. *3* Slide 821/54. *7* Slide 821/18. *8* Slide 821/1. *10* Slide 821/41, x 0.9. *4* Slide 821/14, x 0.9. *5* Slide 821/27. *6* Slide 821/46, x 0.9, South Yakutia, Byuk-715 borehole, depth 1923–1929 m, Yudomian, Kursov Formation. *9* Slide 602/8, x 0.9, Verkhny Vilyuchan-602 borehole, depth 2256 m, Yudomian, Bochugunor Formation

Plate 68. *1–5 Renalcis polymorphus* Masl. *1,4* Slide 8–0,5/76. *2* Slide 8–1,5/76. *3,5* Slide 8–0,6/76, colony section showing contrasting black envelope and light internal part, x 0.9, Anabar Massif, Kotui River, exposure 8 in front of the mouth of the Kotuikan River, Nemakit-Daldyn Horizon, 15 m below Korila Member. *6 Korylophyton inopinatum* Vor. Slide 9–1,0/76, x 0.9, Kotui River, Nemakit-Daldyn Horizon

Plate 69. *1–4 Renalcis polymorphus* Masl. Slide 345, 367, colony section with multilayered envelope (centre of photograph), x 0.9, Sukharikha River, upper Sukharikha Formation. *2 R. polymorphus* Masl. Slide 422/3, x 0.9, exp. 422, 12–15 m below the Korila Member, from bioherms. *3 R. gelatinosus* Korde. Sample 19/9 x 0.9, Rassokha River, Manykai Formation, Korila Member. *5 Gemma inclusa* Luch. Slide 89-52/74. Dense black envelopes with lobate processes, x 0.9, Anabar Massif, B. Kuonamka River, Manykai Formation, Korila Member

Plate 70. *1–3 Gemma inclusa* Luch. Slide 63–15/74. *2,3* Slide 89–57/74, x 0.9, Anabar Massif, B. Kunamka River, Manykai Formation, Korila Member, *1,3* Small, round bodies inside of the envelope imparting a cellular aspect. *2* Feebly torn edge of the envelope; *lower left, Korilophyton inopinatum* (Vor.). *4 Protoortonella flabellata* (Vor.). Slide 3–54.0/72, x 0.9, Kotui River, Nemakit-Daldyn Horizon, Korila Member. *5 Botomaella zelenovii* Korde. Slide 41/72, x 0.9, Anabar Massif, Rassokha River, Manykai Formation, Korila Member. *6–9 Korillophyton inopinatum* Vor. *6* Slide 9–1, 0/76, short filaments or more often their transverse section, circular in shape, Kotui River, Nemakit-Daldyn Horizon, Korila Member. *8* Slide 53/72, 35/72, thallus section resembling bush, dichotomous branching of short threads is seen, Rassokha River, Manykai Formation, Korila Member

Plate 71. *1–3 Obruchevella parva* Reitlinger. Slide No. YaM-1/4, x 0.9, Lena River, above Nokhtuisk Village, Tinnovka Formation. *4 Catenuata solida* Yakshin. Holotype, slide No. YaM-1/6, x 0.9, Lena River, above Nokhtuisk Village, Tinnovka Formation

Plate 72. *1–4* Cluster of trichomes not displaying cellular structure. *1* Slide No. YaM-1/5, x 0.9, *2* Slide No. YaM-1/10, x 0.9, Lena River, above Nokhtuisk Village, Tinnovka Formation. *2* Catenuata solida Yakshin. Slide No. YaM-1/5, x 0.9, Lena River, above Nokhtuisk Village, Tinnovka Formation. *3a,b Oscillatoriopsis nochtuica* Yakshin. Holotype, slide No. YaM-1/10, x 0.9, Lena River, above Nokhtuisk Village, Tinnovka Formation. *3a* Image is focussed onto filament surface, septae between cells are visible. *3b* The same filaments focussed onto trichome profile, sawlike pattern due to compression. *5 Catenuata implicata* Yakshin. Holotype, slide No. YaM-1/4, x 0.9, Lena River, above Nokhtuisk Village, Tinnovka Formation

Plate 73. *1–4 Fistularia volubila* Yakshin. *1* Holotype, slide No. YaM-1/13. *2–4* Slide No. Yam-1/3, x 0.9, Lena River, above Nokhtuisk Village, Vendian, Tinnovka Formation

Plate 74. *1,2 Contextuopsis zonata* sp. nov. *1* Specimen No. 1083/3, x 0.9. *2* Holotype No. 1082/1, x 0.9, Dniester Basin of Podolia, Zinkov Beds. *3,6,7 Volyniella valdaica* Schep. Slide No. 1046/3, *3,6* x 0.9; *7* x 0.9, Dniester Basin of Podolia, Lyadova Beds. *4 Volyniella* sp. Specimen No. 669a, x 0.9, Dniester Basin of Podolia, Lyadova Beds. *5 Volyniella jampolica* As. Specimen No. 1046/3, x 0.9, Dniester Basin of Podolia, Lyadova Beds. *8 Leiothrichoides typicus* Herm. Specimen No. 1046/3, x 0.9, Dniester Basin of Podolia, Lyadova Beds. *9 Siphonophicus* sp. Specimen No. 753, x 0.9, Dniester Basin of Podolia, Komarov Beds. *10 Eoholynia* sp.(?). Specimen No. 643/1, x 0.9, Dniester Basin of Podolia, Zinkov Beds. *11* Filamentous thallus with poorly preserved cellular structure. Specimen No. 1066/2, x 0.9, Dniester Basin of Podolia, Bernashevka Beds. *12 Botoobia wernadskii* (Schep.) Specimen No. 1053, x 0.9; Dniester Basin of Podolia, Lyadova Beds

Plate 75. *1,9,10 Zinkovioides perforata* sp. nov. *1* Holotype, No. 1082/1, x 0.9. *9* Specimen No. 1083/3, x 0.9. *10* Specimen No. 1083/3, x 0.9, Dniester Basin of Podolia, Zinkov Beds. *2–4 Zinkovioides inclusis* sp. nov. *2* Specimen No. 1083/3, x 0.9. *3* Holotype, specimen No. 1083/3, x 0.9. *4* No. 1083/3, x 0.9, Dniester Basin of Podolia, Zinkov Beds. *5,11 Ljadovia exasperata* sp. nov. *5* Holotype, No. 718/1, x 0.9. *11* Specimen No. 718/1, x 0.9, Dniester Basin of Podolia, Lyadova Beds. *6 Leiothrichoides gracilis* Pjat. Specimen No. 696/1, x 0.9, Dniester Basin of Podolia, Kalyus Beds. *7 Siphonophycus kestron* Schopf. Specimen No. 1046/1, x 0.9, Dniester Basin of Podolia, Lyadova Beds. *8,12* Fragments of filamentous algae of cellular structure. *8* Specimen No. 1083/3, x 0.9. *12* Specimen No. 741, x 0.9, Dniester Basin of

Podolia, Zinkov Beds. *13* Spirally coiled trichome. Specimen No. 1083/4, x 0.9, Dniester Basin of Podolia, Zinkov Beds. *14* A fragment of empty, tubular thallus, strongly plicated. Specimen No. 1079/3, x 0.9, Dniester Basin of Podolia, Zinkov Beds. *15,16 Oscillatorites wernadskii* Schep. Slide 736/1. *15* x 0.9. *16* x 0.9, Dniester Basin of Podolia, Sokolets Beds

Plate 76. *1–3 Collumnaefacta usatica* Schenf. Holotype, No. 310/13, North Baikal Highland (Chaya River), Yudomian, Usaty Formation. *1* Longitudinal section of columns, x 0.9, slide. *2* Microstructure, x 0.9, slide. *3* Longitudinal section, x 0.9, polished section. *4–6 Jurusania aldanica* Schenf. Holotype No. 310/9, Aldan Shield (Aldan River), Yudomian, Yudoma Formation. *4 column margins*, x 0.9, slide. *5* Microstructure, x 0.9, slide. *6* Microstructure, x 0.9, slide.

Plate 77. *1,2 Jurusania tuructachica* Schenf. Holotype, No. 310/10, Aldan Shield (Olekma River), Yudomian, Porokhtakh Formation. *1* Microstructure, x 0.9. *2* Character of lateral termination of columns, x 0.9 slide. *3,4 Jurusania sibirica* (Jak.). Holotype, GIN No. 3590/3, Aldan Shield (Aldan River), Yudomian, Yudoma Formation. *3* Character of *column margins*, x 0.9, slide. *4* Microstructure, x 0.9, slide

Plate 78. *1,2 Dgerbia grumulosa* Dol. IGU sample 85/51, longitudinal section of columns, Zherba Formation, Patom Highland, Zhuya River. *1* Character of *column margins*, lamination, microstructure, x 0.9, slide. *2* Character of microstructure, x 0.9, slide

Plate 79. *1 Collumnaefacta minuta* Dol. Holotype, IGU, No. 48, longitudinal section, character of lateral termination, laminae, microstructure, slide, x 0.9, Usolka Formation of the Baikal area, Sokhatiny Creek (tributary to the Lena River). *2 Boxonia zharcovi* (Korol.). Topotype, IGU, sample 339/31, longitudinal section of columns. Shape of columns, character of *column margins* and laminae, microstructure, x 0.9, slide, Patom Highland, Chuya River below Berezovy Creek, lower member of Nokhtuisk Formation

Plate 80. *1 Boxonia knjasevi* Dol. Holotype, IGU, sample 583/159, longitudinal section of columns, character of lateral termination and laminae, microstructure, x 0.9, slide, Patom Highland, Lena River near Nokhtuisk Village, lower member of the Nokhtuisk Formation. *2 Boxonia zharcovi* (Korol.). Topotype, IGU, sample 339/31, longitudinal section of columns, character of laminae and microstructure, x 0.9, slide, distribution and age as above

Plate 81. *1 Dgerbia grumulosa* Dol. Shape of columns, Patom Highland, Zhuya River, Zherba Formation. *2 Collumnaefacta minuta* Dol. Shape of columns, Usolka Formation of the Baikal area, Sokhatiny Creek (tributary to the Lena River). *3 Boxonia knjasevi* Dol. Shape of columns, North Baikal Highland, Chaya River, lower member of the Usaty Formation. *4 Boxonia zharcovi* Dol. Shape of columns, Patom Highland, Chaya River below Berezovy Creek, lower member of the Nokhtuisk Formation

Plate 82. *1*–5 Conophyton gaubitza Kryl. Chichkan Horizon, Lesser Karatau Range, Chbakty Basin (*1,5*) and Ichkele Tau Ridge, Tien Shan (*2–4*); *1–4* x 0.9, *5*, demagnified. *1* Sample 3570/11, longitudinal section of columns (*dark*, chert; *light*, carbonate). *2–4* Sample 4285/24. *2* Longitudinal section of columns. *3* Transverse section; column is composed of chert. *4* Intricate relief on lower surface of column. *5* Large cnophytons in outcrop

Plate 83. *1 Linella avis* Kryl. Sample 73/62, longitudinal section of columns, x 0.9, Tien Shan, Lesser Karatau Range, Chichkan Horizon. *2–5 Patomia ossica* Kryl. Sample 3599/16, Tien Shan, Ichkele tau Ridge Chichkan Horizon. *2,5* Longitudinal section of columns, x 1; *3* Ibid, x 4. *4* Transverse section of columns, x 1

Plate 84. *1,2 Aldania sibirica* (Jak.) Kryl. Topotype, GIN, No. 3599/7, K-265; Aldan River near Uugun Village, Yudoma Formation. *1* Shape and microstructure of laminae, x 0.9 *2 column margins*, x 0.9. *3,4 Patomia aldanica* Kryl. Holotype, GIN, No. 3599/5, K-266. Distribution and age as above. *3* Shape and microstructure of columns, x 0.9. *4 column margins*, x 0.9

Plate 85. *1–3 Jurusania tumuldurica* Kryl. Holotype, GIN, No. 3599/1, K-268, longitudinal section of columns, Aldan River near Tumuldur Island, Yudoma Formation. *1* General view of columns, x 0.9. *Column margins*, x 0.9. *3* Column structure, x 0.9

Plate 86. *1–3 Collumnaefacta vulgaris* Sid. *1* Holotype, sample 7564, slide x 0.9, Lena River, Rysiya Village, Pestrotsvet Formation. *2* Typical specimen, sample 317, slide, x 0.9, Lena River, Nokhtuisk Village, Pestrotsvet Formation. *3* Typical specimen, sample 317, slide, x 0.9, Lena River, Nokhtuisk Village, Pestrotsvet Formation

Plate 87. *1–4 Collumnaefacta elongata* Korol. Holotype, below bauxite unit of Bokson Formation, East Sayan. *1,2* Longitudinal section of columns, x 0.9. *3* Transverse section, x 0.9. *4* Longitudinal section. x 0.9, slide

Plate 88. *1–5 Redkinia spinosa* Sokolov. PIN No. 3993C/1-5, 5x10, Nepeitsino borehole, depth 1417–1426 m, Ust Pinega Formation, Valdai Series, Vendian

Plate 89. All specimens are from Asha Series of the Basa Formation, Gornaya Bashkiria, Ryauzyak Basin. CGM, No. 1/11406. *1 Protodipleurosoma paulus* Becker, x 0.9, sample 176/86. *2 Pseudorhizostomites howchini* Sprigg, x 0.9, sample 354^b/86. *3 Medusinites* sp., x 0.9, sample 358/86. *4 Paliella patelliformis* Fed., x 0.9, sample 354^b/86. *5 Neonereites uniserialis* Seil., x 0.9, sample 354^b/86

References

Aalto KR (1971) Glacial marine sedimentation and stratigraphy of the Toby conglomerate (Upper Proterozoic), southeastern British Columbia, northeastern Idaho and northeastern Washington. Can J Earth Sci 8-7:733-787

Abaimova GP (1978) Anabaritids, the most ancient fossils having carbonate skeleton. In: Sukhov SV (ed) New data on stratigraphy and paleontology of Siberia (in Russian). Tr SNIIGGIMS (Nauka, Novosibirsk) 260:77-83

Adegoke OS (1967) A probable pogonophoran from the Early Oligocene of Oregon. J Paleontol 41-5:126-214

Aksenov EM (1967) On the Vendian complex in the eastern Russian platform (in Russian). Izv Akad Nauk SSSR Ser Geol 9:81-91

Aksenov EM (1974) Main features of stratigraphy of the Upper Precambrian of the northern Russian plate. In: Keller BM (ed) Meeting on the Upper Precambrian (Riphean) of the Russian platform (in Russian). Abstracts. Nauka, Moscow, pp 96-97

Aksenov EM, Igolkina NS (1969) On the recognition of the Redkino Formation (Valdai Series) in the northern Russian platform (in Russian). Izv Vuzov Geol Razvedka 8:22-25

Aksenov EM, Vlasov VV, Volkova SA (1970) On the paleogeography of the Valdai Series of the north-east of the Volga-Kama region. In: Lobov VA and Valeev RN (eds) New data on the geology and oil and gas content of the Volga-Kama region (in Russian). Tr Geol Ins (Kazan) 30:37-52

Aksenov EM, Bukatchuk PD, Drumya AV et al. (1974) On the Vendian deposits of the cis-Dobruja trough (in Russian). Dokl Akad Nauk SSSR 218:163-166

Aksenov EM, Keller BM, Sokolov BS, Solontsov LF, Shulga PL (1978) The general scheme for the Upper Precambrian stratigraphy of the Russian platform (in Russian). Izv Akad Nauk SSR Ser Geol 12:17-34

Akulcheva ZA, Galperova EM, Drobkova EL et al. (1981) The Moty sediments and their equivalents in the Irkutsk Amphitheater. In: Meshkova NP, Nikolaeva IV (eds) Precambrian-Cambrian boundary deposits of the Siberian platform (in Russian). Tr IGiG SF Akad Nauk SSSR (Nauka, Novosibirsk) 475:65-139

All-Union Symposium on Paleontology of the Precambrian and Early Cambrian. Abstracts (1965) In: Sokolov BS (ed) (in Russian). IGiG SF Akad Nauk SSSR, Novosibirsk, 115 pp

Anderson MM (1972) A possible time span for the Late Precambrian of the Avalon Peninsula, southern Newfoundland in the light of worldwide correlation of fossils, tillites and rock units within the succession. Can J Earth Sci 9-12:1710-1726

Anderson MM, Misra SB (1968) Fossils found in the Precambrian Conception Group in southeastern Newfoundland. Nature 220:680-681

Anderson MM, Conway Morris S (1982) A review with description of four unusual forms of the soft-bodied fauna of the Conception and St John's Groups (Late Precambrian), Avalon Peninsula, Newfoundland. In: Proceedings of the Third North American Paleontological Convention, vol 1. Montreal, p 1-8

Andreeva EM (1960) Upper Proterozoic and Lower Paleozoic spore assemblages of the Russian platform. In: XXI IGC Session: Papers presented by Soviet geologists (in Russian). Problem 6, pp 172-175

Andreeva EM (1966) Precambrian plant microfossil assemblages of the USSR (in Russian). Tr VSEGEI (Leningrad) 141:10-17

Andreeva EM (1973a) Guide to the Upper Proterozoic and Lower Paleozoic plant microfossil assemblages of the Russian platform (in Russian). Tr VSEGEI (Leningrad) 195:188–195
Andreeva EM (1973b) Upper Proterozoic and Lower Paleozoic plant microfossils in some areas of the Russian platform. In: Vozzhennikova TF (ed) Microfossils in the most ancient deposits (in Russian). Nauka, Moscow, pp 42–44
Anon (1962) Resolutions of the meeting on the stratigraphy of Upper Vendian deposits of Siberia and the Far East. In: Sokolov BS (ed) (in Russian)IGiG SF Akad Nauk SSSR, Novosibirsk, 9 pp, supplement
Anon (1965) Resolutions of the interdepartmental meeting on the development of unified stratigraphic schemes for the Upper Precambrian and Paleozoic of the Russian platform, 1962. In: Keller BM et al. (eds) (in Russian) VSEGEI, Leningrad, 80 pp
Anon (1972a) Resolutions of the meeting on stromatolites. In: Sokolov BS, Khomentovsky VV (eds) Meeting on the stratigraphy of Vendian deposits of the Siberian platform and its fringe: Resolutions (in Russian). IGiG SF Akad Nauk SSSR, Novosibirsk, pp 27–32
Anon (1972b) Resolutions of the meeting on the stratigraphy of Vendian deposits of the Siberian platform and its fringe. In: Sokolov BS, Khomentovsky VV (eds) (in Russian). IGiG SF Akad Nauk SSSR, Novosibirsk, 44 pp
Anon (1975) Resolutions of the All-Union meeting on microphytolites, stromatolites and microfossils (in Russian). IGiG SF Akad Nauk SSSR, 37 pp
Anon (1979) Resolutions and recommendations of the All-Union meeting on microphytolites of reference sections of the Upper Precambrian of Central Siberia (in Russian). IPI, Irkutsk, 58 pp
Anon (1983) Resolutions of the All-Union stratigraphic meeting on the Precambrian, Paleozoic and Quaternary System of Central Siberia. Part 1. Upper Proterozoic and Lower Paleozoic. In: Krasnov VI (ed) (in Russian). Novosibirsk, 215 pp, supplement
Arén B (1968) Tility wschodniej Polski (in Polish). Przegl Geol 5:221–226
Arkhangelsky AD (1940) On the problem of structure of the Russian platform (in Russian). Dokl Akad Nauk SSSR 2152–154
Arutyunov SL, Gorodnichev VI, Drobot DI et al. (1982) New data on the Late Precambrian stratigraphy of the Anabar-Lena gas field (in Russian). Geol Geofiz 3:41–43
Aseev AA (1974) The most ancient continental glaciations of Europe (in Russian). Nauka, Moscow, 319 pp
Aseeva EA (1968) Palynological grounds for stratigraphic subdivision of the Precambrian and Lower Cambrian terrigenous strata of the Dniester Basin of Podolia. In: Abstracts of the papers presented at the V conference of the young geologists of the Ukraine (in Russian). Naukova Dumka, Kiev, pp 73–74
Aseeva EA (1971) Microfossils of the Yaryshev Formation (Vendian) of Podolia. In: Problems of palynology (in Russian). Naukova Dumka, Kiev, pp 204–212
Aseeva EA (1974) On spiral- and ring-like forms in the Upper Precambrian deposits of Podolia (in Russian). Paleontol Sbornik 11-2:95–98
Aseeva EA (1976) Microfossils and algae from the Upper Precambrian of Volyn-Podolia. In: Ryabenko VA et al. (eds) Paleontology and stratigraphy of the Upper Precambrian and Lower Paleozoic of the south-western part of the East European platform (in Russian). Naukova Dumka, Kiev, pp 40–63
Aseeva EA (1981) On regional micropaleophytological studies of the Upper Precambrian deposits in Volyn-Podolia. In: Byvsheva TN et al. (eds) Biostratigraphic aspects in palinology. Abstracts of papers of the IV All-Union palynological conference (in Russian). ZapSib NIGNI, Tyumen, pp 11–12
Aseeva EA (1983a) Vendian microphytofossils of the Ukraine. In: Velikanov VA et al. (eds) Vendian of the Ukraine (in Russian). Naukova Dumka, Kiev, pp 102–127
Aseeva EA (1983b) Stratigraphic significance of the Late Precambrian microfossils of the south-western part of the East European Platform. In: Ryabenko VA (ed) Stratigraphy and formations of the Precambrian of the Ukraine (in Russian). Naukova Dumka, Kiev, pp 148–156
Aseeva EA, Bratslavsky PF, Velikanov VA et al. (1979) Characteristics of the Vendian sections in the northern Podolia reference region (Ushitsa and Yuzhny Bug River Interfluve). In: Ryabenko VA (ed) Problems of Precambrian geology and stratigraphy of the Ukraine (in Russian). Naukova Dumka, Kiev, pp 112–157
Asklund B (1960) Studies in the thrust region of the southern part of the Swedish Mountain chain. XXI IGC Session. Norden Guide Excurs Nos A24, C19, p 60

Atlas of the lithofacies maps of the Russian platform, parts 1, 2 (1952). In: Barkhatova VP et al. (eds) (in Russian). VNIGRI, Leningrad (Russian)
Atlas of the lithopaleogeographic maps of the Russian platform and its geosynclinal margins. Part 1: Late Precambrian and Paleozoic (1961). In: Vinogradov AP (ed) (in Russian). Gosgeoltekhizdat, Moscow-Leningrad, 56 pp
Atlas of the lithopaleogeographic maps of the USSR, vol 1 (1968). In: Vinogradov AP (ed) (in Russian). Nedra, Leningrad
Avdeeva OD, Dragunov VI, Kovtun AS et al. (1978) Late Precambrian stratigraphy of the Botuobiya gas field. In: Khomentovsky VV (ed) New data on stratigraphy and paleontology of the Late Precambrian of the Siberian platform (in Russian). IGiG SF Akad Nauk SSSR, Novosibirsk, pp 76-84
Avdeeva VI (1982) New data on stratigraphy of oil- and gas-bearing Vendian and Lower Cambrian deposits of south-western Yakutia. In: Geology and mineral resources of Yakutia (in Russian). Yakut Knizh Izd, Yakutsk, pp 9-13
Awramik SM (1976) Gunflint stromatolites: microfossil distribution in relation to stromatolite morphology. In: Walter MR (ed) Stromatolite developments in sedimentology. Elsevier, Amsterdam, pp 311-320
Awramik SM, Semikhatov MA (1979) The relationship between morphology, microstructure and microbiota in three vertically integrading stromatolites from the Gunflint Iron Formation. Can J Earth Sci 16-3(1):484-495
Babkova MS (1951) A new formation in a section of ancient strata of the Sayany area (in Russian). Tr Irkutsk Univ Ser Geol 5-2:100-103
Balashova MM, Golub IN, Denisov AI (1973) On biostratigraphic subdivision of the Upper Vendian of the Kama River basin (in Russian). Tr VNIGRI (Kama Branch) 123:151-154
Banks NL (1970) Trace fossils from the Late Precambrian and Lower Cambrian of Finnmark, Norway. Geol J, Spec Issue 3:19-34
Basharin AK (1967) East Verkhoyanie epicratonic geosyncline (in Russian). Nauka, Moscow, 200 pp
Bassot JP, Bonhomme M, Roques M, Vachette M (1963) Mesures d'ages absolus sur les séries precambriennes et paléozoiques du Sénégal oriental. Bull Soc Geol Fr 5:401-405
Becker YR (1968) Late Cambrian molasse of the southern Urals (in Russian). Nedra, Leningrad, 158 pp
Becker YR (1975) On the recognition of the Bakeevo deposits in the Riphean stratotype (in Russian). Izv Akad Nauk SSSR Ser Geol 6:125-129
Becker YR (1977) The first paleontological findings in the Riphean of the Urals (in Russian). Akad Nauk SSSR Ser Geol 3:90-100
Becker YR (1978) Urals-Timan fold area (in Russian). Tr IGiG SF Akad Nauk SSSR 420:92-117
Becker YR, Keller BM, Kozlov VI, Rotar AF (1979) Riphean stratotype. In: Keller BM and Semikhatov MA (eds) Upper Proterozoic stratigraphy of the USSR (Riphean and Vendian) (in Russian). Nauka, Leningrad, pp 71-85
Bednarczyk W (1970) Trilobites fauna of the lower Paradoxides oelandicus stage from the Brzechow area in the western part of the Swietokrzyskie Mts. Bull Acad Pol Sci Sér Géol 18:29-35
Beklemishev VN (1964) Principles of comparative anatomy of invertebrates, 3rd edn, vols 1 and 2. Univ Chicago Press, Chicago (English translation in 1969 of Russian original)
Bell M (1976) Phylogenetic implications of ontogenetic development in the class Edrioasteroidea (Echinodernata). J Paleontol 50-6:1001-1019
Bergström J (1973) Organization, life and systematics of trilobites. Fossils Strata 2:1-69
Bergström J (1979) Morphology of fossil arthropods as a guide to the phylogenetic relationship. In: Gupta AP (ed) Arthropod phylogeny. Van Nostrand Reinhold New York, 762 pp
Bertrand-Sarfati J, Coby R (1974) Précisions sur l'âge précambrien terminal (vendien) de la série carbonatée à stromatolites du groupe d'Eleonora Bay (Groenland oriental). C R Acad Sci D 728-18:2267-2270
Bessonova VY, Chumakov NM (1968) On glacial deposits in the Upper Precambrian of Byelorussia (in Russian). Dokl Akad Nauk SSSR 178-4:68-71
Bessonova VY, Chumakov NM (1969) Upper Precambrian glacial deposits of the western areas of the USSR (in Russian). Litologiya Poleznye Iskopaemye 2.73-89
Bessonova VY, Chumakov NM (1979) Orsha depression and western Moscow syneclise. In: Keller BM, Rozanov AY (eds) Stratigraphy of Upper Precambrian and Cambrian deposits of the western East European platform (in Russian). Nauka, Moscow, pp 69-96

Bessonova VY, Narozhnykh LI (1970) Microphytolites of the Upper Precambrian dolomite-terrigenous strata of Byelorussia (in Russian). Dokl Akad Nauk SSSR 14-1:48-52
Bessonova VY, Veretennikov NV (1966) Some features of psephitic material in tilloid deposits of the Mogilev Formation (Upper Proterozoic of East and Central Byelorussia). In: Manykin SS (ed) Stratigraphy, lithology and mineral resources of the BSSR (in Russian). Nauka Tekhnika, Minsk, pp 103-117
Bessonova VY, Gnilovskaya MB, Piskun LV (1979) Fossils of Precambrian-Cambrian boundary deposits of Byelorussia. In: Sokolov BS (ed) Paleontology of the Precambrian and Early Cambrian (Proceedings of the All-Union Meeting, Leningrad, 1976) (in Russian). Nauka, Leningrad, pp 166-173
Bessonova VY, Sinichka AM et al. (1972) Borehole No. 2 of the Orsha depression (in Russian). Nauka Tekhnika, Minsk, 232 pp
Bidgood DRT, Harland WB (1961) Paleomagnetism in some East Greenland sedimentary rocks. Nature 189-4765:633-634
Biju-Duval B, Gariel O (1969) Nouvelles observations sur les phênomènes glaciaires "éocambriens" de la bordure nord de la synéclise de Taoudeni, entre le Hank et le Tanezrouft, Sahara occidental. Palaeogeogr Palaeoclimatol Palaeoecol 6-4:283-315
Birket-Smith SJR (1981a) Is Praecambridium a juvenile Spriggina? Zool Jahrb Anat 106:233-235
Birket-Smith SJR (1981b) A reconstruction of the Precambrian Spriggina. Zool Jahrb Anat 105:237-258
Bjørlykke K, Englund JO, Kirkbusmo LA (1967) Latest Precambrian and Eocambrian stratigraphy of Norway. Nor Geol Unders 251:5-17
Bjørlykke K, Elvsborg A, Hoy T (1976) Late Precambrian sedimentation in the central Sparagmik basin of South Norway. Nor Geol Tidsskr 56:233-290
Blagodatsky AV, Vladimirov AI, Galimova TF et al. (1980) On the correlation of the Upper Proterozoic deposits in the junction zone of the Sayany area and southern Yenisei Range. In: Khomentovsky VV (ed) Reference junction areas of the unified scheme for the Late Precambrian of Central Siberia (in Russian). Nauka, Novosibirsk, pp 113-127
Bland BH (1984) *Arumberia* Glaessner and Walter, a review of its potential for correlation in the region of the Precambrian – Cambrian boundary. Geol Mag 121:625-633
Bobrov AK (1976) On Upper Precambrian and Cambrian stratigraphy of southern Yakutia. In: Bodunov EI (ed) Prospecting and exploration of oil fields in the Yakutian ASSR (in Russian). Yakut Knizhn Izd, Yakutsk, pp 70-83
Bogdanov AA (1968) Editorial notes. In: Keller BM (ed) Upper Proterozoic of the Russian Platform (Riphean and Vendian) (in Russian). MGU, Moscow, 102 pp
Borovko NG (1967) Vendian and Lower Paleozoic of the Polyudov Range, North Urals (in Russian). VSEGEI, Leningrad, 30 pp
Borovko NG, Golub IN (1966) On the age of fossiliferous strata of the Polyudov Range, North Urals (in Russian). Dokl Akad Nauk SSSR 169-4:895-897
Borovko NG, Kell GN, Smirnov YD (1964) Stratigraphy, depositional environment and diamond content of the Churochnaya Formation (North Urals). In: Vinogradskaya GM (ed) Materials on geology of the Urals (in Russian). Tr VSEGEI (Leningrad) 109:83-98
Bozhko NA, Piekhova NG, Raaben MS (1974) Biostratigraphy of the Upper Precambrian in Africa (in Russian). Dokl Akad Nauk SSSR 214:643-646
Brookfield ME (1987) Lithostratigraphic correlation of Blaine Formation (Late Proterozoic, Lesser Himalaya, India) with other Late Proterozoic tillite sequences. Geol Rundsch 76-2:477-484
Bruns EP (1957) Stratigraphy of the ancient pre-Ordovician deposits of the western Russian platform (in Russian). Sov Geol 59:3-18
Bruns EP (1963) North-western, central and northern areas. In: Keller BM (ed) Stratigraphy of the USSR, vol 2. Upper Precambrian (in Russian). Gosnauchtekhizdat, Moscow, pp 22-46
Bubnoff S (1926) Geologie von Europa. Bd 1. Berlin, 322 S
Bubnoff S (1930) Geologie von Europa. Bd 2. Berlin, 690 S
Bukatchuk PD (1967) Pre-Silurian stratigraphy of Moldavia and adjacent areas of the USSR. In: Paleontology, geology and mineral resouces of Moldavia (in Russian). RIO Akad Nauk MSSR, Kishinev, pp 62-74
Bukatchuk PD (1974) Riphean-Vendian of Moldavia and western Black Sea area. In: Keller BM (ed) Meeting on the Upper Precambrian (Riphean) of the Russian platform. Abstracts (in Russian). Nauka, Moscow, pp 59-64

Bukatchuk PD (1975) Upper Precambrian formations of the sedimentary cover of the south-western margin of the East European platform. Thesis of Cand's dissertation (in Russian). IGN USSR, Kiev, 28 pp

Bulck M (1933) The algal sediments of Andres Island, Bahamas. Philos Trans R Soc Lond B Biol Sci 122:165–192

Butakov EP, Karpinsky RB, Ustalov VV (1975) Vendian of the Teya-Chapa area (Yenisei Range) (in Russian). Geol Geofiz 8:67–77

Butakov EP, Viks EG, Skorobogatykh PP (1978) Yudomian of the south-western Siberian platform. In: Khomentovsky VV (ed) New data on stratigraphy and paleontology of the Late Precambrian of the Siberian platform (in Russian). IGiG SF Akad Nauk SSSR, Novosibirsk, pp 84–104

Chadwick GH (1930) Subdivision of geologic time. Bull Geol Soc Am 41:47–48

Chapman DM (1966) Evolution of the Scyphistoma. Symp Zool Soc Lond 16:51–75

Chen Jinbiao, Zhang Huimin, Xing Yusheng, Ma Guogan (1981) On the Upper Precambrian (Sinian Suberathem) in China. Precambrian Res 15-314:207-228

Chepikova IK (1966) Paleophytological characteristics of the Upper Precambrian deposits in the eastern Russian platform. In: Neistadt MI (ed) Significance of palynological analysis in stratigraphy and paleofloristics (in Russian). Nauka, Moscow, pp 42–45

Chepikova IK (1971a) On the distribution of acritarchs of the group Minuscula Naum in the Upper Proterozoic of the eastern Russian platform. In: Aliev MM (ed) Biostratigraphy of the Mesozoic and Paleozoic of oil- and gas-bearing areas of Central Asia, western Siberia and Russian platform (in Russian). Nauka, Moscow, pp 87–92

Chepikova IK (1971b) Vendian acritarchs of the Volga-Urals oil- and gas-bearing areas and their stratigraphic significance. In: Aliev MM, Chepikov KR (eds) Spore and pollen in oil and rocks of oil- and gas-bearing areas of the USSR (in Russian). Nauka, Moscow, pp 60–70

Chepikova IK (1971c) On the problem of stratigraphic position of the Upper Bavly deposits in the Volga-Urals area (in Russian) Dokl Akad Nauk SSSR 197-3:671-673

Chepikova IK (1973) On the distribution of the Lower Vendian deposits in the eastern Russian platform. In: Aliev MM (ed) Debatable problems of the stratigraphy of oil- and gas-bearing areas of the Soviet Union (in Russian). Nedra, Moscow, p 114–120

Chibrikova EV (1961) Stratigraphy and age of the pre-Devonian deposits on the eastern margin of the Russian platform. In: Shatsky NS, Rogers D (eds) Cambrian system, its paleogeography and problem of the lower boundary. XX IGC Session. Proceedings (in Russian). Izd Akad Nauk SSSR, Moscow, pp 27–32

Chistyakov VG, Kalmykova NA, Nesov LA, Suslov GA (1984) On the presence of Vendian deposits in the middle Onega River and on the possible presence of Tunicana: Chordata (in Russian). Vestn LGU 6:11–19

Chumakov NM (1965) On some Precambrian tilloid rocks in the USSR (in Russian). Izv Akad Nauk Ser Geol 2:83-101

Chumakov NM (1968) On the character of the Late Precambrian glaciation of Spitsbergen (in Russian). Dokl Akad Nauk SSSR Ser Geol 180:1446-1449

Chumakov NM (1970) On the character of the Upper Precambrian glaciation in Eurasia (in Russian). Byull MOIP, Otd Geol 4:170

Chumakov NM (1971) Vendian glaciation in Europe and the North Atlantic (in Russian). Dokl Akad Nauk SSSR 198-2:419–422

Chumakov NM (1972a) International Symposium on the Precambrian glacial deposits (in Russian). Vestn Akad Nauk SSSR 4:101-102

Chumakov NM (1972b) Late Precambrian glaciation in Europe and some related problems. In: XXIV IGC Session: Papers presented by the Soviet geologists. Problem 1. Precambrian geology (in Russian). Nauka, Leningrad, pp 6–17

Chumakov NM (1974) Laplandian glaciation. In: Yanshin AL (ed) Essay of stratigraphy (in Russian). Nauka, Moscow, pp 1-96

Chumakov NM (1978a) On stratigraphy of Precambrian upper horizons in the southern Urals (in Russian). Izv Akad Nauk SSSR Ser Geol 121:95–107

Chumakov NM (1978b) Precambrian tillites and tilloids (in Russian). Nauka, Moscow, 202 pp

Chumakov NM (1981) Upper Proterozoic glacigenic rocks and their stratigraphic significance. Precambrian Res 13-3/4:373–395

Chumakov NM (1984) Main glacial events of the past and their geological significance (in Russian). Izv Akad Nauk SSSR Ser Geol 7:35–53

Chumakov NM (1985) Glacial events of the past and their geological significance. Palaeogeogr Palaeoclimatol, Palaeoecol 51-1/4:319-346
Chumakov NM, Semikhatov MA (1981) Riphean and Vendian of the USSR. Precambrian Res 15-3/4:229-253
Claesson S (1976) The age of the Ottfjallet dolerites of Särv Nappe, Swedish Caledonides. Geol Foren Stockh Forh 98:370-374
Clark RB (1969) Systematics and phylogeny: Annelida, Echiura, Sipuncula. Vol4. In: Florkin M, Scheer BT (eds) Chemical zoology. Academic Press, London, 68 pp
Clarke IM (1902) *Parapsonema cryptophya*, a peculiar echinoderm from the Intumescenszone (Portage Beds) of western New York. State Mus Bull (NY) 8-39:172-178
Clauer N (1976) Geochemie isotopique du strontium des milieux sedimentaires. Application a la géochronologie de la couverture Ouest-Africaine. Thèse, Strasbourg, 256 pp
Cloud PE (1968) Pre-Metazoan evolution and the origins of the Metazoa. In: Drake ET (ed) Evolution and environment. Yale Univ Press, pp 1-72
Cloud PE, Dardenn M (1973) Proterozoic age of the Bambui Group in Brazil. Bull Geol Soc Am 84:1673-1676
Cloud PE, Glaessner MF (1982) The Ediacarian period and system: Metazoa inherit the Earth. Science 217-4562:783-792
Cloud PE, Semikhatov MA (1969) Proterozoic stromatolite zonation. Am J Sci 267:1017-1061
Cloud PE, Wright J, Clover L (1976) Traces of animal life from 620-million-year-old rocks in North Carolina. Am Sci 64-4:396-406
Coats RP, Preiss WV (1980) Stratigraphic and geochronological reinterpretation of Late Proterozoic glaciogenic sequences in the Kimberley region, western Australia. Precambrian Res 13:181-208
Conway Morris S (1979) Middle Cambrian polychaetes from Burgess Shale of British Columbia. Philos Trans R Soc Lond B Biol Sci 285:227-274
Conway Morris S (1981) Parasites and the Fossil record. Parasitology 82:489-509
Cowie YW (1978) IUGS/IGCP Project 29. Precambrian-Cambrian boundary Working Group in Cambridge. Geol Mag 115:151-152
Cowie YW, Cribb SY (1978) The Cambrian system in contributions to the geologic time scale AAPG. Stud Geol Pol 6:355-362
Cowie YW, Rozanov AY (1973) Report of the International Working Group at the symposium on the Cambrian-Precambrian boundary (in Russian). Izv Akad Nauk SSSR Ser Geol 12:72-82
Cowie YW, Rozanov AY (1982) On the problem of choosing the stratotype of the Precambrian-Cambrian boundary (in Russian). Izv Akad Nauk SSSR, Ser Geol 12:87-93
Cramer FH, Diez MCR (1977) Late Arenigian (Ordovician) acritarchs from cis-Saharan Morocco. Micropaleontology (NY) 23-3:339-360
Crimes TP, Germs GJB (1982) Trace fossils from the Nama Group (Precambrian Cambrian) of southwest Africa (Namibia). Paleontology (Lond) 56-4:890-907
Daily B (1976) New evidence for the Cambrian in southern Australia (in Russian). Izv Akad Nauk SSSR Ser Geol 3:45-52
Datsenko VA, Zhuravleva IT, Lazarenko NP et al. (1968) Cambrian biostratigraphy and fauna of the north-western Siberian platform (in Russian). Nedra, Leningrad, 213 pp
Debrenne F, Naud G (1981) Méduses et traces fossiles supposées précambriennes dans la formation de San Vito, Sarrabus, Sud-West de la Sardaigne. Bull Soc Geol Fr 7 (XXIII)-1:23-31
Delsman HC (1923) Beiträge zur Entwicklungsgeschichte von Porpita. Treubia, Bd III, S 243-266
Demokidov KK, Lazarenko NP (1964) Upper Precambrian and Cambrian stratigraphy and Lower Cambrian trilobites of Central Siberian and Soviet Arctic islandS (in Russian). Tr NIIGA (Nedra, Leningrad) 137:288
Deynoux M, Trompette R, Clauer N, Sougy J (1978) Upper Precambrian and Lowermost Paleozoic correlations in West Africa and in the western part of Central Africa. Probable diachronism of the Late Precambrian tillite. Geol Rundsch 67-2:615-630
Dikenshtein GK (1957) Paleozoic deposits of the south-western Russian platform (in Russian). Gostoptekhizdat, Moscow, 154 pp
Diver WL, Peat CJ (1979) On the interpretation and classification of Precambrian organic-walled microfossils. Geology (Boulder) 7-8:401-404
Dolnik TA (1978) Stromatolites of the Akitkan Series of the north-western Baikal area. In: Raaben ME (ed) Lower boundary of the Riphean and stromatolites of the Aphebian (in Russian). Tr GIN Akad Nauk SSSR (Nauka, Moscow) 312:106-110

Dolnik TA, Vorontsova GA (1972) New data on the age of deposits of the Karagas Series (Sayany area) (in Russian) Dokl Akad Nauk SSSR 204-2:426-429
Dolnik TA, Vorontsova GA (1974) Biostratigraphy of the Upper Precambrian and Lower Cambrian horizons of the North Baikal Upland (in Russian). East Sib, Irkutsk, 96 pp
Dolnik TA, Faizulina ZK, Lysova AL et al. (1982) Correlation of the deposits in the junction zone of the Baikal area and Irkutsk Amphitheater. In: Arutyunov SA (ed) Correlation of the pre-Usolka deposits in the central and southern areas of the Lena-Tunguska oil- and gas-bearing province (in Russian). Irkutsk, pp 27-30
Dow DB, Gemuts I (1969) Geology of the Kimberley region, W. Australia. The East Kimberley. Bull Bur Mines Res Geol Geophys 106:135-146
Downie C, Evitt W, Sarjeant W (1963) Hystricosphaeres and the classification of the acritarchs. Stanford Univ Publ Geol Sci 7-3:1-16
Dzevanovsky YK (1943) Does the Proterozoic occur to the east of the Aldan plate? (in Russian). Izv Akad Nauk SSSR Ser Geol 3:38-51
Edwards MB (1975) Glacial retreat sedimentation in Smalfjord Formation, Late Precambrian, North Norway. Sedimentology 22-1:75-94
Edwards MB (1984) Sedimentology of the Upper Proterozoic glacial record, Vestertana Group, Finnmark, North Norway. Nord Geol unders Bull 394:76
Egorova LI, Savitsky VE (1969) Cambrian stratigraphy and biofacies of the Siberian platform (western Anabar area) (in Russian). Tr SNIIGGiMS Ser Paleontol Stratigr (Nedra, Moscow) 43:408
Eichwald E (1854) Paleontology of Russia, ancient period. 1 Flora of the graywacke, calcareous and copper shale formations of Russia (in Russian). St Petersburg, 245 pp
Equivalents of the Vendian Complex in Siberia (1975) In: Sokolov BS, Khomentovsky VV (eds) Proceedings of meeting on the stratigraphy of Vendian deposits of the Siberian platform (in Russian). Tr IGiG SF Akad Nauk SSSR (Nauka, Moscow) 232:251
Evans AM, Ford TD, Aller JRL (1968) Precambrian rocks. In: Sylvester-Bradley PC, Ford TD (eds) The geology of the East Midlands. Leicester Univ Press, 400 pp
Evseev GF (1955) On the stratigraphy of the pre-Gotland Paleozoic of Podolia (in Russian). Uchen Zap Kharkov Univ 12:67-81
Fairchild IJ, Hambrey MJ (1984) The Vendian succession of northeastern Spitsbergen: petrogenesis of a dolomite-tillite association. Precambrian Res 26-1:111-167
Faizulina ZK, Popova GP (1986) Acritarchs of the Gissyusa Formation. In: Biostratigraphy to the geologic map, scale 1:50 000 (in Russian). SNIIGGiMS, Irkutsk, pp 17-18
Faizulina ZK, Lysova LA, Treshchetenkova AA et al. (1981) Biostratigraphy of the Late Precambrian and Early Cambrian of the Nepa-Botuobiya area (in Russian). Izv Akad Nauk SSSR Ser Geol 2:13-26
Faizulina ZK, Treshchetenkova AA (1979) Algal microfossils of the Lower Paleozoicum (analogous to the Motskaya and Usolskaya suites) of the Batuobian Region. In: Sokolov BS (ed) Paleontology of the Precambrian and Early Cambrian (in Russian). Nauka, Leningrad, pp 163-165
Fedonkin MA (1976a) Traces of multicellular organisms from the Valdai Series (in Russian). Izv Akad Nauk SSSR Ser Geol 4:129-132
Fedonkin MA (1976b) Paleoichnology of the Precambrian and Early Cambrian. In: Sokolov BS (Ed) All-Union Symposium on Paleontology of the Precambrian and Early Cambrian. Abstracts (in Russian) IGiG SF Akad Nauk SSSR, Novosibirsk, pp 23-26
Fedonkin MA (1977) Precambrian-Cambrian ichnocoenoses of the East European platform. In: Crimes TP, Harper JC (eds) Trace fossils 2. Geol J Spec Issue 9:183-194
Fedonkin MA (1978) A new occurrence of non-skeletal Metazoa in the Vendian of the Zimny Coast (in Russian) Dokl Akad Nauk SSSR 239-6:1423-1426
Fedonkin MA (1979) Paleoichnology of the Precambrian and Early Cambrian. In: Sokolov BS (ed) Paleontology of the Precambrian and Early Cambrian (in Russian). Nauka, Leningrad, pp 183-192
Fedonkin MA (1980a) Trace fossils of the Precambrian Metazoa (in Russian). Izv Akad Nauk SSSR Ser Geol 1:39-46
Fedonkin MA (1980b) New representatives of the Precambrian coelenterates in the northern Russian platform (in Russian) Paleontol Zh 2:7-15
Fedonkin MA (1980c) Vendian fauna of the northern Russian platform. In: XXVI IGC Session. Papers presented by the Soviet geologists. Paleontology. Stratigraphy (in Russian). Nauka, Moscow, pp 21-29

Fedonkin MA (1981a) Vendian White Sea biota (Precambrian non-skeletal fauna of the northern Russian platform) (in Russian). Tr GIN Akad Nauk SSSR (Nauka, Moscow) 342:98
Fedonkin MA (1981b) The largest occurrence of the Precambrian fauna (in Russian). Priroda 5:94–102
Fedonkin MA (1982a) A new name of the Precambrian coelenterates (in Russian). Paleontol Zh 2:137
Fedonkin MA (1982b) Precambrian non-skeletal fauna and the earliest stages of metazoan evolution. J Paleontol 56-2:9
Fedonkin MA (1982c) Precambrian soft-bodied fauna and the earliest radiation of invertebrates. In: Third North Amer Paleontol Convention Proc Montreal 1:165–167
Fedonkin MA (1983a) Organic world of the Vendian. In: Scientific and technical achievements. Stratigraphy. Paleontology Vol 12 (in Russian) VINITI, Moscow, 128 pp
Fedonkin MA (1983b) Non-skeletal fauna of the Podolian Dniester basin. In: Velikanov VA et al. (eds) Vendian of the Ukraine (in Russian). Naukova Dumka, Kiev, pp 128–139
Fedonkin MA (1984) Promorphology of Vendian Radialia. In: Ivanovsky AB, Ivanov IB (eds) Stratigraphy and paleontology of the most ancient Phanerozoic (in Russian). Nauka, Moscow, pp 30–58
Fedonkin MA (1985) Promorphology of Vendian Bilateria and the problem of origin of metamerism in Articulata. In: Sokolov BS, Zhuravleva IT (eds) Late Precambrian and Paleozoic problematics (in Russian). Nauka, Moscow, pp 79–92
Firsov LV (1978) Examples of ambiguous isotope dating on glauconites from the Lower Cambrian (in Russian). Geol Geofiz 6:24–30
Firsov LV, Nikolaeva IV, Lebedev YN, Solntseva SM (1971) Composition, origin and absolute age of micaceous minerals in Lower Cambrian blue clays of the Baltic area. In: Yanshin AL (ed) Glauconite in recent, Lower Paleozoic and Precambrian deposits (in Russian). Tr IGiG SF Akad Nauk SSSR (Nauka, Moscow) 144:165–192
Ford TD (1958) Precambrian fossils from Charnwood Forest. Proc Yorkshire Geol Soc 31:211–217
Ford TD (1963) The Precambrian fossils of Charnwood Forest. Trans Leicester Lit Philos Soc 57:57–62
Føyn S (1937) The Eo-Cambrian series of the Tana District, northern Norway. Nor Geol Tidsskr 17:65–164
Føyn S (1967) Dividal-gruppen ("Hyolithus sonen") i Finmark og dens forhold til de eokambrisk-kambriske formasjoner. Nor Geol Unders 249:1–84
Føyn S, Glaessner MF (1979) Platysolenites, other animal fossils, and the Precambrian-Cambrian transition in Norway. Nor Geol Tidsskr 59:25–46
Frakes LA (1979) Climates throughout geologic time. Elsevier, Amsterdam, 310 pp
Fundamentals of paleontology. Sponge, archaeocyatheans, coelenterates, worms (1962) In: Sokolov BS (ed) (in Russian). Izd Akad Nauk SSSR, Moscow, 486 pp
Garris MA (1977) Magmatic and metamorphic stages in the pre-Jurassic history of the Urals and cis-Urals (in Russian). Nauka, Moscow, 235 pp
Garris MA (1983) Late Precambrian geochronology of the western slope of the South Urals and cis-Urals. In: Keller BM, Khomentovsky VV (eds) Riphean stratotype (Stratigraphy. Geochronology) (in Russian). Tr GIN Akad Nauk SSSR (Nauka, Moscow) 377:143–169
Garris MA, Kazakov GA, Keller BM et al. (1964) Geochronologic scale for the Upper Proterozoic (Riphean and Vendian). In: Afanasiev GD (ed) Absolute age of geologic formations (in Russian). Nauka, Moscow, pp 431–455
Garris MA, Postnikov DV, Popov YN et al. (1967) On the age of gabbrois in the Volga-Urals area. In: Afanasiev GD (ed) Proceedings of the XIV Session of the Commission on Absolute age dating of geologic formations (in Russian). Nauka, Moscow, pp 36–44
Geisler AN (1959) On the problem of stratigraphic subdivision and correlation of the Lower Cambrian deposits on the north-western Russian platform (in Russian). Inf Sbornik VSEGEI (Leningrad) 11:5–16
Geologic formations on the sedimentary cover of the Russian platform (1981) In: Igolkina NS (ed) (in Russian). Tr VSEGEI (Nedra, Leningrad), Nov Ser 296:33–34
Germs GJB (1973) Possible Sprigginid worm and a new trace fossil from the Nama Group, South West Africa. Geology (Boulder) 10:69–71
Glaessner MF (1958) New fossils from the base of the Cambrian in South Australia. Trans R Soc S Aust 81:185–188
Glaessner MF (1959) Precambrian Coelenterate from Australia, Africa and England. Nature 183:4673:1472–1473

Glaessner MF (1960) Precambrian fossils from South Australia. In: XXI IGC Session. Proceedings, Copenhagen, pp 59–64
Glaessner MF (1961) Precambrian animals. Sci Am 204-3:72–78
Glaessner MF (1962) Precambrian fossils. Biol Revs 37:467–494
Glaessner MF (1963) Zur Kennthis der Nama-Fossilien Süd-West Africa. Ann Naturhist Mus Wien 66:113–120
Glaessner MF (1969) Trace fossils from the Precambrian and basal Cambrian. Lethaia 2(4): 369–393
Glaessner MF (1971a) Geographic distribution and time range of the Ediacara Precambrian fauna. Geol Soc Am Bull 82-2:509–514
Glaessner MF (1971b) The genus *Conomedusites* Glaessner et Wade and the diversification of the Cnidaria. Palaeontol Z 45-1-2:7–17
Glaessner MF (1972) Precambrian paleozoology. Spec Pap Univ Adelaide Cent Precambrian Res 1:43–52
Glaessner MF (1976) A new genus of Late Precambrian polychaete worms from South Australia. Trans R Soc S Aust C 3:169–170
Glaessner MF (1979a) An echiurid worm from the Late Precambrian. Lethaia 12-2:121–124
Glaessner MF (1979b) Precambrian. In: Robison RA, Teichert C (eds) Treatise on invertebrate paleontology. Geol Soc Am, Univ Kansas Press, pt A, pp A79–A118
Glaessner MF, Daily B (1959) The geology and Late Precambrian fauna of the Ediacara fossil reserve. Rec S Aust Mus (Adelaide) 13-3:369–401
Glaessner MF, Wade M (1966) The Late Precambrian fossils from Ediacara, South Australia. Palaeontology (Lond) 9-4:599–628
Glaessner MF, Wade M (1971) *Precambridium* – a primitive arthropoda. Lethaia 4:71–77
Glaessner MF, Walter MR (1975) New Precambrian fossils from the Arumbera Sandstone, Northern Territory, Australia. Alcheringa 1-1:59–69
Glaessner MF, Preiss WV, Walter MR (1969) Precambrian columnar stromatolites in Australia: morphological and stratigraphic analysis. Science 164-3883:1056–1058
Gnilovskaya MB (1971) The most ancient Vendian water plants on the Russian platform (in Russian). Paleontol J 3:101–107
Gnilovskaya MB (1974) On the stratigraphic implication of vendotaenid. In: Timofeev BV (ed) Microphytofossils of the Proterozoic and Early Paleozoic of the USSR (in Russian). Nauka, Leningrad, pp 58–59
Gnilovskaya MB (1975) New data on the nature of vendotaenid (in Russian). Dokl Akad Nauk SSSR 221-4:953–955
Gnilovskaya MB (1976) The most ancient Metaphyta. In: XXV IGC Session: Papers presented by the Soviet geologists. Paleontology (in Russian). Marine geology. Nauka, Moscow, pp 10–14
Gnilovskaya MB (1979a) The Vendian Metaphyta. Bull Cent Rech Explor. Prod Ell-Aquitaine 1-2:611–618
Gnilovskaya MB (1979b) Vendotaenids. In: Keller BM, Rozanov AY (eds) Paleontology of Upper Precambrian and Cambrian deposits of the East European platform (in Russian). Nauka, Moscow, pp 39–48
Gnilovskaya MB, Mens KA, Paškevičiene LT, Pirrus EA (1979) Horizon subdivision of the Upper Vendian in the type locality. In: Keller BM, Semikhatov MA (eds) Stratigraphy of the Upper Proterozoic of the USSR: Riphean and Vendian (in Russian). (Proceedings of the meeting on the Precambrian geology. Ufa, 1977). Nauka, Leningrad, pp 117–121
Gnilovskaya MB, Istchenko AA, Kolesnikov CM, Korenchuk LV, Udaltsov AP (1988) Vendotaenids of the East European platform (in Russian). Nauka, Leningrad, 140 pp
Gollerbach MM, Kosinskaya EK, Polyansky VI (1953) Blue-green algae, 2nd edn (in Russian). Sov Nauka, Moscow, 651 pp
Golovenok VK, Belova MY (1980) Precambrian microorganisms in cherts of the Anabar uplift (in Russian). Dokl Akad Nauk SSSR 261-3:713–715
Golub IN (1974) Acritarchs from the Upper Bavly deposits of the Cis-Urals area at Perm (in russian). Tr IGiG SF Akad Nauk SSSR 81:23–27
Golub IN (1979) A new group of problematic microforms in the Vendian deposits of the Orsha Depression (Russian platform). In: Sokolov BS (ed) Paleontology of the Precambrian and Early Cambrian (in Russian). Nauka, Leningrad, pp 147–155

Gorokhov IM, Semikhatov MA (1984) Rb and Sr behaviours during formation of sedimentary rocks. Contribution 2. Rb and Sr behaviour in the course of diagenesis, katagenesis and initial metamorphism (in Russian). Litologiya Poleznye Iskopaemye 2:87–109
Gorokhov IM, Clauer N, Varshavskaya ES, Kutyavin EP, Drannik AS (1979) Studies of clay minerals in the Upper Precambrian series of the Ovruch Range related to dating technique of sedimentary rocks. In: Ovchinnikov LN (ed) Isotope geochronology of the Precambrian (in Russian). Ufa, pp 94–96
Gorokhov IM, Clauer N, Kutyavin EP et al. (1982) Rb-Sr systematics in the Vendian-Cambrian boundary deposits on the northwestern East European platform. In: Shcherbak NP (ed) Problems of isotope dating of volcanic and sedimentary processes (in Russian). Naukova Dumka, Kiev, pp 38–40
Gorozhanin VM, Kutyavin EP, Kaliganov BA (1987) Isotope age from glauconite of the Uk Formation. In: Bibikova EV, Glebovitsky VA (eds) Precambrian magmatism, metamorphism and geochronology of the East European platform related to large-scale mapping. Abstracts (in Russian). Petrozavodsk, pp 166–167
Grabau AW (1922) The Sinian system. Bull Geol Soc China 1-1/4:44–88
Grigelis AA (1971) Some problems of regional stratigraphy of the Baltic area (in Russian). Sov Geol 3:65–78
Gritsik VV (1969) Lithostratigraphic section of the Markha reference borehole. In: Zhuravleva IT (ed) Biostratigraphy and paleontology of the Lower Cambrian of Siberia and the Far East (in Russian). Nauka, Moscow, pp 186–201
Guan B, Wu R, Hambrey MJ, Geng W (1986) Glacial sediments and erosional pavements near the Cambrian-Precambrian boundary in western Henan Province, China. J Geol Soc (Lond) 143-2:311–323
Gudymovich SS (1970) On the biostratigraphy, paleontology and lithology of the Anastasievo Series of the Mana trough (NW East Sayany) (in Russian). Thesis of Cand's dissertation. TPI, Tomsk, 23 pp
Guide to excursion. Examination of the Sinian and Sinian-Cambrian boundary, stratigraphic section, Eastern Yangzegorge, Hubei Province (1982) Hubei Geol Bur, 14 pp
Gürich G (1930a) Die bislang ältesten Spuren von Organismen in Südafrica. In: XV IGC Session. C.r. 2:670–680
Gürich G (1930b) Über den Kuibisquartzit in Südwest-Afrika. Z Geol Ges 82:637
Gürich G (1933) Die Kuibis-Fossilien der Nama-Formation von Südwest-Afrika. Palaeontol Z 15:137–154
Hahn G, Pflug HD (1980) Ein neuer Medusen-fund aus dem Jung-Präkambrium von Zentral-Iran. Senckenb Lethaea 60-4/6:449–461
Haller J (1971) Geology of the East Greenland Caledonides. Intersci, NY, 465 pp
Hambrey MJ (1982) Late Precambrian diamictites of northeastern Svalbard. Geol Mag 119-6:527–551
Hambrey MJ (1983) Correlation of the Late Proterozoic tiltites in the North Atlantic region and Europe. Geol Mag 120-13:209–320
Hambrey MJ et al. (eds) (1981) Earth's pre-Pleistocene glacial record. Cambridge Univ Press, Cambridge, 1004 pp
Häntzchel W (1975) Trace fossils and problematica. In: Haas WH et al. (eds) Treatise on invertebrate paleontology, Pt W, vol 1. Boulder & Lawrence, 269 pp
Harland WB (1965) Critical evidence for a great infra-Cambrian glaciation. Geol Rundsch 54-1:45–61
Harland WB (1983) The Proterozoic glacial record. Geol Soc Am Mem 161:279–288
Harland WB, Bidgood DET (1960) Paleomagnetism in some Norwegian sporagmites and the Late Precambrian Ice Age. Nature 184:1860
Harland WB, Cox AV, Glewellyn PG et al. (1982) A geologic time scale. Cambridge Univ Press, Cambridge, 140 pp
Hermann TN (1974) Findings of mass accumulations of trichomes in the Riphean. In: Timofeev BV (ed) Proterozoic and Paleozoic microfossils of the USSR (in Russian). Nauka, Moscow, pp 6–10
Hermann TN (1981) Occurrences of filamentous algae in the Miroedikha Formation (Upper Precambrian) (in Russian). Paleontol J 4:118–121
Heinzilin J de (1964/1965) Pogonophores fossils? Bull Soc belge Geol Paleontol Hydrol 73-3:501–510
Hessler RR, Newman WA (1975) A trilobitomorph origin for the Crustacea. Fossils Strata 4:437–459
Hofmann HJ, Aitken JD (1979) Precambrian biota from the Little Dal Group, Mackenzie Mountains, north-western Canada. Can J Earth Sci 16-1:150–166

Holtedahl O (1961) The "Sporagmite Formation" (Kjerulf) and "Eocambrian" (Brögger) of the Scandinavian Peninsula. In: Shatsky NS, Rogers D (eds) Cambrian system, its paleogeography and problem of the lower boundary, vol 3 (in Russian). Izd Akad Nauk SSSR, Moscow, pp 9–43
Howell BF (1940) Late Proterozoic and Early Cambrian climates. In: XVII IGC Session (1937). Proceedings, vol 6 (in Russian). Gostoptekhizdat, Moscow, pp 7–10
Huckriede R (1967) Archaenectris benderi n. gen. n. sp. (Hydrozoa), eine Chondrophore von der Wende Ordovicium/Silurium aus Jordanien. Geol Palaeontol 1:101–109
Hyman LH (1940) The invertebrates: Protozoa through Ctenophora. McGraw-Hill Book, NY, 726 pp
Igolkina NS (1956) On the age of sand-clay rocks of Zimny Coast of the White Sea. In: Tolstikhina MM (ed) Materials on geology of the European part of the USSR (in Russian). Tr VSEGEI (Leningrad), Nov Ser 14:169–173
Igolkina NS (1959) On the recognition of the Baltic Complex (Lower Cambrian) to the north of the Russian platform (in Russian). Inf Sborn VSEGEI (Leningrad) 11:17–23
Igolkina NS (1961) Precambrian deposits of the sedimentary cover to the north of the Russian platform (in Russian). Inf Sborn VSEGEI (Leningrad) 43:3–10
Ioganson AK, Kropachev AP (1979) Structural features of Vendian deposits of south-eastern Yakutia (Maya-Kyllakh zone) (in Russian). Geol Geofiz 10:28–38
ISC Resolution on the general stratigraphic scale for the Precambrian adopted at the Ufa Conference in 1977 (1981) In: Sokolov BS (ed) Resolutions of the Interdepartmental Stratigraphic Committee (ISC) and its standing commissions (in Russian). 19. Leningrad, pp 5–11
Ishchenko AA (1982) New data on the occurrence of vendotaenian flora in the Dniester Basin (in Russian). Geol J 42–4:109–116
Ishchenko AA, Korenchuk LV (1979) New data on the flora and lithology of the Nagoryany Formation (Upper Precambrian) of Podolia (in Russian). Geol J 39:1–111–117
Isotta Carlos AL, Rocha-Campos AC, Yoshida R (1969) Striated pavement of the Upper Precambrian glaciation in Brazil. Nature 222–5192:466–468
Ivanov AV, Mamkaev YV (1973) Ciliophore worms (Turbellaria), their origin and evolution: essays of phylogeny (in Russian). Nauka, Leningrad, 221 pp
Ivanov AV, Polyansky YI, Strelkov AA (1981) Practical research on the zoology of invertebrates (protozoans, coelenterates, ctenophorans, flatworms, nemertines, round worms), 3rd edn (in Russian). Vyssh Shkola, Moscow, 504 pp
Ivanova EP, Veselovskaya MM, Kirsanov VV (1965) On occurrence of the Volyn Series on the Russian Platform (in Russian). Byul MOIP Otd Geol XL-5:137–146
Ivanovskaya TA, Smelov SB (1976) Materials on isotope age and mineralogy of glauconite of the Upper Riphean of the South Urals (Minjar area) (in Russian). Litologiya Poleznye Iskopaemye 1:147–151
Iwanoff PP (1933) Die embryonale Entwicklung von *Limulus mollucanus*. Zool Jahrb Anat 56:163–348
Jankauskas TV (1980) The most ancient plant microfossils from the Vendian of the Urals in Bashkiria (*Serga microbiota*) (in Russian). Dokl Akad Nauk SSSR 250–6:1434–1436
Jenkins RJF (1981) The concept of an "Ediacaran Period" and its stratigraphic significance in Australia. Trans R Soc S Aust 105–4:179–194
Jenkins RJF, Gehling IG (1978) A review of the frond-like fossils of the Ediacara assemblage. Rec S Aust Mus 17:347–359
Jones D, Tompson I (1977) Echiura from the Pennsylvanian of northern Illinois. Lethaia 10:317–325
Jonin M, Vidal P (1975) Etude geochronologique des granitoids de la Mancellia, Massif Armoricain, France. Can J Earth Sci 12:920–927
Juhner NR, Hirst R (1946) The geology and hydrogeology of the Voltaian basin. Mem Geol Surv 8:207
Kabankov VY, Lopatin BG, Tabunov SM (1970) The Proterozoic group of the southern and eastern slopes of the Anabar uplift. In: Markov FG (ed) Geology of the USSR, v. XVIII. Western part of the Yakutian ASSR (in Russian). Nedra, Moscow, pp 83–90
Kalberg ES (1940) Geologic description of the Onega Peninsula (in Russian). Tr Severn Geol Upravl 4:25–29
Karlova GA (1987) First findings of skeletal fauna in the Turkut Formation of the Olenek uplift (in Russian). Dokl Akad Nauk SSSR 292–1:204–205
Kats KR (1964) Stratigraphy of the Late Precambrian of East Greenland. In: Sheinmann YM (ed) Geology of the Arctic (in Russian). Mir, Moscow, pp 181–212
Kazakov GA, Knorre KG (1970) Geochronology of the Siberian platform, Uchur-Maya region. Ecol Geol Helv 63–1:173–183

Kazakov GA, Knorre KG (1975) Upper Precambrian geochronology of the Uchur-Maya area of the Siberian platform. In: Afanasiev GD (ed) Geologic-radiometric interpretation of discrepancy in age values (in Russian). Nauka, Moscow, pp 192–205
Kazakov GA, Knorre KG, Strizhov VP (1967) New data on the age of the lower formations of the Lower Bavly Series of the Volga-Urals area (in Russian). Geokhimiya 4:482–486
Keller BM (ed) (1963) Upper Precambrian. Stratigraphy of the USSR (in Russian). GONTI, Moscow, 716 pp
Keller BM (1966a) Units of the unified stratigraphic scale for the Precambrian (in Russian). Dokl Akad Nauk SSSR 171-6:1405–1408
Keller BM (1966b) Vendian Complex of the Urals (in Russian). Sov Geol 5:58–69
Keller BM (1968) Upper Proterozoic of the Russian platform (Riphean and Vendian) (in Russian). Izd MGU, Moscow, 101 pp
Keller BM (1969) Imprint of an unknown animal from the Valdai Series of the Russian platform. In: Rozanov AY (ed) The Tommotian stage and the problem of the lower boundary of the Cambrian (in Russian). Tr GIN Akad Nauk SSSR 206:175–176. English translation edited by M.E. Raaben published 1981 for the U.S. Department of the Interior and the National Science Foundation (Washington, D.C.) by Amerind, New Delhi, India, 1981, 359 pp
Keller BM (1973) Vendian, Yudomian and terminal Riphean (Vendomian) (in Russian). Izv Akad Nauk SSSR Ser Geol 1:86–92
Keller BM (1974) Stratotype of the Vendomian in the South Urals (in Russian). In: Yanshin AL (ed) Essays of stratigraphy. Nauka, Moscow, pp 97–101
Keller BM (1982) On the Vendian-Cambrian boundary (in Russian). Izv Akad Nauk SSSR Ser Geol 2:5–12
Keller BM (1983a) Terminal Riphean (Kudash) and lower boundary of the Vendian (in Russian). Izv Akad Nauk SSSR Ser Geol 6:67–78
Keller BM (1983b) Systems of the Upper Precambrian (in Russian). Izv Akad Nauk SSSR Ser Geol 12:54–84
Keller BM, Chumakov NM (eds) (1983) Riphean stratotype. Stratigraphy. Geochronology (in Russian). Tr GIN Akad Nauk SSSR (Nauka, Moscow 377:184
Keller BM, Fedonkin MA (1976) New findings of fossils in the Valdai Series (Precambrian) on the Syuzma River. Izv Akad Nauk SSSR Ser Geol 3:38–44
Keller BM, Jankauskas TV (1980) Microfossils of the Riphean stratotype of the South Urals (in Russian). Izv Akad Nauk SSSR Ser Geol 12:58–68
Keller BM, Krasnobaev AA (1983) Late Precambrian geochronology of the European part of the USSR. Geol Mag 120-4:381–389
Keller BM, Semikhatov MA (1968) Riphean reference sections on continents. In: Stratigraphy and paleontology (in Russian) Nauka, Moscow, pp 5–108
Keller BM, Shulga PL (1978) Explanatory notes to the scheme showing the Upper Precambrian stratigraphy of the Russian Platform (in Russian). Inst Geol Nauk Akad Nauk USSR (Kiev) 4:36
Keller BM, Kazakov GA et al. (1960) New data on stratigraphy of the Riphean Group (Upper Proterozoic) (in Russian). Izv Akad Nauk SSSR Ser Geol 12:26–41
Keller BM, Korolev VG, Semikhatov MA, Chumakov NM (1968) Major features of Late Proterozoic paleogeography of the USSR. In: XXIII IGC Session. Papers presented by the Soviet geologists. Problem 4 (in Russian). Nauka, Leningrad, pp 82–94
Keller BM, Aksenov EM et al. (1974a) Vendomian (terminal Riphean) and its regional subdivisions (in Russian). VINITI, Moscow, 128 pp
Keller BM, Menner VV, Stepanov VA, Chumakov NM (1974b) New findings of Metazoa in the Vendomian of the Russian Platform (in Russian). Izv Akad Nauk SSSR Ser Geol 12:130–134
Keller BM, Krats KO, Mitrofanov FP et al. (1977) Progress in development of the general stratigraphic scale for the Precambrian of the USSR (in Russian). Izv Akad Nauk SSSR Ser Geol 11:16–21
Keller BM, Semikhatov MA, Chumakov NM (1984) Type sections of the upper erathem of Proterozoic. In: XXVI IGC Session. Papers presented by the Soviet geologists. Problem C.05. Precambrian geology (in Russian). Nauka, Moscow, pp 56–76
Khomentovsky VV (1974) Principles of recognition of the Vendian as a system of the Paleozoic. In: Yanshin AL (ed) Essays of stratigraphy (in Russian). Nauka, Moscow, pp 33–70
Khomentovsky VV (1976) Vendian system (in Russian). Nauka, Novosibirsk, 272 pp

Khomentovsky VV (1978) The state of the art of the stratigraphic base for the Late Precambrian in Central Siberia. In: Khomentovsky VV (ed) New data on stratigraphy and paleontology of the Late Precambrian of the Siberian Platform (in Russian). IGiG SF Acad Nauk SSSR, pp 3–21

Khomentovsky VV (1980) Some problems of the Late Precambrian stratigraphy, reference sections and junction areas of the unified scheme for Central Siberia. In: Khomentovsky VV (ed) Reference junction areas of the unified scheme for the Late Precambrian of Central Siberia (in Russian). Nauka, Novosibirsk, pp 3–11

Khomentovsky VV (1984a) Vendian. In: Khomentovsky VV et al. (eds) Phanerozoic of Siberia, vol 1 (in Russian). Vendian, Paleozoic. Nauka, Novosibirsk, pp 5–34

Khomentovsky VV (1984b) Baikalian Complex and Baikalides (in Russian). Geol Geofiz 8:33–40

Khomentovsky VV (1986a) On the stratotype of the lower Cambrian boundary in the Aldan River Basin (in Russian). Geol Geofic 10:3–9

Khomentovsky VV (1986b) The Vendian system of Siberia and a standard stratigraphic scale. Geol Mag 123-4:333–348

Khomentovsky VV, Karlova GA (1986) On the lower boundary of the Pestrotsvetnaya Formation in the Aldan River Basin. In: Khomentovsky VV (ed) The Late Precambrian and Early Paleozoic of Siberia. Siberian platform and outer zone of the Altai-Sayan foldbelt (in Russian). IGiG SF Akad Nauk SSSR, Novosibirsk, pp 3–22

Khomentovsky VV, Pyatiletov VG (1978) On the stratigraphic significance of microfossils of the Siberian platform and fringe areas. In: Khomentovsky VV (ed) New data on stratigraphy and paleontology of the Late Precambrian of the Siberian platform (in Russian). IGiG Siber Branch, Akad Nauk SSSR, pp 49–69

Khomentovsky VV, Repina LN (1965) Lower Cambrian of the type section of Siberia (in Russian). Nauka, Moscow, 200 pp

Khomentovsky VV, Trofimov VR (1980) Vendian of the western Anabar area. In: Khomentovsky VV (ed) New data on stratigraphy of the Late Precambrian of the western Siberian platform and its mountain fringe (in Russian). IGiG SF Akad Nauk SSSR, pp 3–30

Khomentovsky VV, Yakshin MS (1973) On the influence of different conditions of sedimentation on morphological features of microphytolites. In: Betekhtina OA, Zhuravleva IT (eds) Environment and life in the geologic past (in Russian). Nauka, Novosibirsk, pp 140–146

Khomentovsky VV, Shenfil VY, Yakshin MS (1967) Yudomian Complex of Central Siberia. In: All-Union Meeting on Precambrian-Cambrian boundary deposits. Abstracts (in Russian). Ufa, pp 34–37

Khomentovsky VV, Shenfil VY, Yakshin MS (1969a) On the Yudomian Complex of Siberia (in Russian). Geol Geofiz 3:25–33

Khomentovsky VV, Shenfil VY, Yakshin MS, Butakov EP (1969b) Equivalents of the Yudomian Complex in the Sayany area and in the outer belts of the Baikal-Patom fold area (in Russian). Tr IGiG SF Akad Nauk SSSR 51:56–72

Khomentovsky VV, Shenfil VY, Yakshin MS, Butakov EP (1972) Reference sections of Upper Precambrian and Lower Cambrian deposits of the Siberian Platform (in Russian). Nauka, Moscow, 356 pp

Khomentovsky VV, Shenfil VY, Gibsher AS et al. (1978) Geologic structure of the Mana trough and its position in the Sayan-Altai Baikalides (in Russian). Nauka, Novosibirsk, 223 pp

Khomentovsky VV, Didenko AN, Pyatiletov VG (1982) General conclusions on Vendian stratigraphy of the western Anabar area. In: Khomentovsky VV (ed) New data on stratigraphy of the Late Precambrian of Siberia (in Russian). IGiG SF Akad Nauk SSSR, Novosibirsk, pp 3–20

Khomentovsky VV, Shenfil VY, Pyatiletov VG (1987) Main problems of stratigraphy of pre-Usolka deposits of inland areas of the Siberian Platform (in Russian). Geol Geofiz 11:3–11

Kiderlen H (1937) Die Conularien. Über Bau und Leben der ersten Scyphozoa. Neues Jahrb Miner Geol Paleontol Abh (B) 77:113–169

Kielan -Jaworowska Z (1966) Polychaete jaw apparatuses from the Ordovician and Silurian of Poland and a comparison with modern forms. Palaeont Pol 16:152

Kirsanov VV (1966) Riphean and Vendian trappean magmatism in the central and eastern Russian platform (in Russian). Dokl Akad Nauk SSSR 167–1:162–165

Kirsanov VV (1968) On the problem of stratigraphy and correlation of Vendian Complex deposits on the eastern margin of the Russian platform (in Russian). Izv Akad Nauk SSSR Ser Geol 6:86–103

Kirsanov VV (1972) On evidence of Vendian Glaciation on the central Russian platform. In: Zubov IP (ed) Lithology and stratigraphy of Paleozoic and Mesozoic deposits of the Russian platform (in Russian). Nauka, Moscow, pp 39–46

Kiryanov VV (1978) Silurian acritarchs of Volyn-Podolia (in Russian). Naukova Dumka, Kiev, 116 pp

Klevtsova AA (1965) On the problem of correlation of a section of Upper Precambrian deposits of the Russian platform and Urals (in Russian). Dokl Akad Nauk SSSR 165-6:1371–1374

Klevtsova AA (1968) Stratigraphic relations of sections of Upper Precambrian deposits of the Russian platform (in Russian). Izv vuzov. Geol Razvedka 9:10–20

Klevtsova AA (1972a) Main features of evolution of the Russian platform in the Vendian (in Russian). Izv vuzov. Geol Razvedka 8:3–15

Klevtsova AA (1972b) Vendian Complex of the Russian platform and its problematic fossils. In: Zubov IP (ed) Lithology and stratigraphy of Paleozoic and Mesozoic deposits of the Russian and Scythian platforms (in Russian). Nauka, Moscow, pp 4–14

Kokoulin ML (1974) Pre-Mesozoic rocks of the Yakutian uplift (in Russian). Geol Geofiz 6:16–23

Kolosov PN (1975) Upper Precambrian stratigraphy of southern Yakutia (in Russian). Nauka, Novosibirsk, 156 pp

Kolosov PN (1977) Ancient oil-and-gas-bearing strata of the south-eastern Siberian Platform (in Russian). Nauka, Novosibirsk, 89 pp

Kolosov PN, Avdeeva VI (1979) Stratigraphic scheme for the Upper Precambrian of the gas-bearing area in south-western Yakutia (in Russian). Yakut Department of SF Akad Nauk SSSR, Byull, pp 3–7

Kolosov PN, Avdeeva BI, Bakin VE et al. (1980) Stratigraphy of Upper Precambrian and Lower Cambrian deposits of the Botuobiya gas-bearing area in south-western Yakutia. In: Kolosov PN (ed) Stratigraphy of oil and gas deposits of Yakutia (in Russian). Yakut Knizhn Izd, Yakutsk, pp 3–35

Komar VA (1964) Riphean columnar stromatolites of the northern Siberian platform (in Russian). NIIGA, Leningrad, pp 84–105

Komar VA (1966) Upper Precambrian stromatolites of the northern Siberian platform and their stratigraphic significance (in Russian). Tr GIN Akad Nauk SSSR (Nauka, Moscow) 154:122

Komar VA (1979) Classification of stromatolites according to microstructures. In: Sokolov BS (ed) Paleontology of the Precambrian and Early Cambrian (in Russian). Nauka, Leningrad, pp 42–45

Komar VA, Furdui RS (1969) On Cambrian and Yudomian deposits of the Kolyma uplift (in Russian). Dokl Akad Nauk SSSR 184-4:915–916

Komar VA, Raaben ME, Semikhatov MA (1965) Riphean conophytons of the USSR (in Russian), Nauka, Moscow, 125 pp

Komar VA, Krylov IN, Raaben ME, Semikhatov MA, Serebryakov SN, Zhuravleva ZA (1977) Phytolites in Precambrian stratigraphy. In: Sidorenko AV (ed) Correlation of the Precambrian (in Russian). Nauka, Moscow, pp 40–64

Konopleva NG (1976) Subdivision and correlation of the Upper Riphean and Vendian deposits of the Kola Peninsula (in Russian). Izv Akad Nauk SSSR Ser Geol 11:50–61

Kopeliovich AV (1965) Epigenesis of ancient strata of the south-western Russian platform (in Russian). Tr GIN Akad Nauk SSSR (Nauka, Moscow) 121:310

Korde KB (1958) On some species of fossil blue-green algae. In: Sharov AG (ed) Materials on fundamentals of paleontology (in Russian). Moscow: An SSSR 2:113–118

Korde KB (1961) Cambrian algae of the south-eastern Siberian platform (in Russian). Tr PIN Akad Nauk SSSR (Nauka, Moscow) LXXXIX: 146

Korde KB (1973) Cambrian algae (in Russian). Tr PIN Akad Nauk SSSR (Nauka, Moscow) 139:273

Korolev VG, Ogurtsova RN (1982) Correlation of the Vendian-Lower Cambrian boundary deposits in the Talas-Karatau zone (Lesser Karatau Ridge) with reference sections of the East European and Siberian Platforms (in Russian). Izv Akad Nauk SSSR Ser Geol 6:27–36

Korolev VG, Maksumova RA (1984) Precambrian tillites and tilloids of the Tien Shan (in Russian). Ilim, Frunze, 189 pp

Korolev VG, Maksumova RA, Ogurtsova RN (1979) Relations between the Vendian and terminal Riphean (Vendomian) in the Tien Shan and southern Kazakhstan. In: Keller BM, Semikhatov MA (eds) Stratigraphy of the Upper Proterozoic of the USSR: Riphean and Vendian (in Russian). Nauka, Leningrad, pp 184–189

Korolev VG, Kiselev VV et al. (1985) Riphean of the Tien Shan. In: Regular features in crustal structure underlying the Tien Shan and its mineralization (in Russian). Ilim, Frunze, pp 14–23

Korolyuk IK (1960a) Lower Cambrian and Proterozoic stromatolites of the Irkutsk Amphitheater. In: Teodorovich GI (ed) Geologic and geochemical studies of oil- and gas-bearing areas in the USSR (in Russian). Izd Akad Nauk SSSR, Moscow, pp 112-161
Korolyuk IK (1960b) Subdivision of the Cambrian and Precambrian of eastern Siberia from stromatolites. In: Shatsky NS (ed) Stratigraphy of the Late Precambrian and Cambrian (in Russian). Izd Akad Nauk SSSR, Moscow, pp 118-124
Korolyuk IK (1966) Microproblematics of the Riphean and Lower Cambrian of the Baikal area and the Angara-Lena trough. In: Problems of micropaleontology (in Russian). 10:174-198
Korolyuk IK, Sidorov AD (1969) Stromatolites of the Moty Formation in the southern Baikal area and south-eastern Sayan area (in Russian). Dokl Akad Nauk SSSR 183-3:669-671
Korolyuk IK, Sidorov AD (1971) Stromatolites of the southern Baikal area (in Russian). Izv Akad Nauk SSSR Ser Geol 11:112-123
Kotyk VA, Markovsky BM (1974) Upper Proterozoic deposits of the submerged part of the Volyn-Podolia margin of the Russian Platform. In: Keller BM (ed) Meeting on the Upper Precambrian (Riphean) of the Russian platform. Abstracts (in Russian). Nauka, Moscow, pp 70-73
Kozhemyakina IA (1968) Mogilev deposits of north-eastern Byelorussia and their possible use for underground gas storage (in Russian). Tr Soyuzburgas 6:202-217
Kozhemyakina IA, Chumakov NM (1969) Buried relief at the base of the Volyn Series of Byelorussia (Upper Precambrian) (in Russian). Dokl Akad Nauk SSSR 188-4:885-888
Kozlov VI (1982) Upper Riphean and Vendian of the South Urals (in Russian). Nauka, Moscow, 128 pp
Kozlowski R (1956) Sur quelques appareils masticateurs des Annélidês et Polychètes ordoviciens. Acta Palaeontol 1-3:165-210
Kozlowski R (1967) Sur certains fossiles ordoviciens à test organique. Acta Palaeontol Pol 12-2:221-233
Kozlowski R (1968) Nouvelles observations sur les Conularies. Acta Palaeontol Pol 8:497-529
Krandievsky VS, Ishchenko TA, Kiryanov VV (1968) Paleontology and stratigraphy of the Lower Paleozoic of Volyn-Podolia (in Russian). Naukova Dumka, Kiev, 124 pp
Krasheninnikova OV (1956) Ancient formations of the western slope of the Ukrainian crystalline shield (in Russian). Izd Akad Nauk USSR, 194 pp
Krasilshchikov AA (1973) Stratigraphy and paleotectonics of the Precambrian-Early Paleozoic of Spitsbergen (in Russian). Tr NIIGA (Nedra, Leningrad) 172:121
Krasilshchikov AA, Biterman IM (1970) Proterozoic group of the Olenek uplift. In: Markov FG (ed) Geology of the USSR, vol 18. Western part of the Yakutian ASSR (in Russian). Nedra, Moscow, pp 91-100
Krasnobaev AA (1986) Zircon as an indicator of geologic processes (in Russian). Nauka, Moscow, 147 pp
Krasnobaev AA, Semikhatov MA (1986) Geochronological scale for the Upper Proterozoic (Riphean and Vendian of the USSR): present state of the art. In: Shukolyukov YA, Bibikova EV (eds) Methods of isotope geology and geochronological scale (in Russian). Nauka, Moscow, pp 159-183
Krasnobaev AA, Ronkin YL, Lepikhina OP, Mezentseva VN (1979) Rb-Sr dating of Vendian deposits. In: Ovchinnikov LA (ed) Isotope geochronology of the Precambrian (in Russian). Ufa, pp 78-79
Krasnobaev AA, Stepanov AI, Ronkin YL, Lepikhina OP (1984) Age of granitoids of the Troitsky Massif and the upper radiometric boundary of Precambrian in the Urals. Izv Akad Nauk SSSR Ser Geol 11:128-131
Krasnobaev AA, Ronkin YL, Lepikhina OP (1986) Rb-Sr dating of clay sediments (in Russian). Izv Akad Nauk SSSR Ser Geol 6:3-17
Krivtsov AI (1958) Lower Paleozoic of the eastern slope of the Baltic shield (in Russian). Izv Akad Nauk SSSR Ser Geol 4:63-75
Kröner A (1974) The Gariep Group. Pt. 1: Late Precambrian formations in the western Richtersveld, Norden Cape province. Precambrian Res. Univ Cape Town Bull 13:115
Kröner A (1977) Non-synchroneity of Late Precambrian glaciations in Africa. J Geol 85:289-300
Krylov IN (1967) Riphean and Lower Cambrian stromatolites of the Tien Shan and Karatau (in Russian). Tr GIN Akad Nauk SSSR (Nauka, Moscow) 171:190
Krylov IN (1968) The oldest traces of life on Earth (in Russian). Priroda 11:44-54
Krylov IN (1972) At the outset of life (in Russian). Nauka, Moskow, 104 pp
Krylov IN (1975) Stromatolites of the USSR Riphean and Phanerozoic (in Russian). Nauka, Moscow, 243 pp

Krylov IN, Komar VA (1965) On stromatolites of the Vendian Complex: In: Sokolov BS (ed) All-Union Symposium on paleontology of the Precambrian and Early Cambrian Abstracts (in Russian). Novosibirsk, pp 44–45
Krylov IN, Korolyuk IK, Sidorov AD (1969) Stromatolites. In: Rozanov AY (ed) The Tommotian stage and the problem of the lower boundary of the Cambrian (in Russian). Tr GIN Akad Nauk SSSR 206:195–223 English translation edited by ME Raaben published in 1981 for the U.S. Department of the Interior and the National Science Foundation (Washington, D.C.) by Amerind, New Delhi, India, 1981, 359 pp
Krylov IN, Shapovalova IG, Kolosov PN, Fedonkin MA (1971) Riphean deposits at the lower Lena River (in Russian). Sov Geol 7:85–95
Kulling O (1934) The Hecla Hoek Formation round Hinlopenstredet. Geogr Ann Stockh 16-4:101–162
Kulling O (1942) Grunddragen ov fjallkedjerandens bergbyggnad inom Vasterbottens lan. Sver Geol Unders Arsb Ser C Avh Uppsatsel 36-445:1–320
Kurman IM, Zonov NG, Larin NI (1932) On the problem of formation of the Podolian phosphorite deposits (in Russian). Agron Ores USSR 2:11–24
Kurochka VP (1959) Cycles of sedimentation of ancient Paleozoic deposits in the Dniester area and facies association of clay minerals, microelements and some mineral deposits. Thesis of Cand's dissertation (in Russian). Minsk, 39 pp
Kuteinikov NS, Kuteinikov ES, Voronova LG, Missarzhevsky VV (1973) New data on stratigraphy of Proterozoic-Paleozoic boundary beds on the flanks of the Anabar crystalline massif. In: Zhuravleva IT (ed) Problems of paleontology and biostratigraphy of the Lower Cambrian of Siberia and the Far East (in Russian). Nauka, Novosibirsk, pp 215–219
Lagutenkova NS (1953) Evidence of volcanic activity in the Upper Bavly Series of the Kama Basin at Perm and of the Bashkirian ASSR (in Russian). Dokl Akad Nauk SSSR 150-6:1352–1353
Lagutenkova NS, Chepikova IK (1976) Upper Precambrian deposits of the Shkapovo District and their micropaleophytological characteristics. In: Aliev MM, Chepikov KR (eds) Use of palynology in petroleum geology (in Russian). Nauka, Moscow, pp 31–37
Lamansky VV (1897) Translator's notes to the book by M. Neumayer. History of the Earth. vol 2 (in Russian). St. Petersburg, pp 2–3
Larin NI, Svetozarova TA (1932) Stratigraphy of the Silurian sand-shale sequence in Podolia (in Russian). Agron Ores USSR 1-2:31–38
Leloup E (1954) A propos des Siphonophores. Inst R Sci Nat Belg Mem 2(N 4):643–699 (vol. jubil. Victor van Straelen, 1925–1954)
Loeblich ARJ (1970) Morphology, ultrastructure and distribution of Paleozoic acritarchs. In: Proc North Am Paleontol Convention (Chicago 1969). Lawrence, Kans., pp 705–788
Lu Songnian, Ma Guogan, Gao Zhenjia, Lin Weixing (1983) Classification of ice ages and glacial sedimentary facies – areas of the Sinian in China. In: Zhang Huimin, Chen Jinbiao (eds) International Symposium on Late Precambrian Geology, China. Abstracts.Chin Acad Geol Sci, pp 55–58
Lu Songnian, Ma Guogan, Gao Zhenjia, Lin Weixing (1985) Sinian ice age and glacial sedimentary facies – areas in China. Precambrian Res 29-1:53–63
Luchinina VA (1969) Renalcis polymorphum Maslov from the Yudomian Complex on the Sukharikha River. In: Zhuravleva IT (ed) Biostratigraphy and paleontology of the Lower Cambrian of Siberia and the Far East (in Russian). Nauka, Moscow, pp 184–185
Luchinina VA, Kiryanov VV, Gnilovskaya MB (1978) On the correlation of the Precambrian-Cambrian boundary deposits of the East European and Siberian platforms (in Russian). Dokl Akad Nauk SSSR 240-5:1184–1187
Lungershausen LF (1938) Evolutionary stages of the Podolian platform and its Black Sea slope. In: Porfiriev VB (ed) Proceedings of Petroleum Conference (in Russian). Izv Akad Nauk USSR, Kiev, pp 107–148
Lungershausen LF (1939) Evidence of glacial climate in the Lower Paleozoic of Podolia (in Russian). Dokl Akad Nauk SSSR 22-6:318–321
Lungershausen LF (1947) On the facies and emplacement conditions of ancient formations of the Bashkirian Urals (in Russian). Sov Geol 18:35–74
Lungershausen LF, Nikiforova OI (1942) On stratigraphic relations between the Silurian beds of Podolia and equivalent beds in some other parts of western Europe (in Russian). Dokl Akad Nauk SSSR Nov Ser 34-2:69–79

Mackie GO (1959) The evolution of Chondrophora (Siphonophora-Disconanthae): new evidence from behavioural studies. Proc Trans R Soc Can Ser 3, 53:7–20
Macurda DB (1964) The blastoid Pentremites Say – a trimerous mutant and some new occurrences. J Paleontol 38:705–710
Makhnach AS, Veretennikov NV, Shkuratov VI (1974) Stratigraphic scheme for the Upper Proterozoic of Byelorussia (in Russian). Dokl Akad Nauk BSSR XVIII-5:453–456
Makhnach AS, Veretennikov NV, Shkuratov VI, Bordon VE (1976) Riphean and Vendian of Byelorussia (in Russian). Nauka Tekhnika, Minsk, 360 pp
Malcuit RJ, Winters RR (1980) The Late Proterozoic glaciations: possible product of the evolution of the Earth-Moon system. In: XXVI IGC, Abstracts, Paris, 11:600
Manton SM (1941) On the Hydrozhira and Claspers of the hydroid *Myriothela cocksi* (Vigurs). J Mar Biol Assoc UK 25:143–150
Manton SM (1967) The polychaete *Spinther* and the origin of the Arthropoda. J Nat Hist 1:1–22
Manton SM (1969) Evolution and affinities of Onychophora, Myriapoda, Hexapoda and Crustacea. In: Moore RC (ed) Treatise of invertebrate paleontology. Boulder & Lawrence, Geol Soc Am Univ Kansas, pp 15–56
Manuilova MM (ed) (1968) Precambrian geochronology of the Siberian platform and the adjacent fold areas (in Russian). Nauka, Leningrad, 332 pp
Marchese HG (1974) Estromatolitos "Gymnosolenidos" en el lado oriental de Minas Gerais. Bras Rev Bras Geol 4-4:257–271
Martin H (1965) The Precambrian geology of South-West Africa and Namaqualand. PRU, Cape Town Univ, Wunberg, Rustica Press, 159 pp
Maslov VP (1937) On the distribution of carbonate algae in East Siberia. In: Problems of paleontology, vol 2/3 (in Russian). Izv MGU, Moscow, pp 327–342
Maslov VP (1950) Systematic position and difference between stromatolites and oncolites and fossil calcareous algae (in Russian). Byull MOIP, Otd Geol XXV-4:66–76
Maslov VP (1952) Carbonate nodules of organic origin (in Russian). Byull MOIP, Otd Geol XXVII-4:28–45
Mazarovich AN (1947) On the basic units of geochronology (in Russian). Dokl Akad Nauk SSSR LVIII-3:443–446
Melnikov NV (1966) Litho-stratigraphic subdivision of the Riphean, Vendian and Cambrian on the lower Lena River. In: Trofimuk AA (ed) Geologic structure and oil and gas content of the eastern Siberian platform and adjacent areas (in Russian). Yakut Knizhn Izd, Yakutsk, pp 20–24
Melton W, Scott HW (1973) Conodont-bearing animals from the Bear Gulch limestone, Montana. Geogr Soc Am Spec Pap, 141 pp
Menner VV, Streiss NA (1971) On tectonic aspects of the geochronological scale. In: Peive AV (ed) Problems of theoretical and regional tectonics (in Russian). Nauka, Moscow, pp 300–309
Mens KA, Pirrus EA (1971) On the stratigraphy of Vendian-Cambrian boundary beds on the north-western Russian platform (in Russian). Izv Akad Nauk SSSR Ser Geol 11:93–103
Mens KA, Pirrus EA (1979) The southern slope of the Baltic shield. In: Keller BM, Rozanov AY (eds) Stratigraphy of Upper Precambrian and Cambrian deposits of the western East European platform (in Russian). Nauka, Moscow, pp 7–41
Meshkova NP, Zhuravleva IT, Luchinina VA (1973) Lower Cambrian and Lower Middle Cambrian of the Olenek uplift. In: Zhuravleva IT (ed) Problems of paleontology and biostratigraphy of the Lower Cambrian of Siberia and the Far East (in Russian). Nauka, Novosibirsk, pp 194–214
Meshkova NP, Nikolaeva IV, Kulikov YP et al. (1976) Stratigraphy of Precambrian-Cambrian boundary deposits of the northern Anabar uplift. In: Zhuravleva IT (ed) Stratigraphy and paleontology of the Lower and Middle Cambrian of the USSR (in Russian). Tr IGiG SF Akad Nauk SSSR (Nauka, Novosibirsk) 296:3–22
Meshkova NP, Luchinina VA, Karpov GP, Sidorov SD (1979) Section of the Lower Cambrian on the Bolshaya Kuonamka River (north-eastern slope of the Anabar uplift). In: Zhuravleva IT, Meshkova NP (eds) Biostratigraphy and paleontology of the Lower Cambrian of Siberia (in Russian). Tr IGiG SF Akad Nauk SSSR (Nauka, Novosibirsk) 406:76–84
Mezentsev AV, Nesenenko AP, Sukhorukov VI, Yan-Zhin-shin VA (1978) New data on the structure and correlation of the Yudoma Series of the Kyllakh uplift (in Russian). Geol Geofiz 3:19–28
Misra SB (1981) Depositional environment of the Late Precambrian fossil-bearing rocks of south-eastern Newfoundland, Canada. J Geol Soc India 22-8:375–382

Missarzhevsky VV (1969) Hyolithes, gastropods, and groups of uncertain systematic origin. In: Rozanov AY (ed) The Tommotian stage and the problem of the lower boundary of the Cambrian (in Russian). Tr GIN Akad Nauk SSSR 206:93–175 English translation edited by ME Raaben published in 1981 for the U.S. Department of the Interior and the National Science Foundation (Washington DC) by Amerind, New Delhi, India, 1981, 359 pp

Missarzhevsky VV (1974) New data on the most ancient Early Cambrian fossils of the Siberian platform. In: Zhuravleva IT, Rozanov AY (eds) Biostratigraphy and paleontology of the Lower Cambrian of Europe and North Asia (in Russian). Nauka, Moscow, pp 179–189

Missarzhevsky VV (1980) On Precambrian-Cambrian boundary strata on the western slope of the Olenek uplift (Olenek River) (in Russian). Byull MOIP, Otd Geol 55-4:23–34

Missarzhevsky VV (1982) Subdivision and correlation of the Precambrian-Cambrian boundary strata based on some oldest groups of skeletal organisms (in Russian). Byull MOIP Otd Geol 57-5:52–67

Missarzhevsky VV (1983) Stratigraphy of the most ancient Phanerozoic strata of the Anabar Massif (in Russian). Sov Geol 9:62–73

Missarzhevsky VV, Rozanov AY (1975) On the problem of correlation of the Precambrian-Cambrian boundary strata on the Siberian platform (in Russian). Tr IGiG SF Akad Nauk SSSR 232:101–111

Mladshikh SV, Ablizin BD (1967) Stratigraphy of the Upper Precambrian of the western slope of the Central Urals (in Russian). Izv Akad Nauk SSSR Ser Geol 2:67–80

Monty C (1967) Distribution and structure of recent stromatolites algal mats, eastern Andros Island, Bahamas. Ann Soc Géol Belg 90-3:55–99

Moore RC (ed) (1956) Treatise on invertebrate paleontology. Pt F: Coelenterata. Boulder & Lawrence, Geol Soc Am Univ Kansas, 498 pp

Moore RC, Harrington HJ (1956) Scyphozoa, Conulata. In: Moore RC (ed) Treatise on invertebrate paleontology. Boulder & Lawrence, Geol Soc Am Univ Kansas, pp 27–38, 54–66

Morozov SG, Revenko EA (1969) On the age of carbonate strata of the Bavly sediments in Bashkiria (in Russian). Dokl Akad Nauk SSSR 184-4:917–926

Morris WA (1977) Paleolatitude of glaciogenic Upper Precambrian Rapitan Group and the use of tillites as chronostratigraphic marker horizon. Geology (Boulde) 5-2:85–88

Moskvitin IE, Nuzhnov SV, Yakshin MS (1980) Results of revisional stratigraphic studies on the Upper Precambrian of the north-western slope of the Aldan shield. In: Khomentovsky VV (ed) Reference junction areas of the unified scheme for the Late Precambrian of Central Siberia (in Russian). Nauka, Novosibirsk, pp 12–42

Naumova SN (1951) Spores in ancient formations on the western slope of the South Urals (in Russian). Tr MOIP Otd Geol, pp 183–187

Naumova SN (1960) Spore and pollen assemblages in Riphean and Lower Riphean deposits of the USSR. In: XXI IGC Session. Papers presented by Soviet geologists. Problem 8 (in Russian). Izd Akad Nauk SSSR, Moscow, pp 109–117

Naumova SN (1968) Zonal assemblages of Precambrian and Lower Cambrian plant microfossils of Eurasia and their stratigraphic significance. In: Sokolov BS (ed) Stratigraphy of the Lower Paleozoic of Central Europe (in Russian). Nauka, Moscow, pp 30–39

Naumova SN (1974) Precambrian and Lower Cambrian plant microfossils of Eurasia and their stratigraphic significance. In: Naumova SN (ed) Palynology of proterophytes and paleophytes (in Russian). Proceedings of III Intern Palyn Conf (Nauka, Moscow), pp 7–12

Nechaev SV (1974) Geochronology of Late Precambrian strata on the south-western slope of the Russian platform. In: Keller BM (ed) Meeting on the Upper Precambrian (Riphean) of the Russian platform. Abstracts (in Russian). Moscow, pp 40–48

Nekrasova OI (1955) Lithology of Lower and Middle Cambrian deposits in a section of Amga key borehole (in Russian). Gosgeoltekhizdat, Moscow, 68 pp

Nevolin BS, Potapov SV, Stavtsev AL (1978) Upper Proterozoic (Riphean) and Lower Cambrian of the south-eastern margin of the Siberian platform, Yudoma-Maya trough, and the Okhotsk median mass. In: Khomentovsky VV (ed) New data on stratigraphy and paleontology of the Late Precambrian of eastern and northern Siberia (in Russian). Novosibirsk, pp 21–63

Nuzhnov SV (1967) Riphean deposits of the south-eastern Siberian platform (in Russian). Nauka, Moscow, 159 pp

Nuzhnov SV, Potapov SV, Rabotnov VT (1975) Stratigraphic position and subdivision of the Vendian (Yudomian) deposits in the Uchur-Maya area (in Russian). Tr IGiG SF Akad Nauk SSSR 232:152–161

Nystuen JP (1976) Late Precambrian Moevl Tillite deposited on a discontinuity surface associated with a fossil ice wedge, Rendalen, southern Norway. Nor Geol Tidsskr 56:29–56

Odin GS, Gale NH, Auvray B et al. (1983) Numerical dating of Precambrian-Cambrian boundary. Nature 301-5895:21–23

Oftedahl Ch (1945) Om tillitene i det Central-Norske sparagmitomrade. Nor Geol Tidsskr 25:285–294

Ogurtsova RN (1975a) Findings of Lontova acritarchs in Tommotian deposits of the Olenek uplift (in Russian). Izv Akad Nauk SSSR Ser Geol 11:84–89

Ogurtsova RN (1975b) Upper Vendian and Lower Cambrian acritarchs of the Lesser Karatau Range in southern Kazakhstan (section on the Kyrshakty River) (in Russian). Izv Akad Nauk KSSR 5: 19–27

Olli AI (1948) Ancient deposits on the western slope of the Urals (in Russian). Izd Saratov Univ, Saratov, 412 pp

On the Sinian geochronological scale of China based on isotopic ages for the Sinian strata in the Yenshan Region, North China (1977). Sci Sin 20-6:818–834

Öpik AA (1961) Alimentary caeca of agnostids and other trilobites. Palaeontology 3:410–438

Ozhiganova LD (1959) Petrographical-mineralogical studies of ancient deposits in western Bashkiria. In: Timergazin KR (ed) Ancient deposits of western Bashkiria (in Russian). Ufa, pp 28–87

Paasikivi LB (1966) Geologic structure and history of evolution of the Khanya-Loknya and Mynisto uplifts. In: Problems of exploration geophysics vol 5 (in Russian). Gostoptekhizdat, Leningrad pp 86–99

Paleogeography and lithology of the Vendian and Cambrian of the western East European platform (1980). In: Keller BM, Rozanov AY (eds) Nauka, Moscow, 119 pp

Paleontology of Upper Precambrian and Cambrian deposits of the East European platform (1979) In: Keller BM, Rozanov AY (eds) (in Russian). Akad Nauk SSSR, Moscow, 108 pp

Paliy VM (1976) Remains of non-skeletal fauna and trace fossils from Upper Precambrian and Lower Cambrian deposits of Podolia. In: Ryabenko VA (ed) Paleontology and stratigraphy of the Upper Precambrian and Lower Paleozoic of the south-western part of the East European platform (in Russian). Naukova Dumka, Kiev, pp 63–77

Paliy VM, Posti E, Fedonkin MA (1979) Soft-bodied Metazoa and Vendian and Early Cambrian trace fossils. In: Keller BM, Rozanov AY (eds) Paleontology of Upper Precambrian and Cambrian deposits of the East European platform (in Russian). Nauka, Moscow, pp 49–82

Paškevičiene LT (1976) Micropaleophytological characteristics of Vendian and Lower Cambrian deposits of the Baltic area. In: Grigelis AA (ed) Materials on stratigraphy of the Baltic area (in Russian). Vilnius, pp 30–31

Paškevičiene LT (1980) Acritarchs from Vendian-Cambrian boundary beds of the western East European platform (in Russian). Nauka, Moscow, 76 pp

Pflug HD (1970a) Zur Fauna der Nama-Schichten in Südwest-Afrika. I. Pteridinia, Bau und systematische Zugehörigkeit. Palaeontogr A 134:226–262

Pflug HD (1970b) Zur Fauna der Nama-Schichten in Südwest Afrika. II. Rangeidae, Bau und systematische Zugehörigkeit. Palaeontogr A 135-3/6:198–231

Pflug HD (1971a) Neue Fossilfunde im Jungpräkambrium und ihre Aussagen zur Entstehung der Höheren Tiere. Geol Rundsch 60-4:1340–1350

Pflug HD (1971b) Neue Zeugnisse zum Ursprung der Höheren Tiere. Naturwissenschaften 58-7:348–352

Pflug HD (1972a) Systematik der jung-präkambrischen Petalonamae Pflug 1970. Palaeontol Z 46-1/2:56–57

Pflug HD (1972b) The Phanerozoic-Cryptozoic boundary and the origin of Metazoa. In: XXIV IGC Session. Sect 1, Montreal, pp 58–67

Pflug HD (1972c) Zur Fauna der Nama-Schichten in Südwest-Afrika. III. Erniettomorpha, Bau und Systematik. Palaeontogr A 139-4/6:134–170

Pflug HD (1973) Zur Fauna der Nama-Schichten in Südwest-Afrika. IV. Mikroskopische Anatomie der Petalo-Organismen. Palaeontogr A 144-4/6:166–202

Pisarchik YK (1963) Lithology and facies of Cambrian deposits of the Irkutsk Amphitheater as related to oil and gas and salt content (in Russian). Gostoptekhizdat, Leningrad, 347 pp

Polevaya NI (ed) (1978) Catalogue of definitions of age of rocks by radiometric methods. Russian platform (crystalline basement and sedimentary and volcanic cover). Leningrad, 398 pp

Porfirieva NA (1977) Planarias of Lake Baikal (in Russian). Nauka, Novosibirsk, pp 1–207

Postnikov DV (1974) New geochronological data on the Upper Precambrian of the Russian platform. In: Keller BM (ed) Meeting on the Upper Precambrian (Riphean) of the Russian platform. Abstracts (in Russian). Nauka, Moscow, p 48
Postnikova IK (1972) Correlation of Upper Precambrian sections on the western slope of the Urals and the East European platform (in Russian). Byull MOIP Otd Geol 47-5:86-102
Postnikova IK, Revenko EA (1969) New data on the Vendian Complex of the Volga-Urals area (in Russian). Dokl Akad Nauk SSSR 188-5:1123-1125
Poulsen C (1930) Contributions to the stratigraphy of the Cambro-Ordovician of East Greenland. Medd Gronland 74-XII:297-316
Precambrian-Cambrian boundary in the geosynclinal areas (1982) Rozanov AY (ed) Salany Gol reference section, Mongolia. Joint Soviet-Mongolian Paleontological Expedition, Transactions, vol 18 (in Russian). Nauka, Moscow, pp 110-118
Preiss WV (1973) The systematics of South Australian Precambrian and Cambrian stromatolites. Pt 1. Trans R Soc S Aust 97-2:91-125
Preiss WV (1976) The biostratigraphic potential of Precambrian stromatolites. In: XXV IGS Session. Abstracts. Canberra, vol 1. s.a., p 36
Preiss WV, Forbes BG (1981) Stratigraphy, correlation and sedimentary history of Adelaidean (Late Proterozoic) Basins in Australia. Precambrian Res 15-3/4:255-304
Preiss WV, Krylov IN (1980) Upper Precambrian stratigraphy and algal fossils of South Australia (in Russian). Izv Akad Nauk SSSR Ser Geol 7:61-74
Pringle JR (1973) Rb-Sr age determinations on shales associated with the Varanger Ice Age. Geol Mag 109-6:465-472
Pritula YA, Nechaev ZK, Frolov BM et al. (1962) Geologic structure and potential oil and gas content in the Angara-Lena fold zone (in Russian). Gostoptekhizdat, Leningrad, 147 pp
Pyatiletov VG (1979) On the occurrence of blue-green algae in the Yudomian deposits of Yakutia (in Russian). Dokl Akad Nauk SSSR 249-3:714-716
Pyatiletov VG (1980a) Yudomian microfossil assemblage of southern Yakutia (in Russian). Geol Geofiz 7:8-20
Pyatiletov VG (1980b) Microfossils from Late Precambrian deposits penetrated by the Vanavara Borehole. In: Khomentovsky VV (ed) New data on stratigraphy of the Late Precambrian of the western Siberian platform and its mountain fringe (in Russian). IGiG SF Akad Nauk SSSR, Novosibirsk, pp 71-76
Pyatiletov VG, Karlova GA (1983) Microphytological characteristics of the Chistyakov Formation of the Taseevo Series of the Yenisei Mountain Range. In: Khomentovsky VV (ed) Stratigraphy of the Late Precambrian and Early Paleozoic of Central Siberia (in Russian). IGiG SF Akad Nauk SSSR, Novosibirsk, pp 95-99
Pyatiletov VG, Luchinina VA, Shenfil VYu, Yakshin MS (1981) New data on ancient algae (in Russian). Dokl Akad Nauk SSSR 261-4:982-984
Pykhova NG (1973) Precambrian acritarchs of the South Urals, Siberia, the East European platform and their stratigraphic significance. In: Vozzhennikova TF (ed) Microfossils in the most ancient deposits (in Russian). Nauka, Moscow, pp 5-17
Raaben ME, Zabrodin VE (1972) Upper Riphean algal problematics (stromatolites, oncolites) (in Russian). Tr GIN Akad Nauk SSSR (Nauka, Moscow) p 217
Rabotnov VT (1977) Stratigraphy of the Upper Precambrian of the Okhotsk massif (in Russian). Dokl Akad Nauk SSSR 234-1:148-151
Rabotnov VT, Narozhnykh LI (1961) Stratigraphy of the Lower Cambrian of the Uchur-Maya area. In: Kobelyatsky IA (ed) Materials concerning the geology and mineral resources of the Yakutian ASSR (in Russian). Yakut Knizhn Izd, Yakutsk VII:107-113
Rabotnov VG, Narozhnykh LI, Komar VA (1971) On stratigraphy of the Upper Precambrian of the Sayany area (in Russian). Dokl Akad Nauk SSSR 200-6:1407-1410
Reading HG (1965) Eocambrian and Lower Paleozoic geology of the Digermul Peninsula, Tanafjord, Finnmark. Nor Geol Unders 234:167-191
Reading HG, Walker RG (1966) Sedimentation of Eocambrian tillites and associated sediments in Finnmark, Northern Norway. Palaeogeogr Palaeoclimatol Palaeoecol 2:177-212
Reitlinger EA (1948) Cambrian foraminifers of Yakutia (in Russian). Byull MOIP Otd Geol 23-2:77-81
Reitlinger EA (1959) Atlas of microscopic fossils and problematics of ancient strata of Siberia (in Russian). Izd Akad Nauk SSSR, Moscow, 55 pp

Repina LN, Lazarenko NP, Meshkova NP et al. (1974) Lower Cambrian biostratigraphy and facies of Kharaulakh (Tuora-Sis Range) (in Russian). Tr IGiG SF Akad Nauk SSSR (Nauka, Moscow), 300 pp

Research Group of Sinian glaciation (1976) Sinian glacial deposits of China. Peking, 15 pp

Richter R (1975) Die ältesten Fossilien Süd-Afrikas. Senckenb Lethaea 36-3/4:243-289

Ritchie A, Gilbert-Tomlinson J (1977) First Ordovician vertebrates from the Southern Hemisphere. Alcheringa 1:351-368

Rozanov AY (1966) Problem of the lower boundary of the Cambrian. In: Scientific achievements. Geology: general geology. Stratigraphy (in Russian). VINITI, Moscow, pp 91-111

Rozanov AY (1976) Precambrian-Cambrian boundary. In: Keller BM (ed) Boundaries of geologic systems (in Russian). Nauka, Moscow, pp 31-53

Rozanov AY, Missarzhevsky VV (1966) Biostratigraphy and fauna of lower horizons of the Cambrian (in Russian). Tr GIN Akad Nauk SSSR (Nauka, Moscow) 148:126

Rozanov AY, Sokolov BS (1980) Precambrian-Cambrian boundary: recent state of knowledge. In: XXVI IGC Session. Papers presented by Soviet geologists (in Russian). Precambrian. Nauka, Moscow, pp 159-164

Rozanov AY, Sokolov BS (1982) Precambrian-Cambrian boundary: recent state of knowledge. Precambrian Res 17:125-131

Rudavskaya VA (1971) Acritarchs of the Moty Formation of the Markovo oil field in East Siberia. In: Samoilovich SR (ed) Palynology in petroleum geology (in Russian). Tr VNIGRI (Nauka, Leningrad) 296:93-100

Rudavskaya VA (1973) Acritarchs of Riphean-Cambrian boundary deposits in south-eastern Siberia. In: Vozzhennikova TF (ed) Microfossils in the most ancient deposits (in Russian). Nauka, Moscow, pp 17-21

Rudavskaya VA (1980) Acritarchs of Late Precambrian oil-and-gas-bearing horizons and oils in East Siberia. In: Timoshina NA (ed) Microfossils in petroleum geology (in Russian). Nauka, Leningrad, pp 130-140

Rudavskaya VA, Frolov BM (1974) First findings of acritarchs in lower horizons of reference sections of the Yudoma Formation (in Russian). Tr IGiG SF Akad Nauk SSSR 81:11-13

Rudavskaya VA, Vasilieva NI (1985) Acritarchs and skeletal problematics near the Vendian, Tommotian and Atdabanian. In: Kokoulin ML, Rudavskaya VA (eds) Stratigraphy of the Late Precambrian and Early Paleozoic of the Siberian platform (in Russian). Leningrad, pp 51-58

Rudavskaya VA, Kossovoi LS, Tyunegin SP (1977) Stratigraphic subdivision of the Upper Precambrian in the Mezen depression based on microfossil assemblages (in Russian). Tr VNIGRI 398:7-15

Ruedemann R (1916) Note on *Paropsonema cryptophya* Clarke and *Discophyllum peltatum* Hall. Bull NY State Mus 189:22-27

Rukhina EV (1960) Lithology of till deposits (in Russian). Izd LGU, Leningrad, 148 pp

Runnegar B (1982) Oxygen requirements, biology and phylogenetic significance of the Late Precambrian worm Dickinsonia and the evolution of burrowing habit. Alcheringa 6:223-239

Rutland RWR, Parker AJ, Pitt GM et al. (1981) The Precambrian of South Australia. In: Hunter DR (ed) Precambrian of the Southern Hemisphere. Elsevier, Amsterdam, pp 309-360

Rzhevsky VF, Chekhovich KM (1978) Stratigraphy of the Precambrian of the Igarka area. In: Khomentovsky VV (ed) New data on stratigraphy and paleontology of the Late Precambrian of eastern and northern Siberia (in Russian). Novosibirsk, pp 101-114

Sagyndykov KS, Cherepanova VA (1977) On the effusive-sedimentary nature of jaspilites in the mid-Tien Shan. In: Adyshev M (ed) Main problems of sedimentary ore formation (in Russian) Ilim, Frunze, pp 168-186

Sakalauskas V (1968) The Vendian of Lithuania. In: Grigelis A (ed) Stratigraphy of the Baltic Lower Paleozoic and its correlation with other areas (in Russian). Mintis, Vilnius, pp 9-21

Savitsky VE (1962) On the relationships between the Cambrian and Upper Precambrian on the Anabar shield. In: Sokolov BS (ed) Meeting on stratigraphy of Late Precambrian deposits of Siberia and the Far East. Abstracts (in Russian). Novosibirsk, pp 53-54

Savitsky VE (1978) The Precambrian-Cambrian boundary problem in Siberia and some general problems of stratigraphy (Preliminary Report). Geol Mag 11-2:127-130

Savitsky VE, Shabanov YY, Evtushenko VM, Shishkin BB (1972) Reference section of the Cambrian of the Khorbusuonka facies area. In: Sukhov SV, Krasnov VI (eds) Materials on stratigraphy and paleontology of Siberia (in Russian). Novosibirsk, pp 53-60

Schmidt F (1897) Excursion durch Estland. In: Guide des excurs du VIIme. Congr Geol Intern, St. Petersbourg XII:26

Schopf JM (1968) Microflora of the Bitter Springs Formation, Late Precambrian, Central Australia. J Paleontol 42-3:925-960

Schopf JW, Haugh BN, Molnar RE, Satterthwait DF (1973) On the development of Metaphytes and metazoans. J Paleontol 47-1:1-9

Scrutton CT (1979) Early fossil Cnidarians. In: House MR (ed) Systematics Association spec vol 12: The origin of major invertebrate groups. Academic Press, London, pp 161-207

Scupin H (1928) Ostbalticum (1 Th.), Algonkium, Paläozoicum und Mesozoicum. Berlin, 126 pp

Sedgwick A (1884) On the origin of the metameric segmentation and some other morphological questions. Q J Microsc Sci 24:96-112

Seilacher A (1984) Late Precambrian and Early Cambrian Metazoa: Preservational or real extractions? In: Patterns of change in Earth evolution (Dahlem Konferenzen, 1984). Springer, Berlin, Heidelberg New York, pp 159-168

Semikhatov MA (1962) Riphean and Lower Cambrian of the Yenisei Range (in Russian). Nauka, Moscow, 236 pp

Semikhatov MA (1974) Vendomian of the Siberian platform and adjacent troughs. In: Menner VV (ed) Vendomian (terminal Riphean) and its regional units. Scientific and technical achievements. Stratigraphy and paleontology (in Russian). VINITI, Moscow 5:26-40

Semikhatov MA (1979) New stratigraphic scale for the Precambrian of the USSR: analysis and lessons (in Russian). Izv Akad Nauk SSSR Ser Geol 11:5-22

Semikhatov MA, Serebryakov SN (1983) Siberian hypostratotype of the Riphean (in Russian). Tr GIN Akad Nauk SSSR (Nauka, Moscow) 367:224

Semikhatov MA, Komar VA, Serebryakov SN (1967) New data on stromatolites of the Yudoma Formation and its equivalents (in Russian). Dokl Akad Nauk SSSR 175-5:1119-1122

Semikhatov MA, Komar VA, Serebryakov SN (1970) Yudomian Complex of the type locality (in Russian). Nauka, Moscow, 207 pp

Semikhatov MA, Aksenov EM, Becker YR, Velikanov VA, Dolnik TA, Kabankov VY, Solontsov LF (1979) Subdivision and correlation of the USSR Riphean. In: Keller BM, Semikhatov MA (eds) Stratigraphy of the Upper Proterozoic (Vendian and Riphean) (in Russian). Nauka, Leningrad, pp 5-38

Semikhatov MA, Gorokhov IP, Ivanovskaya TA, Rublev AG, Kutyavin EP, Yakovleva OV, Bizunok MB (1987) Rb-Sr and K-Ar age of Riphean and Cambrian globular layered silicates: Materials for geochronometric dating (in Russian). Litol Polezn Iskop 6:78-96

Shatsky NS (1932) Main features of tectonics of the Siberian platform (in Russian). Byul MOIP, Otd Geol 10-34:476-509

Shatsky NS (1945) Outlines of the tectonics of the Volga-Ural oil region and adjacent part of the west slope of the South-Urals (in Russian). MOIP, 131 pp

Shatsky NS (1952a) On the most ancient deposits of the sedimentary cover of the Russian platform and its structure in the Early Paleozoic (in Russian). Izv Akad Nauk SSSR Ser Geol 1:17-32

Shatsky NS (1952b) On the Paleozoic-Proterozoic boundary and Riphean deposits of the Russian platform (in Russian). Izv Akad Nauk SSSR Ser Geol 5:36-49

Shatsky NS (1963) On the relations between the Cambrian and Proterozoic and on Baikalian folding. Selected works (in Russian). Izd Akad Nauk SSSR Moscow 1:581-587

Shenfil VY (1965a) The influence of habitat on the shape of stromatolite structures. In: Sokolov BS (ed) All-Union Symposium on paleontology of the Precambrian and Early Cambrian. Abstracts (in Russian). IGiG SF Akad Nauk SSSR. Novosibirsk, pp 22-27

Shenfil VY (1965b) Some new data on stromatolites of Late Precambrian and Early Cambrian deposits of the Baikal area, Patom Highland, and southern Yakutia. In: Sokolov BS (ed) All-Union Symposium on Paleontology of the Precambrian and Early Cambrian. Abstracts (in Russian). IGiG SF Akad Nauk SSSR, Novosibirsk, pp 38-40

Shenfil VY (1968) Stromatolites of Upper Precambrian deposits of the southern fringe of the Siberian platform. Thesis of Cand's dissertation (in Russian). IGiG SF Akad Nauk SSSR, Novosibirsk, 20 pp

Shenfil VY (1975a) Stromatolites of the Yudomian Complex in Siberia. In: Sokolov BS, Khomentovsky VV (eds) Equivalents of the Vendian Complex in Siberia (in Russian). Nauka, Moscow, pp 123-131

Shenfil VY (1975b) A new group of stromatolites from the Precambrian-Cambrian boundary deposits of the Siberia platform. In: Sokolov BS, Khomentovsky VV (eds) Equivalents of the Vendian Complex in Siberia (in Russian). Nauka, Moscow, pp 141-142

Shenfil VY, Yakshin MS (1975) New data on stratigraphy of the Late Precambrian of the Tokko River basin (in Russian). Tr IGiG SF Akad Nauk SSSR (Novosibirsk) 232:162–168
Shenfil VY, Yakshin MS (1982) On stratigraphy of Riphean deposits of the western slope of the Anabar Massif. In: Khomentovsky VV (ed) New data on stratigraphy of the Late Precambrian of Siberia (in Russian). IGiG SF Akad Nauk SSSR, Novosibirsk, pp 31–42
Shenfil VY, Yakshin MS, Butakov EP (1975) On correlation of deposits of the Yudoma Formation in the lower Belaya River Basin (in Russian). IGiG SF Akad Nauk SSSR, Novosibirsk, 232:146–151
Shenfil VY, Dolnik TA, Sukhanova NV (1980a) On the problem of correlation of Upper Precambrian deposits in the junction zone of the Baikal and Sayany areas. In: Khomentovsky VV (ed) Reference junction areas of the unified scheme for the Late Precambrian of Central Siberia (in Russian). Nauka, Novosibirsk, pp 93–113
Shenfil VY, Bragin SS, Didenko AN, Karlova GA (1980b) Late Precambrian reference section in the eastern areas of the central Yenisei Range. In: Khomentovsky VV (ed) New data on stratigraphy of the Late Precambrian of the western Siberian platform and its mountain fringe (in Russian). IGiG SF Akad Nauk SSSR, Novosibirsk, pp 30–56
Shenfil VY, Arutyunov SL, Dolnik TA, Korobeinikov NI, Permyakov SA, Pyatiletov VG (1984) Correlation of Upper Precambrian deposits of the Baikal area and the Irkut Sayan area. In: Khomentovsky VV (ed) Late Precambrian and Early Paleozoic of Siberia. Problems of subdivision and correlation (in Russian). IGiG SF Akad Nauk SSSR, Novosibirsk, pp 3–17
Shepeleva ED (1960) Finding of blue-green algae in Lower Cambrian deposits of the Leningrad District (in Russian). Tr VNIGRI (Leningrad) 27:170–172
Shepeleva ED (1963) Spore assemblages(?) from deposits of the Bavly Formation in the Volga-Urals area (in Russian). Tr VNIGRI (Leningrad) 37:7–17
Shepeleva ED (1966) New data on the age of the Upper Bavly Series of the north-eastern margin of the Russian platform (in Russian). Tr VNIGRI (Leningrad) 51:20–22
Shepeleva ED (1967) Phytoplankton in pre-Devonian deposits of the Pachelma trough (in Russian). Tr VNIGRI (Leningrad) 52:3–9
Shepeleva ED (1969) On stratigraphic range of some microfossils (acritarchs) in Vendian and Lower Cambrian deposits (in Russian). Tr VNIGRI (Leningrad) 74:60–65
Shepeleva ED (1973) Subdivision of the Vendian on the Russian platform from acritarchs. In: Vozzhenikova TF (ed) Microfossils in the most ancient deposits (in Russian). Nauka, Moscow, pp 13–15
Shepeleva ED (1974) Stratigraphic subdivision of Vendian deposits of the central Russian platform from acritarchs (in Russian). Tr IGiG SF Akad Nauk SSSR (Novosibirsk) 81:13–23
Shepeleva ED, Timofeev BV (1963) On the micropaleophytological characteristic of the Pachelma Series and its stratigraphic equivalents (in Russian). Dokl Akad Nauk SSSR 153-5:1158–1159
Shishkin BB (1975) On stratigraphy of Precambrian-Cambrian boundary deposits of the Igarka area. In: Sokolov BS, Khomentovsky VV (eds) Equivalents of the Vendian Complex in Siberia (in Russian). Nauka, Moscow, pp 192–195
Shishkin BB (1978) On the problem of stratigraphic subdivision of Upper Precambrian deposits in the western Anabar area. In: Vozin VF (ed) Stratigraphy of the Upper Precambrian and Lower Cambrian of the eastern Siberian platform (in Russian). Yakut Knizhn Izd, Yakutsk, pp 66–75
Shishkin BB (1981) New data on stratigraphy of the Early Cambrian of the north-eastern Anabar area (Bolshaya Kuonamka River). In: Meshkova NP, Nikolaeva IV (eds) Precambrian-Cambrian boundary deposits of the Siberian platform (in Russian). Nauka, Novosibirsk, pp 20–27
Shishkin BB, Stepanova MB (1978) Aim Formation and its paleontological characteristics (in Russian). SNIIGGiMS (Nauka, Novosibirsk) 260:27–36
Shpunt BR, Shapovalov IG, Shamshina EA et al. (1979) Proterozoic of the north-eastern margin of the Siberian platform (in Russian). Nauka, Novosibirsk, 213 pp
Shubnikov AV (1940) Symmetry (in Russian). Izd Akad Nauk SSSR, Moscow-Leningrad, 176 pp
Shubnikov AV (1951) Symmetry and antisymmetry of finite figures (in Russian). Izd Akad Nauk SSSR, Moscow, 172 pp
Shulga PL (1952) On the age of the oldest arkose sequence on the south-western margin of the Russian platform. Dopovidi Akad Nauk Ukr SSR Ser B 4:320–323 (Ukrainian)
Sinelnikov DV (1973) Comparison of tectonic evolution of western margins of the African and Siberian platforms in the Late Precambrian (in Russian). Izv Vuzov Geol Razv 12:13–20
Sinian Suberathem in China (1980) Tianjin Sci Technol, Tianjin, 407 pp

Sin Yu-Sheng, Lin Kui-chin (1973) On Sinian micropalaeoflora in Yenliao region of China and its geological significance. Acta Geol Sin 6:1-64
Sin yu-Sheng, Lin Kui-chin (1976) Microplantflora from the Sinian subera of W. Hupch and its stratigraphic significance. Peking, 23 pp
Sivertseva IA, Stankovsky AF (1979) Microfossils of Upper Precambrian deposits of the Arkhangelsk District. In: Sokolov BS (ed) Paleontology of the Precambrian and Early Cambrian (Proceedings of the All-Union Symposium, Leningrad, 1976) (in Russian). Nauka, Leningrad, pp 157-159
Sivertseva IA, Verichev EM, Grib VP, Stankovsky AF (1981) Upper Precambrian microfossils of the south-eastern White Sea region: Lithology and paleogeography (in Russian). Izd LGU, Leningrad, pp 133-148
Skorobogatykh PP, Yanshina NA (1982) Correlation of Vendian-Lower Cambrian terrigeneous deposits of the western Siberian platform. In: Arutyunov SA (ed) Correlation of pre-Usolka deposits in the central and southern areas of the Lena-Tunguska oil-and-gas-bearing province (in Russian). Irkutsk, pp 33-35
Sobolevskaya RF (1959) Stratigraphy of Sinian deposits of the Anabar uplift. In: Tkachenko BV (ed) Stratigraphy of Sinian and Cambrian deposits of the north-eastern Siberian platform (in Russian). Gostoptekhizdat, Leningrad, pp 34-36
Sokolov BS (1952) Age of the most ancient sedimentary cover of the Russian platform (in Russian). Izv Akad Nauk SSSR Ser Geol 5:21-31
Sokolov BS (1956) Comparative characteristic of pre-Eifelian deposits on the central and eastern Russian platform (in Russian). Tr VNIGRI (Leningrad) 95:35-88
Sokolov BS (1958a) Problem of the lower boundary of the Paleozoic and the most ancient deposits of the pre-Sinian platforms in Eurasia (in Russian). Tr VNIGRI (Leningrad) 3:5-67
Sokolov BS (1958b) Le problème de la limite inferieure du paléozoique et les dépôts les plus anciens sur les plates-formes antesiniennes de l'Eurasie. In: Les relations entre precambrien et cambrien. Problèmes de séries intermediaires. P.: Coll CNRS, pp 103-128
Sokolov BS (1964a) The Vendian and the problem of the boundary between the Precambrian and the Paleozoic Group. In: XXII IGC Session. Rep Pt X: Archean and Precambrian Geology. New Delhi, pp 288-304
Sokolov BS (1964b) The Late Precambrian. In: Nalivkin VD (ed) History of geologic evolution of the Russian platform and its fringe (in Russian). Nedra, Moscow, 252 pp
Sokolov BS (1965a) The most ancient Early Cambrian deposits and sabelliditis. In: Sokolov BS (ed) All-Uniom Symposium on Paleontology of the Precambrian and Early Cambrian. Abstracts (in Russian). IGiG SF Akad Nauk SSSR, Novosibirsk, pp 78-91
Sokolov BS (1965b) Precambrian paleontology and the organic world at the beginning of the Phanerozoic. In: Sokolov BS (ed) All-Union Symposium of Paleontology of the Precambrian and Early Cambrian. Abstracts (in Russian). IGiG SF Akad Nauk SSSR, Novosibirsk, pp 3-7
Sokolov BS (1967) The most ancient pogonophores (in Russian). Dokl Akad Nauk SSSR 177-1:201-204
Sokolov BS (1971) Vendian of northern Eurasia (in Russian). Geol Geofiz 6:13-22
Sokolov BS (1972a) The Vendian Stage in Earth history. In: XXIV IGC Session. Sect 1: Precambrian Geol, Montreal, pp 78-84
Sokolov BS (1972b) Vendian and Early Cambrian sabelliditida (Pogonophora) of the USSR. In: Proc IPU. XXIII IGC (Prague, 1968) W-wa, pp 79-84
Sokolov BS (1973) Vendian of North Eurasia. In: Pitcher MG (ed) Arctic Geology. Tulsa (Okla), pp 204-218
Sokolov BS (1974a) Vendian of the Russian platform: Boundaries of its subdivision and stratigraphic equivalents. In: Keller BM (ed) Meeting on the Upper Precambrian (Riphean) of the Russian platform. Abstracts (in Russian). Nauka, Moscow, pp 3-10
Sokolov BS (1974b) Principal results of the International Symposium on the Precambrian-Cambrian boundary (in Russian). Geol Geofiz 4:18-26
Sokolov BS (1974c) Problem of the Precambrian-Cambrian boundary (in Russian). Geol Geofiz 2:3-29
Sokolov BS (1975) On the Vendian (instead of a foreword). In: Sokolov BS, Khomentovsky VV (eds) Equivalents of the Vendian Complex in Siberia (in Russian). Tr IGiG SF Akad Nauk SSSR (Nauka, Moscow) 232:5-10
Sokolov BS (1976a) The Earth's organic world on the path toward Phanerozoic differentiation (in Russian). Vestn Akad Nauk SSSR 1:126-143

Sokolov BS (1976b) Precambrian Metazoa and the Vendian-Cambrian boundary (in Russian). Paleontol Zh 1:3-18
Sokolov BS (1976c) Metazoa of the Precambrian and the Vendian-Cambrian boundary. In: XXV ICC Session. Abstracts, vol 1. Canberra, pp 37-38
Sokolov BS (1977a) Organic kingdom of the Earth on the way to the Phanerozoic differentiation. In: 250 years to the Academy of Science of the USSR (in Russian). Nauka, Moscow, pp 423-444
Sokolov BS (1977b) Perspectives on the biostratigraphy of Precambrian (in Russian). Geol Geofiz 11:54-70
Sokolov BS (1978) Some problems of stratigraphy of the Upper Precambrian and the position of the Vendian. In: Milanovsky EE, Dobruskina IA (eds) Problems of stratigraphy and history of geology (in Russian). Nauka, Moscow, pp 20-29
Sokolov BS (1979) Vendian: principles of recognition of boundaries and position in the scale. In: Keller BM, Semikhatov MA (eds) Stratigraphy of the Upper Proterozoic of the USSR. Riphean and Vendian. Proceedings of the 5th Session of the Scientific Council on the geology of the Precambrian (Ufa 1977) (in Russian). Nauka, Leningrad, pp 42-61
Sokolov BS (1980) The Vendian System: Eocambrian geologic environment. In: XXVI IGC Session. Papers presented by Soviet geologists. Paleontology, Stratigraphy (in Russian) Nauka, Moscow, pp 9-21
Sokolov BS (1983) Paleontology, biochronology and stratigraphy of the Precambrian (Precambrian geobiology). In: Tatarinov LP (ed) Main problems of paleontological studies in the USSR (in Russian). Nauka, Moscow, pp 79-89
Sokolov BS (1984) Vendian period in the Earth's history (in Russian). Priroda 12:3-18
Sokolov BS, Aleksandrova EP (1952) Sinian, Cambrian, Ordovician and Silurian system. In: Barkhatova VP (ed) Atlas of the lithofacies maps of the Russian platform (in Russian). VNIGRI, Leningrad
Sokolov BS, Khomentovsky VV (1980) Age of the oil-and-gas-bearing strata of the south-western Siberian platform (in Russian). Sov Geol 5:45-56
Solontsov LF (1974) Upper Precambrian formations of the Ryazan-Saratov trough. Thesis of Cand's dissertation (in Russian). Geol Inst, Kazan, 27 pp
Solontsov LF, Aksenov EM (1970) On the stratigraphy of the Valdai Series of the East European platform. In: Lebedev LD (ed) Transactions of higher educational institutions. Geology and exploration (in Russian). Izd Geol Razv Inst, Moscow, pp 3-13
Solontsov LF, Khomentovsky VV, Aksenov EM (1975) Main features of stratigraphy of the Upper Precambrian of the East European platform and its possible equivalents on the Siberian platform. In: Sokolov BS, Khomentovsky VV (eds) Equivalents of the Vendian Complex in Siberia (in Russian). Nauka, Moscow, pp 207-224
Spencer AM (1971) Late Precambrian glaciation in Scotland. Mem Geol Soc Lond 6:100
Spjeldnaes N (1963) A new fossil (*Papillomembrana* sp.) from the Upper Precambrian of Norway. Nature 200-4901:53-83
Sprigg RS (1947) Early Cambrian(?) jellyfishes from the Flinders Ranges, South Australia. Trans Soc S Aust 71:212-234
Sprigg RS (1949) Early Cambrian "jellyfishes" of Ediacara, South Australia and Mount John, Kimberley district, Western Australia. Trans R Soc S Aust 73:72-99
Stankovsky AF, Verichev EM, Konstantinov YG et al. (1977) First finding of effusives in the Redkino deposits of the northern Russian platform (in Russian). Dokl Akad Nauk SSSR 234-3:661-664
Stankovsky AF, Verichev EM, Grib VP et al. (1979) A new type of magmatism in the Vendian of the northern Russian platform (in Russian). Dokl Akad Nauk SSSR 247-6:1459-1460
Stankovsky AF, Verichev EM, Grib VP, Dabeiko IP (1981) Vendian of the south-eastern White Sea region (in Russian). Izv Akad Nauk SSSR Ser Geol 2:78-87
Stanley SM (1976) Fossil data and the Precambrian-Cambrian evolutionary transition. Am J Sci 276:56-76
Stashchuk MF (1957) On the stratigraphy of the Early Paleozoic of the middle Dniester area (in Ukrainian). Geol Zh 2:38-48
Stachchuk MF (1958) Lithological features of Early Paleozoic deposits in the middle Dniester area (in Ukrainian). Tr Inst Geol Nauk UkrSSR. Ser Stratigraf Paleontol 21:43
Stasinska A (1960) Velumbella charnoekii n. gen. Acta Palaeontol Pol 5-3:36-44

Stephen AC, Edmonds SJ (1972) The Phyla Sipuncula and Echiura. Brit Mus Nat Hist Lond 717:528
Stormer L (1942) Studies on trilobite morphology. Pt III. The larval development, the segmentation and the sutures, and their bearing on trilobite classification. Nor Geol Tidsskr 21:49–164
Stormer L (1966) Jordens og livets historie En innforing i historisk geologi. Univ Forl, Oslo 275 pp
Stratigraphy of the Ukrainian SSR. II – Riphean-Vendian (1971) (in Ukrainian). Naukova Dumka, Kiev, 273 pp (Ukrainian)
Stratigraphy of Upper Precambrian and Cambrian deposits of the western East European platform (1979) In: Keller BM, Rozanov AY (eds) (in Russian). Akad Nauk SSSR, Moscow, 234 pp
Sysoev VA (1965) Main features of evolution of hyolithid. In: Vozin VF (ed) Paleontology and biostratigraphy of Paleozoic and Triassic deposits of Yakutia (in Russian). Nauka, Moscow, pp 5–20
Tarling DH (1974) A paleomagnetic study of Eocambrian tillites in Scotland. Geol Soc Lond 130:163–177
Termier H, Termier G (1968) Evolution et biocénose. Les invertébrés dans l'histoire du monde vivant. Masson, Paris, 242 pp
Termier H, Termier G (1976) Ediacaran fauna and evolution of the animal kingdom (in Russian). Paleontol Zh 3:22–29
The Sinian System of China (1976) Publ Inst Geol Miner Res, Chinese Acad Geol Sci, Peking, 15 pp
Thelander T, Kumpulainen R (1980) The Late Precambrian tillites in central Scandinavia and correlation with adjacent areas. In XXVI IGC Session. Abstracts, vol 11. Paris, p 618
Thiel H (1966) The evolution of Scyphozoa: a review. Symp Zool Soc Lond 16:77–117
Thompson DW (1942) On growth and form. Cambridge Univ Press, Cambridge, 1116 pp
Thurakov AM (1932) Proterozoic glaciation in the southern part of central Siberia. Bull Geol Soc Am 43-3:581–602
Timergazin KR (1959) Pre-Devonian formations of western Bashkiria and their oil-and-gas potentials (in Russian). Ufa, 311 pp
Timofeev BV (1952) Early Paleozoic deposits in Moldavia (in Russian). Dokl Akad Nauk SSSR 86-6:1207–1209
Timofeev BV (1959) The most ancient flora of the Baltic area (in Russian). Gosgeotekhizdat, Leningrad, 320 pp
Timofeev BV (1966) Micropaleophytological studies of ancient formations (in Russian). Nauka, Moscow-Leningrad, 147 pp
Timofeev BV (1968) Precambrian microphytofossils and thier diagnostic and correlational value. In: Krats KO (ed) Geology of the Precambrian (in Russian). Nauka, Leningrad, pp 95–99
Timofeev BV (1973a) Precambrian microphytofossils of the Ukraine (in Russian). Nauka, Leningrad, 59 pp
Timofeev BV (1973b) Plant microfossils from the Proterozoic and Lower Paleozoic. In: Vozzhennikova TF, Timofeev BV (eds) Microfossils in the most ancient deposits (in Russian). Nauka, Moscow, pp 7–12
Timofeev BV (1977) Findings of Vendian microphytofossils in the flysch zone of the Transcarpathian area. In: Afanasiev GD (ed) Experience in correlation of magmatic and metamorphic rocks of Czechoslovakia and some areas of the USSR (in Russian). Nauka, Moscow, pp 223–227
Timofeev BV, Hermann TN, Mikhailova NS (1976) Microphytofossils of the Precambrian, Cambrian and Ordovician (in Russian). Nauka, Leningrad, 106 pp
Titorenko TN, Drobkova EL (1974) Biostratigraphy of the Vendian and Lower Cambrian of the eastern Irkutsk Amphitheater from algae and microphytolites (in Russian). Geol Geofiz 2:30–38
Tommotian stage and the problem of the lower boundary of the Cambrian (1969) In: Rozanov AY (ed) (in Russian). Tr Geol Inst Akad Nauk SSSR 206:380. English translation edited by Raaben ME published 1981 for the US Department of the Interior and the National Science Foundation (Washington, D.C.) by Amerind, New Delhi, India, 1981, 359 pp
Treshchetenkova AA, Faizulina ZKh, Shirobokov IM (1982) Plant microfossils of the Ushakovka Formation of the south-western Baikal area (in Russian). Izv Akad Nauk SSSR Ser Geol 5:116–121
Trofimov VR (1982) On the problem of Late Proterozoic trappean intrusions of the western Anabar area. In: Khomentovsky VV (ed) New data on stratigraphy of the Late Precambrian of Siberia (in Russian). IGiG SF Akad Nauk SSSR, Novosibirsk, pp 43–49
Trompette R (1973) Le Précambrien supérieur et le Paléozoique inférieur de l'Adrar de Mauritanie (bordure occidentale du bassin de Taoudeni, Afrique de l'Quest). Un exemple de sedimentation de craton. Etude stratigraphique et sedimentologique. Trav Lab Sci Terre B 7:574–702

Tyschenco LF (1980) Regional correlation of deposits of the Moty Formation in the Irkutsk Amphitheater. In: Karogodin YN (ed) Debatable problems of lithostratigraphy (in Russian). Nauka, Novosibirsk, pp 149–158
Upper Precambrian. Stratigraphy of the USSR (1963) In: Keller BM (ed) (in Russian). GONTI, Moscow, 716 pp
Urbanek A (1979) Sabelliditid. In: Keller BM, Rozanov AY (eds) Paleontology of Upper Precambrian and Cambrian deposits of the East European platform (in Russian). Nauka, Moscow, pp 88–92
Ushakov PV (1972) Polychaete worms. In: Fauna of the USSR, vol 1, ns, N 102 (in Russian). Nauka, Leningrad, 272 pp
Ushakova ZG (1962) Lower Paleozoic trappean formation of the western Russian platform (in Russian). VSEGEI (Leningrad), Nov Ser 80:108
Vakhrameev VA, Volkova NA (1968) Plant microfossils and the Cambrian-Precambrian boundary on the Russian platform. In: Krats KO (ed) Geology of the Precambrian (in Russian). Nauka, Leningrad, pp 124–127
Valkov AK (1968) On the fauna of the Kessyuse-Yurega Formation of the Olenek uplift. In: Mokshantsev KB (ed) Tectonics, stratigraphy and lithology of sedimentary formations of Yakutia (in Russian). Yakut Knizhn Izd, Yakutsk, pp 115–119
Valkov AK (1975) Cambrian biostratigraphy and hyolithid of the north-eastern Siberian platform (in Russian). Nauka, Moscow, 139 pp
Valkov AK (1982) Biostratigraphy of the Lower Cambrian of the eastern Siberian platform (Uchur-Maya area) (in Russian). Nauka, Moscow, 92 pp
Valkov AK, Sysoev VA (1969) On a new group of animals (angustiochreides) from the oldest Cambrian deposits of Siberia. In: Vasilevskaya ND et al. (eds) XV All-Union Paleontol Soc Session (in Russian). VPO, Leningrad, pp 15–16
Valkov AK, Sysoev VA (1970) Cambrian angustiochreides of Siberia. In: Bobrov AK (ed) Stratigraphy and paleontology of the Proterozoic and Cambrian of the eastern Siberian platform (in Russian). Yakut Knizhn Izd, Yakutsk, pp 94–100
Velikanov VA (1975) On the distribution pattern of phosphoritic nodules in the Kalyus Beds (Vendian) of Podolia (in Russian). Litologiya Polezn Iskop 6:84–90
Velikanov VA (1976) Stratigraphy of the Upper Precambrian of the south-western slope of the Ukrainian shield. In: Ryabenko VA et al. (eds) Paleontology and stratigraphy of the Upper Precambrian and Lower Paleozoic of the south-western part of the East European platform (in Russian). Naukova Dumka, Kiev, pp 15–40
Velikanov VA (1977) Kanilovka Formation (Vendian) of Volyn-Podolia (in Russian). Geol Zh 1:73–83
Velikanov VA (1979a) Vendian of Podolia. In: Keller BM, Rozanov AY (eds) Stratigraphy of Upper Precambrian and Cambrian deposits of the western East European platform (in Russian). Nauka, Moscow, pp 131–150
Velikanov VA (1979b) Structural and facies zonation of the south-western margin of the East European platform in the Riphean and Vendian (in Russian). Geol Zh 6:54–56
Velikanov VA, Zaika-Novatsky VS, Khanisenko OM (1971) Volyn Series of Podolia (in Ukrainian). Vest Kiev Univ Geologiya (Kiev) 12:7–13
Velikanov VA, Aseeva EA, Ivanchenko VY, Korenchuk LV (1979) Improvement of the stratigraphic scheme for a reference section of the Vendian in the Dniester River basin of Podolia (in Russian). Dokl Akad Nauk Ukr SSR 12:987–991
Velikanov VA, Aseeva EA, Fedonkin MA (1983) Vendian of the Ukraine (in Russian). Naukova Dumka, Kiev
Veretennikov NV (1968) Upper Precambrian tilloid rocks of Byelorussia and their genesis: In: Makhnach AS (ed) Lithology geochemistry and mineral resources of Byelorussia and the Baltic area (in Russian). Nauka Tekhnika, Minsk, pp 29–50
Veretennikov NV, Vernikhovsky VN, Klevtsova AA (1972) Tilloids in the Upper Precambrian of Byelorussia and the north-western Ukraine (Volyn) and their implication for the stratigraphy of the Riphean-Vendian boundary beds. In; Bogaits AT, Sviridenko VI (eds) New data on geology and oil-and-gas content of the Ukr SSR (in Russian). Ukr NIGRI Press, (Lvov) 7:3–13
Vidal G (1976) Late Precambrian microfossils from the Visingö Beds in southern Sweden. Fossils Strata 9:1–57
Vidal G (1981a) Lower Cambrian acritarch stratigraphy in Scandinavia. Geol Foren Stockh Forh 103-2:183–192

Vidal G (1981b) Micropaleontology and biostratigraphy of the Upper Proterozoic and Lower Cambrian Sequence in East Finnmark, Northern Norway. Nor Geol Unders Bull 59-362:1-53
Vinogradov VA (1962) Stratigraphy of Sinian and Cambrian deposits of North Kharaulakh. In: Shvedov NA (ed) Papers on paleontology and biostratigraphy (in Russian). Tr NIIGA (Leningrad) 28:5-23
Virzhikivsky PP (1932) Geologic map of the Ukraine: Sheets XXVI-6, XXII-6 (Dniester River area: Mogilev-Yampol) (in Ukrainian). Izd Ukr Geol Razv Trest, Kiev, 226 pp
Vlasov BI (1974) Upper Precambrian of Volyn. In: Keller BM (ed) Meeting on the Upper Precambrian (Riphean) of the Russian platform. Abstracts (in Russian). Nauka, Moscow, pp 67-69
Vlasov BI (1979a) Vendian of Volyn. In: Keller BM, Rozanov AY (eds) Stratigraphy of Upper Precambrian and Cambrian deposits of the western East European platform (in Russian). Nauka, Moscow, pp 119-131
Vlasov BI (1979b) Upper Precambrian of Volyn. Thesis of Cand's dissertation (in Russian). IGN Ukr SSR, Kiev, 26 pp
Vlasov BI, Volovnik BY (1973) Structural affinity and space distribution of facies of the Upper Proterozoic trappean formation of Volyn-Podolia. In: Luchitsky IV, Fremd GM (eds) Evolution of volcanism in Earth's history (in Russian). Nauka, Moscow, pp 221-223
Volkova NA (1962) Precambrian spores of the Dnieper River area (in Russian). Dokl Akad Nauk SSSR 142-4:893-895
Volkova NA (1964) Phytoplankton in the most ancient deposits of the north-western Moscow District and its stratigraphic implication (in Russian). Izv Akad Nauk SSSR Ser Geol 4:74-84
Volkova NA (1968) Precambrian and Lower Cambrian acritarchs of Estonia (in Russian). Tr GIN Akad Nauk SSSR 188:8-36
Volkova NA (1969) Acritarchs of the north-western Russian platform. In: Rozanov AY (ed) Tommotian stage and the problem of the lower boundary of the Cambrian (in Russian). Tr Geol Inst Akad Nauk SSSR 206:224-236 English translation edited by Raaben ME published 1981 for the U.S. Department of the Interior and the National Science Foundation (Washington, D.C.) by Amerind, New Delhi, India, 1981, 359 pp
Volkova NA (1973) Acritarchs and correlation of Vendian and Cambrian on the western Russian platform (in Russian). Sov Geol 4:48-62
Volkova NA (1974a) Acritarchs from Lower-Middle Cambrian boundary beds of West Latvia. In: Zhuravleva IT, Rozanov AY (eds) Biostratigraphy and paleontology of the Lower Cambrian of Europe and North Asia (in Russian). Nauka, Moscow, pp 194-198
Volkova NA (1974b) Types of damage of walls of Precambrian and Cambrian acritarchs (in Russian). Paleontol Zh 4:101-108
Volkova NA (1981) Upper Precambrian acritarchs of south-eastern Siberia (Ust-Kirba Formation) (in Russian). Byull MOIP Otd Geol 56-4:66-75
Volkova NA, Kiryanov VV, Piskun LV et al. (1979) Plant microfossils. In: Keller BM, Rozanov AY (eds) Paleontology of Upper Precambrian and Cambrian deposits of the East European platform (in Russian). Nauka, Moscow, pp 4-39
Volkova NA, Kiryanov VV, Pyatiletov VG et al. (1980) Precambrian microfossils of the Siberian platform (in Russian). Izv Akad Nauk SSSR Ser Geol 1:23-29
Volkova SA (1980) Clay minerals of the Valdai Series on the north-eastern Russian platform. Thesis of Cand's dissertation (in Russian). Geol Inst, Kazan, 22 pp
Vologdin AG (1962) The most ancient algae of the USSR (in Russian). Izd Akad Nauk SSSR, Moscow, 655 pp
Vologdin AG, Maslov AB (1960) On a new group of fossil organisms from the lower Yudoma Formation of the Siberian platform (in Russian). Dokl Akad Nauk SSSR 134-3:691-693
Volovnik BY (1971) Petrology and mineralogy of the trappean formation of Volyn-Podolia. Thesis of Cand's dissertation (in Russian). Lvov, 23 pp
Vorobiev VN, Aleksandrov VV, Arutyunov SL et al. (1982) Pre-Usolka oil-and-gas-bearing deposits of the Nepa-Botuobiya anteclise (in Russian). Geol Geofiz 2:3-13
Voronova LG, Missarzhevsky VV (1969) Findings of algae and columnar worms in Cambrian-Precambrian boundary beds on the northern Siberian platform (in Russian). Dokl Akad Nauk SSSR 184-1:207-211
Voronova LG, Radionova EP (1976) Paleozoic algae and microphytolites (in Russian). Tr GIN Akad Nauk SSSR (Nauka, Moscow) 294:220

Vorontsova GA, Dolnik TA, Sukhanova NV, Fomin NI (1975) Yudomian Complex of the Sayan-Baikal fold area (in Russian). Tr IGiG SF Akad Nauk SSSR 232:169–179
Wade M (1968) Preservation of soft-bodied animals in Precambrian sandstones at Ediacara, South Australia. Lethaia 1-3:238--267
Wade M (1969) Medusae from uppermost Precambrian or Cambrian sandstones, central Australia. Palaeontology 12-3:351–365
Wade M (1971) Bilateral Precambrian chondrophores from the Ediacara fauna, South Australia. Proc R Soc Victoria 84-1:183–188
Wade M (1972a) Hydrozoa and Scyphozoa and other medusoids from the Precambrian Ediacara fauna, South Australia. Palaeontology 15-2:197–225
Wade M (1972b) Dickinsonia: polychaete worms from the Late Precambrian Ediacara fauna, South Australia. Mem Mus 16-2:171–190
Walcott CD (1896) Fossil jellyfishes from the Middle Cambrian terrane. Proc US Nay Mus 18:611–615
Walcott CD (1898) Fossil Medusae. US Geol Surv Monogr 30:201
Walcott CD (1911) Cambrian geology and paleontology. II,2. Middle Cambrian Merostomata. Smithson Misc Collect 57:17–40
Walcott CD (1914) Cambrian geology and paleontology. III,2. Precambrian Algonkian algal flora. Smithson Misc Geol LXIV-2:77–156
Walter MR, Oehler JH, Oehler DZ (1976) Megascopic algae 1300 million years old from the Belt Supergroup, Montana: a reinterpretation of Walcott's Helminthoidichnites. J Paleontol 50-5: 872–881
Walton HS (1980) Cryptarcha and Acritarcha from Late Proterozoic and Cambrian strata of the Colville Lake area, district of Mackenzie, N.W.T., Canada. In: V Intern Palynol Conf. Abstracts. Cambridge, p 422
Wang Fuxing (1987) On the evolution of microbiota during the transitional stage between the Sinian and Cambrian Periods. In: Wang Xiaofeng et al. (eds) Some aspects of the studies on the Late Precambrian macrofossils in China. Abstracts. Intern Symp Precambrian and Cambrian Geology 3. Yichang, Hubei, China, pp 79–80
Wang Yuelem et al. (1980) Subdivision and correlation of the Upper Precambrian in China. In: Sinian Suberathem in China. Tianjin Sci & Technol, Tianjin, pp 1–30
Wells JW, Hill D (1956) Ctenophora. In: Moore RC (ed) Treatise on invertebrate paleontology. Pt F: Coelenterata. Boulder & Lawrence, Geol Soc Am Univ Kansas, Kansas, p 478
Werner B (1966) *Stephanoscyphus* (Scyphozoa, Coronatae) und seine direkte Abstammung von den fossilien Conulata. Helgol Wiss Meeresunters 13:317–347
Werner B (1967a) Morphologie, Systematik und Lebensgeschichte von *Stephanoscyphus* (Scyphozoa, Coronatae) sowie seine Bedeutung für die Evolution der Scyphozoa. Zool Anz Suppl 30: 297–319
Werner B (1967b) *Stephanoscyphus* Allman (Scyphozoa, Coronatae), ein rezenter Vertreter der Conulata? Palaeontol Z 41:137–153
Williams GE (1975) Late Precambrian glacial climate and the Earth's obliquity. Geol Mag 112-5:441–465
Wilson CB, Harland WB (1964) The Polarisbreen Series and other evidences of Late Precambrian Ice Ages in Spitsbergen. Geol Mag 101-3:198–219
Xing Yusheng (1982) Microflora of the Sinian System and Lower Cambrian near Kunming, Yunnan and its stratigraphic significance (in Chinese). Acta Geol Sin 56-1:49–61
Xing Yusheng, Ding Qixiu, Luo Huilin (1982) Biotic characteristics of the Sinian-Cambrian boundary beds in China and the boundary problems. Precambrian Res 17-1:77–85
Yakobson KE (1971) Relations between the Volyn Series and other units of the Upper Precambrian on the western Russian platform (in Russian). Sov Geol 2:66–74
Yakobson KE (1974) New data on copper content in volcano-sedimentary rocks of the Volyn Series (Upper Precambrian) of Byelorussia (in Russian). Dokl Akad Nauk SSSR 217-1:180–183
Yakobson KE, Krylov NS (1977) Lower boundary of the Vendian in its type locality (in Russian). Sov Geol 7:59–70
Yakovlev NN (1934) On a finding of the Murmansk genus Gymnosolen in the Lower Cambrian of East Siberia (in Russian). Dokl Akad Nauk SSSR 2-12:5S4–591
Yakshin MS (1975) Microphytolites of the Yudomian Complex of Siberia. In: Sokolov BS, Khomentovsky VV (eds) Equivalents of the Vendian Complex in Siberia (in Russian). Tr IGiG SF Akad Nauk SSSR (Nauka, Moscow) 232:132–140

Yakshin MS (1978) On the stratigraphic significance of microphytolites. In: Khomentovsky VV (ed) New data on stratigraphy and paleontology of the Late Precambrian of the Siberian platform (in Russian). IGiG SF Akad Nauk SSSR, Novosibirsk, pp 37–48

Yakshin MS (1979) On the oldest microphytolites of Kharaulakh. In: Zhuravleva IT, Meshkova NP (eds) Biostratigraphy and paleontology of the Lower Cambrian of Siberia (in Russian). Tr IGiG SF Akad Nauk SSSR (Nauka, Novosibirsk) 406:71–76

Yakshin MS (1980) On the Precambrian microphytolites from boreholes of the Kuyumba exploration area (in Russian). Geol Geofiz 5:137–138

Yakshin MS, Luchinina VA (1981) New data on fossil algae of the family Oscillatoriaceae. In: Meshkova NP, Nikolaeva IV (eds) Precambrian-Cambrian boundary deposits of the Siberian platform (in Russian). Tr IGiG SF Akad Nauk SSSR (Nauka, Novosibirsk) 475:28–34

Yanishevsky ME (1926) On fossil *Tybicola* in the Cambrian blue clay (in Russian). Paleont Soc Russia J 4:99–III

Yan Zhin-shin (1975) Yudomian Complex of the Sette Daban Ridge. In: Sokolov BS, Khomentovsky VV (eds) Equivalents of the Vendian Complex in Siberia (in Russian). Tr IGiG SF Akad Nauk SSSR (Nauka, Novosibirsk) 232:143–145

Yang Xianle, Chen Yuande (1983) The Sinian system of south-west China. In: Zhang Huimin and Chem Jinbiao (eds) International Symposium on Late Precambrian Geology, China. Abstracts. Chin Acad Geol Sci, Tianjin, pp 26–27

Yarmolyuk VA (1946) Proterozoic of the eastern Aldan plate. In: Materials on geology and mineral resources of the Far East (in Russian). Khabarovsk, pp 58–79

Yuon AA (1978) Staraya Rechka Formation and the Aldan stage on the south-western slope of the Anabar anteclise. In: Khomentovsky VV (ed) New data on stratigraphy and paleontology of the Late Precambrian of eastern and northern Siberia (in Russian). IGiG SF Akad Nauk SSSR, Novosibirsk, pp 88–100

Zaika-Novatsky VS (1965) New problematics from the Upper Precambrian on the Dniester River basin. In: Sokolov BS (ed) All-Union Symposium on Paleontology of the Precambrian and Early Cambrian. Abstracts (in Russian). Nauka, Novosibirsk, pp 98–99

Zaika-Novatsky VS, Bukatchuk PD, Velikanov VA, Zhovinsky EY (1968) Stratigraphic position of volcanics of Podolia and Moldavia. In: Einor OL (ed) Materials on geology, hydrogeology, geophysics, and geochemistry of the Ukraine, Moldavia, Kazakhstan, the Transbaikal area (in Russian). Kiev Univ Press, Kiev, pp 36–40

Zaika-Novatsky VS, Velikanov VA, Khanisenko AM (1972) Grushka Formation of Volyn-Podolia, vol 8 (in Russian). In: Reports of research sector of Kiev Univ, pp 5–12

Zaitsev YA, Kheraskova IN (1979) Vendian of Central Kazakhstan. In: Materials on geology of Central Kazakhstan (in Russian). 14:252

Zaitsev YA, Zykov SI, Krasnobaev AA et al. (1974) On radiometric dating of Precambrian rocks in Central Kazakhstan. In: Abdullin AA (ed) Pre-Paleozoic and Paleozoic of Kazakhstan, vol 1 (in Russian). Alma Ata, pp 25–28

Zang Wenlong (1987) Late Proterozoic and Early Cambrian microfossils from Australia and China: Distribution and correlation. In: Wang Xiaofeng et al. (eds) Some aspects of the studies on the Late Precambrian macrofossils in China. Abstracts, Intern Symp Precambrian & Cambrian Geology, 3. Yichang, Hubei, China, pp 110–111

Zhang Huimin, Chem Jinbiao (eds) (1983) International Symposium on Late Precambrian Geology, China. Abstracts. Chin Acad Geol Sci, Tianjin, 119 pp

Zhang Zi-Chang, Compston W, Page RW (1982) The isotopic age of the Cambrian boundary from the Sinian-Lower Cambrian sequence in South China. In: V Intern Conf geochronol., cosmochronol., isotope geol. Abstracts. Tokyo, pp 411–412

Zharcov MA, Khomentovsky VV (1965) Main problems of stratigraphy of the Lower Cambrian and Vendian of the southern Siberian platform in connection with salt content (in Russian). Byull MOIP, Otd Geol 1-1:100–118

Zhoa Wendjie (1983) Proterozoic eonothem in China. In: Zhang Huimin and Chem Jinbiao (eds) International Symposium on Late Precambrian Geology, China. Abstracts. Chin Acad Geol Sci, Tianjin, pp 1–3

Zhuravleva IT, Meshkova NP, Luchinina VA et al. (1982) Biofacies of the Anabar sea in the Late Precambrian and Early Cambrian. In: Betekhtina OA, Zhuravleva IT (eds) Environment and life in the geologic past (in Russian). Tr IGiG SF Akad Nauk SSSR (Nauka, Novosibirsk) 510:74–103

Zhuravleva IT, Korshunov VI, Luchinina VA et al. (1983) Key boreholes in the Upper Precambrian-Lower Cambrian of the type area on the middle Lena River. In: Meshkova NP (ed) Biostratigraphy and paleontology of the Lower and Middle Cambrian of the Asian part of the USSR (in Russian). Nauka, Moscow, pp 3–44

Zhuravleva ZA (1964) Riphean and Lower Cambrian oncolites and catagraphies of Siberia and their stratigraphic significance (in Russian). Tr GIN Akad Nauk SSSR 114:119

Zhuravleva ZA (1968) Diagnostic features of oncolites and catagraphies and their distribution in Riphean section of the South Urals. In: Keller BM (ed) Problematics of Riphean-Cambrian boundary beds on the Russian platform, Urals and Kazakhstan (in Russian). Nauka, Moscow, pp 83–100

Zhuravleva ZA (1974) On the age of Upper Precambrian deposits of Kharaulakh (in Russian). Izv Akad Nauk SSSR Ser Geol 11:127–131

Zhuravleva ZA (1980) Distribution of oncolites in Riphean deposits of the South Urals (in Russian). Izv Akad Nauk SSSR Ser Geol 2:38–58

Zhuravleva ZA, Chumakov NM (1968) Catagraphies, oncolites and stromatolites from the Late Precambrian of East Byelorussia (in Russian). Dokl Akad Nauk SSSR 178-3:668–671

Zhuravleva ZA, Krylov IN, Postelnikov ES (1969) On stratigraphy and organic remains of the Dashka Formation (Oslyanka Series, Upper Precambrian of the Yenisei Range). (in Russian). Izv Akad Nauk SSSR Ser Geol 7:125–130

Zlobin MN, Golovanov NP (1970) Essay of stratigraphy of Upper Precambrian deposits on the western slope of the Anabar uplift (Kotuikan River). In: Tkachenko BV (ed) Reference section of Upper Precambrian deposits on the western slope of the Anabar uplift (in Russian). NIIGA, Leningrad, pp 6–20

Zlobin MN, Golovanov NP, Kobankov VY et al. (1970) Reference section of Upper Precambrian deposits on the western slope of the Anabar uplift (in Russian). Nauka, Leningrad, 146 pp

Zoricheva AI (1956) Stratigraphy of Paleozoic deposits of the northern Russian platform. In: Tolstikhina MM (ed) Materials on geology of the European part of the USSR (in Russian). Tr VSEGEI (Leningrad), Nov Ser 14:153–168

Zoricheva AI (1963) North of the Russian platform. In: Zoricheva AI (ed) Geology of the USSR, vol 2. Arkhangelsk, Vologda districts and Komi ASSR (in Russian). Geoltoptekhizdat, Moscow, pt 1:87–99

Zubtsov EI (1972) Precambrian tillites of the Tien Shan and their stratigraphic significance (in Russian). Byull MOIP, Otd Geol 47-1:42–56

MIX
Papier aus verantwortungsvollen Quellen
Paper from responsible sources
FSC® C105338

If you have any concerns about our products,
you can contact us on
ProductSafety@springernature.com

In case Publisher is established outside the EU,
the EU authorized representative is:
Springer Nature Customer Service Center GmbH
Europaplatz 3, 69115 Heidelberg, Germany

Printed by Libri Plureos GmbH
in Hamburg, Germany